AF578229

ASTRONOMICAL DATA ANALYSIS SOFTWARE AND SYSTEMS I

A SERIES OF BOOKS ON RECENT DEVELOPMENTS IN ASTRONOMY AND ASTROPHYSICS

Printed by BookCrafters, Inc.

First published 1992

Library of Congress Catalog Card Number: 92-81835
ISBN 0-937707-44-9

D. Harold McNamara, Managing Editor of Conference Series
408 ESC Brigham Young University
Provo, UT 84602

ASTRONOMICAL SOCIETY OF THE PACIFIC
CONFERENCE SERIES

Volume 25

ASTRONOMICAL DATA ANALYSIS SOFTWARE AND SYSTEMS I

Edited by
Diana M. Worrall, Chris Biemesderfer, and Jeannette Barnes

A SERIES OF BOOKS ON RECENT DEVELOPMENTS IN ASTRONOMY AND ASTROPHYSICS

Vol. 1-Progress and Opportunities in Southern Hemisphere Optical Astronomy: The CTIO 25th Anniversary Symposium
ed. V. M. Blanco and M. M. Phillips ISBN 0-93707-18-X

Vol. 2-Proceedings of a Workshop on Optical Surveys for Quasars
ed. P. S. Osmer, A. C. Porter, R. F. Green, and C. B. Foltz ISBN 0-937707-19-8

Vol. 3-Fiber Optics in Astronomy
ed. S. C. Barden ISBN 0-937707-20-1

Vol. 4-The Extragalactic Distance Scale: Proceedings of the ASP 100th Anniversary Symposium
ed. S. van den Bergh and C. J. Pritchet ISBN 0-937707-21-X

Vol. 5-The Minnesota Lectures on Clusters of Galaxies and Large-Scale Structure
ed. J. M. Dickey ISBN 0-937707-22-8

Vol. 6-Synthesis Imaging in Radio Astronomy: A Collection of Lectures from the Third NTAO Synthesis Imaging Summer School
ed. R. A. Perley, F. R. Schwab, and A. H. Bridle ISBN 0-937707-23-6

Vol. 7-Properties of Hot Luminous Stars: Boulder-Munich Workshop
ed. C. D. Garmany ISBN 0-937707-24-4

Vol. 8-CCDs in Astronomy
ed. George H. Jacoby ISBN 0-937707-25-2

Vol. 9-Cool Stars, Stellar Systems, and the Sun
ed. G. Wallerstein ISBN 0-937707-27-9

Vol. 10-The Evolution of the Universe of Galaxies. The Edwin Hubble Centennial Symposium
ed. Richard G. Kron ISBN 0-937707-28-7

Vol. 11-Confrontation Between Stellar Pulsation and Evolution
ed. C. Cacciari and G. Clementini ISBN 0-937707-30-9

Vol. 12-The Evolution of the Interstellar Medium
ed. L. Blitz ISBN 0-937707-31-7

Vol. 13-The Formation and Evolution of Star Clusters
ed. K. Janes ISBN 0-937707-32-5

Vol. 14-Astrophysics with Infrared Arrays
ed. R. Elston ISBN 0-937707-33-3

Vol. 15-Large-Scale Structures and Peculiar Motions in the Universe
ed. D. W. Latham and L. A. N. da Costa ISBN 0-937707-34-1

Vol. 16-Atoms, Ions and Molecules: New Results in Spectral Line Astrophysics
ed. A. D. Haschick and P. T. P. Ho ISBN 0-937707-35-X

Vol. 17-Light Pollution, Radio Interference, and Space Debris
ed. D. L. Crawfoed. D. L. Crawford ISBN 0-937707-36-8

Vol. 18-The Interpretation of Modern Synthesis Observations of Spiral Galaxies.
ed. M. Duric and P. C. Crane ISBN 0-937707-37-6

Vol. 19-Radio Interferometry: Theory, Techniques, and Applications, IAU Colloquium 131
ed. T. J. Cornwell and R. A. Perley ISBN 0-937707-38-4

Vol. 20-Frontiers of Stellar Evolution, celebrating the 50th Anniversary of McDonald Observatory
ed. D. L. Lambert ISBN 0-937707-39-2

Vol. 21-The Space Distribution of Quasars
ed. D. Crampton ISBN 0-937707-40-6

Vol. 22-Nonisotropic and Variable Outflows from Stars
ed. Laurent Drissen, Claus Leitherer, and Antonella Nota ISBN 0-937707-41-4

Vol. 23-Astronomical CCD Observing and Reduction Techniques
ed. Steve B. Howell ISBN 0-937707-42-4

Vol. 24-Cosmology and Large-Scale Structure in the Universe
ed. Reinaldo R. de Carvalho ISBN 0-937707-43-0

Inquiries concerning these volumes should be directed to the:
Astronomical Society of the Pacific
CONFERENCE SERIES
390 Ashton Avenue
San Francisco, CA 94112-1722

TABLE OF CONTENTS

♦ SECTION 4: DECONVOLUTION AND IMAGE RESTORATION

PREFACE

Astronomical research has been revolutionized by the rapid growth in computing and networking and strongly influenced by the trend towards distributed data processing. We have reached a stage where astronomers expect to exchange data, intercompare multidisciplinary data sets, access archives, and re-analyze data, for the most part without leaving their offices and using common software which can be applied regardless of the heritage of the data. These expectations necessitate effective communication between scientists and programmers concerned with algorithms, software, and software systems employed in the reduction and analysis of astronomical data.

A series of annual conferences, to be held in North America, was proposed with the aim of fostering such communication. The conferences will be held in different locations, with a basic format that is expected to change only slowly from year to year. The present volume documents the first meeting, which was held in Tucson, from the 6th to 8th of November 1991, and was hosted by the National Optical Astronomy Observatories. The second meeting is planned for Boston in November 1992 and will be hosted by the Smithsonian Astrophysical Observatory. Responses from a questionnaire sent by electronic mail to participants of the Tucson meeting are being used to help plan the second conference.

The first annual conference on "Astronomical Data Analysis and Software Systems" attracted 312 registered participants, of whom 92 represented 20 countries outside the United States. The conference comprised invited and contributed talks, poster papers, software demonstrations, and "birds of a feather" (BOF) sessions, some of which were arranged by the conference organizers, and some of which resulted from an invitation to potential participants to suggest and organize BOFs. The conference highlighted four special topics: all sky data, deconvolution and image restoration, error analysis and statistics, and high energy imaging, and invited additional contributions on the topics of software systems and image and spectral analysis. BOF sessions were held on the following topics: data acquisition software with on-line and pipeline processing, IRAF in the undergraduate environment, the use of IDL in astronomy, and installation and maintenance of the IRAF, STSDAS and PROS software. Immediately following the conference there was an IRAF status report and panel discussion, and user-group meetings for the NOAO, STSDAS and PROS analysis packages. The organization of this volume largely follows that of the conference, except in one respect. In line with the multidisciplinary trends of the subject, there is no separate section on high energy imaging.

The conference was sponsored by the National Aeronautics and Space Administration, the National Optical Astronomy Observatories, the National Science Foundation, the Smithsonian Astrophysical Observatory, and the Space Telescope Science Institute, and their ongoing interest is greatly appreciated.

The Program Organizing Committee comprised Carol Christian, Dennis Crabtree, Bob Hanisch, Steve Ridgway, Doug Tody and Diana Worrall.

Members of the Local Organizing Committee were Jeannette Barnes, Chris Biemesderfer, Lindsey Davis, Mike Fitzpatrick, Steve Grandi, Suzanne Jacoby, Dyer Lytle, Steve Ridgway, Doug Tody, and Frank Valdes. We are particularly grateful to Steve Grandi for his able assistance in setting up the complex hardware arrangements at the meeting. Many others of our colleagues helped considerably: we especially mention Ian Gatley, Michelle Nickol, Nigel Sharp, and Mark Stevens from NOAO and Debbie Brissenden, Tricia Buckley, and Su Tuttle from SAO.

Finally, we thank those who helped us edit these proceedings: Mark Birkinshaw and Mark Reid for software useful in producing the subject index; and, most especially, Michelle Nickol for invaluable help with reformatting some papers and for assistance in assembling the final manuscript.

Diana M. Worrall
Smithsonian Astrophysical Observatory

Chris Biemesderfer
National Optical Astronomy Observatories

Jeannette Barnes
National Optical Astronomy Observatories

March 23, 1992

Cover illustration: Hubble Space Telescope images of R136 before and after Richardson-Lucy restoration, provided by R. L. White (see his paper, page 176).

PARTICIPANT LIST

Timothy M. C. Abbott ⟨ tmca@astro.as.utexas.edu ⟩
McDonald Observatory/University of Texas, Austin, TX

Jeff Achtermann ⟨ jeffrey@astro.as.utexas.edu ⟩
Astronomy Department, University of Texas, Austin, TX

Hans-Martin Adorf ⟨ hmadorf@eso.org ⟩
ST-ECF/ESO, Garching, GERMANY

Rudolf Albrecht ⟨ ralbrech@eso.org ⟩
Space Telescope European Coordinating Facility, Garching, GERMANY

Peter Allan ⟨ 19457::pma ⟩
STARLINK, SERC, Oxfordshire, UNITED KINGDOM

Ron Allen ⟨ rjallen@stsci.edu ⟩
Space Telescope Science Institute, Baltimore, MD

Edwin Anderson ⟨ eanderson@noao.edu ⟩
NSO/GONG, Tucson, AZ

Lorella Angelini ⟨ heasrc::angelini ⟩
HEASARC/NASA, Greenbelt, MD

David Axon ⟨ dja@star.jb.man.ac.uk ⟩
NRAL, Jodrell Bank, Chesire, ENGLAND

Thomas R. Ayres ⟨ hyades::ayres ⟩
University of Colorado (CASA), Boulder, CO

Dietrich Baade ⟨ esomc1::baade ⟩
European Southern Observatory, Garching, GERMANY

Karin Babst ⟨ sdcdcl::cobecl::babst ⟩
STX-COBE, Greenbelt, MD

Roberto Baglioni ⟨ baglioni@arcetri.astro.it ⟩
Dipartimento Di Astronomia, Firenze, ITALY

Pascal Ballester ⟨ pballest@eso.org ⟩
European Southern Observatory, Garching, GERMANY

Klaus Banse ⟨ esomc1::klaus ⟩
European Southern Observatory, Garching, GERMANY

Irene Barg ⟨ ibarg@as.arizona.edu ⟩
Steward Observatory, Tucson, AZ

Jeannette Barnes ⟨ jbarnes@noao.edu ⟩
NOAO, Tucson, AZ

Paul Barrett ⟨ grossc::barrett ⟩
USRA-NASA/GSFC, Greenbelt, MD

John P. Basart ⟨ jpbasart@iastate.edu ⟩
Iowa State University, Ames, IA

Suzanne Bauer ⟨ ssdo::bauer ⟩
BDC/STX, Washington, DC

David Bazell ⟨ bazell@stsci.edu ⟩
Space Telescope Science Institute, Baltimore, MD

Stēphane Bēland ⟨ beland@cfht.hawaii.edu ⟩
Canada-France-Hawaii Telescope Corp., Kamuela, HI

Eileen F. Berman ⟨ berman@fnal.fnal.gov ⟩
Fermilab, Batavia, IL

Chris Biemesderfer ⟨ cbiemes@noao.edu ⟩
NOAO, Tucson, AZ

Teresa Bippert-Plymate ⟨ teresa@mips3.as.arizona.edu ⟩
Steward Observatory, Tucson, AZ

James H. Blackwell, Jr. ⟨ stars::hrsblackwell ⟩
CSC IUE/GHRS, Lanham-Seabrook, MD

Jonathan Bland-Hawthorn ⟨ hawthorn@pegasus.rice.edu ⟩
Rice University, Houston, TX

Jeffrey Bloch ⟨ 103283@essdp1.lanl.gov ⟩
Los Alamos National Laboratory, Los Alamos, NM

Fred Blundell ⟨ blundell@cfa.bitnet ⟩
Harvard-Smithsonian Center for Astrophysics, Cambridge, MA

Bruce Bohannan ⟨ bruce@noao.edu ⟩
Kitt Peak National Observatory, Tucson, AZ

Edouard Boily ⟨ boily@phy.ulaval.ca ⟩
Universite Laval, Ste-Foy, Quebec, CANADA

Michael Bolte ⟨ bolte@helios.ucsc.edu ⟩
Lick Observatory, Santa Cruz, CA

Jerry Bonnell ⟨ jerry@iuesn1.gsfc.nasa.gov ⟩
CSC, Greenbelt, MD

Tj. Romke Bontekoe ⟨ romke@guspace.rug.nl ⟩
European Space Agency, Noordwijk, THE NETHERLANDS

Angela Bragaglia ⟨ looney::bragaglia ⟩
Lunar and Planetary Lab, Tucson, AZ

Daniel Shenon Briggs ⟨ dbriggs@nrao.edu ⟩
NRAO Socorro/NMIMT, Socorro, NM

Roger J. Brissenden ⟨ rbrissenden@cfa.harvard.edu ⟩
SAO - Center for Astrophysics, Cambridge, MA

Edward W. Brugel ⟨ brugel@puppis.colorado.edu ⟩
CASA, University of Colorado, Boulder, CO

Patricia A. Buckley ⟨ adass@cfa.harvard.edu ⟩
SAO Center for Astrophysics, Cambridge, MA

Luciano Burderi ⟨ ipamat::exosat ⟩
Universita Di Palermo, Palermo, ITALY

Howard Bushouse ⟨ 6890::bushouse ⟩
CSC/IUE, Lanham-Seabrook, MD

Ivo C. Busko ⟨ 47556::busko ⟩
INPE, Sao Jose dos Campos-SP, BRAZIL

George A. Carlson
Citrus College, Glendora, CA

Ed Cheng ⟨ ec@cobi.gsfc.nasa.gov ⟩
NASA/GSFC, Greenbelt, MD

Kwang-Ping Cheng ⟨ iue::cheng ⟩
NASA/GSFC/NRC, Greenbelt, MD

Eric Chipman ⟨ chipman@grossc.gsfc.nasa.gov ⟩
GRO Science Support Ctr/CSC, Greenbelt, MD

Carol Christian ⟨ carolc@ssl.berkeley.edu ⟩
EUVE/CEA/UCB, Berkeley, CA

Riccardo Claudi ⟨ 39003::claudi ⟩
Osservatorio Astronomico Padova, Padova, ITALY

Maureen Conroy ⟨ mo@cfa.harvard.edu ⟩
Smithsonian Astrophysical Observatory, Cambridge, MA

Robin Corbet ⟨ corbet@astro.psu.edu ⟩
Penn State, University Park, PA

Mark E. Cornell ⟨ cornell@astro.as.utexas.edu ⟩
McDonald Observatory, Austin, TX

Tim Cornwell ⟨ tcornwel@nrao.edu ⟩
NRAO/VLA, Socorro, NM

David Cottingham ⟨ dc@cobi.gsfc.nasa.gov ⟩
USRA, Greenbelt, MD

Dennis R. Crabtree ⟨ crabtree@dao.nrc.ca ⟩
DAO/CADC, Victoria, BC, CANADA

Eric R. Craine ⟨ craine@noao.edu ⟩
Western Research Co., Inc., Tucson, AZ

Gabriele Cremonese ⟨ 39003::cremonese ⟩
Osservatorio Astronomico Padova, Padova, ITALY

Deborah Crocker ⟨ crock@okra.astr.ua.edu ⟩
Univ. of Alabama, Tuscaloosa, AL

Cindy Cunningham ⟨ ccc@nereid.sal.ists.ca ⟩
Institute for Space and Terrestrial Science, Concord, Ontario, CANADA

Francesca D'Alessandro
Telespazio (I), Rome, ITALY

Tim Davidge ⟨ davidge@cfht.hawaii.edu ⟩
CFHT, Kamuela, HI

Lindsey Davis ⟨ davis@noao.edu ⟩
NOAO, Tucson, AZ

Charles Day ⟨ heavax::day ⟩
LHEA, NASA GSFC, Greenbelt, MD

Michele De La Peña ⟨ 6890::delapena ⟩
CSC/IUE Obs., Lanham-Seabrook, MD

Susana Delgado-Marante ⟨ iac::sdm ⟩
Instituto de Astrofisica de Canarias, Tenerife, SPAIN

Janet DePonte ⟨ janet@cfa.harvard.edu ⟩
Smithsonian Astrophysical Observatory, E. Taunton, MA

E.R. Deul ⟨ deul@maas.leidenUniv.nl ⟩
Sterrewacht Leiden, Leiden, THE NETHERLANDS

Daniel Dewey ⟨ dd@space.mit.edu ⟩
MIT, Cambridge, MA

Carl Dobson ⟨ cad@ssl.berkeley.edu ⟩
U.C. Berkeley, Center for EUV Astrophysics, Berkeley, CA

Edmund G. Dombrowski
BDC/SFA, Alexandria, VA

Gitta Domik ⟨ domik@cs.colorado.edu ⟩
University of Colorado, Boulder, CO

Carlos José Donzelli ⟨ donzelli@famaf.edu.ar ⟩
Observatorio Astronomico Cordoba, Cordoba, ARGENTINA

John Dorband ⟨ dorband@nibbles.gsfc.nasa.gov ⟩
NASA Goddard Space Flight Center, Greenbelt, MD

Daniel Durand ⟨ durand@dao.nrc.ca ⟩
CADC/DAO, Victoria, BC, CANADA

Jonathan Eisenhamer ⟨ eisenham@stsci.edu ⟩
Space Telescope Science Institute, Baltimore, MD

R. Mark Elowitz ⟨ elowitz@astro.as.arizona.edu ⟩
Steward Observatory, Tucson, AZ

Martin Elvis ⟨ elvis@cfa.harvard.edu ⟩
Harvard-Smithsonian CFA, Cambridge, MA

Darrel T. Emerson ⟨ demerson@nrao.edu ⟩
NRAO, Tucson, AZ

John R. Engel ⟨ craine@noao.edu ⟩
Western Research Co., Inc., Tucson, AZ

Robert E. Eplee, Jr. ⟨ stars::eplee ⟩
General Sciences Corporation, Greenbelt, MD

S. Allen Farris ⟨ farris@stsci.edu ⟩
Space Telescope Science Institute, Baltimore, MD

Fabio Favata ⟨ estsa2::ffavata ⟩
ESTEC, Noordwijk, THE NETHERLANDS

Eric Feigelson ⟨ edf@astro.psu.edu ⟩
Penn State, University Park, PA

Anthony J. Ferro ⟨ tony@quasar.la.asu.edu ⟩
Arizona State University, Tempe, AZ

Michael Fitzpatrick ⟨ fitz@noao.edu ⟩
NOAO, Tucson, AZ

Dale Fixsen ⟨ fixsen@cobecl.gsfc.nasa.gov ⟩
USRA, Savage, MD

John Flanagan ⟨ johnf@ssl.berkeley.edu ⟩
U.C. Berkeley Center for EUV Astrophysics, Berkeley, CA

Al Fowler ⟨ afowler@noao.edu ⟩
NOAO, Tucson, AZ

I. Freedman ⟨ cobecl::freedman ⟩
STX-COBE, Greenbelt, MD

Henry Freudenreich ⟨ cobecl::freudenreich ⟩
STX-COBE, Greenbelt, MD

Lynn Friesen ⟨ dirac2::friesen ⟩
The Aerospace Corp., Los Angeles, CA

Tom Fuller ⟨ fuller@voodoo.ucsb.edu ⟩
UCSB, Santa Barbara, CA

Marc Gagné ⟨ gagne@jove.physast.uga.edu ⟩
University of Georgia, Athens, GA

Michael Garcia ⟨ garcia@cfa.harvard.edu ⟩
Smithsonian Astrophysical Observatory, Cambridge, MA

Thomas L. Garrard ⟨ tlg@citsrl.caltech.edu ⟩
Caltech Space Radiation Lab, Pasadena, CA

Nick Gautier ⟨ nick@ipac.caltech.edu ⟩
IPAC, Jet Propulsion Lab, Pasadena, CA

Richard Gelderman ⟨ rg2v@virginia.edu ⟩
University of Virginia, Charlottesville, VA

Douglas B. George ⟨ ottr@cc.sfu.ca ⟩
Carleton University, Kanata, Ontario, CANADA

Pedro Gigoux ⟨ pgigoux@noao.edu ⟩
NOAO (CTIO), La Serena, CHILE

Paolo Giommi ⟨ 29617::giommi ⟩
ESA/ESIS, Frascati, ITALY

Frank Giovane ⟨ fgiovane@nhqvax.span.nasa.gov ⟩
NASA Headquarters, Washington, DC

John Glaspey ⟨ glaspey@cfht.hawaii.edu ⟩
CFHT, Kamuela, HI

Paul Gogan
CNRS-France, Marseille, FRANCE

Gabriel Cruz Gonzalez ⟨ gabriel@sunpuide1.puide.unam.mx ⟩
Instituto de Astronomia, U.N.A.M., Mexico, D.F., MEXICO

John Good ⟨ jcg@ipac.caltech.edu ⟩
IPAC, Pasadena, CA

Anil Gore ⟨ anil@ccit.arizona.edu ⟩
Dept. of Stats., Univ. of AZ, Tucson, AZ

Steve Grandi ⟨ grandi@noao.edu ⟩
NOAO, Tucson, AZ

Richard Green ⟨ rgreen@noao.edu ⟩
Kitt Peak National Observatory, Tucson, AZ

Ted Groner ⟨ ted@galileo.sao.arizona.edu ⟩
Smithsonian Astrophysical Observatory, Tucson, AZ

Preben Grosbøl ⟨ pgrosbol@eso.org ⟩
European Southern Observatory, Garching, GERMANY

Pasi Hakala ⟨ pahakala@finfun.bitnet ⟩
University of Helsinki, Helsinki, FINLAND

Masaru Hamabe ⟨ hamabe@c1.mtk.nao.ac.jp ⟩
Kiso Observatory, Nagano, JAPAN

Robert Hanisch ⟨ hanisch@stsci.edu ⟩
Space Telescope Science Institute, Baltimore, MD

Paul Harding ⟨ pharding@noao.edu ⟩
NOAO, Tucson, AZ

F. Rick Harnden, Jr. ⟨ frh@cfa.harvard.edu ⟩
Harvard-Smithsonian CfA, Cambridge, MA

Dan Harris ⟨ harris@cfa.harvard.edu ⟩
Smithsonian Astrophysical Observatory, Cambridge, MA

R. Lee Hawkins ⟨ lhawkins@lucy.wellesley.edu ⟩
Wellesley College, Wellesley, MA

Susan G. Hazelton ⟨ swg@cfa.harvard.edu ⟩
Smithsonian Astrophysical Observatory, Westford, MA

Steve Heathcote ⟨ sheathcote@noao.edu ⟩
CTIO, La Serena, CHILE

Jim Hecht ⟨ dirac2::hecht ⟩
Aerospace Corp., Los Angeles, CA

Arne A. Henden ⟨ henden@mps.ohio-state.edu ⟩
Ohio State University, Columbus, OH

Leon Herreid ⟨ herreid@xenium.gsfc.nasa.gov ⟩
ST Systems Corporation, Silver Spring, MD

Frank Hill ⟨ fhill@noao.edu ⟩
NOAO/NSO/GONG, Tucson, AZ

Richard Hill
Warner and Obs., Tucson, AZ

Kerry C. Hilldrup ⟨ hilldrup@jmh.gsfc.nasa.gov ⟩
NASA/GSFC, Greenbelt, MD

Ignacio Hirata
INAOE, Mexico D.F., MEXICO

Philip E. Hodge ⟨ hodge@stsci.edu ⟩
Space Telescope Science Institute, Baltimore, MD

Jay B. Holberg ⟨ looney::holberg ⟩
Lunar and Planetary Lab, Tucson, AZ

Jan M. Hollis ⟨ hollis@amarna.gsfc.nasa.gov ⟩
NASA, Greenbelt, MD

Richard Hook ⟨ rhook@eso.org ⟩
ST-ECF/ESO, Garching, GERMANY

Sethanne Howard ⟨ howard@ssl.msfc.nasa.gov ⟩
MSFC, Huntsville, AL

Steve Howell ⟨ 5470::howell ⟩
Planetary Science Institute, Tucson, AZ

Jin-Chung Hsu ⟨ hsu@stsci.edu ⟩
Space Telescope Science Institute, Baltimore, MD

B. R. Hunt ⟨ hunt@ece.arizona.edu ⟩
University of Arizona, Tucson, AZ

Juhani Huovelin ⟨ huovelin@cc.helsinki.fi ⟩
University of Helsinki, Helsinki, FINLAND

David G. Iadevaia ⟨ diadevaia@east.pima.edu ⟩
Pima College - East Campus, Tucson, AZ

Shin-ichi Ichikawa ⟨ ichikawa@c1.mtk.nao.ac.jp ⟩
National Astronomical Observatory, Mitaka, Tokyo, JAPAN

Richard Isaacman ⟨ isaacman@stars.gsfc.nasa.gov ⟩
General Sciences Corp., Greenbelt, MD

Takashi Isobe ⟨ ti@space.mit.edu ⟩
CSR, MIT, Cambridge, MA

Peter D. Jackson ⟨ cobecl::jackson ⟩
STX-COBE, Greenbelt, MD

Paul Jacobs ⟨ jacobs@rosserv.gsfc.nasa.gov ⟩
STX-NASA/GSFC, Greenbelt, MD

George Jacoby ⟨ gjacoby@noao.edu ⟩
Kitt Peak National Observatory, Tucson, AZ

Suzanne Jacoby ⟨ sjacoby@noao.edu ⟩
NOAO, Tucson, AZ

H. Jager ⟨ rieksj@sron.ruu.nl ⟩
Space Research Utrecht, Utrecht, THE NETHERLANDS

Stuart Jefferies ⟨ sjefferies@noao.edu ⟩
Bartol Research Institute, Tucson, AZ

Vicki Johnson ⟨ vjohnson@quake.stanford.edu ⟩
Stanford University, Stanford, CA

Todd Karakashian ⟨ todd@cfa.harvard.edu ⟩
Smithsonian Astrophysical Observatory, Cambridge, MA

Michael Keane ⟨ mk@helios.ucsc.edu ⟩
Lick Observatory, Santa Cruz, CA

Phillip B. Keegstra ⟨ keegstra@csdr2.gsfc.nasa.gov ⟩
STX-COBE, Greenbelt, MD

Bill Keel ⟨ keel@okra.astr.ua.edu ⟩
University of Alabama, Tuscaloosa, AL

Phil Kelton ⟨ pwk@astro.as.utexas.edu ⟩
Univ. of Texas McDonald Observatory, Austin, TX

Ajit Kembhavi ⟨ ajit@incaa.ernet.in ⟩
Inter University Centre for Astron. and Astrophys., Pune, INDIA

John M. Kerr III ⟨ john@cfht.hawaii.edu ⟩
CFHT, Kamuela, HI

Do Kester ⟨ do@guspace.rug.nl ⟩
Space Research Organization Netherlands, Groningen, THE NETHERLANDS

Joao Luiz Kohl Moreira ⟨ 47556::kohl ⟩
CNPq Observatorio Nacional, Rio de Janeiro-RJ, BRAZIL

Leonid Kokourine ⟨ iaas@node.ias.msk.su ⟩
INASAN, USSR, Moscow, USSR

Alan T. Koski ⟨ akoski@as.arizona.edu ⟩
Steward Observatory, Tucson, AZ

Tobias J. Kreidl ⟨ tjk@lowell.edu ⟩
Lowell Observatory, Flagstaff, AZ

John Eric Krist ⟨ krist@stsci.edu ⟩
Space Telescope Science Institute, Baltimore, MD

Michael Kurtz ⟨ kurtz@cfa.harvard.edu ⟩
Center for Astrophysics, Cambridge, MA

Wayne Landsman ⟨ stars::landsman ⟩
ST Systems Co., Greenbelt, MD

Michael LaValley ⟨ mlv@stat.psu.edu ⟩
Penn State, University Park, PA

Denis A. Leahy ⟨ leahy@iras.ucalgary.ca ⟩
University of Calgary, Calgary, Alberta, CANADA

Dawn D. Lenz ⟨ elroy::lenz ⟩
Univ. Colorado (CASA), Boulder, CO

Michael Lesser
Steward Observatory, Tucson, AZ

Zoltan G. Levay ⟨ levay@stsci.edu ⟩
STScI/SCARS, Baltimore, MD

Ti-pei Li ⟨ litp@vxihep.ihep.cern.ch ⟩
Institute of High Energy Physics, Beijing, CHINA

Don J. Lindler ⟨ hrs::hrslindler ⟩
ACC, Inc., Potomac, MD

Philip Lubin ⟨ lubin@voodoo.bitnet ⟩
Physics Dept. UC Santa Barbara, Santa Barbara, CA

Tom Lutz ⟨ tlutz@wsumath.bitnet ⟩
Washington State University, Pullman, WA

Dyer Lytle ⟨ lytle@noao.edu ⟩
NOAO, Tucson, AZ

Maria Concetta Maccarone ⟨ 39386::maccarone ⟩
IFCAI/CNR, Palermo, ITALY

Ronald J. Maddalena ⟨ rmaddale@nrao.edu ⟩
NRAO, Green Bank, WV

Barry Madore ⟨ barry@ipac.caltech.edu ⟩
Caltech, Pasadena, CA

Oleg Yu. Malkov ⟨ iaas@node.ias.msk.su ⟩
INASAN USSR, Moscow, USSR

Eric Mandel ⟨ eric@cfa.harvard.edu ⟩
Smithsonian Astrophysical Observatory, Cambridge, MA

Paolo Marcucci ⟨ marcucci@astrts.astro.it ⟩
Astronomical Observatory of Trieste, Trieste, ITALY

Hugo G. Marraco ⟨ ssl::psi%iafe::hmarraco ⟩
IAFE, Buenos Aires, ARGENTINA

Phil Massey ⟨ pmassey@noao.edu ⟩
KPNO/NOAO, Tucson, AZ

Chris Mayer ⟨ cjm@kekoa.jach.hawaii.edu ⟩
Joint Astronomy Center, Hilo, HI

J. D. Mayfield
Ball Aerospace, Boulder, CO

Mark McCaughrean ⟨ mjm@as.arizona.edu ⟩
Steward Observatory, Tucson, AZ

James E. McGaha
Grasslands Observatory, Tucson, AZ

Thomas A. McGlynn ⟨ mcglynn@grossc.gsfc.nasa.gov ⟩
GRO SSC/CSC, Greenbelt, MD

Rick McGonegal ⟨ sysrick@noao.edu ⟩
GEMINI 8-M Telescopes Project, Kamucla, III

John T. McGraw ⟨ jmcgraw@as.arizona.edu ⟩
Steward Observatory, Tucson, AZ

Brian K. McIlwrath ⟨ 19457::bkm ⟩
STARLINK, SERC, Oxfordshire, UNITED KINGDOM

Justin F. McNeill, Jr. ⟨ jfm059@ipl.jpl.nasa.gov ⟩
Jet Propulsion Lab/CALTECH, Pasadena, CA

Stephen Meatheringham ⟨ sjm@moggie.anu.edu.au ⟩
Mt. Stromlo Observatory, Private Bag, AUSTRALIA

John Menzies ⟨ jwm@saao.ac.za ⟩
SAAO, Observatory, SOUTH AFRICA

William J. Merline ⟨ merline@hindmost.lpl.arizona.edu ⟩
Planetary Science Institute, Tucson, AZ

Mike Merrill ⟨ mmerrill@noao.edu ⟩
NOAO, Tucson, AZ

Doug Mink ⟨ mink@cfa.harvard.edu ⟩
Center for Astrophysics, Cambridge, MA

Kenneth Mitchell ⟨ cobecl::mitchell ⟩
GSC COBE Group, Greenbelt, MD

Kazuhisa Mitsuda ⟨ mitsuda@astro.isas.ac.jp ⟩
ISAS, Kanagawa-ken, JAPAN

Jeffrey S. Morgan ⟨ morgan@orca.astro.washington.edu ⟩
University of Washington, Seattle, WA

Heather Morrison ⟨ heather@vegemite.noao.edu ⟩
KPNO/NOAO, Tucson, AZ

Fionn Murtagh ⟨ fionn@dgaeso51.bitnet ⟩
Space Telescope - European Coordinating Facility, Garching, GERMANY

Maria Barbara Negri ⟨ asirom::negri ⟩
Agenzia Spaziale Italiana, Roma, ITALY

Michael V. Newberry ⟨ newberry@astro.as.arizona.edu ⟩
Steward Observatory, Tucson, AZ

Tom Nicinski ⟨ nicinski@fnal.fnal.gov ⟩
Fermilab, Batavia, IL

Michelle Nickol ⟨ mnickol@noao.edu ⟩
NOAO, Tucson, AZ

Jorge Núñez ⟨ d3fajnm0@eb0ub011.bitnet ⟩
Universitat de Barcelona, Barcelona, SPAIN

Earl J. O'Neil, Jr. ⟨ draco::oneil ⟩
NOAO, Tucson, AZ

Eric C. Olson ⟨ ericco@ssl.berkeley.edu ⟩
Center for EUV Astrophysics, UC Berkeley, Berkeley, CA

Jeffrey Orszak ⟨ orszak@cfa.harvard.edu ⟩
Smithsonian Astrophysical Observatory, Cambridge, MA

Ludwig F. Oster ⟨ loster@nsf.gov ⟩
National Science Foundation, Washington, DC

Clive G. Page ⟨ 19838::cgp ⟩
XRA Group, Leicester University, Leicester, ENGLAND

A. N. Parmar ⟨ exosat::arvind ⟩
ESA, Noordwijk, THE NETHERLANDS

Fabio Pasian ⟨ fpasian@eso.org ⟩
ST-ECF/ESO, Garching, GERMANY

L. Pásztor ⟨ h2295pas@ella.hu ⟩
Loránd Eötvös University, Budapest, HUNGARY

Jeffrey A. Pedelty ⟨ pedelty@amarna.gsfc.nasa.gov ⟩
NASA/GSFC, Greenbelt, MD

José H. Peña ⟨ astros@unamvm1.bitnet ⟩
UNAM - INAOE, Mexico D.F., MEXICO

William Pence ⟨ pence@tetra.gsfc.nasa.gov ⟩
HEASARC/GSFC, Greenbelt, MD

Rosario Peniche ⟨ astros@unamvm1.bitnet ⟩
UNAM - INAOE, Mexico D.F., MEXICO

Fidel Cruz Peregrino ⟨ fidel@alfa.astroscu.unam.mx ⟩
Instituto de Astronomia, UNAM, Mexico D.F., MEXICO

Michael Pfeiffer
Institute for Astronomy University Munich, Muenchen, GERMANY

Jeff Pier ⟨ jrp@lowell.edu ⟩
USNO Flagstaff, Flagstaff, AZ

Robert Pierson
Applied Technology Associates, Albuquerque, NM

Benoit Pirenne ⟨ eso::pirenne ⟩
Space Telescope - European Coordinating Facility, Garching, GERMANY

Donald Platt ⟨ Platt_don@ctstateu.bitnet ⟩
Western Ct. St. University, Newton, CT

David Plummer ⟨ dplummer@cfa.harvard.edu ⟩
Smithsonian Astrophysical Observatory, Cambridge, MA

V. F. Polcaro ⟨ 40058::polcaro ⟩
C.N.R. Istituto di Astrofisica Spaziale, Frascati, ITALY

Alain Porter ⟨ aporter@noao.edu ⟩
KPNO/NOAO, Tucson, AZ

Tony Poyner ⟨ apoyner@as.arizona.edu ⟩
MMTO/Steward Observatory, Tucson, AZ

Bill Press ⟨ wpress@cfa.harvard.edu ⟩
Harvard-Smithsonian Center for Astrophysics, Cambridge, MA

C. Pritchet ⟨ pritchet@orca.phys.uvic.ca ⟩
University of Victoria, Victoria, BC, CANADA

Mauro Pucillo ⟨ pucillo@astrts.astro.it ⟩
Astronomical Observatory of Trieste, Trieste, ITALY

Anne C. Raugh ⟨ cobecl::raugh ⟩
STX-COBE, Greenbelt, MD

Shirley M. Read ⟨ cobecl::read ⟩
STX-COBE, Greenbelt, MD

Christian Reile
Institut fur Astronomie und Astrophysik, Munchen, GERMANY

Steve Ridgway ⟨ sridgway@noao.edu ⟩
NOAO, Tucson, AZ

Fred Ringwald ⟨ fred.ringwald@dartmouth.edu ⟩
Dartmouth College, Hanover, NH

Wendy Roberts ⟨ roberts@cfa.harvard.edu ⟩
Smithsonian Astrophysical Observatory, Cambridge, MA

Peter R. Roelfsema ⟨ pjotr@afrodite.rug.nl ⟩
Space Research Organization Netherlands, Groningen, THE NETHERLANDS

William Romanishin ⟨ romanishin@phyast.nhn.uoknor.edu ⟩
University of Oklahoma, Norman, OK

Fred Romelfanger ⟨ fred@stsci.edu ⟩
Space Telescope Science Institute, Baltimore, MD

Steve Rooke ⟨ srooke@noao.edu ⟩
NOAO, Tucson, AZ

Corinne Rossi ⟨ 40058::rossi ⟩
Università di Roma, Roma, ITALY

Arnold H. Rots ⟨ arots@xebec.gsfc.nasa.gov ⟩
USRA/NASA/GSFC, Greenbelt, MD

Vincent Saba ⟨ vsaba@ssl.berkeley.edu ⟩
EUVE, U.C. Berkeley, Berkeley, CA

Bruno Sacco ⟨ 39405::sacco ⟩
IFCAI/CNR, Palermo, ITALY

Skip Schaller ⟨ skip@as.arizona.edu ⟩
Steward Observatory, Tucson, AZ

Eric M. Schlegel ⟨ eric@heasfs.gsfc.nasa.gov ⟩
NASA-GSFC/USRA, Greenbelt, MD

Timothy J. Schneeberger
Logicon/RDA, Albuquerque, NM

J. Schuurmans ⟨ jaaps@sron.ruu.nl ⟩
Space Research Utrecht, Utrecht, THE NETHERLANDS

Ian C. Scott-Fleming ⟨ icsf@as.arizona.edu ⟩
MMTO/Steward Observatory, Tucson, AZ

Paul Scowen ⟨ uk1@spacsun.rice.edu ⟩
Rice University, Houston, TX

Rob Seaman ⟨ seaman@noao.edu ⟩
NOAO (CCS), Tucson, AZ

Ajai Sehgal ⟨ ajai@sce.carleton.ca ⟩
Carleton University, Gloucester, Ontario, CANADA

Miquel Serra-Ricart ⟨ iac::msr ⟩
Instituto de Astrofisica de Canarias, Tenerife, SPAIN

Nigel Sharp ⟨ nsharp@noao.edu ⟩
NOAO, Tucson, AZ

Richard A. Shaw ⟨ shaw@stsci.edu ⟩
Space Telescope Science Institute, Baltimore, MD

Patrick L. Shopbell ⟨ pls@pegasus.rice.edu ⟩
Rice University, Houston, TX

David Silva ⟨ silva@noao.edu ⟩
NOAO/KPNO, Tucson, AZ

John Arthur J. Skard ⟨ cobecl::skard ⟩
STX-COBE, Greenbelt, MD

Giselle Sleiman ⟨ sleiman@noao.edu ⟩
NOAO, Tucson, AZ

Larry Smarr ⟨ lsmarr@ncsa.uiuc.edu ⟩
National Center for Supercomputing Applications, Champaign, IL

Oleg M. Smirnov ⟨ iaas@node.ias.msk.su ⟩
INASAN USSR, Moscow, USSR

Eric P. Smith ⟨ hrs::ericsmith ⟩
LASP/GSFC, Greenbelt, MD

William A. Snyder ⟨ 11334::snyder ⟩
Naval Research Laboratory, Cheverly, MD

Luigi Stella ⟨ astmib::stella ⟩
Osservatorio Astronomico di Brera, Milano, ITALY

David M. Stern ⟨ idl@boulder.colorado.edu ⟩
Research Systems, Inc., Boulder, CO

Peter Stetson ⟨ stetson@dao.nrc.ca ⟩
Dominion Astrophysical Observatory, Victoria, BC, CANADA

Mark Stevens ⟨ mstevens@noao.edu ⟩
NOAO, Tucson, AZ

J. M. Stewart ⟨ jms@star.roe.ac.uk ⟩
Royal Observatory Edinburgh, Edinburgh, SCOTLAND

Thomas Stiff ⟨ tstiff@nereid.sal.ists.ca ⟩
Simcoe County Board of Education, Midhurst, Ontario, CANADA

Elizabeth B. Stobie ⟨ stobie@stsci.edu ⟩
Space Telescope Science Institute, Baltimore, MD

L.G. Taff ⟨ lgtaff@stsci.edu ⟩
Space Telescope Science Institute, Baltimore, MD

Muriel A. Taylor ⟨ stars::taylor ⟩
Applied Research Corporation, Greenbelt, MD

Randall Thompson ⟨ iue::rthompson ⟩
CSC, Greenbelt, MD

Doug Tody ⟨ dtody@noao.edu ⟩
NOAO, Tucson, AZ

Gary Toller ⟨ cobecl::toller ⟩
G.S.C., Columbia, MD

L. V. Tóth
Eötvös University, Budapest, HUNGARY

Francesco Tribioli ⟨ tribioli@arcetri.astro.it ⟩
Osservatorio di Arcetri, Firenze, ITALY

Massimo Trifoglio ⟨ 38045::massimo ⟩
CNR-TESRE, Bologna, ITALY

Ryland J. Truax ⟨ truax@mps.ohio-state.edu ⟩
The Ohio State University, Columbus, OH

Hiroshi Tsunemi ⟨ tsunemi@jpnoskfm.bitnet ⟩
Osaka University, Toyonaka, Osaka, JAPAN

Benoit Turgeon ⟨ turgeon@nereid.sal.ists.ca ⟩
ISTS/Space Astrophysics Lab and York Univ., Concord, ON, CANADA

Susan J. Tuttle ⟨ logan@cfa.harvard.edu ⟩
SAO Center for Astrophysics, Cambridge, MA

Suzanne Tyc-Dumont
CNRS FRANCE, Marseille, FRANCE

Stephen Unger ⟨ swu@starlink.astronomy.cambridge.ac.uk ⟩
RGO, Cambridge, UNITED KINGDOM

Frank Valdes ⟨ fvaldes@noao.edu ⟩
IRAF Group/NOAO, Tucson, AZ

Thijs van der Hulst ⟨ vdhulst@rug.nl ⟩
Kapteyn Astronomical Institute, Groningen, THE NETHERLANDS

Frans van Hoesel ⟨ hoesel@hrdksw5.bitnet ⟩
Kapteyn Observatory, Roden, THE NETHERLANDS

Roland Vanderspek ⟨ roland@space.mit.edu ⟩
MIT, Cambridge, MA

Luis A. Martinez Vazquez ⟨ lamb@alfa.astroscu.unam.mx ⟩
Instituto de Astronomia-UNAM, Mexico D. F., MEXICO

Peter W. Vedder ⟨ pwv@ssl.berkeley.edu ⟩
UC Berkeley/SSL/CEA, Berkeley, CA

Wolfgang Voges ⟨ mpe::whv ⟩
MPI fuer Extraterrestrische Physik, Garching, GERMANY

Michael H. Ward ⟨ ward@astro.as.utexas.edu ⟩
McDonald Observatory, Ft. Davis, TX

Rein H. Warmels ⟨ rwarmels@eso.org ⟩
European Southern Observatory, Garching, GERMANY

Ron Watkins ⟨ ron@argus.lpl.arizona.edu ⟩
LPL (Lunar and Planetary Lab), Tucson, AZ

M. G. Watson ⟨ 19838::mgw ⟩
University of Leicester, Leicester, ENGLAND

Ann E. Wehrle ⟨ aew@ipac.caltech.edu ⟩
IPAC/Caltech, Pasadena, CA

Nicholas Weir ⟨ weir@deimos.caltech.edu ⟩
Caltech, Pasadena, CA

Rick Wenk ⟨ raw@physics.att.com ⟩
AT&T Bell Laboratories, Murray Hill, NJ

N. E. White ⟨ lheavx::white ⟩
HEASARC/GSFC, Greenbelt, MD

Richard A. White ⟨ sdcdcl::rwhite ⟩
NASA/GSFC, Greenbelt, MD

Richard L. White ⟨ rlw@stsci.edu ⟩
STScI/JILA, Boulder, CO

T. N. Wilkins ⟨ tnw@ast-star.cam.ac.uk ⟩
Institute of Astronomy, Cambridge, UNITED KINGDOM

Winifred E. Williams ⟨ wwilliams@noao.edu ⟩
NOAO/NSO, Tucson, AZ

Ramon L. Williamson II ⟨ williamson@stsci.edu ⟩
Space Telescope Science Institute, Baltimore, MD

Daryl Willmarth ⟨ willmart@noao.edu ⟩
NOAO/KPNO, Tucson, AZ

George W. Wolf ⟨ gww836f@vma.smsu.edu ⟩
SW Missouri State University, Springfield, MO

Richard Wolff ⟨ rwolff@noao.edu ⟩
NOAO, Tucson, AZ

Andrew Woodsworth ⟨ wdswrth@dao.nrc.ca ⟩
Canadian Astronomy Data Centre, DAO, Victoria, BC, CANADA

Diana M. Worrall ⟨ dmw@cfa.harvard.edu ⟩
Smithsonian Astrophysical Observatory, Cambridge, MA

Nailong Wu ⟨ bmabao@ica.beijing.canet.cn ⟩
Peking Astronomical Observatory, Peking, CHINA

Ryan J. Wyatt ⟨ rjw@spacsun.rice.edu ⟩
Rice University, Houston, TX

Chengda Yang ⟨ chengda@convx1.ccit.arizona.edu ⟩
University of Arizona, Dept. of Statistics, Tucson, AZ

Daryl J. Yentis ⟨ yentis@xip.nrl.navy.mil ⟩
Naval Research Laboratory, Washington, DC

Wesley Young ⟨ wyoung@nrao.edu ⟩
NRAO, Socorro, NM

Wan Yuzhi
Shanghai Observatory/NOAO, Tucson, AZ

Nelson Zarate ⟨ zarate@stsci.edu ⟩
Space Telescope Science Institute, Baltimore, MD

Martin V. Zombeck ⟨ mvz@cfa.harvard.edu ⟩
Harvard/Smithsonian Center for Astrophysics, Cambridge, MA

W. Zwitser ⟨ zwitser@rug.nl ⟩
Kapteyn Astronomical Institute, Groningen, THE NETHERLANDS

First Annual Conference on Astronomical Data Analysis Software and Systems
November 6-8, 1991 Tucson, AZ

SECTION 1

DATA COMPRESSION AND FILE FORMATS

WAVELET-BASED COMPRESSION SOFTWARE FOR FITS IMAGES

WILLIAM H. PRESS
Harvard-Smithsonian Center for Astrophysics, 60 Garden Street, Cambridge, MA 02138, USA

ABSTRACT FITSPRESS is a program for compressing astronomical FITS images. It is capable of functioning in both lossless (reconstructed file byte identical to original) and high-compression filtered (truncated wavelet representation) modes. A brief description of the program's design and implementation is given, and sample output is shown. The program is available in beta release by anonymous ftp.

INTRODUCTION

The FITS (Flexible Image Transport System) format, originally proposed by Wells, Greisen & Harten (1981), has rapidly become the standard by which astronomers exchange machine-readable data, notably including astronomical images. Designed for maximal portability to a wide variety of machine architectures, the FITS standard avoids, insofar as possible, any processing or encoding of the packaged data. In essence it "giftwraps" the input data in a simple format that can be unwrapped at the receiving end.

Because astronomical images contain large numbers of pixels, FITS image files are very large. Currently, images of 800 by 800 by 16 bits deep are common, implying a minimum file size on the order of 1.2 Mbyte; in the near future, one will begin to see 4096 by 4096 by 16 bit images, whose size is on the order of 33.5 Mbyte. All-sky digital surveys will produce many thousands of such images. The problem of distributing such libraries of images, and of making them available on-line to interested researchers, is not trivial.

As a feasibility exercise, we have developed software for the compression of FITS image data. A program, called FITSPRESS, is available, currently in a version 0.8 beta release, by anonymous ftp from Internet address 128.103.40.79 (file pub/fitspress08.tar.Z). This paper gives a brief introduction to the design considerations embodied in FITSPRESS, and shows examples of its use.

The FITSPRESS algorithm is not a final optimized proposal for a standard. Below, we will indicate some "roads not taken" where different techniques might improve the algorithm. It is intended, however, that FITSPRESS provide a lower bound for the cabilities that a good astronomical compression standard should have, and that it should stimulate discussion, both from the user community (regarding what features they collectively want) and from the algorithm community (regarding what techniques are best in this application).

DESIDERATA FOR ASTRONOMICAL IMAGE COMPRESSION

It is a mathematical fact that independently random, uniformly white-noise data values cannot be compressed *at all.* This fact puts fundamental limits on what one can expect from a *lossless* compression algorithm, one where the original data can be exactly reconstructed from the compressed file. For example, if a 16 bit image has, in each pixel, 5 low-order bits of uniform white noise (whether from sky background, dark current, A-to-D conversion, or whatever), then the image *cannot possibly* be compressed to a size less than 5/16 of the original, that is, with a compression ratio of greater than 3.2:1.

The alternative to lossless compression is what we will call "filtered" compression. If an image has a truly uncorrelated white-noise background, then the pixel-by-pixel values of that noise are themselves uninteresting, and should not be stored at all (except possibly in some statistical summary form). Rejecting the noise in this manner allows very much higher compression ratios; we will see that a range from 20:1 to 50:1 is feasible. However, (1) the reconstructed image is not byte for byte identical to the original input; (2) there is no unique way to positively identify true noise, so that real algorithms will inevitably introduce artifacts into the image; and (3) there is not an observer in the world who will be willing to run his or her precious data into an irreversible filtering algorithm designed by theorists!

In FITSPRESS we approach this confluence of mathematical and sociological issues in the following way: The output of a FITSPRESS compression consists of two files. The first, a "FITSPRESS summary file" can by itself be reconstructed to be a filtered version of the original image. The degree of filtering is a parameter set by the user at the time of compression: stronger filtering gives a shorter summary file (e.g., 50:1), weaker filtering gives a longer summary file (e.g., 10:1), which reconstructs to an image that is arithmetically closer to the original. The second file, a "FITSPRESS residual file" contains, in an only slightly compressed form, all the data that the summary file regards as noise. The two files together are shorter than the original image by a factor between 2 and 5, and reconstruct exactly to the original (lossless compression).

In practice one might distribute the two files through different channels. For example, the summary files of an all-sky survey might be kept on-line, allowing them to be immediately accessed in rapid succession, while the residual files might be off-line on CD ROM. In many applications, if the filtering parameter ("compression level") is properly chosen at compression time, all required detail will be present in the summary file, and one may rarely or never need to use the residual file.

In practice, an algorithm decides what noise to throw away by instead deciding what non-noise information to keep. Unfortunately, there is as much art as science in this. Results are highly dependent on the nature of the images, and on what their user community thinks "looks right". Discussions with a number of optical and radio observers lead, as the choice embodied in FITSPRESS, to a two-pronged strategy: First, a certain fraction of pixels are "important" and are stored exactly in the summary file. While more sophisticated choices are probably desirable, FITSPRESS version 0.8 takes the brightest pixels to be the important ones. The fraction of all pixels thus stored is determined by the user-chosen compression level: level 1 keeps 1% exactly; level 2, 2%, etc.

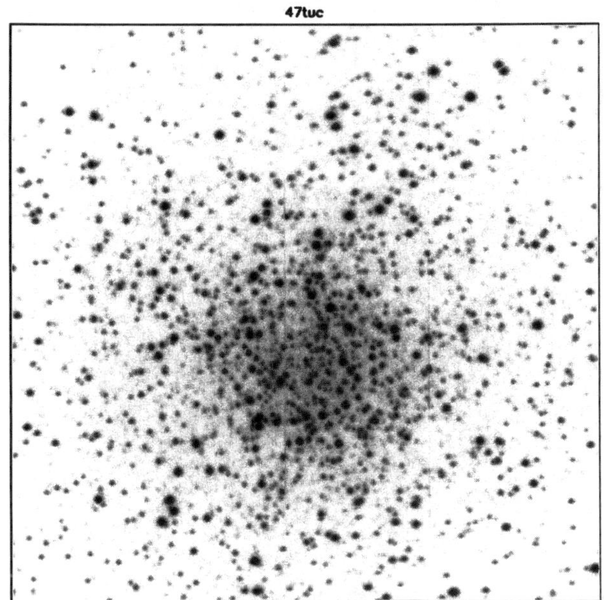

FIGURE I Original CCD image of globular cluster 47 Tuc (FITS image courtesy of J. Grindlay).

Second, it is desirable to keep features of low surface brightness that are statistically significant because they extend over many pixels. Wavelet transforms (Daubechies 1992; Press et al. 1992; Strang 1989; Mallat 1989) provide a very natural and computationally efficient means for doing this. Here, the user-chosen compression level specifies what fraction of the image's complete set of wavelet coefficients are stored: 1% at level 1, etc. (Choices must also be made as to how the coefficients are truncated and discretized.) For a given compressed size, there is obviously a tradeoff between how much to allocate for important pixels versus how much to allocate for important wavelets. We have made a fixed choice on the basis of a particular set of test images, described below. An obvious improvement would be to make an adaptive choice, based on the image.

Some additionally imposed design constraints are these: 1. All output files should themselves be minimally conforming FITS files consisting of valid FITS headers followed by streams of bytes (in multiples of 2880). 2. All program I/O should correspondingly be done in blocks of 2880 bytes. 3. The code must port correctly to machines with differing byte orderings of 16 bit integers. 4. The command line syntax should be "safe" against accidentally overwriting an original FITS data file.

Before describing the algorithm in more detail, let us get oriented with a concrete example. Figure I shows a raw 800 by 800 by 16 bit CCD image of globular cluster 47 Tuc. In fact, this is quite a difficult image to compress, because it has a very large number of resolved and partially resolved stars, and it has structure on all spatial scales. Compression artifacts will thus be particularly acute for this image. In lossless compression mode, FITSPRESS (version 0.8) achieves a compression ratio of only 1.5:1. Figure II shows the

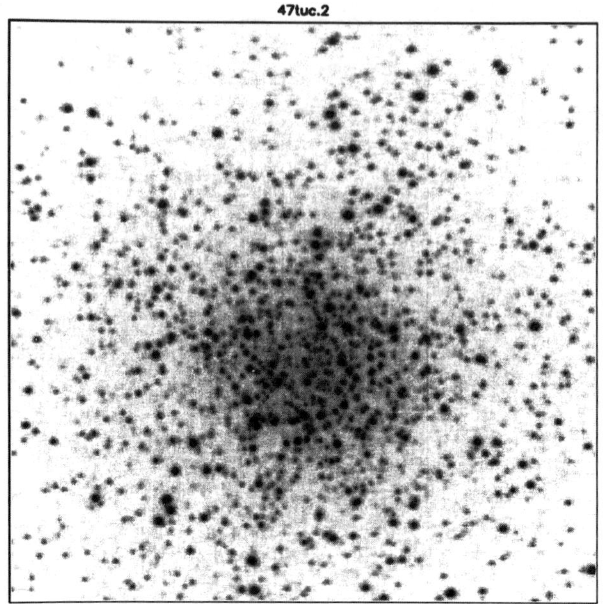

FIGURE II 47 Tuc reconstructed from FITSPRESS compression level 2, with compression ratio 12:1. This is probably a useful image for most purposes.

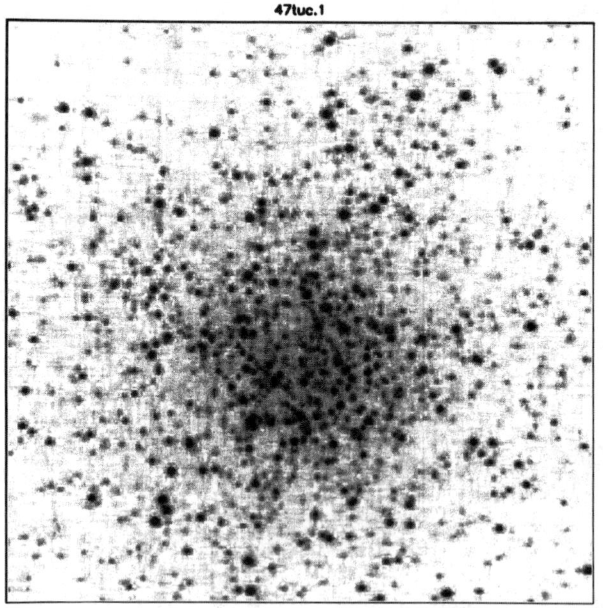

FIGURE III 47 Tuc reconstructed from FITSPRESS compression level 1, with compression ratio 21:1. This is a marginally useful image.

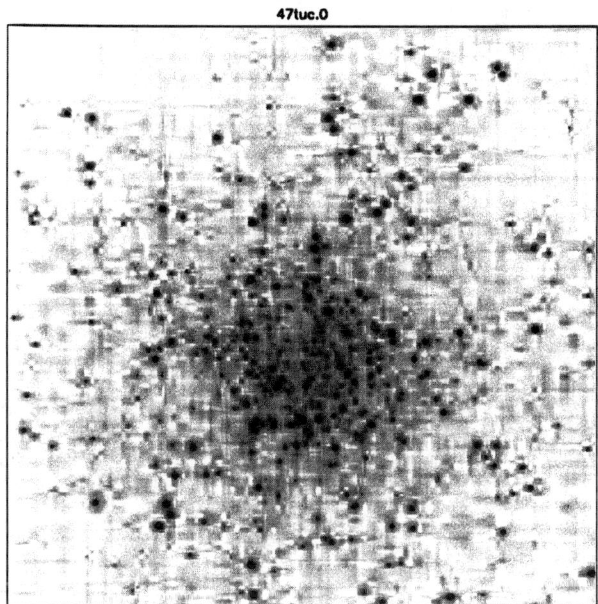

FIGURE IV 47 Tuc reconstructed from FITSPRESS compression level 0, with compression ratio 37:1. This is probably not a useful image.

reconstruction from filtered compression level 2, whose compression ratio is 12:1. Artifacts are clearly present – one sees incipient signs of "wavelet plaid" –, but the overall image quality is not significantly degraded. Figures III and IV show compression levels 1 and 0, shows compression ratios are respectively 21:1 and 37:1. One sees the wavelet plaid artifact (which is due to discretization and truncation of the wavelet coefficients) becoming increasingly visible. One would probably not recommend that these images be used for astronomical measurements, though the image at 21:1 might be useful in a survey catalog. Note that the smoothed luminosity profile of the cluster remains accurately represented even at the highest compression: when stars are lost as individually resolved objects, their luminosity is transferred to *some* scale of local wavelet.

FITSPRESS IMPLEMENTATION

The FITSPRESS program is presently implemented in portable C source code for UNIX systems, and has been run successfully on Sun Sparc and DEC RISC workstations. There should be no intrinsic difficulty in porting the code to VMS or other operating systems. Ports to IBM/PCs are more problematic because of the relatively large memory requirements; with a suitable DOS extender, and a lot of memory, such ports are probably possible.

Figure V shows the data flow within the algorithm, as follows: The length of the input FITS file's ASCII header is noted, and the header is transferred uncompressed to the output buffer. (This is so that the original header of a compressed summary file can be examined with any text editor, a desirable

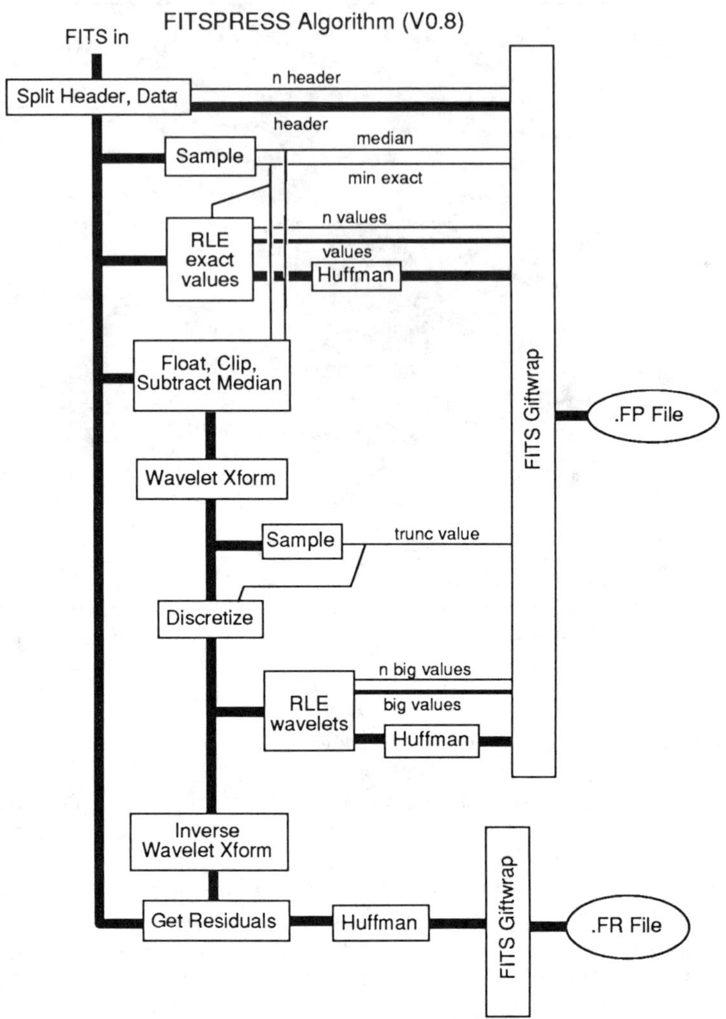

FIGURE V Data flow in the FITSPRESS algorithm. See text for details.

feature.) The image data is then sampled (random pixels chosen) to estimate its median, and – based on the input compression level – the minimum brightness level that should be stored exactly. (Random sampling is much faster than processing all pixels, and is accurate enough.)

We now need to save the exact pixel values and their locations. The values (and their number) are put to the output buffer in uncompressed form. (One could achieve minor performance improvements by compressing at this stage, but we have not done this.) The locations are run-length encoded, and then Huffman coded into the output buffer.

The image, with its median value subtracted and its bright pixels clipped

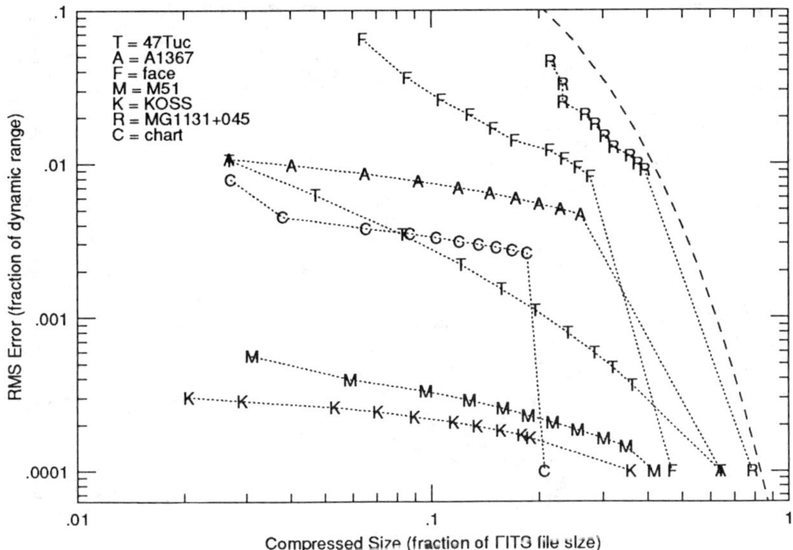

FIGURE VI Accuracy of reconstructed images as a function of FITSPRESS compression level for 7 sample images. The plotted letters indicate the accuracy and size obtained using FITS compression levels 0 through 9 (left to right along each curve). The tenth (rightmost) point on each curve shows the size of the compressed file (summary plus residual) in lossless compression mode. The error is arbitrarily plotted as 0.0001 for these points, but the reconstruction is in fact exact. The dashed curve shows the degree of compression and accuracy that would obtain from simply truncating low-order bits on random 16-bit data. See text for further discussion.

to the minimum exact brightness level, is now transferred to a floating point buffer. A two-dimensional wavelet transform is taken, with transform coefficients replacing the real image values. The transform coefficients are sampled, and a truncation level and unit for discretization is chosen. The wavelet image is then discretized to integer values, most of which are zero by choice of the truncation level.

The nonzero discretized wavelet coefficients are now run-length encoded by a procedure almost identical to that used for the bright pixels. One difference is that the values $0, \pm 1$ (truncation units) are encoded as part of the run-length code, so that only values with magnitude > 2 (truncation units) are transferred directly to the output buffer, where they are stored in discretization units. We choose the discretization unit to be 8 times smaller than the truncation unit, so that values > 2 (which are presumed to be the most statistically significant) are stored to at least 3 bit accuracy. There is some room for further experimentation and improvement in this procedure.

The contents of the output buffer is at this point "giftwrapped" into a FITS

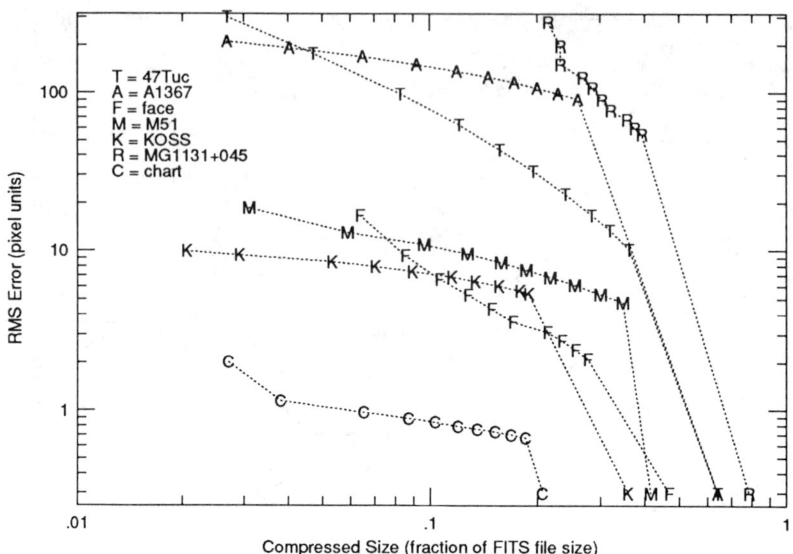

FIGURE VII Same as Figure VI, but errors now shown as raw counts. For those curves that are relatively flat, one sees the level of their white noise backgrounds: ≈ 1 count for image C, ≈ 10 counts for images K and M, ≈ 100 counts for image A (a digitized photographic plate). These images are highly compressible, i.e., at high compression there is very little lost except noise. Image T, a globular cluster, is genuinely difficult to compress, since it has a lot of true structure on all scales. See text for discussion of image R.

byte-data format, and it becomes the FITSPRESS summary file, with filename extension ".FP".

To calculate the residual file, the truncated and discretized wavelet coefficients are inverse transformed, exactly as will be done in decompression mode. The results are subtracted from the original (input) image values, and Huffman coded to the output buffer, which is FITS giftwrapped and becomes the FITSPRESS residuals file (filename extension ".FR").

PERFORMANCE OF THE ALGORITHM

Figures VI and VII show the performance of the algorithm (version 0.8) on a set of test images collected from a variety of sources. The test images are as follows: "T" is the globular cluster 47 Tuc (Figure I, above); "M" is a CCD image of M51 (Figure VIII, below); "K" is a segment of the KOSS (Kirshner et al. 1992) CCD drift scan survey (Figure X, below); "A" is a section of digitized photographic plate from the Hubble Space Telescope Guide Star Survey containing the central region of the galaxy cluster Abell 1367 (Figure XII, below); "R" is a VLA radio image of the gravitational lens MG1131+045 (not shown); "C" is a supernova

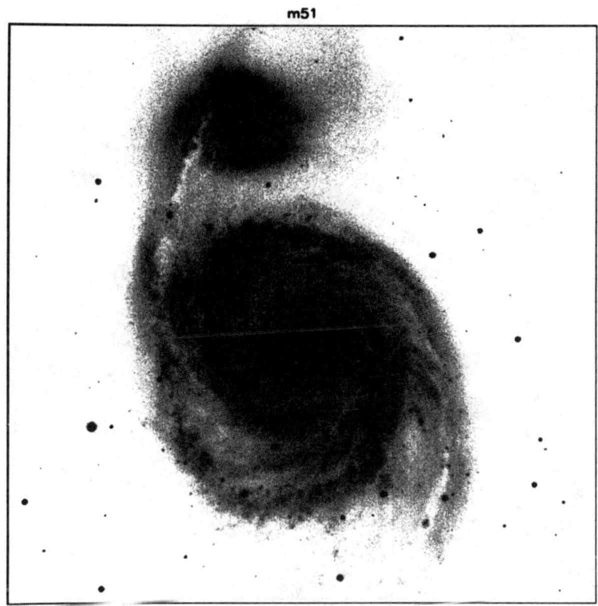

FIGURE VIII Original CCD image of M51 (FITS image courtesy of J. Tonry).

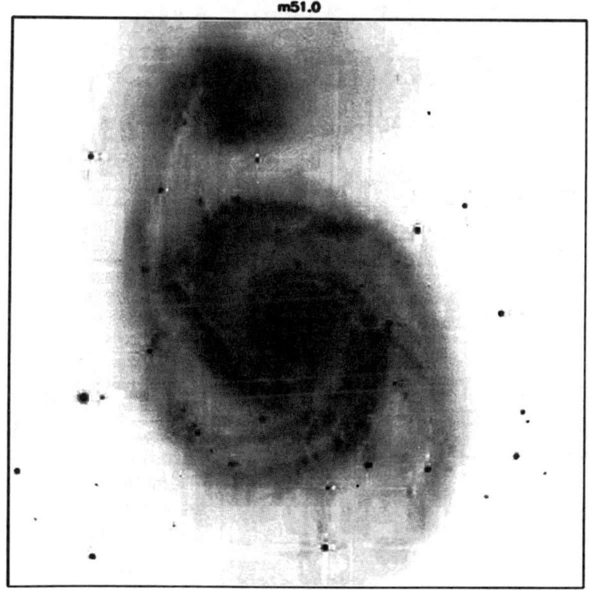

FIGURE IX M51 as reconstructed from compression level 0 (32:1 compression).

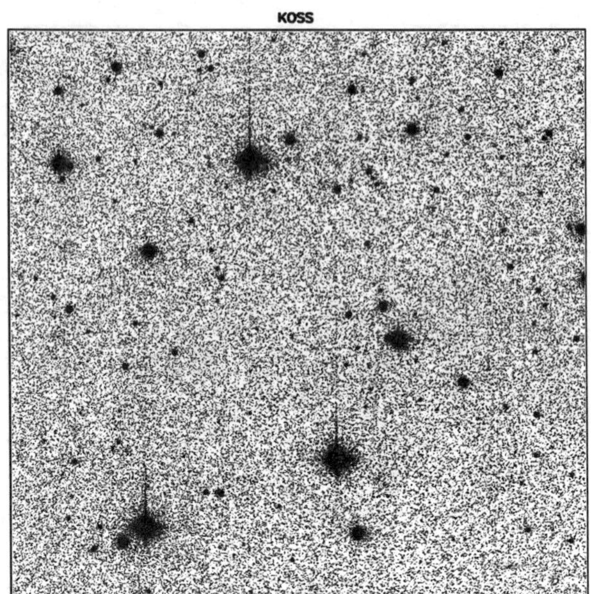

FIGURE X Original KOSS CCD drift scan, detail (FITS image courtesy of R. Kirshner).

finding chart showing a cluster of galaxies and including some lettering (not shown); "F" is a standard Kodak test image of a human face (not shown).

In general, images A, C, M, and K compress beautifully. The flat curves in the Figures demonstrate a very clean separation of signal from noise, so that the reconstruction accuracy degrades hardly at all even at the highest compressions. The common accuracy of the compression levels represents the actual noise level of the images. For these images, arguably, the compressed image is "better" than the original: it has been stripped of noise and very little else. Figures VIII through XIV show originals and sample reconstructed images. In general, one compression level larger than the level shown becomes practically indistinguishable from the original; that is, our examples are chosen to show some compression artifacts.

Images T, F, and R do not compress well, though for differing reasons. We have already discussed image T above. Image F (photographic image of face) has a combination of structure on many scales (e.g., fine texture in the female model's hair) and also grain noise that is poorly modeled as independent white noise. Image R is the most incompressible, but for reasons that are, in practice, avoidable: The signal to noise ratio of the brightest feature of the ring is only about 5, and the image is normalized to the largest 16 bit value. Furthermore, the ring occupies only the central 1% (by area) of the image. Thus, virtually all the bits in the image are noise, and it is not surprising that the points in Figure VI fall close to the dashed curve, which represents compression by simply truncating low-order bits. With sensible cropping and scaling, image R could be made about as compressible as any of the more favorable cases.

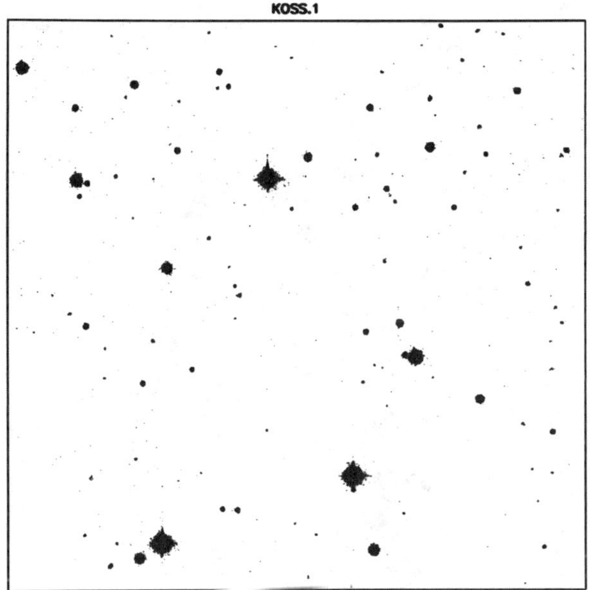

FIGURE XI KOSS image as reconstructed from compression level 1 (34:1 compression).

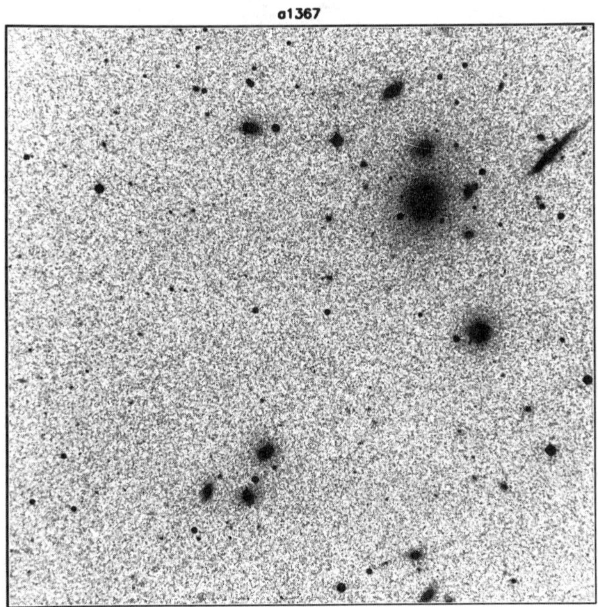

FIGURE XII Digitized photographic GSS image of Abell 1367, detail (FITS image courtesy of R.L. White).

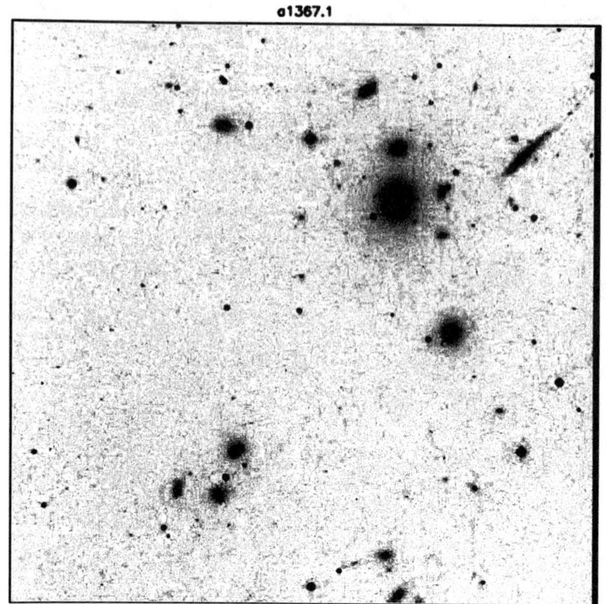

FIGURE XIII Abell 1367 as reconstructed from compression level 1 (25:1 compression).

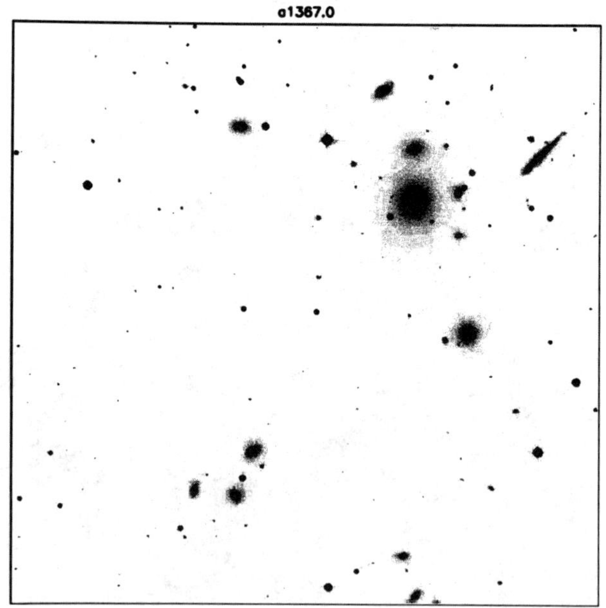

FIGURE XIV Abell 1367 as reconstructed from compression level 0 (37:1 compression).

DO'S AND DON'TS FOR OBSERVERS

One can now readily give a set of "do's and don'ts" for observers who want to collect images that are easily compressible to high compression ratios:

1. Don't store many bits deep of sky noise or dark current. (At issue is not the mean noise level, which can be anything, but rather the pixel-to-pixel noise fluctuations.) Three bits (variations ± 4 counts) are more than enough. It is a mathematical fact that low surface brightness features are *not* significantly degraded by quantization, even when the quantization steps are larger than the feature brightness (the so-called Dithering Theorem).

2. Don't invert the FITS brightness scale. A compression algorithm should be able to rely on large positive values being "bright". Likewise, do use 16 bit two's-complement arithmetic (-32768 to 32767) as required by the FITS standard. Don't use unsigned integer format (0 to 65535).

3. Don't border images with extreme pixel values. It only wastes precious compression bits. Do fill in formatted blank spaces with a constant value that is not too different from the median value of the image.

4. Do crop images so that the object of interest fills the frame.

OTHER APPLICABLE TECHNIQUES

With the exception of wavelets, which are new and trendy, we have chosen in FITSPRESS version 0.8 to use a rather standard set of tools: run-length encoding and Huffman coding. Buried in the code there is also some use of range compression (logarithmic encoding), and Monte Carlo sampling.

A primitive type of wavelet, the Haar transform, has previously been used by Fritze et al. (1977) and by R.L. White at STScI (unpublished). We have done some experiments that suggest that the simplest Daubechies wavelets ("D4 coefficients") are significantly more efficient. We also find that wavelets strongly dominate simple linear prediction schemes (including delta-mod and previous line prediction).

On the other hand, it may well be that the quad-tree encoding schemes advocated by Huang & Bijaoui (1991) and by White (see Dewdney 1989 for an introduction) are more efficient than the run-length encoding that we have used. This will be investigated for later FITSPRESS versions.

Arithmetic coding (see Press et al. 1992 for an introduction) is another alternative to the combination of run-length and Huffman coding. At present, it seems to be too slow for the slight increase in compression that it might give.

We have not exploited the techniques of pre-whitening or range-scaling, as means for rendering the actual noise closer to white. This could be quite important in cases where the noise is actually $\sqrt{N}$ counting noise, or $\sqrt{N}$+constant. We will investigate this in a later FITSPRESS version.

CONCLUSION

In its present version, the FITSPRESS program, available by anonymous ftp (see Introduction), seems able to compress a variety of astronomical FITS images to compression ratios of between 10:1 and 30:1 with little loss of astronomical utility. There is perhaps an additional factor of two to be had (at the same quality levels) by refinement of the general algorithm. However, there is also a need for dialogue between users and algorithm developers, so that observers can both become accustomed to the kind of artifacts that compression algorithms may introduce and can learn to gather images that are more easily compressible.

ACKNOWLEDGMENTS

I thank Seth Finkelstein for his professional programming assistance in translating the FITSPRESS algorithm into portable C. I thank John Tonry, Robert Kirshner, Jacqueline Hewitt, Richard L. White, and Phil Pinto for supplying FITS images, and for discussions of user needs; and Alberto Accomazzi and Sara L. Press for assistance in producing hardcopy images.

REFERENCES

Daubechies, I. 1992, Wavelets (Philadelphia: Society for Industrial and Applied Mathematics)

Dewdney, A. K. 1989, The Turing Omnibus (Rockville, MD: Computer Science Press), §44

Fritze, K., Lange, M., Möstl, G., Oleak, H., & Richter, G. M. 1977, Astron. Nachr., 298, 189

Huang, L., & Bijaoui, A. 1991, Experimental Astron., 1, 311

Kirshner, R. P., Oemler, A., Schechter, P. L., & Shectman, S. A. 1992, in preparation

Mallat, S. G. 1989, IEEE Trans. on Pattern Analysis and Machine Intelligence, 11, 674

Press, W. H., Teukolsky, S. A., Vetterling, W. T., & Flannery, B. P. 1992, Numerical Recipes: The Art of Scientific Computing, Second Edition (New York: Cambridge University Press)

Strang, G. 1989, SIAM Review, 31, 614

Wells, D. C., Greisen, E. W., & Harten, R. H. 1981 A&AS, 44, 363

DEFINITION OF A DISTRIBUTION FORMAT FOR X-RAY DATA

M.A. CONROY

M.R. GARCIA, E.G. MANDEL, J. ROLL, AND D.M. WORRALL
Smithsonian Astrophysical Observatory, Cambridge, MA 02138

ABSTRACT X-ray data from the ROSAT satellite will be distributed to Guest Observers on FITS tapes. The precise contents and format of the tape were chosen bearing the following issues in mind:
1. The primary data component for an X-ray observation is a list of photon events, and the user requires additional data for selection and analysis of the primary data. The introduction of the FITS bintable extension has provided a standard format that makes the storage of photon events plausible and efficient.
2. The photon events have associated properties, such as position, time, energy, quality, but the precise list of properties is experiment-dependent. This issue has been addressed by defining minimum elements for each record type, but allowing self-defining record extensions, specific to individual projects.
3. Filtering and selection of photons will be a user's first step in the scientific analysis of the data. This issue has been addressed by the inclusion of multiple *filters* to permit the storage of all photon events, while providing a simple mechanism for extracting a nominal subset that includes only *good* events.

These solutions in defining the contents and formats of the FITS files should be useful for other event-based data sets.

INTRODUCTION

The goal of the US Rosat Science Data Center (RSDC) is to allow observers to analyze their observations at their home institution. This requires two things: a portable software analysis system and a set of machine independent data files. The first item is the focus of the Post-Reduction Off-line Software (PROS) project (Worrall et al. 1991) which is developing a layered package to run within the Image Reduction and Analysis Facility (IRAF; Tody 1986). The Flexible Image Transport Format (FITS; Wells, Greisen & Harten 1981) is the natural choice for the second item, since it has been used for years to transport astronomical data between institutions.

X-RAY DATA AND EVENT LISTS

X-ray Data and Standard FITS

Many of the ROSAT standard processing system output files are well-suited to the FITS image format. In particular, all data maps are formatted as images. These include a low resolution map of the screened events, a model of the field background and an exposure map recording the relative amount of observing time each image pixel was exposed. Many of the derived files can be easily formatted into FITS table extension (Harten et al. 1988). These include analysis results such as detected sources and known catalog objects in the field of view.

X-ray Events and BINTABLE FITS

The actual event list produced by an X-ray detector is not well suited to either the image or table extension FITS format. Special properties of X-ray data which make both inappropriate are as follows:

Sparse For ROSAT, the Position Sensitive Proportional Counter (PSPC) has 15360 x 15360 pixels and the High Resolution Imager (HRI) has 8192 x 8192 pixels. Most pixels are empty. Therefore a FITS image format would be unwieldy and expensive in terms of data storage, and the dynamic range of the data tends to be too small for optimum use of image-compression techniques, such as those described by Press (1992).

Copius While the pixel arrays may be sparsely filled, a modest 15000 second exposure with an average five counts per second counting rate will produce a 1.5 Megabyte file, even in the optimized native machine representation. An ASCII table representation would increase this size by at least a factor of three.

Precious Experience has shown that there is no such thing as a *bad* photon. No observer is willing to allow any photons to be automatically and permanently discarded. Therefore, all photons from an observation must be archived, with provisions for easy selection of data subsets, such as photons used in the ROSAT standard processing.

Non-Uniform X-ray data are acquired under a set of constantly changing conditions. The position of the target on the detector is not constant, and observing conditions, such as background rate, aspect quality, and bright object contamination change as a function of satellite position and telemetry quality. Auxiliary data must be stored to record these conditions as well as to maintain time- and energy-tags for each photon.

For these reasons the newly proposed bintable extension to the FITS format (Cotton & Tody 1991) is an ideal choice for recording both the master X-ray event list as well as the support files that record the observing conditions.

QPOE/EVENT IMPLEMENTATION OF BINTABLE FITS

The bintable extension allows a self-defining record structure and so in principle any simple fixed-length binary-record format can be converted to a FITS representation. However, a FITS format is only as useful as the system-dependent data structure into which it is converted for analysis. FITS image files and (ASCII) table extensions have well-defined and well-understood counterparts in most major astronomical systems, (*e.g.*, AIPS, MIDAS and IRAF). The FITS bintable format for the ROSAT event data is supported in IRAF by the Quick (*i.e.*, prototype) Position Ordered Event (QPOE) data structure. The design of the ROSAT bintable FITS file is driven by the QPOE implementation. The required extension name is EVENT.

Required EVENT Elements

The only required elements in the event definition are:

x	x position	2-byte integer	(strictly positive)
y	y position	2-byte integer	(strictly positive)

These two entries allow compatibility with IRAF image format. This upward compatibility guarantees access to all existing software which accesses data in the standard IRAF image format.

Optional EVENT elements

All other record elements are optional and self-defining, and the QPOE structure will accommodate whatever data-structure attributes are defined. Entries defined for ROSAT data are:

time	arrival time in seconds	8-byte floating point
pi	energy	2-byte integer
pha	pulse-height amplitude	2-byte integer
dx	detector x	2-byte integer
dy	detector y	2-byte integer
corx	corrected x	2-byte integer
cory	corrected y	2-byte integer

The IRAF/PROS package contains several special-purpose tasks which access time and/or energy (or pulse-height) information.

QPOE IMPLEMENTATION OF AUXILIARY BINTABLE FITS

In addition to the primary event list, the QPOE structure allows the inclusion of other user-defined binary data structures. Additional bintable extensions are stored here to record the observing conditions. There are currently two well-defined and supported extensions.

Good Time Intervals

The time-intervals during which all observing conditions were determined to meet the *standard* acceptable conditions are denoted as the standard *good time intervals.* This FITS bintable extension is named *GTI* and has the format:

start	start time in seconds	8-byte floating point
stop	stop time in seconds	8-byte floating point

Temporal Status Intervals

The record which stores the status of all the defined observing conditions is called a *temporal status interval.* The FITS bintable extension is named *TSI* and it has both a required and optional section. The required portion of the record is:

tstart	start time in seconds	8-byte floating point
failed	reasons failed	4-byte integer (=0 if accepted)

The definition of observing conditions is extremely mission specific, and so the rest of the record is self-defining. Some of the attributes defined for the ROSAT mission are:

logicals	bit-encoded on/off indicators
hibk	high background level
hvlev	high voltage setting
vg	viewing geometry level
aspstat	aspect quality status

IRAF/QPOE FILTERING

The FITS bintable extensions defined above can be converted into an IRAF QPOE data structure by the PROS software. In turn the IRAF/QPOE data structure comes with the following filtering support:

Command Line/Runtime Filtering

The QPOE data structure supports automatic filtering using any of the defined event attributes. The filtering is specified by appending the filter to the filename. Where no filter is specified, all photons are taken.

QPOE Default Filters

The QPOE data structure supports the definition of a default time filter. This allows the inclusion of all photons in the event list, while defining a default time filter that excludes the photons that did not pass the standard screening criteria. Additional time filters can be specified by users. This support requires that the *time* event attribute be included in the event definition and that it have the 8-byte floating point representation.

Coordinate System Change

The user may select detector coordinates in place of x,y by appending a simple string to the filename.

Extended TSI Filtering

The PROS package is being developed to support additional filtering based on the attributes defined in the *TSI* extension. The syntax will be modeled after the existing event-attribute filtering syntax. It will then be possible to define the desired conditions in terms of limits on any of the defined TSI attributes. The PROS software will use the TSI records to convert the TSI filter to the corresponding time attribute filter for use by IRAF. The IRAF system will then automatically retrieve just the photons that meet the stated criteria.

APPLICATIONS AND MISSIONS

A committee of representatives from the Goddard Space Flight Center (RSDC, HEASARC, NSSDC) and SAO (RSDC) is currently working to finalize the details of the FITS formats used for ROSAT and the archival X-ray data. Meanwhile FITS formats currently defined and implemented for the ROSAT project have been used successfully for the *Einstein* SLEW-survey CD-ROM (Plummer et al. 1991), the *Einstein* HRI Merged Photons CD-ROM (Plummer et al. 1992). and the AXAF mirror-test data system. The ASTRO-D project has also used this format as part of its analysis system (Corbet et al. 1992).

REFERENCES

Corbet, R. H. D., Larkin, C., & Nousek, J. A. 1992, this volume

Cotton, W., & Tody, D. "Binary Table Extension to FITS: A Proposal" Draft 20 September 1991, available on fits.cx.noao.edu

Harten, Grosbøl, Greisen, & Wells 1988, Astron. Astrophys. Suppl, 73, 365

Plummer, D., Schacter, J., Garcia, M., & Elvis, M. 1991, "The Einstein Observatory IPC Slew Survey (FITSA3D/CD-ROM Version)", Smithsonian Institution, Astrophysical Observatory

Plummer, D., Manning, K., Schacter, J., Primini, F., & Garcia, M. 1992, "Einstein HRI Merged Photons CD-ROM", Smithsonian Institution, Astrophysical Observatory

Press, W. 1992, this volume

Tody, D. 1986, in Instrumentation in Astronomy VI, SPIE, 627, part 2

Wells, D. C., Greisen, E. W., & Harten, R. H. 1981, Astron. Astrophys. Suppl, 44, 363

Worrall, D. M., et al. 1991, in Data Analysis in Astronomy IV, eds. V. Di Gesu et at., Plenum Press, in press

FITSIO AND FITS FILE UTILITY SOFTWARE

W. D. PENCE
HEASARC, NASA/GSFC, Greenbelt, MD 20771

ABSTRACT A Fortran subroutine package called FITSIO has been written to provide a standard programming interface to easily read and write data files in FITS format. FITSIO provides a machine independent interface which insulates programmers from having to deal directly with the complicated FITS file formats. Programmers at the HEASARC are now developing a set of general utility programs using FITSIO to manipulate the contents of FITS files.

FITSIO SUBROUTINE INTERFACE LIBRARY

FITSIO is a machine-independent Fortran-77 subroutine interface for reading or writing data files in the FITS (Wells, Greisen & Harten 1981) format. This package was written to provide an easy to use interface for accessing FITS files which insulates the programmer from having to deal with the complicated formating details in the FITS file. The FITSIO package contains about 15000 lines of ANSI-standard Fortran-77 code most of which is machine independent, however, a small set of machine specific subroutines are also provided for most popular computers including VAX/VMS, DECstations, Sun workstations, IBM PC's, and IBM mainframes. It supports all the commonly used types of FITS files, including 'random grouped' data, and ASCII and binary table extensions. One of the main features of FITSIO is that it provides random access I/O capabilities which allow users to read or write information in the FITS file in any desired order. Because of this, the package is oriented to FITS files on magnetic disk and does not support magnetic tape devices. Subroutines are provided in the FITSIO package to perform all the required types of operations on FITS files, such as: open and close files, read, write, modify, or delete header keyword values, and read or write any element of the associate data array or table. The package also contains extensive error checking to help verify that the file has a valid FITS format.

The FITSIO software is freely available from the HEASARC via anonymous ftp from tetra.gsfc.nasa.gov (or 128.183.8.77) in subdirectory pub/fitsio3. Alternatively, the FITSIO files may be copied over the SPAN network from:

NDADSA::HEASARC:[EXOSAT.XANADU.FITSIO.VERSION3].

FITS FILE UTILITY SOFTWARE

The HEASARC is starting a software development project to produce a generic set of software utilities to manipulate FITS files. Each utility program will be designed to perform one simple task, such as extracting specified columns or rows from a FITS table, or sorting the elements in a column, or printing the contents of a FITS file. These tasks may be strung together in scripts to perform more complex operations. For those familiar with the STSDAS software package, this set of FITS utilities will be analogous to the TTOOLS set of tasks except that they will operate on FITS table or binary table extensions instead of SDAS tables. These FITS utilities will be designed to run as standalone tasks in a Unix or VMS environment, as well as run as tasks in the IRAF environment.

A particularly exciting aspect to this project is that many of these tasks will be layered on top of a SQL interface to FITS binary tables. SQL (Structured Query Language) is a database query language that has become the defacto industry standard. This may open up whole new possibilities for using FITS binary tables in database applications, and, at least for some simple applications, could eliminate the need to purchase an expensive commercial database software package.

REFERENCES

Wells, D. C., Greisen, E. W., & Harten, R. H. 1981, A&AS, 44, 363

FITS KERNEL INTERFACE IN IRAF

NELSON R. ZARATE
Space Telescope Science Institute

INTRODUCTION

A FITS kernel to the IRAF imio interface is being developed by the STSDAS group at STScI. This kernel will allow any IRAF task that handles images to also access or create images that are on disk in the FITS standard format (Draft Standard. NSDSSO 100-0.1, December 1990). The FITS kernel will use a logical record size of 2880 bytes and a physical record length determined by the host operating system.

The FITS kernel is expected to provide several advantages over the other formats supported by IRAF, especially when dealing with large images. Currently, a user must convert the FITS file using some sort of FITS reader and then use a program to analyze the data; this requires large amounts of disk space and time to perform the conversions. A clear advantage to the FITS kernel is that it will access remote FITS files, regardless of the architecture of the machine on which the file resides, and will write output in FITS files to any machine on which the user has an account.

The FITS kernel probably will not support access to tapes since the imio does have an interface for them; nor FITS extensions (XTENSION) under the imio because the kernel contains no clear definition as to the format of a disk file created by the FITS extensions.

FITS FILES ON DISK

A FITS image is a binary file on disk, consisting of two parts. The first part is a header made up of 2880-byte blocks. Each line in the header is an ASCII record of 80 characters, with the last line containing only the string "END" in columns 1-3. The rest of the header should be padded with blanks to fill it to the required 2880 characters (or multiple of 2880).

The second part of the image is the binary data, which immediately follows the last header block. This is a stream of bytes whose size is defined by the information in header keywords NAXISn and BITPIX. The total number of bytes is calculated as (NAXIS1 * NAXIS2 * .... NAXISn) * Nbytes. (Nbytes is BITPIX/8). The file must be padded with zeros after the last pixel to fill the 2880-byte block length. A value of NAXIS=0 is used to indicate that there is no data portion.

In the current implementation, the kernel ignores any extensions in the input FITS file.

FITS DATATYPE SUPPORT

The FITS kernel was designed to optimize processing speed, the assumption being that the less computation done on the pixels by the kernel, the better. The best results are obtained when the input datatype requires no scaling calculation, but if scaling is required, the FITS IEEE datatype is preferred.

The following datatypes are supported for input files:

```
8-bit unsigned binary integers
16-bit signed integers (scaled)
32-bit signed integers (scaled)
32-bit FITS IEEE single floating
       point (no scaled)
64-bit FITS IEEE double precision
       floating point (no scaled)
```

The FITS kernel will produce only the following datatypes:

```
16-bit signed integers (no scaled)
32-bit signed integers (no scaled)
32-bit FITS IEEE single floating point
       (no scaled)
64-bit FITS IEEE double precision
floating point. (no scaled)
```

SCALED FITS DATA

The input and output datatype supported by the FITS kernel can be characterized as either *scaled* or *unscaled.* Scaled data from an input FITS file will have keywords BSCALE and BZERO with values other than 1.0 or 0.0, and will force real pixel values to be computed. (Values of 1.0 and 0.0 will cause the kernel to simply check input data for possible byte swapping.)

The FITS kernel does not yet support any scaled FITS IEEE files, i.e., a FITS file with BITPIX values of -32 or -64 and BSCALE values other than 0.0 or 1.0.

TYPICAL FITS IMAGE HEADER

```
SIMPLE  =                 T
BITPIX  =                32 / Bits per pixel
NAXIS   =                 2 / Number of axes
NAXIS1  =               512 / First axis length
NAXIS2  =               512 / Second axis length
BSCALE  =               1.0 / REAL=TAPE*BSCALE+BZERO
BZERO   =               0.0
```

FIT KERNEL STANDARD EXTENSION NAME

The IKI/FIT extension has been set to the string 'fit', i.e., a filename such as 'wfpc.fit' is a legal IRAF image filename.

SUMMARY

The FITS kernel is expected to provide several advantages over the other formats supported by IRAF, such as accessing remote FITS files, regardless of the architecture of the machine on which the files reside, and writing output in FITS files to any machine on which the user has an account.

The FITS kernel will only support in its first release the primary Header and Data unit; in other words, there is no capability to handle any FITS EXTENSION at this time.

IRAF DATA FORMAT CONVERSION ISSUES

NELSON R. ZARATE
Space Telescope Science Institute

INTRODUCTION

Passing binary data between different types of computers using different data formats is a challenging problem. Now, with the iraf networking capabilities it is possible to write simple conversion programs that would read a local or remote file, do the binary conversion for each pixel and write out to the local or remote machine. There is no need to use time consuming transfer to tape using FITS format and read it back on another machine with possible loss of data precision.

At STScI we have developed a set of programs following the schema above to transfer and convert Iraf and Geis images and stsdas tables between Sun and Vax architecture. If the iraf networking is not available at any point, the programs are still useful since they can do the conversion in the local machine after the files have been transferred using FTP for example.

SUN2VAX, VAX2SUN AND TCONVERT

These are the main applications now running under iraf at STSci to transfer images from a SUN machine to a VAX one and vice versa; sun2vax and vax2sun accept both Geis and iraf image types and tconvert accepts Stsdas tables. As with any application that uses the iraf networking an account on the remote machine is necessary. An entry in the user's local ".irafhosts" in the local machine with the remote account name if is different from the local name. For example if a user on a VAX wants to transfer a set of iraf images to a SUN machine with name "aurora" in the directory /au/data1/ then the usual command line is:

```
cl> vax2sun jan*.imh aurora!/au/data1/
```

The hidden point here is where the pixel file go. The default here is to send it to the directory indicated in the command line.

If an image has been transferred using FTP to a Sun machine, then the command to convert the VAX formatted iraf image to a SUN formatted iraf image would be:

```
cl> vax2sun jan1.imh jan1_sun.imh
```

For vax2sun running on the VAX, the application converts the local binary format for each pixel to the SUN format and writes to the remote SUN machine. The same thing is done for vax2sun running on the SUN, but now the VAX data is the remote one. Sun2vax on the SUN will convert the data to the VAX format before writing it to the remote machine.

Tconvert uses a parameter to specify whether to convert stsdas tables from VAX to SUN or from SUN to VAX format; otherwise, it works the same way as vax2sun and sun2vax.

All of these applications make use of the `mii` (machine independent routines) to convert the local data format and pass it to remote machine, where the same kind of routines are used to unpack the data to the remote format.

IRAF NETWORKING

At Stsci the iraf networking is set up to work under the TCP/IP protocol; which means both VAXes and Suns run TCP/IP and the transfer of files is guaranteed for the VAX side. In other words a user on a VAX machine can run vax2sun and sun2vax, reading and creating images on the SUNs; whereas the transfer of files to a Vax machine for a user on a SUN machine is not guaranteed if the VAX is running Wollongong TCP/IP for example.

Here at the Institute we are currently running MULTINET which supports the iraf network calls from a SUN machine.

IRAF KERNEL SERVER

The iraf kernel server `irafks.e` program running on a remote node is the one that provides low level services such as opens, reads and writes for the application on a local node. This means that all the conversion and packing of the data should be done by the application while the server is waiting for input. This schema has worked fine for machines that have the same processing power, and probably the user does not mind or does not have a better alternative but to wait for a transfer to finish.

An idea developed while building application to transfer data to the CRAY machine is to have the irafks server do most of the work of converting, packing and unpacking data to and from the supercomputer environment. Under this situation, where the pixel format is so different, a modified or improved server that would do any complex computation for the application running it would greatly increase the efficiency of a network program.

OTHER POSSIBLE APPLICATIONS

We can extend the vax2sun idea to any other architecture that runs iraf; in fact we have working versions of `cray2sun` and `sun2cray` for our local version of iraf on the CRAY-YMP. The transfer of any binary file would also be possible given the required flexibility on the user interface such as the input datatype, output datatype, dimensionality and byte offset.

THREE-DIMENSIONAL TABLES IN STSDAS

PHILIP E. HODGE AND NELSON ZARATE
Space Telescope Science Institute

ABSTRACT Proposed extensions to STSDAS tables are described that will allow table entries to be arrays rather than just single values.

INTRODUCTION

A 2-D table is a set of values arranged into rows and columns, where the values are single numbers or text strings. In a 3-D table the entry at a given row and column may be an array. Prototype code has been written which implements this feature for STSDAS tables.

APPLICATIONS

Group-format images (e.g. GEIS images) map naturally into 3-D tables, where the different image groups are separate table rows, the data arrays are stored in a fixed-length array column, and the group parameters are in ordinary single-element columns. Users often have a large number of related spectra, which could be stored in separate rows in an array column. Associated wavelengths or coefficients of a fit to the dispersion solution would be kept in a second array column.

Variable-length arrays could be used for coefficients of a fit to data. References, history information, or comments could be stored in a variable-length character string. This is especially appropriate for comments in a catalog, where some rows would have extensive comments, and others would have none at all. Locations of reseau marks on the Faint Object Camera photocathode are currently being stored in single rows with many columns, one pair of columns (X and Y) for each reseau mark. Two array columns, one for X positions and one for Y, would be more reasonable.

FITS

The prototype 3-D table I/O routines were used in `stwfits`, the STSDAS FITS writer, to create a FITS binary table (BINTABLE) extension. Since the table routines calling sequence is the same as for 2-D tables (in the prototype), the `stwfits` code did not need many changes, but rather include the necessary new routines to write binary table data according to the proposal by Cotton and Tody (Draft September 20 1991).

The STSDAS write and read FITS tasks support all the 3-D table datatypes. For FITS binary table, the 3-D floating point data have to be converted to the FITS IEEE standard, which is not a problem for a SUN machine where this is the standard already and the byte ordering is the same as IEEE; but on a VAX machine, the `stwfits` task will convert the host machine floating point representation and change the byte ordering in memory before creating a FITS record on tape or disk.

For a GEIS image with multiple groups on a 3-D table format, the FITS writer will take the table columns description and produce the FITS extension header description consisting of column name, units, null value (if any), and the datatype description of the form (nD); where 'n' is the number of elements in the vector and 'D' is a one-character code representing the datatype. Then for each table row it will read the vector in each column and produce the data portion of the FITS file.

IMPACT ON TASKS

Although current 2-D tables will be upward compatible with 3-D tables, some existing tasks are incompatible with the concept of array entries, some should be enhanced to take advantage of this option, and some tasks as currently implemented will actually fail on 3-D tables.

In tasks such as `tsort` and `tselect`, the entry at a given row and column must have a definite value. It would be possible to assign a value to an array by using the average, median, minimum or maximum depending on context, but the confusion would probably outweigh the benefit. These tasks will therefore give an error message when asked to operate on a array-type column, but they will be able to sort or select 3-D tables based on values in single-element columns.

Tasks that create or display tables will need to be enhanced. For `tcreate`, a special symbol could be defined to indicate the end of a list of items for a variable-length entry. We would like `tcalc` to be able to operate on arrays, but within reasonable limits. For example, `tcalc` might choke when asked to add arrays of different lengths, but it certainly should be possible to add a constant to an array. Array entries in separate columns could be related, such as wavelength and flux, and one might want to sort the elements of two columns at one row based on the values in one of those columns. Because such enhancements to `tcalc` and `tsort` could take considerable programming time and effort, a simple script-based workaround would first be implemented. This would extract columns at a single row into a separate table, expanding array entries into multiple rows, then would call a task such as `tsort` or `tcalc`, and then would write the values back into array entries in the original table.

SPECIFYING ARRAY ENTRIES

The issue here is not just how to specify table entries which are arrays (when plotting with `sgraph`, for example), but also how to specify a column of ordinary elements, or to restrict it to a range of rows within a column. It would also be nice if the syntax were similar to that for images and ascii files. The following suggestions have been made for tables:

syntax	*interpretation*
`test.tab[wl]`	column "wl" in table "test.tab"
`test.tab[x 1,5,31:35]`	rows 1, 5 and 31-35 in column "x"
`test.tab[flux 27]`	row 27 of array-type column "flux"

DISPLAYING ARRAY ENTRIES

Tasks such as **tedit** and **tprint** display the contents of a 2-D table in tabular form (no surprise), writing as many columns as will fit on the page at one time. One option for displaying a 3-D table would be to print one row at a time, with the array entries printed in columns, possibly with element numbers. This could be very wasteful of space, and it would make it awkward to compare values in different rows. Another option is to first print non-array columns in the usual tabular format, and then print the array columns separately, formatted into blocks to take up less space.

When editing a table using a workstation, the task could display rows and columns in the usual way but with a special value (e.g. "array") for entries in an array column, and it could use a separate window when the cursor was in such a column. An alternative that would work on a plain terminal is to display only one array element at a time, also showing the array index. When the cursor is in that column, a (new) command would be available to move to a different index.

IMPLEMENTATION

One requirement for any extension to STSDAS tables is that there be minimum impact on the current format. The column descriptor already includes an item for the number of elements in an entry. Fixed-length arrays can be implemented in a straightforward manner by storing the values in consecutive locations in the table, with the length item in the column descriptor set appropriately.

Variable-length arrays are handled in a completely different manner. The arrays are read and written to a separate, machine-independent binary file by routines in the IRAF file management I/O (FMIO) package. This file has the same root name as the table but a different extension. For a variable-length column, the table file itself contains an ordinary integer-valued column. These numbers are FMIO lfile numbers, and they are assigned and stored in the table only when needed. Thus there is minimal overhead when a variable-length column contains values in only some of the rows. Although prototype code has been written which implements variable-length arrays in this manner, substantial additional work remains to be done.

For fixed-length arrays we intend to support the same data types as for 2-D tables: real, double, integer, boolean, and character strings. For character strings that would mean 2-D arrays, e.g. `ctype[SZ_CTYPE,NUM_CTYPE]`. For variable-length arrays, short and byte arrays would also be supported, and instead of arrays of character strings we would have 1-D strings of arbitrary length.

SECTION 2

DATABASES AND ANALYSIS OF ARCHIVAL DATA

OVERVIEW OF THE ASTROPHYSICS DATA SYSTEM (ADS)

J. C. GOOD
Astrophysics Data System, Infrared Processing and Analysis Center, California Institute of Technology Pasadena, CA, 91125

ABSTRACT The Astrophysics Data System (ADS) has been built with the aim of unifying all the datasets and data systems currently available to the users of NASA astrophysics data. It is a complete distributed geographic information system using the latest information systems technologies, while at the same time being very much a science-driven project.

INTRODUCTION

The Astrophysics Division of NASA has built a geographically- and logically-distributed heterogeneous information system for coordinated multispectral analysis of data from Astrophysics missions. This system, called the Astrophysics Data System (ADS), is a truly distributed system in which the user, data and processing tools can be widely separated. It's architecture is driven by the existing geographical distribution of data and users in Astrophysics and is made possible by the rapid advances in workstation and networking technology over the past few years.

To accommodate the anticipated growth and changes in both requirements and technology, the ADS employs a client/server architecture in which services and data can be added or replaced without having to change the basic architecture or interfaces. The future evolution of the system will be driven in large part by user services mounted both by the ADS Project itself and by members of the astrophysics community.

ASTROPHYSICS DATA SYSTEM PHILOSOPHY

Astrophysicists today have a bewildering array of powerful NASA resources to call upon. Among these are the data centers for the High Energy Astrophysics Observatories (HEAO's), the International Ultraviolet Explorer (IUE), the Infrared Processing and Analysis Center (IPAC), and the Hubble Space Telescope Science Institute (STScI), as well as the National Space Science Data Center (NSSDC). Unfortunately, the services provided by these centers are essentially independent and some are only accessible through mission-unique hardware and software. Furthermore, they can generally only be accessed directly through the centers themselves.

The Final Report of the Astrophysics Data System Study, March 1988, characterized the data environment of the Astrophysics community at that time and defined for the future an "... *architecture for a data system which will serve the astrophysics community in multi-spectral research through the decade of the 1990's.*" One of its recommendations was that each data set, and the human expertise which supports the data, be maintained at the same physical location. Moreover, these multiple locations should be linked together and to researchers by means of high speed communications networks. Finally, to allow true multi-spectral research, the various data sets should be accessible through a common set of tools.

The Astrophysics Data System (ADS) Project is NASA's response to the Data System Study. For the science investigator, the ADS makes NASA's current and future data holdings more broadly and efficiently accessible, and makes the data itself more interpretable. For NASA, it provides a common information system infrastructure for science analysis, thereby reducing duplication of effort while increasing the scientific return from missions.

ADS SOFTWARE ARCHITECTURE

Conceptually, the software can be divided into two categories. Core Services are those which provide the basic system functionality upon which user applications or User Services can be built. These Core Services will be built primarily by the ADS Project itself. User Services are other applications which reside on the ADS and provide the science support and analysis functions required by investigators. These may be built by the Project in response to community demand or by individuals or groups.

In its initial release, the ADS provides basic Core and User Services. These will be expanded and supplemented in the future based on feedback from the Astrophysics community.

INITIAL CORE SERVICES

The MESSAGE PASSING SERVICE (MPS) enables remote inter-operability and data transfer/translation among the ADS services. It provides homogeneity across networks and operating systems for process requests and responses. While this service is largely invisible to the end user, it provides the application programmer (and the ADS system programmers) what the User Interface provides for the science end user: an environment in which to access the services of the ADS without needing to know which servers will execute the program. At a minimum, the MPS implements the functionality of:

- Remote Procedure Call (RPC): a mechanism by which the subroutines of a program can call each other as if they were executing on a single server when, in fact, they may be segmented across several machines;
- Remote Process Invocation (RPI): a mechanism by which a program on one server can spawn programs on other servers;
- Remote Inter-Process Communication (RIPC): a mechanism whereby two or more programs running concurrently on different servers can communicate among themselves.

In the initial release, the ADS Message Passing Service conforms to the Advanced Network System Architecture (ANSA) Testbench and the Open Software Foundation / Distributed Computing Environment (OSF/DCE).

USER INTERFACES are the services which provide the local working environment (including its look and feel) to the end user. They provide a means to access all other services within the ADS. The structure of the ADS is such that there can be any number of user interface programs tailored to the constraints, hardware limitations, and preferences of specialized segments of the community.

The ADS currently has user interface capabilities that will work with any character terminal supported under the UNIX "termcap" facility and will shortly release an X-windows GUI version. The ADS User Interface provides the following combined functionality:

- specialized table display and interactive manipulation facilities;
- a first-order complete relational database facility including support for Structured Query Language (SQL) and Query by Example (QBE);
- menu bar/pulldown menus and multiple windows (split-screen on the character terminal interface);
- context-sensitive help and a dynamic tutorial facility;
- full-featured text management facility that supports browsing, plain-text inquiry and cut-paste editing of selected files.

Through this interface, investigators can locate data sets (descriptions and catalogs) of interest, access these data sets either individually or in combination, and select from among them one or more subsets that they wish to import into a local working environment for further analysis of their own choosing. More importantly, it allows the investigator to accomplish this without requiring him to know which servers execute the services he invokes. In the initial release, the Knowledge Dictionary System (KDS) (tm) is being used to implement the functionality of the User Interface.

The DATA INDEPENDENT ACCESS MODEL (DIAM) is a service which provides inter-operability and language conversion among the various distributed Database Management Systems (DBMS's) in the ADS. These DBMS's include the following: Ingres and Sybase (UNIX), RDB (VMS), and IM/DM (CDC NOS/VE). The DIAM is required in the ADS in order to provide the user with a uniform relational view of all these distributed catalogs even though some of the catalogs are still maintained using DBMS's that do not yet support the relational SQL standard. The functionality of the DIAM is:

- to accept SQL statements generated by the User Interface (either directly or indirectly using the QBE translator);
- to convert them to the language supported by the DBMS that hosts the referenced catalog(s);
- to RPI a process to submit them to the DBMS for processing and to collocate the resulting table(s);

• to return those results to the User Interface that issued the request.

A generic SQL Server program has been written to implement the functionality of the DIAM.

Besides providing uniform access to the multiplicity of existing DBMS's, the ADS provides a complete internal distributed DBMS which allows further processing with special features not available in most commercial DBMS's.

FUTURE CORE SERVICES

There are several Core Services which are important and anticipated in the near future, but which were not included in the initial release of the Astrophysics Data System. Among these are the User/Service Authentication Service and the Transaction Management Service. Other Core Services will be added on as they are required and become available.

The USER/SERVICE AUTHENTICATION SERVICE will provide the capability to automatically verify the authorization of a user to access specific services and data within the system. It will be provided through an implementation of the KERBEROS software on the ADS System. Currently, KERBEROS is only being used to provide initial user authentication.

The TRANSACTION MANAGEMENT SERVICE (TMS) will provide the process and resource management protocol for client-defined transactions. It assures successful execution, synchronization, and release of all services and resources used in a transaction.

INITIAL USER SERVICES

The initial User Services available through the ADS have usually been collectively referred to as Directory Services, with components that provide access to catalog data and to documentation about data holdings for specific Astrophysics archives, without requiring the user to know where the data physically resides. These Directory Services include Documentation Browse and a Catalog Data Retrieval and Processing Service.

The DOCUMENTATION BROWSE SERVICE provides a simple means for users to access and organize directories and files through the functionality of the User Interface described above.

The DATA RETRIEVAL AND PROCESSING SERVICE provides the capability to retrieve cataloged data and perform data base management processing on that data through the query, text management, and relational data base functionality of the DIAM and User Interface services described above.

FUTURE USER SERVICES

It is expected that other User Services will be made available as they are requested by the Astrophysics user community and integrated into the ADS. Because of the flexible modular nature of the ADS architecture, such integrations should be relatively straightforward and can be implemented in a variety of ways, dependent on how the prospective User Service is coded.

Specific user services which are planned for the near future include Document Location and Retrieval and Archive Access.

The DOCUMENT RETRIEVAL SERVICE provides uniform, subject-matter indexing (and English-language querying) across all the data in the ADS, regardless of whether data are highly structured in databases, or unstructured. For the initial release, the Document Retrieval Service is implemented by the Factor Space Access Method (tm) as an integral part of the Knowledge Dictionary System (tm). The various components of the Document Retrieval Service will be exported as discrete services through the Message Passing Service, through which they can be generally accessed by Remote Procedure Calls.

The Factor Space (FS) is an n-dimensional Euclidian space, the axes of which are statistically constructed to account for the variance and covariance in expert judgments made by astrophysicists about the relevance of ADS data items to different subject matter contexts. The functions of the Document Retrieval service are:

- to scan all data entered in the ADS and compute its subject matter profile as a sequence of one or more vectors in the Factor Space;
- to similarly analyze natural language requests and to search the Factor Space for relevant data items;
- to monitor the distribution of vectors in Factor Space for clusters (i.e., undifferentiated data items);
- to periodically generate, disseminate, collect and synthesize questionnaires to obtain additional relevance judgments needed to increase the data resolution;
- to factor-analyze these additional relevance judgments and modify the number and/or orientation of axes in the Factor Space accordingly.

These functions also support the generation of new Factor Spaces, either to accommodate new subject matter or to accommodate personalized perspectives on existing subject matter.

The ARCHIVE ACCESS service is a logical extension of the current Data Retrieval Service. However, whereas the current service is limited to catalog data in textual form, the Archive Access service will allow retrieval of all forms of data (images, documents, spectra, tables, etc.).

ADS PHYSICAL ARCHITECTURE

An overview of the ADS physical architecture is presented in the figure below. In its initial instantiation, the ADS consists of six physically distributed primary nodes which are interconnected via NASA Science Internet (NSI). These six nodes are the following:

- Center for Astronomy and Space Astrophysics (CASA), Boulder, Colorado

- Infrared Processing and Analysis Center (IPAC), Pasadena, California
- International Ultraviolet Explorer (IUE), Greenbelt, Maryland
- National Space Science Data Center (NSSDC), Greenbelt, Maryland
- Smithsonian Astrophysical Observatory (SAO), Boston, Massachusetts
- Space Telescope Science Institute (STScI), Baltimore, Maryland

PLATFORMS AND OPERATING SYSTEMS

Each of these nodes above has important astrophysics catalogs for which it has principal responsibility. In general, these catalogs are maintained in different types of DBMS's on different types of machines, each of which runs a unique operating system. Components of the ADS Core Services will be resident on a server at each node and connected by the ADS Message Passing Service software. The mapping of services to servers in the ADS is unconstrained: some servers simultaneously provide several (simple) services while some (complex) services are segmented across several servers. In addition, the numbers and kinds of services that can be mounted on the ADS are also unconstrained.

USER SCENARIO

The prospective user of the Astrophysics Data System should first obtain a copy of the ADS User's Guide (contact the administrative node at the address at the end of this paper), which will give specific details of how to obtain access to the ADS System. Generally a user will be assigned to one of the ADS Primary Nodes for user support.

It is assumed that the user is a novice and is trying to do a very simple, well-defined task but one which requires access to multiple astrophysics data centers and their on-line databases. In this scenario, the user will locate and subset two independent datasets and then intercompare the results. Specifically, the problem is to correlate measurements of galaxies which have been observed to have both a significant infrared flux (indicative of a large amount of cold dust), and an X-ray or ultraviolet flux (indicative of hot gas).

The user first asks about the existence of 100 micron (infra-red) data, and ultimately finds his/her way to the IRAS Point Source Catalog. This user then wants to know if a subset of IRAS sources (e.g. galaxies that lie above 30 degrees galactic latitude with 100 micron flux greater than 10 Janskys and 60/100 micron flux ratios indicative of temperatures less than 40K) have been observed in the X-ray (Einstein Catalog).

Further, for those sources which were observed both in the infrared and with Einstein list the ones detected in IR and X-ray and plot the IR flux versus the X-ray flux for these objects.

The specific steps are outlined below:

Locating Infrared Datasets

If we consider all the datasets potentially available in astrophysics (not just the 'standard' products of NASA missions), the potential size and diversity of these datasets become quite large. Without data location tools, locating the correct dataset for our investigation would be at least as difficult, if not more so, than the processing we will do once the data is found. In most cases accessing this data would be much more difficult than processing it.

One of the primary services of the initial ADS implementation is the factor space documentation location method described above. With this we can pose a factor space query of the type "I am interested in long wavelength infrared measurements of galaxies, particularly 100 micron measurements, in the form of catalogs. Color temperatures, specifically ones derived from long wavelength measurements, will be correlated with X-ray data to try and determine the relationship between cold dust emission and that from hot gas".

The factor space query will return a list of documents, some of which might be journal articles about similar research, some descriptions of catalogs or other data (e.g., images), some mission descriptions, etc. Somewhere near the top of this list is likely to be the description of the IRAS Point Source Catalog. Unless we have been distracted by some of the other documents, we will read this and decide that this catalog is indeed what we want for our IR data. By poking around the documentation related to this catalog (a simple browse mechanism through our documentation) we also determine what fields in the catalog are important for our investigation.

The documents used in the search are maintained at the same institutions that bear responsibility for the data itself.

Locating the X-ray Datasets

The X-ray data (in particular the Einstein database) would be found the same way. One of the rules of the ADS is that catalog datasets are not made accessible through the system until appropriate documentation is also on-line.

Extracting the IR Data

We have determined the correct catalog and even have some idea as to which fields are appropriate for our endeavor. The next step is to actually query the database where the data resides and get our subset. The basis for such queries is SQL, as described in the section on our DIAM. Our query will look something like:

```
select *
from ipac.fsc2x
where fnu4 > 10
and fnu3 < fnu4
and (glat>30 or glat<-30)
```

In addition to a direct SQL query capability, ADS provides a more intuitive query by example (QBE) mechanism where the user puts constraint information directly into a template of the catalog in question.

The infrared data of interest is located at the Infrared Processing and Analysis Center at the California Institute of Technology in Pasadena, California on a CDC Cyber mainframe in the IM/DM database.

Extracting the relevant X-ray Data

To compare sources which both have an X-ray flux in the Einstein catalog and correspond to the sources obtained from our infrared query, requires the use of SQL for a database join. This is complicated somewhat by the need to potentially transform coordinates from one system or representation and then to perform the join based on spherical distance proximity. This process can also be performed using SQL directly or through the QBE mechanism described above.

Selection of the data must be done at the site where the data resides. However, the join process can be done at either data site or at the user's local site. Optimization can be done to try and minimize the amount of data transfer but in practice the user usually wants to see the results of the two 'selects' before joining them and therefore will transfer the data to the local site anyway.

The X-ray data will be at the Harvard-Smithsonian Astrophysical Observatory in Cambridge, Massachusetts in INGRES on a SUN workstation.

Comparison of IR and X-ray Data

Once the join has been completed, the result is a single table in which are combined IR and X-ray measurements. Proximity on the sky was the deciding factor in matching sources and so we may well want to edit this table in a spreadsheet manner before proceeding. The final step in this example is simply to plot one column in this table versus another.

Table processing and plotting will be performed locally to the user.

SYSTEM DEVELOPMENT AND ADMINISTRATION

The design of the ADS is motivated by the fact that it exists in a dynamic science and information system environment, and therefore it must be a dynamic system. Thus, even as the ADS is released, development continues on Core Services, User Services, Network Connectivity, and the addition of new nodes. System Management is responsible for technical oversight of prototyping research, selection criteria, development of standards and conventions, and all aspects of systems integration, test, and operation of new or revised ADS services.

SYSTEM ADMINISTRATION

The administration of the Astrophysics Data System is shared by NASA Headquarters, Astrophysics Division (Programmatic), the ADS Project Office and ADS Administrative Node (Project), and the other primary nodes making up the ADS.

Further information on the ADS can be obtained by contacting Dr. Good at

Internet	jcg@ipac.caltech.edu
BITnet	jcg%ipac@HAMLET.BitNet
Telemail	[JGOOD/NASA]NASAMAIL/USA
uucp	(cit-vax,trwrb!csula-ps)!ipac!jcg
SPAN	ROMEO:: "jcg%ipac"

TEL: (818) 584-2939
FAX: (818) 584-9945

THE EINSTEIN ON-LINE SYSTEM

T. KARAKASHIAN, D. E. HARRIS, M. GARCIA, G. FABBIANO, J. M. FLANAGAN, C. P. STERN
Smithsonian Astrophysical Observatory, 60 Garden St., Cambridge, MA 02138

ABSTRACT The Einstein On-Line System (EOLS) is a collection of data, software, and documentation from the Einstein Observatory project and is located on a computer at the Smithsonian Astrophysical Observatory. The interface is menu-driven, terminal-independent, and is available via the Internet, SPAN, or telephone dial-up. Using EOLS, astronomers have access to 4 Gigabytes of Einstein processed data and documentation. Products may be retrieved via VMS copy (over SPAN), ftp (over Internet), or electronic mail. In addition, all numerical databases are available via NASA's Astrophysics Data System (ADS). The EOLS project is actively evolving in response to users' comments. Over the next two years, more than 30 Gigabytes of additional Einstein data will be made available in the EOLS.

INTRODUCTION

The increasing decentralization of scientific computing is making it feasible for astronomers to perform data analysis at their home institutions instead of having to travel to centralized data centers. The Einstein On-Line System (EOLS) and its user-interface EINLINE were created to facilitate the selection and transfer of archived Einstein data from the Einstein Data Center to the astronomical community. Since various on-line systems with different styles of user interface are beginning to proliferate, it was felt that the EINLINE interface should be as simple and easy to use as possible. In this way the burden of having to learn yet another on-line system is lessened.

THE EINLINE PROGRAM

The user interface to the EOLS is a program called EINLINE, and it is invoked when a user remotely logs in as user "einline" on the host computer at SAO (see below for information about connecting to EINLINE). EINLINE is organized as a hierarchy of menus. On-line help is available at all menu levels. The program does not require any particular kind of terminal. EINLINE is written in the C language, and consists of about 10,000 lines of code (excluding comments).

EINLINE USER-INTERFACE PHILOSOPHY

We felt that a menu-driven system was easier than a command-driven interface for a new user. At each menu level, the user is presented with a limited series of options, each of which is displayed before the prompt. This is in contrast to a command-based interface which is more complex and may require the user to peruse documentation extensively before productive work can commence. However, a menu system can slow down more experienced users, who are familiar with the menu hierarchy, Therefore we also provide "short-cuts": that is, the user can type several consecutive menu choices on the command line, and thereby step down several menu levels in one step.

We also felt that it was important to make the program independent of terminal type. We were concerned that relying on a specific keyboard layout or graphics character set would make applications dependent on rapidly changing hardware and inconvenience users who do not have the standard terminals. We believe that any loss of convenience resulting from this decision is small compared to the advantage of access from any terminal or workstation.

Finally, we felt that it was important that EINLINE provide a context-sensitive help at every menu level. This facility is important if users are to use EINLINE without having to read any documentation beforehand.

AVAILABLE DATA PRODUCTS

Numerical databases

The numerical data in the EOLS resides in an INGRES database management system (DBMS). We provide tables from the Einstein 2E catalog (e.g., source lists, identifications), the IPC and HRI standard processing, the Slew Survey, the Observation Pointing Catalog ("Yellow Book") and various derived tables (such as quasars and supernova remnants). Also included are certain radio catalogs and selected ROSAT information (such as the mission timeline). New tables are added from time to time; up-to-date listings are available within EINLINE.

Access to these numerical tables is accomplished through either of two services in EINLINE: QQ ("Quick Queries") or DBQ ("Database Queries"). In QQ, users have a choice of numerous "prefabricated" queries. These queries retrieve specific data fields in preset formats; the user fills in only the search parameters. QQ is a simple and fast way to get useful data out of the database. In DBQ, the user is stepped through the process of building a query. There is more flexibility in selecting data to be retrieved and output formats. In addition, joins of tables and multiple search conditions are supported.

In both QQ and DBQ, the user's query is converted to SQL[1] and delivered to the DBMS. The resulting data are then retrieved, formatted (converting to astronomical units if appropriate) and then displayed to the user. If desired, results of queries can be saved and transferred back to the user's home computer.

---

[1]SQL (Structured Query Language) is the standard language used with relational DBMS.

Text documents

A large number of text documents are available in the DOC menu in EINLINE. Documentation is available concerning the Einstein mission (including the Revised Users Manual), all data tables, the EOLS itself, ROSAT bulletins, back issues of HEAO newsletters, and more. Users can read documentation and select it for transfer back to their home computers.

Also provided is a "News" bulletin board containing the latest information about the EOLS, and other selected topics such as the ROSAT mission status.

Images

The Einstein CD-ROMS are available for downloading FITS images: The Einstein Observatory Catalog of X-ray Sources (almost all of the IPC fields); Einstein HRI fields (photon arrays); and the Einstein Slew Survey. These images are provided for astronomers who do not have access to the CD-ROM sets at their home institution.

TRANSFERRING DATA

Text documents, the results of searches, and images can be downloaded over the networks to the user's home computer. This is accomplished by using the **ftp** command or VMS **COPY** command from their home computer. In addition, from within EINLINE users may e-mail themselves copies of documents and the results of their database queries.

HOW TO CONNECT TO EINLINE

- **Telephone and modem** (300/1200/2400 baud)
  dial (+1) 617 495-7047 for use < 20 minutes
  495-7048 for longer
  use odd parity, stop bits = 1, data bits = 7
- **Internet** (TCP/IP using TELNET)
  telnet einline.harvard.edu or
  telnet cfa204.harvard.edu or
  telnet 128.103.40.204[2]
- **SPAN** (DECnet)
  set host cfa204 or
  set host 6714

ACKNOWLEDGMENTS

This work was supported by NASA contract NAS8-30751.

---

[2]Note: this number is subject to change without notice; using the name is preferred. If the name does not work for you, speak to your system administrator about installation of domain nameserving at your site.

THE NASA/IPAC EXTRAGALACTIC DATABASE

BARRY F. MADORE, GEORGE HELOU, HAROLD G. CORWIN JR., MARION SCHMITZ, XIUQIN WU, JUDY BENNETT
NASA/IPAC Extragalactic Database
Infrared Processing and Analysis Center
Jet Propulsion Laboratory 100-22
California Institute of Technology, Pasadena, CA 91125

ABSTRACT The NASA/IPAC Extragalactic Database (NED) is an on-line, free-of-charge, database service that organizes a broad range of published extragalactic data into a computer-based central archive. It is designed for fast, and flexible query via electronic networks by the astronomical community world-wide.

INTRODUCTION

Since its inception in the summer of 1990 the usage of NED (as measured by the number of remote logins per month) has been continuously growing. NED is now handling well over 1,300 interactive sessions every month, with logins coming from across the United States and all other parts of the world, including regular usage from Australia, Canada, Germany, Great Britain, Holland, Italy, Japan, Mexico, and most recently Singapore.

The Fall 1991 version of NED provides POSITIONS, NAMES, and BASIC DATA for over 150,000 extragalactic objects, as well as some 200,000 bibliographic references to 16,000 published papers, including notes from catalogs and other publications. NED supports searches for objects and for references, and upon request NED will automatically forward to the user those files containing data retrieved during a session. NED also allows users to view the contents of several major catalogs, and to browse through 5,500 abstracts of articles of extragalactic interest that have appeared since 1988 in five major journals (A&A, AJ, ApJ, MNRAS, and PASP, including their *Letters* and *Supplements*).

The latest version of NED also supports BATCH SEARCHES on input files (`Option 8` in `Main Menu`). Work is underway to collect DATA from the current literature and from catalogs, and provide more versatile tools for searching through these entries. Meanwhile, additional catalogs are being folded into NED continually, with detailed cross-identifications established. The most recent additions have been the *Einstein Medium Sensitivity Survey, Case Galaxies, IRAS Faint Sources, Abell Clusters*, and *Kiso UV-Excess Galaxies*, as well as basic data from the *RC3* and redshifts from the 1990 release of *ZCAT*.

NED is also now preparing for on-line access the THESIS ABSTRACTS of recent doctoral dissertations on extragalactic topics. As soon as a thesis has been accepted in its final form at an institution, NED plans to accept the abstract for immediate on-line publication.

While our archivists are actively maintaining and up-dating the database, the NED programmers have simultaneously been porting NED away from the CYBER mainframe and on to a SYBASE DBMS working on a SUN server. In-house testing shows this new version of NED to be significantly faster than its predecessor. Outside usage of this new implementation is expected to commence early in the new year, pending the completion of the IPAC-wide computer-system reconfiguration.

ACCESS TO NED

Users may access the NED service over INTERNET or NSI-DECNET (previously SPAN), provided they have a VT100 terminal, or VT100 emulation software. (i) On INTERNET, a connection to IPAC may be set up with the command: `telnet ipac.caltech.edu`. (ii) From a node on NSI-DECNET, use the command: `set host IPAC` (the absolute address is 5.857). Once connected to IPAC and prompted for a "login", simply respond with `NED`; no password is needed. From this point on, the system is self-documenting, especially through the `HELP` utilities and the "control-H" key. First-time users may want to read the `TUTORIAL` in the first screen presented by the interface. A complete set of HELP FILES can also be downloaded via e-mail upon request.

The `COMMENTS` option in the NED interface can be invoked at any time in a session to leave messages or suggestions. Users are encouraged to take advantage of this mechanism for pointing out incompleteness or inaccuracies that they might detect in NED, and also for suggesting new services and/or additions to the database. Questions will be answered by return e-mail directly by one of the staff within a few working days. In the event of problems requiring special attention, users may contact any of the following: Barry F. Madore, G. Helou, M. Schmitz or H.G. Corwin Jr. at (818) 584-2903.

ACKNOWLEDGMENTS

This work is carried out by the Jet Propulsion Laboratory, California Institute of Technology, under contract with the National Aeronautics and Space Administration (Astrophysics Division, Science Operations Branch).

REFERENCES

Helou, G., Madore, B. F., Bicay, M. D., Schmitz, M., & Liang, J. 1990, in Windows on Galaxies, (G. Fabbiano, J. S. Gallagher III, & A. Renzini, eds.) Dordrecht: Kluwer Academic, p. 109

Helou, G., Madore, B. F., Schmitz, M., Bicay, M. D., Wu, X., & Bennett, J. 1991, in Databases and On-Line Data in Astronomy, (M. Albrecht, D. Egret, eds.) Dordrecht: Kluwer Academic, p. 89

A SHORT REVIEW OF SOURCES OF PUBLIC DOMAIN SOFTWARE

FIONN MURTAGH[1]
Space Telescope – European Coordinating Facility, ESO, Karl-Schwarzschild-Str. 2, D-8046 Garching, Germany

ERIC D. FEIGELSON
Dept. of Astronomy & Astrophysics, Penn State University, University Park PA 16802 USA

INTRODUCTION

Software production for observational astronomy has become more efficient during the last decade with the production of large centralized systems like IRAF, MIDAS or AIPS. However, virtually all of the code is produced by astronomers, with few algorithms and little software obtained from outside the astronomical community. This wastes the limited skilled labor resources of astronomers (see Voigt & Smith 1989), and unnecessarily restricts data analysis to familiar methods. A major reason for the failure to use preexisting methods developed for other applications is the difficulty in locating the relevant software. There is no central clearinghouse for scientific software. In a very preliminary effort, we provide here an overview of some of the software resources available to astronomers from "public domain" (PD) sources. PD here means that the source code is available for scholarly use without restriction, and is either free or provided at very low cost (usually to defray distribution expenses). Our overview is restricted to software available from wide-area networks like Internet, from government repositories, and from journals. In particular, we omit the substantial body of PD software associated with scholarly monographs, bulletin boards, and we omit commercial software.

ON-LINE OR NETWORK SOURCES

Here we outline software depositories, and related on-line services such as software discussion groups, available on wide-area networks.

Netlib: A primary network source of high- and medium-quality numerical analysis software (see Dongarra & Grosse 1987 for a description). The one-line command `send index`, sent to *netlib@research.att.com*, gives sufficient information to bootstrap oneself.

Statlib: The same one-line bootstrapping message can be sent to *statlib@temper.stat.cmu.edu*, which houses extensive statistical algorithms, including many from the journal *Applied Statistics* and many in the *S* language.

---

[1]Affiliated to Astrophysics Division, Space Science Department, European Space Agency.

Archie: A very large Internet archive server at *quiche.cs.mcgill.ca* (132.206. 2.3), user name 'archie'. It has a list of about 800 Internet ftp, archive, and Unix sites. It includes a a Software Description Database (SDD), a text database containing names and short descriptions of about 3500 software packages, documents, and datasets available on the Internet.

Usenet newsgroups and listings: Of the ~1000 newsgroups on the Internet, astronomers may be particularly interested in *sci.astro*, *sci.space*, *sci.optics*, *comp.lang.idl-pvwave*, *sci.math.stat*, *comp.graphics.research*, *comp.graphics. visualization*, *comp.text.tex*, *comp.binaries*, and *comp.sources*. The newsgroups *comp.misc* and *comp.sources.wanted* post a list of about 900 sites with software available by anonymous ftp, which is updated monthly (also available by anonymous ftp from *pilot.njin.net*, 128.6.7.38).

Neural networks and artificial intelligence: On-line discussions on neural network software are very active. Send subscription requests to *connectionists-request@cs.cmu.edu* and *neuron-request@cattell.psych.upenn.edu*. A large machine-learning database is accessible through anonymous ftp at *ics.uci. edu*, and a bibliographic service on AI can be found by sending `help` to *lido@cs.uni-sb.de*.

UNIX and C software: GNU ("GNU's Not Unix") is a large collection of high-quality systems and C software. Contact *gnu@prep.ai.mit.edu*. Anonymous ftp to *ftp.uu.net* is worthwhile. These archives contain window, security, mailer, compiler, word processing and similar systems.

MS-DOS software: A large collection is available from SIMTEL20 *wsmr-simtel20.army.mil* (192.88.110.20).

Macintosh software: A large collection is accessible by anonymous ftp to *sumex-aim.stanford.edu* (33.44.0.6). Other large PD collections are available on CD-ROM, including ClubMac, MegaROM, PD-ROM and BCS PD-CD. Costs are typically $50-100. Reviews, addresses and prices can be found in the *Microcomputer Index*, 11:131 (1990).

TeX: One of the largest TeX archives is located at *ymir.claremont.edu* (134.173.4.23).

Mail exploders: The Working Group on Astronomical Software organized by the AAS maintains a mail exploder. Send subscription request *sub wgas-l* your-name to *listserv@scfvm.bitnet*. A list of mail exploders is maintained in directory *info/interest-groups* at *nnsc.nsf.net* (128.89.1.178).

JOURNALS

Computer Physics Communications: Journal of computational physics providing source code for ≥1000 programs, most with 10^3 lines of Fortran. Available by email or other medium, typically costing $30–200.

Applied Statistics: Journal with ~300 short refereed codes, mostly in Fortran, providing statistical analysis.

ACM Transactions on Mathematical Software: Journal describing the *Collected Algorithms of the ACM* (Association for Computing Machinery), ≥600 programs of $10^2 - 10^4$ Fortran lines.

Computers & Geosciences: Journal of algorithms and computer program listings for geology and related fields. Most consist of $10^2 - 10^3$ lines of

Fortran.

Dr. Dobb's Journal, "Software Tools for the Professional Programmer": Trade magazine with C source code accompanying articles, also available on diskette for $15.

GOVERNMENT SOURCES

National Technical Information Service: Provides access to ~2,000 software and related products produced under U.S. and foreign government R&D funding. Some are actual packages, while others are bibliographies or descriptions of other software libraries. The software is published in the annual *Federal Software Exchange Catalog*, and the entire NTIS database can be scanned on CD-ROM or on-line through the DIALOG and BRS commercial bibliographic services. Software products include large mathematical and scientific libraries from Federal agencies, graphics systems, differential equation integrators, etc. NTIS also has important pointers to other software sources such as GAMS (Guide to Available Mathematical Software, 2400 programs), and much more.

SUMMARY

The items covered above represent only a fraction of the scientific PD software that is available, and we encourage others to help construct a more complete catalog. A less abbreviated version of this survey, giving examples of available software available from each source, appears in Feigelson & Murtagh (1992). We emphasize that some of the sources outlined here are libraries of libraries of software, so that we have not scanned the actual software contents. In particular, the astronomical community might delve into the huge Federal NTIS and the network Archie collections for useful software. Now that astronomers have the infrastructure for distributing large specialized software systems like IRAF to the user community, the distribution of appropriate PD software developed in other fields would provide many new data analysis capabilities with relatively little effort.

ACKNOWLEDGMENTS

Statistical research by E.D.F. is supported by NASA grants NAGW-1917 and NAGW-2120, and NSF grant DMS-9007717.

REFERENCES

Dongarra, J. J., & Grosse, E. 1987, "Distribution of mathematical software via electronic mail", Comm. Assoc. Comp. Mach., 30, 403–407

Feigelson, E. D., & Murtagh, F. 1992, submitted for publication

Voigt, S. J., & Smith, K. A. (Eds.) 1989, Software Reuse Issues, NASA CP-3057

A MULTIMISSION DATABASE AND ANALYSIS ENVIRONMENT

N.E. WHITE
High Energy Science Archive Research Center, Code 668, GSFC, Greenbelt, MD, 20771

ABSTRACT In the past, data analysis systems for astrophysics space-based observatories have been custom built for each mission. This has resulted in a large variety of data systems that are difficult to maintain in the long term. However, once the telemetry has been decoded, the underlying functionality of these data systems is the same. The High Energy Science Archive Research Center and the Astro D Guest Observer Facility plan to isolate the data input/output from the data selection function by reformatting the telemetry data from past missions and from Astro D to FITS format files with no loss of information. Since FITS files are self- defining and now a standard in astronomy, the files will be accessible to a wide range of analysis systems with minimum documentation of the data structures. The FITS data will be accessed using a multimission "data selection" program to produce standard data products for further analysis using generic analysis tasks.

A browse tool is described which will allow the seamless connection of selected data in the archive to the user's analysis programs. BROWSE will first be part of the HEASARC online service used to access data from the archive, but later it will be provided to astronomers to provide a local database environment from which they can can write scripts to select data and then pass them transparently to the their own analysis tasks.

INTRODUCTION

The 1990's will see an explosion in the volume of data available to astronomers for archival research. The current and upcoming high energy astrophysics missions ROSAT, GRO, and Astro D will alone produce 1,000 gigabytes of archival data by 1995. The technology now exists to make this available online. The topic of this paper is the analysis and data management of such a large volume of data. The paper naturally falls into two parts, *viz.*, **Multimission Analysis** and **Database Environment**. The first part describes how we can no longer afford to produce mission-specific data processing systems, but instead should use a system which uses a common underlying data format. The second part concentrates on how astronomers must be provided with data management tools that allow the astronomer not only to locate the data but also to make a rapid assessment of its worth so as to avoid needless transport of the data to the user's site. These tools must seamlessly join the data to the analysis tasks.

THE HEASARC

The High Energy Astrophysics Science Archive Research Center, HEASARC, is located at the Goddard Space Flight Center. It was created by NASA in 1990 as a site for X-ray and Gamma-ray archival research. The objectives of the HEASARC are to provide:

- online access to archival X-ray and Gamma-ray data
- scientific expertise in the data holding
- data and software standards
- multimission and multi-wavelength analysis tools

It is a collaboration of Goddard's Laboratory for High Energy Astrophysics, LHEA, and the NSSDC. The LHEA is responsible for the science content of the archive and provides the director. The NSSDC is responsible for the data archive management. The HEASARC data holding will consist of data from past, future, and concurrent missions. The motivation for the HEASARC is to provide a multimission archive for the ROSAT, GRO, BBXRT, Astro D, and XTE missions. But it will also contain data from past domestic missions such as Einstein, HEAO 1, HEAO 3, OSO 8, SAS 2 and 3, Uhuru, and Vela5B. In addition, data from non-US missions will also be made available where possible, e.g., EXOSAT and Ginga. The total data volume will be of the order of 1,000 gigabytes by 1995. The aim is to make these data available online.

MULTIMISSION ANALYSIS

For every mission the data flow is identical. First, the raw data undergoes some form of *data reduction* to produce *data products*, usually a photon list, an image, a spectrum, and/or a lightcurve. These are then *analysed* to produce some *results* which are then *published* (hopefully). While the sequence of events is much the same for each mission, the dilemma facing the HEASARC is that every mission produces a data set in a different format, with a different set of analysis software. This makes the long term support and distribution of a multimission archive problematic, since every mission is a special case. In addition, combining datasets from different missions is non-trivial.

Mission specific formats tend to be used throughout the data processing chain. The raw telemetry data in many cases involves preprocessing and packing of the data on the spacecraft so as to maximize the information transmitted to the ground. There may be multiple telemetry and onboard computer modes which can add to the complexity of reducing the data. The data products produced by the data reduction software are more generic, e.g., a lightcurve is a time and a count rate. However, even for data products, missions typically generate their own formats. A notable exception is that images have recently begun to be distributed in FITS format. The results of each mission are also sometimes kept in mission-specific or vendor-dependent formats, e.g., an INGRES DBMS table.

The data access is limited to a data processing system produced by the project. These tend to be monolithic systems that are not optimal for long term maintenance or general distribution and use by the community. Specific problems are as follows:

- they are custom built for each mission, even though the underlying functions are the same
- there is a failure to modularize and isolate the mission dependent functions
- calibrations and methodology are embedded in the code
- the code is vendor-dependent (e.g., operating system, compiler, DBMS)

The last point is particularly problematic. In the long term it makes maintenance of the data processing system difficult. The code must repeatedly be ported to new hardware and software platforms as technology evolves.

In contrast, the user community is becoming increasingly demanding. They require access to the original raw data, and also want to reduce it from within a familiar analysis environment, e.g., IRAF, IDL or XANADU. Faced with such a dilemma, the HEASARC has only one choice to provide a long lasting archive:

Reformat the data, recycle the software

The root of the problem is that each mission produces data in different formats. The data reduction and analysis functions are basically the same; the driving factor is decoding the different data telemetry. The HEASARC solution is to isolate this function by reformatting the data to a single standard structure. This should be self-describing so that the user need only look in the header to be able to read the file. The FITS standard provides such a capability.

FITS is an IAU and NASA standard for distributing data analysis software, and there are FITS readers within all the popular environments, e.g., IDL, IRAF, and MIDAS. The recent adopting of the *binary table FITS* standard, which allows the byte structure of each column to be defined in the header, has been a real breakthrough. It allows compact table structures to be defined which can mirror the underlying table structures in most data analysis systems such as MIDAS or STSDAS tables.

The HEASARC will distribute all *useful* data as FITS binary tables, including the telemetry. While at first sight it may seem a formidable problem to reformat a complex telemetry stream containing science and housekeeping data, it is actually simpler than having to build from scratch a data reduction and analysis system. By reformatting the data, it forces an isolation of the mission-specific function of decoding the telemetry. The following data reduction tasks will have both mission-specific and mission- independent functions. By reformatting the telemetry, it is simple to recycle the mission-independent functions.

To implement this plan, the HEASARC is taking the following steps: First, the data reduction system for the next high energy astrophysics mission, Astro D, will be constructed so that it forms the basis for a multimission infrastructure. The Astro D telemetry will be reformatted to FITS and all of the mission-dependent and independent bits will be isolated. Second, the HEASARC has begun to reformat existing telemetry and data products from existing missions such as Einstein, HEAO 1, and EXOSAT. The experience learned and FITS file structures defined can be fed into future missions such as XTE.

To enable both the HEASARC and future missions to reformat to FITS, the HEASARC is providing a portable FORTRAN 77 subroutine library to write and read FITS files. This package, called FITSIO, was released earlier this year and has already proved extremely popular (Pence 1992). The HEASARC is also defining mission independent FITS file structures for spectra, lightcurves, and photon lists. These will allow data products to be distributed transparently between different analysis packages.

On top of FITSIO, a Data Selector is being produced by the HEASARC, in collaboration with the Astro D project, to allow Boolean selections from multiple FITS tables. This data selector will form the basis of a multimission data reduction system, and will be very similar in concept to the MIDAS and STSDAS table systems. The major advantage of the HEASARC data selector is that it will directly operate on FITS tables and so the whole system will be fully portable. It will be written in strict FORTRAN 77. The software will isolate the parameter input and validation from the kernel that actually does the task. This will allow the selector to run under different analysis environments. The first version will be built to run under both the IRAF, using the FORTRAN interface, *and* the X-ray ANalysis and Data Utilities, XANADU, package. It will be a trivial matter for other developers to integrate selector into their own analysis environments, so long as their environments have an isolated parameter interface.

The remaining mission-dependent part of any data reduction system is the calibration data. The HEASARC is defining standard formats for distributing calibrations. Like the data itself, calibrations can be divided into raw data, such as a detector energy resolution function, or a telescope point spread function and calibration products such as a detector response matrix or an exposure map. The HEASARC will encourage future developers to externally define all calibration information such that it can be accessed by any data reduction or analysis system.

DATABASE ENVIRONMENT

Before discussing how the data should be accessed, we must first consider what are the reasons for archival research. There are four distinct categories:

1. historical studies,
2. theoretical follow-up,
3. surveys, and
4. assurance

Historical studies are the most obvious archival activity and this comes in several different variants. It could be that an observer discovers a new phenomenon from a recent observation and wants to check earlier data to, e.g., independently confirm its existence and/or track the long term variability of a source. These studies can be perhaps the most difficult, since they will involve combining and/or comparing data sets from different telescopes. A related reason is to use archival data as part of a justification to propose to use a new telescope. The major issue is gaining access to the data as well as cross-instrument calibration.

Theoretical follow-up is the need to test new models against existing data. In many cases, the interpretation of a phenomenon can take many years, with theoreticians repeatedly building models and testing them against the data. Theoreticians currently have to work closely with the original investigator to test their models, or make eyeball fits to published data. In this case, the major issue is that the theoretician does not want or need a detailed understanding of the instrument characteristics or analysis techniques. He or she simply wants a simple data product and associated calibration to test against the model.

Surveys provide the opportunity to combine many observations of a single class of object (e.g., AGN) made using the same telescope and instrument by many different investigators. The current principal investigator approach to allocating observation time means that large uniform samples of particular object types are rarely available to a single observer. Only after the data enters the public domain can a survey of the properties of a particular class be made. The main issue here is ensuring that a user can search an archive for all objects of a particular class.

Assurance is the ability to guarantee both that an observation is analysed (and if appropriate, published) and that unjustified repeat observations are not made. Observation time on satellites is very limited (and expensive). Making the data available after some fixed time ensures that all interested parties in the field get access to that data and that it is eventually looked at. The issue here is that in many cases an observation may never be published because the result is not sufficiently noteworthy. It is essential to provide a simple overview of the main results of the observation.

In all of the above, data availability is not the only key to success. It is also essential to remotely provide a mechanism to visualize and make a quick-look analysis of the data.

Data distribution can be done via online access, and/or by mass distribution, e.g., via CD ROM. Remote access to data is becoming much more common. Online services such as SIMBAD, NED, IUE, EXOSAT, Einline are well known and work well at delivering the data quickly to the user. To date these services have tended to deliver mostly catalog information, but they are becoming increasingly sophisticated with some delivering data products and quick-look analysis.

The disadvantage to these various services is that each one has a different user interface with which the user must become familiar. There are currently two agency-level efforts to provide a uniform online environment which will help the archive researcher locate and connect to the particular data they are interested in. These are:

- NASA's Astrophysics Data System, ADS
- ESA's European Space Information System, ESIS

These systems use a client-server approach to allow remote queries of databases. The archive sites retain control of the archive contents but will rely on a common user interface provided by the central organization.

The HEASARC online service will provide remote access to the HEASARC data holding *and* to data analysis software. The emphasis will be on *browsing* of the data, such that a user can make a quick-look assessment of its worth before exporting it, or part of it, to his or her home site. The heart of the system will be based on the BROWSE program that was developed as part of the EXOSAT database. This is a command driven environment which allows a user to search one or more database tables by coordinates, name, object class, or any other valid parameter combination. The user can then display the selected data or start analysis software on the selected data. The current EXOSAT system has been installed at the HEASARC and is being used to provide the initial online service.

The HEASARC is currently defining a new generation BROWSE based on the current version. The new BROWSE will be a task orientated system with both a command line and graphical user interface. The graphical user interface, GUI, will be installed on the user's workstation, and will use remote procedure calls to communicate with the command line interface. Each BROWSE task will generate SQL calls to the underlying database management system via an application programming interface. There will be BROWSE tasks to allow the selected data to be seamlessly joined to the analysis tasks. These will, e.g., generate IRAF or XANADU script files containing the relevant file names and then run the task. The first build of the new BROWSE (due at the end of 1992) will run at the HEASARC, with the GUI remotely installed on the user's workstation. The next build will allow the users to run BROWSE at their home sites, where they can use it to organise their own *local* archives.

In the meantime, the current BROWSE can be accessed via the HEASARC online service. This can be accessed both by Internet

telnet ndadsa.gsfc.nasa.gov

and Decnet (SPAN)

set host ndadsa or **set host 15761**

In both cases the username is **XRAY**. After logging on, type BROWSE. The following are commonly used BROWSE commands:

```
 Starting browse:
BROWSE ROSOBS       ! start browse to search the rosobs database

 Browse help:
dbhelp              ! description of the current database
help                ! general browse help

 To make a search:

sc                  ! search by coordinates
sn 3c*              ! search for all names starting 3c
sp                  ! search by any indexed parameter
lind                ! list the indexed parameters
mind                ! make an index
sp pi petre*        ! search for all 'pi' names starting petre

 After a search has been made:
```

```
dsam                ! display again the search results
dall 1-5            ! display all the parameters for entries 1 to 5
sif                 ! inclusive filter on a parameter
sef                 ! exclusive filter on a parameter
sort day            ! ascending sort on parameter day
sort/d done         ! descending sort on parameter done
cpd                 ! change (or set) plot device
psam                ! plot one parameter against another

exit                ! quit browse
```

From the HEASARC> prompt:

```
mail browse.log     ! mail home the results of your browse session.
```

Send a note to the account REQUEST to obtain user's guides to the system. Currently there are available catalogs and data products from Einstein, EXOSAT, HEAO 1, Ginga, and ROSAT, plus many popular astronomical catalogs.

REFERENCES

Pence, W. 1992, this volume

THE ESIS CORRELATION ENVIRONMENT PROTOTYPE

P. GIOMMI, S.G. ANSARI, A. CIARLO, P. DONZELLI, H. STOKKE, P. TORRENTE, S. WALKER
ESIS, Information Systems Division, ESA/ESRIN, Frascati, Italy

V. ZAMPOGNARO
Cap Gemini Sesa Italia, Rome, Italy

ABSTRACT This paper gives a brief description of the Correlation Environment Prototype currently under development at ESRIN (Frascati, Italy) as part of the European Space Information System (ESIS) project. The prototype consists of a C program that acts as graphical user-interface, as tasks scheduler and as data-handler within the general ESIS system. This program will provide a uniform view to a number of applications that allow the user to retrieve, inspect and compare Astronomy and Space-Physics data. The prototype architecture is object oriented and open so that new packages can easily be included as they become available. The prototype will run on VAX-VMS and on UNIX systems.

INTRODUCTION

The European Space Information System (ESIS) is an ESA project to access, retrieve and utilize Astronomy and Space Physics data from heterogeneous distributed databases. ESIS consists of a Query Environment (QE), a Correlation Environment (CE), and a number of information services. The purpose of the QE is to allow the user to access and retrieve data from different databases using a single, domain-specific, query language. The ESIS QE is currently under development by a consortium of software companies and scientific institutions and is scheduled to be completed by November 1992. The purpose of the CE is to inspect, compare and correlate data retrieved by means of the QE.

A prototype of the ESIS Correlation Environment is currently under development at ESRIN by the scientific staff of ESIS in collaboration with professional programmers.

QUERY ENVIRONMENT

The Query Environment will make two types of data available:

- Description data (e.g. fluxes, exposure times, observation dates)
- Data sets (e.g. images, spectra, lightcurves)

retrieved from various heterogeneous databases using a homogenized query language, AQL (Astronomical Query Language) and SPQL (Space Physics Query Language).

The description data gives summary information about the data sets. It also includes various parameters on a celestial object, such as its magnitude, spectral type or individual flux values in a wavelength region, which can be used in statistical assessments of a group of objects or events. This type of data is normally found in logs or catalogues and will be made available in ESIS through the Astronomical and the Space Physics Semantic Networks respectively.

Retrieval of the data sets will be facilitated within the Query Language by an internal ESIS format (a FITS format) which is then interfaced to the Correlation Environment for further analysis.

THE PROTOTYPE OF THE CORRELATION ENVIRONMENT

The ESIS CE Prototype consists of a C program (called *SPACE*, Space Physics and Astronomy Correlation Environment) that, through a graphical or a command-line interface, provides a uniform view to a set of software packages that can be used to retrieve and manipulate data. Some of these packages are public-domain programs produced by scientific institutions. The graphical user-interface is based on the OSF-Motif widget set. The packages included in this first version of the prototype are :

- BROWSE, a database management system and data-analysis environment;
- SAOIMAGE, a general purpose image display program;
- XIMAGE, multi-mission X-ray image analysis program;
- XSPEC, multi-mission spectral analysis program;
- XRONOS, multi-mission timing analysis program;
- SPACESTAT, statistical analysis package;
- SPIE, general purpose Space Physics data Inspection Environment.

The ESIS QE software will be interfaced with the prototype as soon as it is available. This will be done by writing a proper driver interface (see below).

The above packages provide most of the necessary capability to retrieve, inspect and compare astronomy and space physics data. Some of the important functions provided by the final system are: selection and comparison of images from different experiments/missions, multi-mission timing and spectral analysis, statistical analysis, generation of spectrograms, multi-panel plots, etc.

The following data types can be handled by the system: images, spectra, light-curves, spectrograms, time-series, etc.

Some of the packages included in the prototype will be modified and improved, as part of the prototyping activity, by the ESIS staff in collaboration with the institutes where the software was originally produced. SPACESTAT and SPIE are under development at ESRIN mainly by assembling and adapting public-domain routines.

The development of the CE prototype started in September 1991 and will continue for several months in parallel to the development of the ESIS QE. A first, limited, version of the prototype will be available in the spring 1992.

THE CE PROTOTYPE ARCHITECTURE

The prototype is based on an object oriented architecture where the objects are software modules. This allows easy maintenance and upgrading of the system. Figure 1 shows an overview of the prototype architecture. *SPACE* includes six modules : a *task manager*, a *communication manager*, a *command interpreter/dispatcher*, a *user interface manager*, a *file manager* , and an *error manager*. *SPACE* is linked to each data handling package through a *command and data dispatcher* and a *driver*. The driver is used to convert the *SPACE* commands into the corresponding instructions in the command language specific to the package. The six modules in *SPACE* work as follows: the Task Manager starts and stops applications and keeps track of their status; the Communication Manager allows the packages to exchange data using VMS mail-boxes or UNIX pipes; the Command Interpreter translates and dispatches commands within the system; the User-Interface Manager handles input/output to and from the user; the File Manager stores and retrieves data on disk and keeps track of configurations files; finally the Error Manager simply handles errors within *SPACE* and from the different packages. This architecture allows the system to grow by adding new software without the need to change the original code. A new program can be added by writing a specific driver which handles all the input-output between *SPACE* and the package, and adding a new set of commands in the system translation table.

All the system-dependent calls are encapsulated and properly handled by conditional compilation flags. This allows the system to run on different hardware platforms (e.g. VAX VMS, SUN/UNIX, etc.).

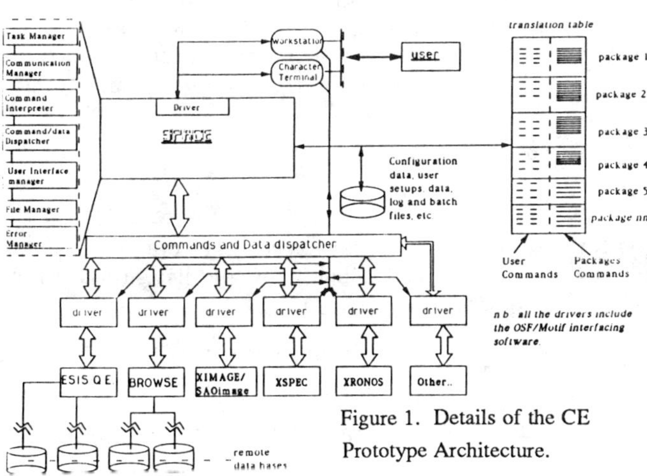

Figure 1. Details of the CE Prototype Architecture.

DISTRIBUTED X-RAY DATA AND DISTRIBUTED ANALYSIS TOOLS

D. PLUMMER, J. SCHACHTER, M. ELVIS, M. GARCIA, M. CONROY
Smithsonian Astrophysical Observatory, 60 Garden St., Cambridge, MA 02138

ABSTRACT X-ray data and analysis packages can now be distributed directly to the desks of the user community using CDROMs and portable, familiar analysis packages (like IRAF/PROS). The Einstein IPC Slew Survey is an example of a complete X-ray data set distributed via CDROMs and is the first to use the new FITS standard for photon event lists (BINTABLE). Users can analyze the Slew data directly off the CDROM using PROS. As an example, we present a recipe for producing a radial profile of the Cygnus Loop using PROS and the Slew Survey CDROM data. CDROMs of the complete Einstein IPC and HRI archive data sets will soon be distributed in BINTABLE format allowing similar analysis with those data.

INTRODUCTION

The bulk of the Einstein data was taken while the satellite was in pointed mode, that is, while the space-craft was pointing in one steady direction. These data can be analyzed with IRAF/PROS tasks that utilize the typical X-ray data files including photon lists, exposure maps, and background maps. Tasks like imcnts account for shadows due to support structures, vignetting effects and do background subtraction by accessing these data files. Slew data was taken while the space-craft was rotating (slewing) from one target to another. These data files differ in format from the pointed data files and require some extra steps to analyze the data.

DISCUSSION

The Slew CDROM contains all the files necessary to analyze the data as well as software to do some basic analysis. The following list summarizes the format of the data files on the Slew Survey CDROM.

1. Photon list data converted to FITS BINTABLE files with an altered version of the xpr2fits task from the IRAF/PROS package.

2. Space-craft Aspect files and Good Time Interval files converted to an ASCII format.

3. Fortran software included on the CDROM to generate binary array exposure files from the Aspect data and Good Time Interval data.

To obtain more information regarding the Slew Survey and to obtain a copy of the CDROM send a request via e-mail to: edpo@cfa.harvard.edu or cfa::edpo or 6699::edpo or edpo@cfa.BITNET or write to Elizabeth Bohlen, EDPO MS 12, High Energy Division, S. A. O., 60 Garden St., Cambridge, MA 02138.

The following is a recipe for producing a radial profile of the Cygnus Loop Super Nova Remnant from within IRAF. The CDROM reader is mounted on the device, "/cdrom".

1. Photon data (in FITS BINTABLE format) from the Slew CDROM is converted to an IRAF/PROS QPOE file.

   cl> xray
   cl> xdataio
   cl> fits2qp /cdrom/data/20h/s2048n30.f3d cygloop

2. Exposure map (as a binary array file) is generated with the software provided on the Slew CDROM. See the documentation on the CDROM and the HEAO newsletter for details on generating the exposure map. The exposure map is converted to an IRAF/PROS Image file.

   cl> ximages
   cl> imcreate cygloop_exp_tmp "390 390" r s2048n30.exp r 0

3. Since the coordinates used by the Slew Survey differ from the IRAF coordinates, the exposure map must be flipped in the y-axis.

   cl> images
   cl> imcopy cygloop_exp_tmp[*,-*] cygloop_exp

4. An image of the count rate is generated by dividing the photon file by the exposure file.

   cl> imcalc "cygloop_flux = cygloop / cygloop_exp"

5. To compensate for the patchy exposure the image is smoothed. A contour map can now be generated.

   cl> xspatial
   cl> imsmooth cygloop_flux cygloop_flux gauss 3.
   cl> xplot
   cl> tvimcontour cygloop_flux_smo

6. A radial profile of the Cygnus Loop can now be produced.

   cl> imcnts cygloop_flux "a 176 247 0 100 n=20"

COBE ASTRONOMICAL DATABASES

I. FREEDMAN, A. C. RAUGH
Hughes STX Corporation, Suite 100, 7601 Ora-Glen Drive, Greenbelt MD 20770

E. S. CHENG
Code 685, NASA Goddard Space Flight Center, Greenbelt MD 20771

ABSTRACT A project to store and convert external astronomical survey maps to the Cosmic Background Explorer (COBE) spacecraft pixelization is described. Established software is reused in order to reduce development costs. The proposed packages and systems include the Image Reduction and Analysis Facility (IRAF), Interactive Data Language Astronomy Library (IDL), the FITSIO data transfer package and the Astronomical Image Processing System (AIPS). The software structure of the astronomical databases, projected conversion schemes, quality assurance procedures and outstanding problems will be discussed.

THRUST

The thrust of this work is that development cost may be reduced by re-use of established software and that heterogeneous software environments promote synergy.

In particular, IDL and IRAF augment each other's capabilities.

INTRODUCTION

The COsmic Background Explorer (COBE), NASA's first satellite devoted to the study of cosmology was launched on 18 November 1989. The main mission covered the time from 21 November 1989 to 21 September 1990. COBE carries three instruments: the infrared experiment DIRBE, the anisotropy experiment, DMR and the spectrum experiment, FIRAS (Wright 1990). To facilitate correlative studies with other surveys at several wavelengths and enable effective foreground subtraction , external survey data will be converted to the COBE sky cube pixelization scheme at the resolution and beam pattern of the COBE instrument beams. The surveys will include point source catalogs and sky maps.

OBJECTIVES

The technical objective is the conversion of external survey data for direct comparison with COBE data at COBE instrument beams in the sky cube pixelization scheme to better than 2% accuracy. The sky cube (White & Stemwedel 1992) is an approximately equal-area projection on the sky of the faces of an inscribed cube. The approximate equal-area property is ensured by a curvilinear coordinate system ruled on each cube face.

Initial plans called for a 2 person-year effort to develop FORTRAN-based software for this conversion—re-use of established software reduced the estimated effort to $\sim$ 1 person-month.

SOFTWARE SYSTEM

The projected software system consists of the following elements : IRAF, the IDL Astronomy Library (Landsman 1989), FITSIO (Pence 1992) AIPS and a small number of project-developed FORTRAN software modules.

COBE astronomical databases are directories of files. Intensity, spectral and polarimetric data are stored in region quadtree maps, with ancillary information in associated header files. Offsets into each map corresponding to each level of resolution are stored in "index" files. Data destined for comparison with maps from the DIRBE experiment will be stored at 9 or more levels of resolution; data intended for FIRAS and DMR will be stored at 6 or more levels of resolution. Data records are fixed length, defined by a Record Definition Language (RDL) file interface to the VAX Common Data Dictionary. RDL and its Record Definition Compiler were developed by the COBE project.

The largest dataset scheduled for conversion is the IRAS all-sky survey ($\sim$ 50 MB)—which constrains the conversion software to execute rapidly.

In order to prevent data proliferation and retain the traceability of map products, only one version of any survey map is destined for storage, converted to the highest resolution COBE instrument beam of lesser resolution than the native beam. Transfer functions will be provided between the beams for each band of each COBE instrument. Templates for the conversion between map coordinates and the COBE sky cube pixelization will be provided together with software to create them.

CONVERSION

If the survey data are not available in FITS format,the conversion process (see Figure I) begins with the conversion of the data from ASCII to FITS format via the IDL-WRITEFITS procedure or the FITSIO subroutine package. Missing data are stored as FITS "BLANKS". The blanking value must be chosen for compatibility between the various software packages used in the conversion process.

Because of the need for synthesized maps from point source catalogs and the individual point sources themselves, there are two point source ingest procedures yielding two different products, the point source sky map and the point source list.

Sky maps are converted from the external survey beam, epoch and pixelization scheme to the COBE instrument beam and pixelization by convolution with a beam transfer function and geometric resampling. IRAF can perform convolution and resampling but does not contain procedures to create the necessary control point grids at this time. These are supplied by IDL Astronomy Library routines supplemented with project-developed FORTRAN software.

An overall transfer function from each external survey to COBE instrument beam is computed via Fast Fourier Transforms (FFT) or the CLEAN algorithm. Since IRAF does not seem to provide an FFT routine, this portion of the work is done by IDL. The AIPS task "VTESS" provides the CLEAN routine. The software is not constrained to require radial symmetry of either the external instrument or COBE beam pattern.

Survey data are convolved with the beam transfer function using IRAF-CONVOLVE which we prefer over IDL-CONVOL or AIPS-CNVL on grounds of accuracy.

Control point grids which specify the correspondence between map coordinates and Ecliptic 2000 sky cube coordinates are separately computed for each cube face. Templates are prepared with IRAF-GEOMAP which models the overall coordinate transformation by piecewise solutions comprising affine transformations of Chebyshev polynomial patches in the form of small text files. This "patch" representation facilitates smooth interpolation and rapid flux-preserving transformation of the convolved data via IRAF-GEOTRAN.

The resulting stack of 6 sky cube faces is reformatted to the standard sky cube pixel list using project-developed FORTRAN software.

Maps observed with instruments whose beam-pattern varies over the sky will be converted initially as a mosaic of overlapping patches. More sophisticated approaches will be used only if this simple technique does not yield sufficient accuracy.

The execution time is less than several seconds for any of these steps on a lightly-loaded DECStation 5000/200 for a 721 by 45 pixel test image.

The databases are fully interfaced with COBEtrieve, a project-developed data access system based on VAX Datatrieve and may be selected by time-range or data set name.

A small number of FORTRAN-based utilities are available to perform primitive operations such as windowing a square or circular region of a map.

QUALITY ASSURANCE

Software validation will require the generation of test patterns with a Nyquist-limited range of spatial frequencies from sharp to diffuse. These patterns (provided by AIPS-PATGN) will be used to *measure* the flux, spectral and geometric distortion over regions from 1 pixel to the full map. The contrast will be stretched at the 2% level to ensure that no moiré pattern or other artifact has been created. The maps may be displayed via the SAOIMAGE tool, IDL or a project-developed Skymap Display Tool (Toller et al. 1992)

Correct software operation will be verified using a small realistic input map. COBE-generated sky map data will be converted from DIRBE to DMR beam patterns to check the numerical accuracy of the conversion process for a realistic range of spatial frequencies and morphologies.

Finally, a large external survey map (e.g. synchrotron survey) will be converted in order to verify that expectations have been met.

The validation of converted data will require the measurement of distortion

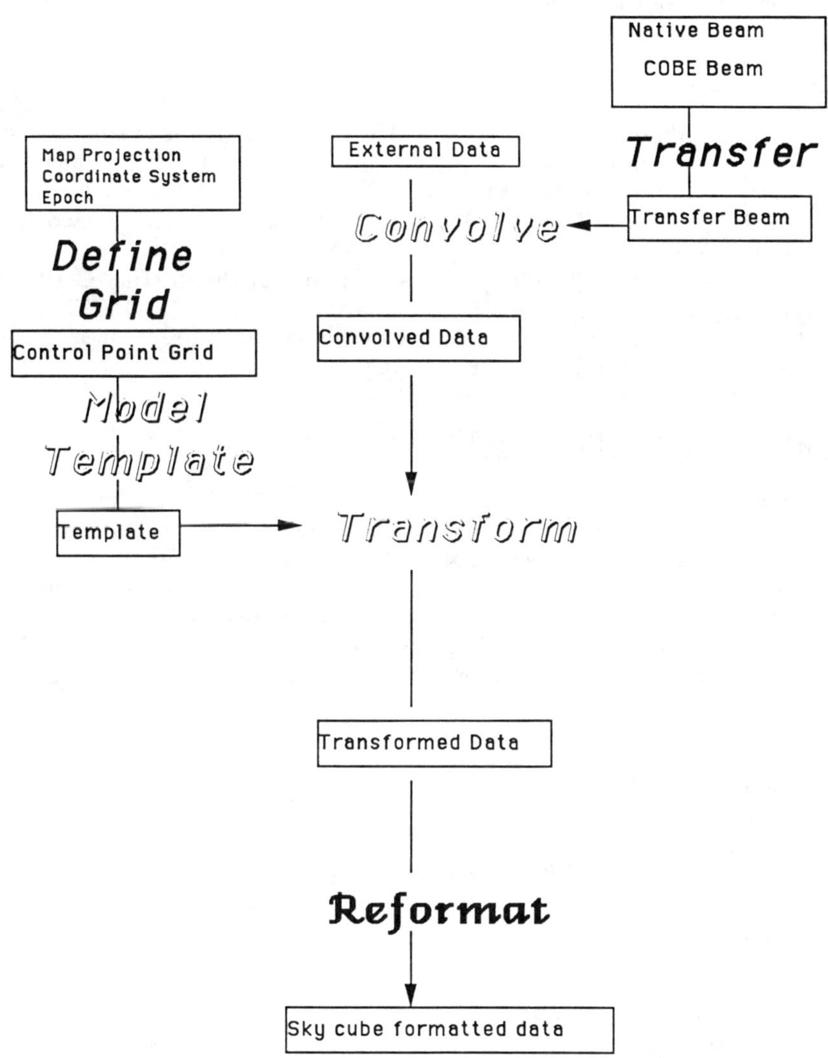

FIGURE I Conversion of external survey data to COBE sky cube format.

using contrast enhancement with particular attention to the invertibility of the coordinate transformations.

PROBLEMS

Correct FITS file transfer between ULTRIX (where IRAF is currently installed) and VMS (the main workhorse) systems is obtained via the File Transfer Protocol (FTP) using the **logical byte** transfer format. The data, which are received at the VMS system in 512–byte blocks are subsequently reformatted to FITS-standard 2880 byte blocks.

Software has been planned in order to reformat the existing COBE instrument beam profile data into text files compatible with IRAF-CONVOLVE. Interpolation of the beam profile to arbitrary resolution is required—super-resolution is to be supported during the sky map ingest.

SUMMARY AND CONCLUSIONS

Astronomical data may be in unique formats with individual beams that vary over the image. COBE needs to transfer these data to instrument beams in skymap format with high accuracy ($\sim 2\%$) for direct comparison with instrument data. The projected approach is highly interactive and requires little programming knowledge but considerable astronomical judgement.

We conclude that re-use of established software may reduce development costs and that a heterogeneous environment of *compatible* software promotes synergy.

ACKNOWLEDGMENTS

This research is supported by the Astrophysics Division of the Office of Space Science and Applications at NASA Headquarters.

REFERENCES

Landsman, W. B. 1989, BAAS, 21, 784

Pence, W. D. 1992, this volume

Toller, G., et al. 1992, this volume

White, R. A., & Stemwedel, S. W. 1992, this volume

Wright, E. L. 1990, COBE, Paper presented at the Texas-ESO-CERN Symposium, 18 Dec 1990

THE BACKGROUNDS DATA CENTER (BDC) VISUAL INTERFACE FOR SPACE AND TERRESTRIAL ANALYSIS (VISTA)

E.G. DOMBROWSKI[1], R. DEMAJISTRE[2], J. WILLIAMS[2], J. HAYNES[1], D.E. ANDERSON[3], AND W.A. SNYDER[4]
Backgrounds Data Center, Space Science Division, Naval Research Laboratory, Washington, D.C. 20375

INTRODUCTION

The VISTA system is currently under development at the BDC for visualization and analysis of atmospheric and celestial backgrounds data. This software system provides the user with the ability to interactively query and manipulate experiment databases through a versatile graphics user's interface (GUI). The object is to allow the user the freedom of reviewing the observing scenario for a particular experiment as well as to define a workable set of data for further processing and analysis.

The VISTA software consists of the union of C (C++), X11 windows (UIM/X), and SQL (INGRES) for the database management. A data pipeline management system will make use of an On-Line Library Manager (OLLM) developed at the BDC (in C) which will handle all data transfer requests. A visual programming language and a text and graphics interface to various analysis packages are also part of this software system. The main feature of VISTA is that it has been designed with modularity so that additional computational algorithms, image processing packages, and in-house data models can be added or withdrawn without loss of functionality.

Overall, VISTA operates as the front-end for the selection, correlation and manipulation of program data, both textual and graphical. The scientific benefits are best represented by the fact that the user can visually query the database, see the physical representation of his query, and interactively manipulate any particular data set item from the query via the same GUI. For astronomical data processing, an example of this would mean that the user could interactively query the data base, view a celestial star map that contains graphical and textual information about selected images, choose a particular image or subset of images, and activate whatever image and data processing tasks that are necessary for further data reduction.

---

[1]Sachs Freeman Associates (SFA)

[2]Hughes/ST systems Corporation (STX)

[3]Computational Physics Incorporated (CPI)

[4]BDC Director, Naval Research Laboratory

I. DEFINING THE QUERYSET

Upon entering the VISTA user interface, the user will first have to define a working data set from the available database. Under the main menu bar, there are several ways in which he can accomplish this. The user can build a new data set via the geophysical and celestial queries (Build QuerySet section), parse an independent catalog source for data set information, or select a previously defined data set that is resident on his workstation or stored elsewhere over the network.

The first choice demonstrates the versatility of the VISTA system. Textual and graphical menus make up the core of the geophysical and celestial queries. For the geophysical selection, the user can specify a time and spectral range via slidebars, and select longitude, latitude, tangent altitude and volume regions in space via two dimensional maps (global) and three dimensional Observer windows (Earthview). The Observer windows give the user a perspective from above the Earth looking down and the view of the instrument line of sight, its field of regard and field of view.

The second choice allows the user to bring up existing catalog information about the available database. This would provide information about each specific program, experiment and observations (images, data, documents, etc.) allowing the user to further restrict his selection of experiments or to focus on a smaller subset of the data. The third selection allows the user to bring up previously defined (through the Build QuerySet section) data sets for further refinement or analysis.

II. EDITING AND VIEWING THE QUERYSET RESULTS

After building his data set, the user can further refine what he has chosen via the Edit QuerySet section. This section plots his results from the query on two and three dimensional graphic windows which provide the visual reference frame for what was observed. In all these windows the user can select and deselect specific images or instrument observations as they appear on a map or projection of various backgrounds. This process defines a smaller subset of the data set which the user can then view or pass on for further data processing.

The windows available for editing the data set are similar to the ones that helped the user build the data set, but actually plot the locations of the data over the Earth and sky backgrounds. The global map is plotted with the continent outlines and latitude and longitude markings with areas drawn for the image overlay. The user can point and click on one of the plotted overlays to get information about the image or observation and to get a quick-look view of the data. The quick-look view capability will most likely work with representative images of a particular scene or compressed versions of real images.

The Observer windows give the user the three dimensional view from above the Earth and from the instrument's perspective. They are interactive by allowing the user to select individual data items, a group of items, manipulate the viewing angle of the observer, or determine information about the satellite orbit or objects within the field of view.

For the user interested in celestial objects, VISTA provides a special section of the user interface that deals only with the celestial background. The VISTA-StarX11 interface graphically plots the user's selected data items over a celestial map in a coordinate system of his choice. Like the other windows, the user can select and deselect the various data items mapped as a result of his query. The user will also have available to him an array of commands with which he can manipulate his celestial view. Such commands would be to change location, magnitude range, constellation and extended object plotting, object labeling, precession date, and more.

III. PROCESSING AND ANALYSIS

Once the user has an active QuerySet (data set) defined, he may want to perform some data or image processing tasks and view the results. The Processing and Analysis section of VISTA comes equipped with a friendly textual and graphical interface to various analysis tools. The design of VISTA is being driven to accommodate interactive links to other analysis tools such as IDL and IRAF. The user can simply take advantage of the VISTA menuing system which will make it easy for him to setup an interactive window for his chosen package while allowing him to remain interactive with the main user interface. Hence he could move to and from his session to accomplish other tasks. VISTA also will provide fast plotting routines for images, graphs, and text.

A graphical visual programmer will be available for the user who wants to set up fast interactive processing tasks for his chosen data set. Through the use of a well defined set of glyphs representing various functions, the user can graphically 'glue' together a chain of tasks to be performed on whatever subset of data he chooses. The visual programmer gives the user the advantage of visually creating programming tasks without ever editing a program.

Overall, the VISTA user interface enhances the analysis process by providing the user with a visual representation of his experimental data in a fashion that is conducive to understanding the context under which the data was observed. This provides the user with important information in determining the quality, quantity, and substances of his data, and allows him to make critical time saving decisions when he is performing his analysis.

ACKNOWLEDGMENTS

The BDC, under the auspices of the Naval Research Laboratory (NRL), Space Science Division (SSD), has been designated by the SDIO Data Management Directive as the prime archive for all backgrounds phenomenology data collected by SDIO programs. In addition, SDIO has directed the BDC to provide data analysis support to allow the SDI research community to conduct research and analysis at the BDC facility.

The VISTA team would like to acknowledge the following for their support and contributions to the on going VISTA development: Ms. Suzanne Bauer, Mr. Richard Dorsey, Mr. Warren Weiss, Mr. Andrew Lawson (Hughes/STX), Dr. Harry Heckathorn (BDC Program Manager/NRL), and Dr. Herb Gursky (SSD Superintendent/NRL).

THE CFHT ARCHIVE SYSTEM

DANIEL DURAND AND DENNIS R. CRABTREE
CADC/DAO, 5071 W. Saanich Rd., Victoria, B.C. Canada V8X 4M6

CAROL CHRISTIAN
EUVE/CEA/UCB, Berkeley, CA 94720

JOHN GLASPEY
CFHT, P.O. Box 1597, Kamuela HI 96743

ABSTRACT The philosophy, design and implementation of the CFHT Archive system will be presented. Data from several CFHT instruments will be archived at the CADC/DAO starting in January 1992. Aspects of the data acquisition system at CFHT which pertain to the archive will be discussed as will the hardware and software being implemented at the CADC for storing and accessing the archive. The CFHT Board of Directors has set a proprietary period of 2 years on data taken at CFHT so that the first data will be available from the archive in 1994.

INTRODUCTION

Currently most data obtained at ground-based observatories is taken home with the observers on magnetic tape and once the data has been reduced, analysed and published it is essentially thrown away. It is usually not possible for another astronomer to find out what data is available or to get access to the data and after a few years the tapes start becoming unreadable so even the original observer can no longer access the data. Also, most ground-based observatories do not have a policy regarding the proprietary period of the data. Unlike space-based missions which specify that the Principal Investigator has exclusive use of the data for a specified period of time before the data becomes publically available, the data taken by ground-based observers essentially belongs to that observer. The need for producing archives from ground-based observatories is growing now that large format CCDs are in use and multi-object spectrographs are capable of producing the same amount of data in one night that used to take several observing runs.

Fortunately, the technology needed for producing a useful archive of ground-based data also exists. It is now possible to get optical media with large capacities (several GB) and lifetimes on the order of 15-20 years. The large capacity is needed as it is now possible for astronomers to produce 1 GB of data each night with the large format CCDs. A lifetime of 20 years for the media is probably longer than what is actually needed as the hardware for reading these disks will not have as long a lifetime. A more likely scenario is one where every 5-7 years the current copy of the archive is copied to a new generation of media which will provide a storage capacity 3-8 times larger.

Storing the data is only one aspect of an archive. The other aspect is cataloging the contents of the archive and searching this catalog to find the data which is of interest. It is now possible to purchase an efficient relational database system (RDBS) along with the CPU to drive it for a reasonable amount of money.

DESIGN CONSIDERATIONS

The CFHT archive system can be logically divided into two sections: the acquisition section which includes the handling of the data at the telescope(including instrument control etc.) and the archive section which includes the actual archiving and retrieval of data.

Ideally, the CFHT and its instruments would have been designed with the idea of a data archive as a goal. Encoding temperatures, various instrument parameters, etc. is the only way to ensure that all of the correct information is available to be included in the archive database. It is much more expensive and in some cases impossible to retrofit a large telescope like the CFHT with all of the encoders necessary to describe the state of the telescope and instrumentation for each exposure. **Hopefully, future large ground-based telescopes will be designed with data archiving in mind.**

The acquisition section has been designed in a way to minimize the impact on the observer doing the observations. One must not forget that the primary purpose of the observing is to acquire the data! While one might wish for more information about the observations in the archive catalog we have relied on information which can be obtained automatically or which is input by the observer during the normal course of observing. All of this information is stored in the FITS headers of the data frames as they are written by the data acquisition software. The information is stored in the FITS headers so that each data frame has as complete a description of observing parameters as is possible. The content of the FITS header is a combination of information from the TCS computer, the data acquisition computer, the instrument and the observer.

The archive section has been designed to build upon the expertise gained at the CADC while dealing with the HST archive and to reuse as much of the hardware and software from that system as possible. Thus the CFHT archive will use the same media as the HST archive. The HST archive currently uses the 2.0 GB LMSI WORM optical disks but will move to Sony 6.5 GB optical disks when DADS is operational. The CFHT archive which will start January 1, 1992 will be completely on the Sony ODs. The CFHT archive will also use the same database system and software (DMF) which has been developed for the HST archive.

ACQUISITION SECTION

The data flow at CFHT is outlined in Figure 1. Once the observer has acquired a data frame at CFHT the data is sent from the mountain down to magnetic disk at the headquarters in Waimea. In Waimea the FITS header of each data frame is stripped off and stored on magnetic disk. The complete FITS file is then written to optical disk in a special device independent format. Backup copies of the data files are kept on Exabyte tape in case of problems with the optical disk. This process of producing copies of the FITS headers and copying the data to OD is handled automatically by daemon software running on the system in Waimea. The optical disks will remain in Hawaii for approximately a year so that the data can be retrieved for engineering purposes or sent to an observer if something happens to the copies which the observer has taken home.

ARCHIVE SECTION

On a regular basis the FITS headers will be retrieved over the network to the CADC where the information in the headers is checked for consistency and selected information is used to update the various tables which form the CFHT catalog (See Figure 2). The catalog of CFHT observations will contain information on filters, gratings, detectors and instruments. Also, the catalog will contain pointers to scanned images of the log sheets and other documentation on the instruments and detectors. The catalog of observations will be available immediately and not subject to the two year proprietary period.

Once the optical disks are received at the CADC they are *ingested.* This process involves scanning the optical disk and cross-checking the contents of the disk with information already stored in the database from the FITS headers. For each file on the optical disk an entry will be made in a special table in the database which will give the diskname and filename for that observation.

Each direct image will be automatically processed by an IRAF script to determine the image quality and this value will be placed in the catalog of observations.

One of our goals is to produce an archive which is as scientifically useful as possible. It is our experience that a sizeable fraction of the data requested by archive researchers is not what was expected for some reason. Either the data was under-exposed, saturated, or simply wasn't the right field. We plan on producing a compressed version of the CFHT archive which is online and which is intended only for previewing the data. The archive researcher will be able to look at the data before actually requesting the data.

For direct images this online archive will consist of spatially block-averaged images which have been scaled to a dimension of 512 pixels or less and then compressed using a lossy compression algorithm. The combination of spatial averaging and special compression will allow us to keep the compressed images online. For example, a data frame from the large 2048 x 2048 CCD is 8MB in size. After spatial averaging and compression the image takes up approximately 70 KB of disk space. This allows us to store over 14,000 images on a 1 GB disk which is 2-3 years of data. This compression process will run automatically in the same IRAF script which calculates the image quality.

A similar procedure will be run for each spectral image as well. For these data frames a quick automatic spectral extraction will be performed and the output of this will be stored on magnetic disk for previewing by an archival researcher.

DATA RETRIEVAL

The system used by an archive researcher to retrieve data is the same used to retrieve HST data. This system which has been developed jointly the the ST-ECF, CADC and STScI is known as STARCAT (Richmond et al. 1987). STARCAT allows the user to search the database with a user friendly screen-oriented interface. The user can qualify any of the fields such and R.A., Declination, object name, etc. and only retrieve those records which match his/her input qualifications (Figure 3). Recent changes to STARCAT made at the CADC allows the user to preview the data if it is available online. The CFHT archive will allow users to preview direct CCD data and Herzberg spectroscopic data using the online archive described in the previous section.

A user can then *mark* any dataset which he/she wishes to have retrieved from optical disk. This data is then sent via the network or copied onto various magnetic media and sent to the researcher by mail.

FUTURE DEVELOPMENTS

In the future we hope to be able to scan the SIMBAD and NED databases in order to include standard object designators in the database to facilitate searching by object. We also hope to provide a real X11 interface to STARCAT to provide a more sophisticated and easier to use interface. Moreover, we also plan to make the CFHT data available on the Astrophysics Data System. The current plan is to archive data from only 4 of the instruments in use at CFHT. However, as new instrumentation is brought online this data will also be archived.

ACKNOWLEDGMENTS

We wish to thank Bob Hanisch and Frank Valdes for making valuable comments on the content and form of the CFHT FITS headers.

REFERENCES

Richmond, A., McGlynn, T., Ochsenbein, F., Romelfanger, F., & Russo, G. 1987, "The Design of a Large Astronomical Database System", Astronomy from Large Databases, eds. F. Murtagh & A. Heck, ESO Conference and Workshop Proceedings No. 28

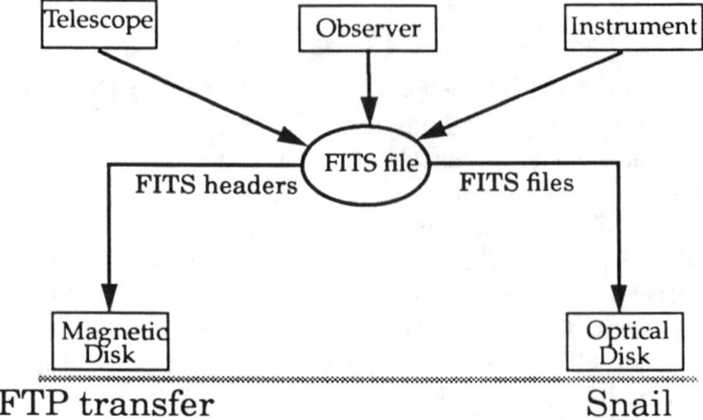

Figure 1: Data Flow at the CFHT

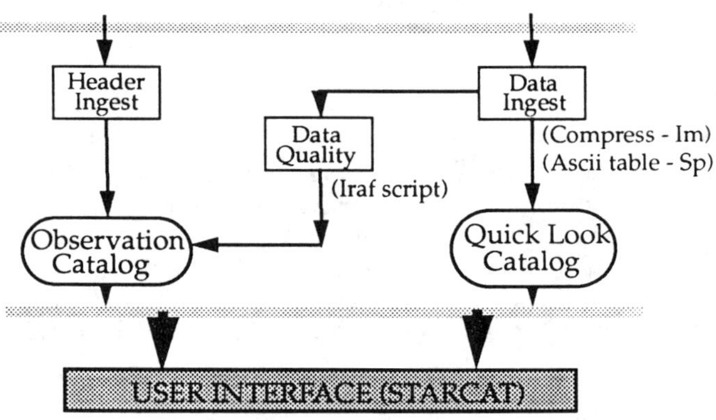

Figure 2: Data Flow at the CADC

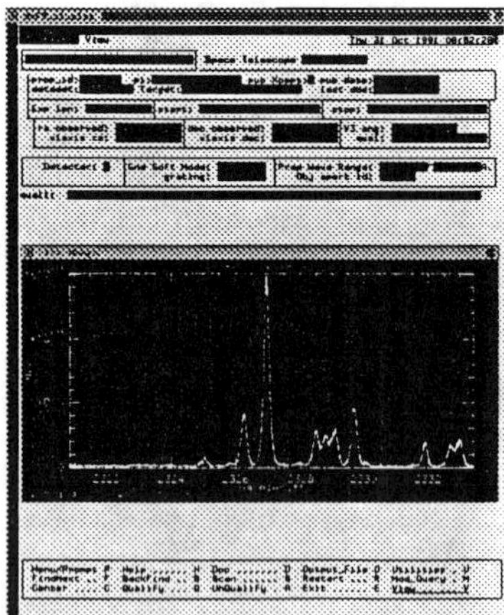

Figure 3: Prototype of the User Interface
Here shown with an HRS spectrum

VARBI DBMS: THE VARIABLE STARS BIBLIOGRAPHICAL DATA BASE MANAGEMENT SYSTEM

O.YU. MALKOV
Institute of Astronomy, USSR AS, 48 Pyatnitskaya st., 109017, Moscow, USSR

INTRODUCTION

The General Catalogue of Variable Stars (GCVS) is a widely used catalogue. It is based on a comprehensive handwritten card catalogue of variable stars data, stored at Shternberg astronomical institute (Moscow), which unfortunately is much less available to the astronomical community. The catalogue consists of data resumes of papers, keeps almost 600 thousand cards about nearly 60 thousand objects and contains, according to rough estimations, about 500 Mb of information.

The main reasons that forced us to start creating the data base are the following:

- to construct a machine readable version of the unique copy of the card catalogue;
- to facilitate the updating and processing of new records in the catalogue;
- to simplify preparation of future versions of the GCVS and the Catalogue of Suspected Variables, them being subsets of the card catalogue.

VARBI DBMS DESCRIPTION

For constructing the data base we needed a tool characterized by a high degree of interactivity while working with large volumes of data both text and complex-structured. A document data model has been chosen in IRIS, an expandable toolkit for information system design (available from "Inforg", a society that develops and distributes the product).

In addition to traditional data types supported by various software, in IRIS documents one can have texts with a variable length, multiple fields containing a variable number of instances of a defined type item, simple and multiple groups of elements, nested group of elements entering other groups and forming a hierarchy up to ten levels deep.

The card catalogue is entered in the following format:

- Designations: GCVS, preliminary designation, designations in the paper.
- Author, source.
- Coordinates: values, accuracy, determination method.

- Observations: number, type, system, date.
- Classification.
- Elements: photometric, spectroscopic, physical.
- Extrema dates.
- Photometric magnitudes and colors.
- Light curve parameters.
- Spectrum, chemical abundance.
- Radial velocity.
- Connection with a nebular, cluster or galaxy membership.
- General information.

Every logical group is supplied with a "comments" field. There are multiple fields ("designation in the paper", "photometric period", etc.) and multiple groups ("coordinates", "observations", etc.). Some fields have initial or succeeded meanings. For another fields (e.g., "type of observation"), a list of possible meanings is proposed. While filling out the document, alphabet, numerical, or another control of some fields (e.g., "declination degrees", "year of publication") is carried out.

The indexed fields in the document are:

- Designation.
- Author, source.
- Card compiler, date.
- Editor, date.

You can use these indexed fields for a document query, using different facilities as full or partial values of fields; logical operators NOT, AND, OR; simultaneous search with several field values as a key. Then you can work with the located documents (sort, correct, print, delete, copy, etc.). The VarBi DBMS allows you to write ASCII files.

To help in the preparation of future GCVS versions, and to solve the cross-identification problems, the "Designations" data base was constructed as well. It is also IRIS-based and will contain lists of designations of variable stars.

The work on entering the card catalogue into the VarBi DBMS is in progress now.

ACKNOWLEDGMENTS

A collaboration with the staff of Variable Object Group of the Institute of Astronomy USSR AS was very fruitful. I am grateful for the help provided by A.V.Orlov and L.Yu.Kokurin in preparation of the demo version of VarBi DBMS. I also acknowledge the Smithsonian Astrophysical Observatory for financial support.

GUIDARES: GUIDE STAR CATALOG DATA RETRIEVAL SOFTWARE

O.YU. MALKOV, L.I. KULKOVA, O.M. SMIRNOV
Institute of Astronomy, USSR AS, 48 Pyatnitskaya st., 109017, Moscow, USSR

INTRODUCTION

The Guide Star Catalog (GSC) was assembled to support the operational requirements of the Hubble Space Telescope (Lasker et al. 1990; Russel et al. 1990; Jenkner et al. 1990) which needs an all-sky set of reference coordinates for pointing and tracking.

The GSC is widely used by the astronomical community for many application as, for instance:

- statistical study of some sky region;
- search for candidate optical counterparts of astronomical objects found at other wavelengths;
- mapping of an object vicinity (in particular, finding charts of variable stars).

The catalog has been made available to the public in the form of a double CD-ROM set. The CD-ROM version was supplied with specialized software (Priou 1990). This allows to specify limits of rectangular area on the sky in equatorial coordinates 2000.0, and to list objects from the GSC on the screen (unfortunately without any possibility to keep an information in a file for further use).

To make the GSC use more comfortable one needs special software for reading the catalogue. We designed the Guide Star Catalog Data Retrieval Software (GUIDARES) to use the GSC on IBM PC as a tool of a sampling of the catalog and creating sky map of desirable area.

GUIDARES DESCRIPTION

GSC retrieval can be made through equatorial, ecliptic, or galactic coordinates at any equinox. The default equinox is the standard equinox in the GSC: 2000.0. The area can be specified as a rectangle or a circle.

The software is supplied with self-explained menu which

- asks you for the logical unit where the GSC files can be found;
- asks you for the shape of the area, system of coordinate, equinox, and coordinates themselves;

- checks coordinates to be in correct limits;
- reorders minimum and maximum declination/latitude when they are inverted;
- displays the evolution in the process and submits a short summary of the retrieval;
- gives short prompt for every action.

The retrieval of objects has to be done in three steps:

- Modification of the limit coordinates to the standard GSC system (J2000.0).
- Determination of GSC files which contain the user area.
- Retrieval of the contents of selected GSC files. All entries are checked if a current object is in the user area. Coordinates of the object are converted to the user system. In a case when the object appears on more than one plate and, therefore, has more than one entry in the GSC (so called few entries object; hereafter FEO) mean weighted values of coordinates are calculated. If the object is in the user area, other parameters are calculated (for FEOs): mean weighted magnitude with an error, and position error.

Two output files are created. One of them is an ASCII table (FITS table soon also) with output data for each object. You can use the other output file when you want to construct the area map. The file contains the MONGO graphic package command lines.

The objects are represented by circles with a size depending on magnitude. As the objects were classified as stellar or non-stellar, different symbols (open or filled circles, correspondingly) on maps are used for object images. FEOs can, in principle, have not the same classification in different entries. These objects were classified as "semi-stars" (in graphical but not astronomical sense) and starred circles represent them on figures.

Comparison with Priou (1990) software shows the following results. Priou's program (including files with limits) occupies near 340 Kb, GUIDARES - about 100 Kb. Performance testing has shown that GUIDARES can examine 4 GSC files in 80 seconds (90 seconds for Priou's software). Other advantages of GUIDARES are listed above.

Our software is written on C and contains about 1800 instructions.

Comments and suggestions are welcome concerning the software.

ACKNOWLEDGMENTS

Support from the Smithsonian Astrophysical Observatory is gratefully acknowledged.

REFERENCES

Jenkner, H., Lasker, B. M., Sturch, C. R., McLean, B. J., Shara, M. M., & Russell, J. L. 1990, AJ, 99, 2081

Lasker, B. M., Sturch, C. R., McLean, B. J., Russell, J. L., Jenkner, H., & Shara, M. M. 1990, AJ, 99, 2019

Priou, D. 1990, Guide Star Catalog Listing Program, Institute Géographique National

Russell, J. L., Lasker, B. M., McLean, B. J., Sturch, C. R., & Jenkner, H. 1990, AJ, 99, 2059

SECTION 3

SOFTWARE SYSTEMS AND ENVIRONMENTS

THE ASSIST USER INTERFACE PROTOTYPE

ERIC MANDEL, JOHN ROLL, AND STEPHEN S. MURRAY
Harvard-Smithsonian Center for Astrophysics, Cambridge, MA 02138

ABSTRACT We believe that the future of astronomical data analysis software lies in a shared effort to make our present systems easier to use and more responsive to the needs of the astronomical community. Our recent prototyping efforts have been directed toward the development of a set of IRAF-compatible interface tools, including an X11 graphical user interface, that can be used by IRAF and non-IRAF programs alike. It is our hope that this AXAF Science Support Interactive Software Tool (ASSIST) will make it possible for our IRAF-based software development group to collaborate with other software groups, without imposing a particular environment on anyone.

INTRODUCTION

In his famous opening to *A Tale of Two Cities*, Charles Dickens described eighteenth century Europe like this:

> "It was the best of times, it was the worst of times, it was the age of wisdom, it was the age of foolishness, it was the epoch of belief, it was the epoch of incredulity ..."

Surely this is how scientists must feel when confronted with the wide array of astronomical analysis systems available today!

On the one hand, we have witnessed the development of several highly successful analysis systems that offer vast functional improvement over past software. Taking advantage of the explosion in hardware computing power over the past ten years, we now offer analysis tools only dreamed of in the days when programs fit into 64 kilobyte address spaces and when 5 megabytes of disk space was all the data storage one could reasonably expect. On the other hand, until very recently we have not had access to standard hardware and software tools that would allow us to evolve our user interfaces to match the functional complexity of our new systems. The result is that our command-line driven systems are often complicated and difficult to master.

All of these analysis systems have special strengths. There are excellent packages available for plotting, for image processing, for spectral and timing analysis, for statistical analysis, and for image display. There are also general systems available that provide a broad range of functionality. But it is clear that no one system can satisfy all of a user's analysis needs. It is inevitably necessary to combine the functions of two or more of these systems in order to

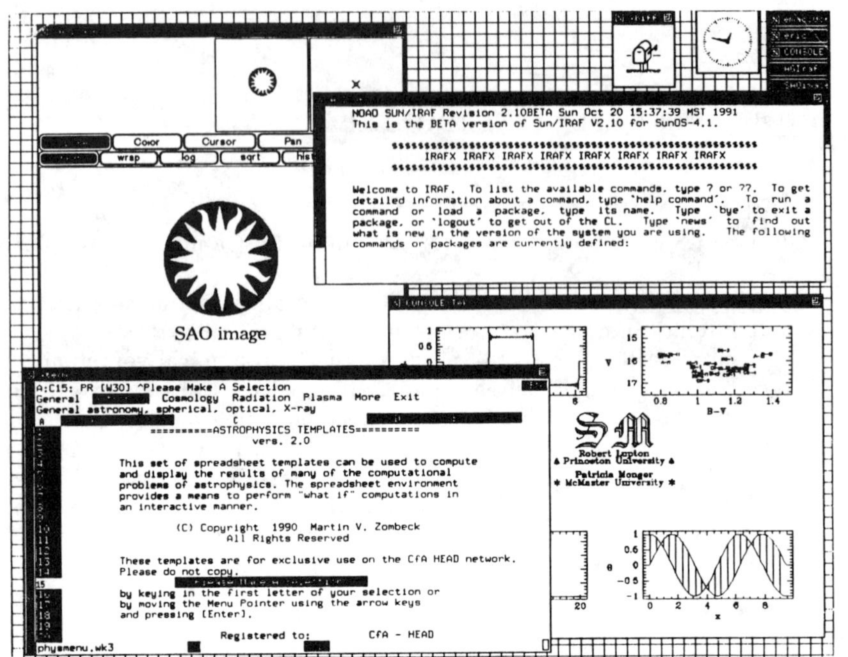

FIGURE I Science analysis often requires use of more than one system.

carry out a research program (see Figure I). Until recently, however, little effort has gone into developing simple interfaces between the different systems. Each one usually supports a special input format, a special command syntax, and a special way of presenting results.

Even in aggregate, these analysis systems cannot provide all of the functionality required by users. It is often necessary for researchers to modify existing functions or to create new functions. Interfacing new code with a given system requires access to that system's data structures and algorithms. Many systems provide such access, but nearly all of them require extensions to be developed within the confines of the given system. There is not much provision to allow a user to interface with these systems using his native environment.

One conclusion that can be drawn from these considerations is that attention can and should be placed on improving various aspects of the "ease of use" of our analysis systems. Indeed, it can even be argued that now is precisely the time for undertaking such work. User interface technology, especially the X11 window system, has matured to the point where we can begin to make real progress in this direction. Our analysis systems also have matured to the point where the development of basic scientific functionality need no longer dominate our efforts. We can begin to look at the question of how to make our systems easier to use, easier to modify, and easier to interface with one another.

A corollary to these conclusions is the belief that collaboration between developers will greatly strengthen the overall software effort in astronomy today. The wide range of successful analysis systems currently available testifies to the fact that there are many talented software developers working in the astronomical community. It is obvious that great mutual benefit will accrue to those groups who are able to work together in the right way.

These considerations and conclusions have shaped the thinking of the AXAF Science Center data system group as we seek to develop a comprehensive set of tools for X-ray analysis. Our fundamental aim is to build a data system that will satisfy the needs of AXAF users. ASC software must be easy to use as well as functionally robust. At the same time, we recognize that no single environment will satisfy all ASC users and that we will need to interface smoothly with other systems and with other developers. Finally, we recognize that users will wish to modify and extend ASC software tools and that facilities must be developed to make this easy to do.

Our initial attempt to balance these various considerations is rather simple: we have chosen IRAF as the basis for our data system, believing, as we do, that IRAF is an environment well-suited to the needs of AXAF users. From the beginning, the ASC will work closely with the IRAF development group at NOAO to develop IRAF to meet the needs of AXAF users. At the same time, we *explicitly* recognize the merits of other analysis systems. We are therefore seeking to develop user and programming interfaces to ASC tools that will encourage the sharing of software without imposing the ASC's chosen environment on anyone. It is our hope that such an "open architecture" will encourage software developers, including the AXAF science instrument groups, individual researchers in the X-ray community, and groups who are developing other astronomical analysis systems, to work with the ASC to develop the best possible software for the AXAF project and for the astronomical community as a whole.

In keeping with our fundamental aim of developing an easy to use data system, one of our first steps has been to explore ways of improving the ease of use of the IRAF analysis environment. In the spring of 1991, work on graphical user interfaces for IRAF was begun under the auspices of the IRAF Technical Working Group. Important contributions were made by John Roll and Eric Mandel at SAO, by Bob Hanisch and his team at STScI, and by Doug Tody at NOAO. Subsequent development efforts were undertaken at SAO on behalf of the AXAF HRC and ASC projects. As the work progressed, we found that the graphical prototypes under investigation had direct bearing on the problem of interfacing ASC software with that of other groups. These issues have become intertwined at SAO and we have been pursuing them as one issue for a number of months. What follows is a progress report on our current efforts to develop a set of ASC interface tools that are straightforward and easy to use, that provide powerful yet familiar functionality, and that provide a possible foundation for cooperation and sharing between astronomical software groups. The discussion of our prototyping efforts will end with an invitation for those interested to join us in our work.

THE ASSIST USER INTERFACE

A central feature of the data system architecture in SAO's 1990 ASC proposal was the development of a user-friendly standard interface for *all* ASC tools. This idea was embodied in the *AXAF Science Support Interactive Software Tool* (ASSIST). The ASSIST originally was conceived as a graphical user interface that would provide a uniform "look and feel" for all ASC tools. It was to

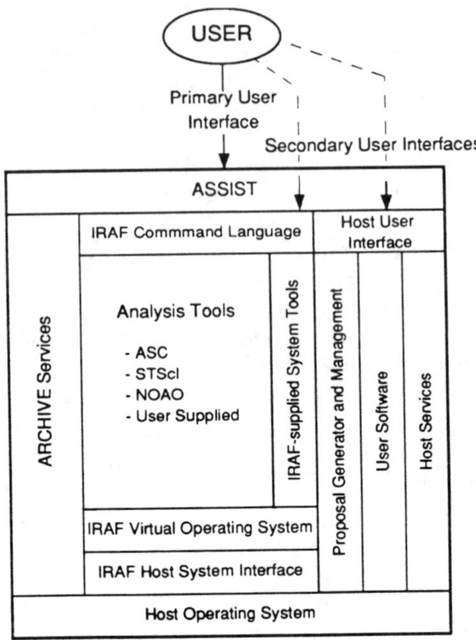

FIGURE II The ASSIST interface is layered on top of ASC tools.

be be layered on top of those tools to allow direct access to the underlying software. The ASC software architecture based on this ASSIST concept is shown in Figure II.

We have continued to develop and refine our concept of the ASSIST since the completion of the ASC proposal in July, 1990. Our expanded concept of the ASSIST is centered on three key design concepts:

- The concept of the ASSIST as a portable interface has been extended to emphasize the use of standards. Thus, for example, we have provisionally chosen the universally-available Xt toolkit as the basis for our graphical interface. We will continue to emphasize standards in order to maximize our compatibility with widely-used systems.

- The concept of the ASSIST as a uniform graphical user interface (GUI) has been extended to include a uniform application programming interface for the underlying tools. In practice, since IRAF is our baseline software environment, we need to make IRAF's parameter and data file I/O capabilities available to non-IRAF programs. This will allow non-ASC programs developed by users in their native environment (*e.g.*, Unix or VMS) to interface with ASC tools.

- The graphical interface will communicate with the underlying tools by *exactly* the same means as would be employed by a user. For example, if the graphical interface is executing an IRAF command, it should send *exactly* the same command to the IRAF window that a user would type at that window. By combining this feature with the ASC parameter-handling interface, we allow the ASSIST to accommodate non-ASC programs that use different command interfaces (*e.g.*, standard Unix or VMS programs).

A prototype of the ASSIST has been developed at SAO based on these central design concepts. The prototype consists of three separate sets of tools:

- an IRAF-compatible parameter interface for "host" programs, *i.e.*, programs that run directly in the native operating system environment rather than in IRAF
- an host interface to the IRAF Virtual Operating System (VOS) that allows non-IRAF programs to call IRAF routines
- an X11 graphical user interface that uses the parameter interface to edit parameters and execute programs

The IRAF parameter interface and the IRAF VOS host interface can be considered "low-level" components of the ASSIST, connecting IRAF and non-IRAF tasks through common parameter and file I/O mechanisms. The graphical interface makes this connection at a higher level by providing a uniform look and feel to tasks that might possibly have different command interfaces. All of these tools are consistent across different platforms, such as Unix and VMS. Together, they allow the ASC to share software with other groups without imposing the ASC environment on those groups.

Each of these tools will be considered separately.

THE IRAF PARAMETER INTERFACE

A prototype parameter-handling interface has been developed for the ASC which implements nearly all of the functionality of the IRAF parameter-handling mechanism. To accomplish this, the low-level C code for parameter handling was extracted from IRAF and modified slightly for the Unix environment. High-level software was then layered on top of the basic IRAF code to produce a host parameter interface that provides IRAF parameter-handling functionality to non-IRAF programs. A full complement of parameter manipulation subroutines are provided in the interface, including routines to open and close named parameter files, read parameters, write parameters, and check for the existence of parameters.

Host programs that make use of the parameter-handling interface have the same look and feel as IRAF programs. Parameters can be of type *auto*, *hidden*, *etc.* They can be input on the command line (either as positional or keyword parameters). The *auto* parameters that are missing from the command line are prompted for, unless the program's *mode* parameter has been set to *hidden*. Limit checking, enumerated types, type conversion, *etc.*, are all implemented. Extended support for indirect parameters (*i.e.*, parameters whose value is actually the value of another parameter) is also implemented.

In addition to the interface itself, special host programs have been developed to allow for the manipulation of parameters at the host command line (*e.g.*, in the Unix shell or in DCL). These programs parallel the functionality afforded by IRAF's command-line parameter manipulation capabilities, but tailored to host environments. The following parameter-handling programs have been developed:

- *pget* - return the value of specified parameter from the named parameter file to the standard output
- *plist* - list the parameters (and values) in the named parameter file
- *pquery* - same as pget, but with a query capability
- *pset* - set the values of parameters in the named parameter file

Host programs that adopt the parameter interface gain IRAF's considerable power in the manipulation of parameters, a functionality that is, moreover, already familiar to IRAF users. The familiarity guarantees that such programs will be easier to use for many people.

THE IRAF HOST INTERFACE

The parameter-handling interface provides a uniform way for IRAF and host programs to manipulate parameters. In order to simplify further the coordination and communication between IRAF and host programs, it also is necessary to allow the latter to access IRAF data files directly. Therefore, a C-language interface to the IRAF libraries has been developed as part of the ASSIST prototype. Using this host interface, C programs can be written to access IRAF data files and other IRAF data structures (such as symbol tables). This prototype was implemented by developing C-language bindings for a sizable subset of IRAF routines, including image I/O, QPOE I/O, PROS region filtering, and STScI table I/O. Because these bindings call the IRAF SPP routines directly, they are guaranteed to work without modification so long as the underlying IRAF interfaces remain unchanged.

The IRAF host interface provides developers with many powerful and familiar capabilities. For example, if a program creates or manipulates QPOE files (X-ray event files), the IRAF host interface automatically provides advanced event filtering capabilities. Similarly, the PROS region capabilities provide sophisticated spatial region filtering. The extraordinary wealth of functionality that is available using this interface would be virtually impossible to duplicate by any individual group.

The host interface prototype is being developed in close cooperation with the IRAF group at NOAO. It is being used to explore the issues surrounding the development of a host interface for C and Fortran that will replace the current IRAF IMFORT interface.

THE GRAPHICAL USER INTERFACE

The graphical user interface completes the ASSIST prototype. Written using the Xt toolkit and layered on top of the interfaces described above, it provides a uniform graphical interface for IRAF and host programs.

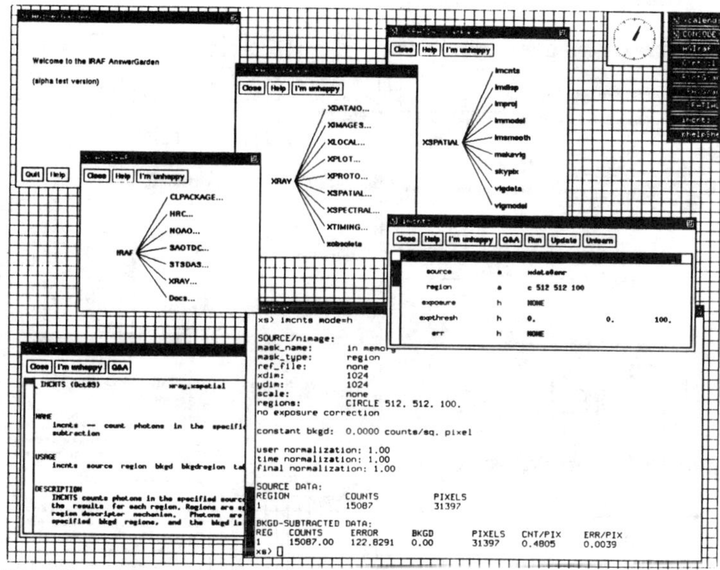

FIGURE III The ASSIST graphical interface consists of a hierarchy of parameter editors that communicate directly with the underlying tools.

The graphical user interface is based on the AnswerGarden, an Xt program that was designed by Mark Ackerman (Ackerman & Malone 1990) at MIT to help groups develop databases of frequently asked questions. The AnswerGarden system includes a branching network of nodes that supports the graphical display of diagnostic questions and informational panes to help users find information. If additional information is needed, or the user does not understand the diagnostic questions or information available, he can ask a question through the system. AnswerGarden automatically sends the question to the appropriate expert, and the answer is returned to the user as well as inserted back into the branching network. Hence, the system grows over time.

For the ASSIST prototype, the AnswerGarden node service has been extended to support the inclusion of *graphical parameter editors* in the node "trees" (see Figure III). An ASC tool that uses IRAF or IRAF-compatible parameter files can be represented by a parameter editor that displays the parameter names, values, data types, modes, limits, and prompt strings. Using the parameter editor, one can edit the values in a given parameter file. User-definable Xt resources control many facets of the parameter editor display (*e.g.*, which fields of a parameter record to display, which fields can be edited, *etc.*). The parameter editor also supports the ability to display help for the program or task, to inspect bug reports, to ask questions, or to run the program.

The parameter editors initiates actions (such as running a program) by sending commands to an xterm window, where these actions are performed. These commands are the *exact* commands that a user would input in order to execute a given function at the command line. For example, to run an IRAF task, the appropriate IRAF task invocation is transmitted. In this way, the parameter editors are layered on top of ASC tools in a modular yet extensible

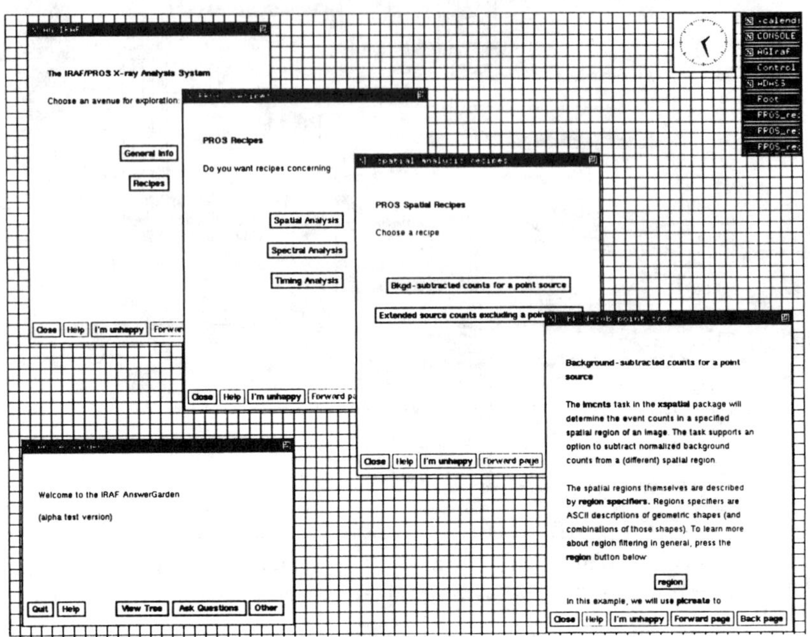

FIGURE IV The ASSIST supports hypertext tutorials and recipes.

way.

Thus, the ASSIST graphical interface consists of hierarchical trees that organize and manage the parameter editor interfaces to ASC software tools. The nodes of each tree are either sub-trees or parameter editor "leaves". For example, the IRAF tree has a node for each sub-package in the IRAF hierarchy. A user can easily traverse the IRAF tree to find the desired tasks, and then bring up the appropriate parameter editors to run these tasks.

The IRAF hierarchy is only one way to organize the package graphs and parameter editor nodes. These same nodes can be organized in any number of ways to suit individual applications. For example, the nodes can be embedded in text windows to produce hypertext tutorials and recipes (see Figure IV). Users can work through instructive examples, bringing up packages and tasks, or even executing demonstration functions directly. As usual, IRAF commands are sent directly to the IRAF window for execution, so that actual results can be viewed.

Finally, the ASSIST graphical interface supports sophisticated means of getting help. Users can view current bug reports, submit new bug and enhancement reports, ask questions, and make comments. As mentioned before, all reports, comments, *etc.* are automatically forwarded to the appropriate expert for resolution. Answers are returned to the user and also are inserted back into the graphical interface "knowledge base".

The first prototype of the ASSIST was created to support the AXAF HRC analysis tools in IRAF. In this prototype, the graphical tree consisted of several

IRAF sub-packages and tasks, including the PROS XRAY package and the STSDAS package.

An expanded alpha version of the ASSIST prototype was released in mid-October, 1991, within SAO. All IRAF tasks can be run from the expanded ASSIST. We are in the process of gathering and evaluating comments on this alpha version. Our plan is to improve the prototype, based on our evaluation, and then to make a beta release available to a small group of sites in the near future. Further plans for a general release will follow evaluation of the beta test sites.

THE ASSIST AND SAOIMAGE

One very important set of software tools for the ASC is that of image display. Currently, the X11-based SAOimage program is one of the most popular image display programs in use by astronomical groups world-wide. Written at SAO between 1985 and 1990, SAOimage employs a "button-box" interface for user input based on the low-level Xlib library.

In order to explore the connection between the ASSIST and future image display tools, we have made two simple modifications to SAOimage:

- SAOimage was modified to read IRAF image files (including QPOE files) directly, without having to go through the IRAF *display* command.
- The SAOimage *cursor/region* commands were modified to allow the option of reading and writing the cursor region descriptors to or from a parameter file (rather than the standard SAOimage region descriptor file).

These two modifications effectively link SAOimage to any program that uses the parameter interface. In particular, the SAOimage cursor facility can be used to specify arbitrary regions of interest graphically, and then communicate these regions of interest to other programs via the parameter interface. Indeed, if a task's region parameter is an indirect parameter pointing to SAOimage's region parameter, the task can be run with new region filters simply by changing the cursors on the SAOimage display, writing the resulting regions to the parameter file, and re-running the task. This facility creates a simple and direct connection between SAOimage and arbitrary analysis programs. Recently, we have been exploring this use of indirect parameters as a means of providing a *common data set interface* between arbitrary analysis programs.

FUTURE DEVELOPMENTS

Much experimentation is underway to develop the functionality of the ASSIST. Among the experiments planned for the coming months are:

- We will further explore the limits of the prototype host interface as a means of planning for a true interface that will be supported by IRAF.
- We will further develop the graphical parameter editor. There are other contexts within the scope of the ASC data system in which this type of

interface is desirable. We intend to generalize the parameter editor to increase its functionality and applicability.

- We will further develop the X-based graphical user interface. The incorporation of user manuals, recipes, and tutorials is in progress for the graphical interface. We also are investigating advanced techniques for organizing and searching documents using the graphical interface.

- We will further develop image display capabilities that are integrated into the ASSIST and that make use of the host interface. Our aim is to develop a successor to SAOimage that is connected both to IRAF and to the graphical interface in a way that allows the user to work with all of these tools as one integrated whole. In addition to new image display capabilities, we will further explore the concept of a common data set interface between separate tasks and programs.

AN INVITATION

As has been mentioned several times, it is our hope that the ASC's "open architecture" will lead to cooperation and sharing between astronomical software development groups. This includes IRAF and non-IRAF developers. We have recently begun to discuss the possibility of such cooperation with a few groups, based on the shared use and development of the graphical interface and the low-level interfaces. If your group is interested in entering discussions with the ASC, please contact us.

We believe that the future of astronomical data analysis software lies in a shared effort to make our present systems easier to use and more responsive to the needs of the astronomical community. We invite you to join us in this endeavor.

ACKNOWLEDGMENTS

Many people have been involved in the development of the ideas presented in this paper. While limited space does not allow us to mention everyone, special thanks must be extended to Doug Tody (NOAO), Bob Hanisch (STScI), Mark Ackerman (MIT), Ralph Swick (DEC), and the IRAF/PROS programming staff at SAO. This work was supported under NASA contracts to the IRAF Technical Working Group (NAGW-1921), the AXAF High Resolution Camera (NAS8-38248), and the AXAF Science Center (NAS8-39073).

REFERENCES

Ackerman, M. S. & Malone, T. W. 1990, in Proceedings of the ACM Conference on Office Information Systems, New York: ACM Press, 31

SOFTWARE DESIGN AND DEVELOPMENT IN A SCIENTIFIC ENVIRONMENT: LESSONS LEARNED DURING THE DEVELOPMENT OF STAR, AN ASTROPHYSICAL ANALYSIS AND VISUALIZATION PACKAGE

GITTA O. DOMIK AND KRISTINA D. MICKUS-MICELI[1]
University of Colorado, Center for Astrophysics and Space Astronomy and Department of Computer Science, CB 430, Boulder, CO. 80309-0430

ABSTRACT Data visualization has become an important component in the analysis of scientific data. This paper describes two common goals aimed at when extending data analysis environments by visualization: offer visualization tools that fit the mind set of the targeted scientists; and integrate visualization tools into the complexity of existing data analysis tools. A case study performed between astrophysicists and computer scientists describes the design and development of STAR, a user-centered and integrated data analysis and visualization system.

INTRODUCTION: THE ROLE OF DATA VISUALIZATION

During the last five years scientific visualization has found a multitude of new applications. With an increase of computing power seen in low-end, mid-range and supercomputing systems, as well as increased data rates from data collecting sensors (higher resolutions, higher transmittal rates, compact storage devices), the availability of scientific data has multiplied. The increased need to interpret large amounts of data and the easy to understand principles (encode numbers into pictures) led to a recent popularity of scientific visualization. Encoding numbers as pictures stimulates mental processes in a different way than the interpretation of numerical values: pictures allow the observer to browse in a large data set to examine large scale structures and to focus at any time to observe small scale structures; pictures emphasize spatial relationships between objects described in a data set; and pictures may suggest patterns not obvious when looking at the representative numbers. Such qualitative judgments are made possible through data visualization, whereas numerical values lead to quantitative judgments of the data to be interpreted.

This paper addresses two common goals aimed at when extending data analysis environments by visualization: offer visualization tools that fit the mind set of the targeted scientists; and integrate visualization tools into the complexity of existing data analysis tools.

[1]CURRENTLY AT NASA AMES RESEARCH CENTER

THE ENVIRONMENT OF THE CASE STUDY

In early 1990, scientists at the Center for Astrophysics and Space Astronomy (CASA) at the University of Colorado, decided to expand CASA's data analysis environment to take better advantage of large amounts of scientific data available to the astrophysical community (Mickus-Miceli & Domik 1991; Domik 1991). The data is described by heterogeneous data structures (zero, one, two and multi dimensional data) and diverse data characteristics, such as varying spatial and spectral resolutions. An improvement in data visualization and interaction techniques held the promise for better exploitation of available data.

At CASA, public domain software packages are used together with in-house developed software. Due to the various characteristics of space and ground sensor data from different wavelengths ranges (such as radio, infrared, visible or x-ray), different software packages are being used. Input and output data streams are not fully standardized, reflecting the lack of standard data formats, and resulting in "islands" of software systems.

CASA's scientists have long appreciated the knowledge gained by utilizing multi-sensor and multi-wavelengths data, but have been hampered by the complex software environment created by the simultaneous use of various software packages and the need to convert between various data formats. The plan to develop multi sensor data sets demanded an integration of existing tools and data.

COLLABORATIVE DESIGN

User needs and resources need to be balanced for a comprehensive and realistic visualization solution. Therefore the design and development of visualization software was to center on the specific users at CASA, scientists with a variety of needs to serve their various research interests but of similar education and doctrines. The "user-model" to drive the new visualization software named STAR (Scientific Toolkit for Astrophysical Research) was to be defined by computer scientists and astrophysicists in a collaborative effort. Real solutions to active problems preceded graphically challenging visualizations.

As a means to initiate discussions between collaborators, a research scenario of limited scope was defined by designers and scientists. It reflected the main tasks of a data analysis and visualization system to solve a specific problem, namely the analysis and display of multi-spectral data: infrared data of various wavelengths responses but of a specific spatial location were searched for in CASA's local data bases; resulting images were visually checked for calibration errors and were preprocessed, if necessary; statistical analysis in form of numerical calculations (e.g. flux integration over an extended area) or graphical representations (e.g. vertical/horizontal profiles in the area of interest) were performed next; finally, three representative data sets were merged into one color image using red-green-blue or hue-lightness-saturation color coding. An initial prototype of STAR was built around a direct manipulation (point-and-click) user interface to reflect the goals of an integrated and user-centered visualization solution and tested against the specific research scenario. With this initial testbed, individual user interviews and general systems demonstrations could start.

One-to-one interviews between designers and users create an interactive discussion, with questions/answers. Individual user interviews resulted in a multitude of changes in the design of STAR, meeting our goal of an iterative design. General demonstrations promoted the existence of STAR at CASA and decreased the general concern about building "another software package that will not interface with the current environment". Besides talks and on-line demonstrations we also provided an easily accessible poster board to describe STAR's user interface and functionality; in cognitive terms this is called a "storyboard technique". Individual user interviews were obviously much richer and useful in their feedback, however, general demonstrations as well as the poster promoted the system to outsiders and initiated talks and discussions.

One of the major problems in a collaborative design with scientists is to get input from scientists as their time is usually already oversubscribed. Another paper by Mickus-Miceli & Domik (1991) describes in more detail issues surrounding a collaborative design to build scientific software systems.

INTEGRATED SOFTWARE SOLUTION

Expanding the analysis cycle through visualization and interactivity is necessary to prepare current data analysis software for use with large sets of diverse scientific data. Many different tasks need to be fulfilled between the acquisition of scientific data and its interpretation by a scientist. These tasks are often represented by different software packages. The scientist trying to solve an individual scientific question is challenged by a complex conglomerate of software systems distributed on various workstations and is thus hampered in his/her scientific work. The development of visualization tools only adds to this complexity.

In order to reduce the complexity as experienced by the user, we first unified existing software packages and tools under a common user interface and then integrated any new tools into the new software environment. The development platform at CASA for in-house developments by scientists and staff is IDL [2]. As STAR was also developed in IDL, integration with in-house developed software could be accomplished with little effort. IDL software packages are collected in source libraries. Depending on the type of data to be analyzed, the user would click one of several round knobs on the user interface indicating the correct data type and thus identify the libraries to activate. Each software package is organized according to the data analysis loop (Domik & Mickus 1991): access, identify, retrieve data; preprocessing; analysis; visualization. Additional menus for "Input/output" and "Environment Options" contain modules to execute administrative tasks. Activating all libraries at the same time will reduce the speed of the execution of individual modules as well as exhibit conflicts between modules of the same name with different functions. The user may switch from one IDL library to another at his/her convenience.

Software packages developed in other languages are not as easy to integrate. Square buttons on the user interface indicate their existence and allow access

[2]Interactive Data Language by Research Systems, Inc.

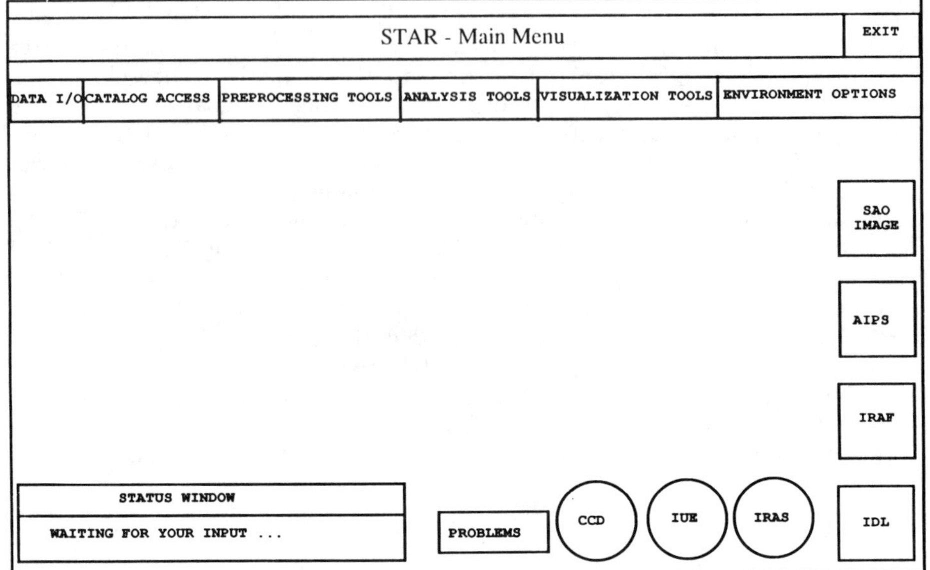

FIGURE I Main user interface level for STAR

to such "foreign" packages. Traveling between different packages demands a common solution to the various data formats. The astronomical community has spent much effort into designing the common data format FITS [3], which has been extended over the years, therefore resulting in several extensions. A suitable FITS definition needed to be chosen as STAR's common data format. Data conversion (to/from FITS format [4]) is handled by the main menu "Data input/output". The user may switch between various foreign analysis packages, entering a new user interface when doing so, and returning to STAR's main user interface after exiting such a software session.

The solution as described above is shared by diverse workstations, as IDL can be implemented on a variety of platforms. This was an additional advantage to choosing IDL as our prototyping/developing language, as the workstation environment at CASA is constantly expanded and new flavors of hardware acquired.

REFLECTIONS ON USER-CENTERED AND INTEGRATED SOLUTIONS

The user interface to STAR as a result of user interviews and demonstrations is reflected in Figure I. A status window is implemented to inform the user

[3] Format Interchange Transform System

[4] Many of the data conversion functions have been made available through the IDL Astronomy Library

about progress of issued commands or the general status of the system. Because problems often occur when the designers/developers are not available, a problem button was implemented to vent the users frustration and document the problem on the spot. The mailing system sends an appropriate message immediately to the designers. The location and size of problem and exit buttons experienced many redesigns.

A surprise to the designers was the demand of visualization tools: interactive tools to analyze and display data (e.g. interactive profiling, interactive manipulation of lookup tables; or interactive mapping of data values from a high dynamic range to 256 available colors) was much more in demand then complex visualization tools (e.g. transparency, or data slicers). The current installation of STAR runs on all graphics workstations at CASA. Not all foreign packages are compatible with the wide range of IDL installations, so that the user interface will not look the same on every hardware.

The current status of STAR reflects two years of work in participatory design and development of a visualization system integrated into a complex data analysis environment. Next steps in the still ongoing project are further expansions of STAR's environment and the addition of more complex visualization tools.

ACKNOWLEDGMENTS

The authors are grateful to all the scientists at the Center for Astrophysics and Space Astronomy, specifically to Dr. Edward W. Brugel, Dr. Thomas R. Ayres, Dr. Bob Stencel and all the graduate students involved, for their collaboration. Scott Merkle and Anthony Veale greatly supported the development of STAR. Prof. Clayton Lewis at the Department of Computer Science was very helpful in the design of the user interface. The project is supported by NASA under grant NAGW-1902. Gitta Domik's efforts were partly also sponsored under NSF grant ASC-9001115.

REFERENCES

Domik, G. 1991, Visualization Techniques to Aid in the Analysis of Multi-Spectral Astrophysical Data Sets, Final Report on NASA Project NAGW-1902

Domik, G., & Mickus, K. D. 1991, Visualization in the Analysis Cycle of Observational Data, Proceeding of Gesellschaft fur Klassifikation, Salzburg, Feb. 25-27, 1991

Mickus-Miceli, K. D., & Domik, G. O. 1991, Participatory Design in the Development of a Scientific Software System, Technical Report CU-CS-561-91, Department of Computer Science, University of Colorado, CB 430, Boulder, CP. 80309

Nielson, G. M. 1991, Visualization in Scientific and Engineering Computing, IEEE Computer, Vol. 24, No. 9, pp. 58 66

XIMAGE A MULTI-MISSION X-RAY IMAGE ANALYSIS PACKAGE

P. GIOMMI[1], L. ANGELINI[2], P. JACOBS[2] AND G. TAGLIAFERRI[3]
[1] ESIS, Information Systems Division, ESA/ESRIN, Frascati, Italy
[2] HEASARC, GSFC, Greenbelt, MD, USA
[3] Astrophysics Division, ESA/ESTEC, Noordwijk, The Netherlands

ABSTRACT XIMAGE is a portable X-ray image analysis and simulation package that supports the analysis of EXOSAT, *Einstein* and ROSAT data. Other X-ray imaging detectors can be supported provided that the proper calibration files are available. The graphic capabilities of XIMAGE are based on PGPLOT which supports most graphic terminals, and on SAOIMAGE which runs on X-Window. XIMAGE supports several data analysis tasks including a slide-cell detect and the simulation of data from present and future detectors. XIMAGE can be run as a standalone program or within the EXOSAT Database System (EXOSAT Observatory team 1991a,b) or from the ESIS Correlation Environment Prototype (Giommi et al. 1992). The XIMAGE code is written in FORTRAN77, runs on VAX/VMS and SUN/UNIX machines and is available on request at no cost.

INTRODUCTION

XIMAGE is a collection of X-ray image manipulation and analysis routines that have been assembled into a coherent and portable software package. Several of the routines were originally written for the EXOSAT interactive analysis system and have been generalized to support other satellites. Currently XIMAGE supports the analysis of EXOSAT CMA, *Einstein* HRI and IPC, ROSAT PSPC and HRI data, as well as some basic analysis of optical, infrared and radio images.

The display capabilities of XIMAGE are based on an extension of the PGPLOT subroutine library (Pearson et al. 1988) and on SAOIMAGE (VanHilst 1991) and support most graphics terminals and workstations.

XIMAGE supports several types of data formats, including a compressed format (Sternberg 1985), ascii files, and FITS format (both normal and photon lists).

MAIN FEATURES

The main functions provided within XIMAGE are

- Image display supporting several plot devices

- Image rotation, specifically written to support sparse images with poissonian background.
- Image sum, subtraction, rescaling, rebinning.
- Image mosaics.
- Image smoothing and contouring.
- Interactive analysis: intensity estimation, upper limits, etc.
- Slide-cell detect routine for images with poissonian background.
- Simulation, including image, spectral, and timing data.
- Maximum Likelihood analysis.
- Point Spread Function estimation.
- Instrument-to-celestial coordinate conversion and vice-versa; coordinate precession.
- Count-rate to flux conversion.
- Interface with spectral and timing analysis packages, i.e. XSPEC (Shafer et al. 1991) and XRONOS (Stella & Angelini 1992).
- Several data formats supported, including a compressed and all FITS formats.

In the following we briefly describe some of the features that are not easily found in other image analysis packages.

THE SIMULATOR

XIMAGE can simulate data from several present and future X-ray imaging instruments. Among the instrument supported are: ROSAT PSPC and HRI, XMM CCD, Jet-X CCD, Astro-D CCD, SAX LEGSPC.

X-ray images including up to 1000 point sources with random or fixed position can be simulated. The distribution of source intensities can follow up to two different logN-logS's; one for AGN and one for stars. Source positions and intensities can also be input from file or interactively. In particular, data from previous measurements (e.g. from published catalogs) can be used to simulate future observations. Background intensity, exposure time and many other parameters can be stored in a parameter file or input from command line. Up to ten different energy spectra can be simulated. All spectral models included in the XSPEC program can be used. The simulated data can be written to a file using standard formats (e.g. a FITS photon file) and analyzed with XIMAGE, XSPEC, and XRONOS.

DETECT

XIMAGE includes a detect routine that has been specifically developed for the analysis of X-ray (i.e. sparse) images. DETECT estimates the background level in several boxes across the image assuming poisson statistics. Sources are searched by means of a slide-cell method. Positions and intensities are estimated in boxes where the signal-to-noise ratio is maximum. The *local* background is used for the calculation of source intensity and statistical significance. When possible, DETECT takes into account of systematic errors which are quadratically added to the statistical uncertainties. DETECT has been optimized for the detection of point sources and is therefore is not particularly efficient with low-brightness extended sources.

FUTURE DEVELOPMENTS

XIMAGE is included in the ESIS Correlation Environment Prototype (Giommi et al. 1992) and will be extended to support tasks that are particularly useful for the correlation of image data. An example of this is the sum of images produced by different detectors even if the original images do not have the same resolution. More satellites will be fully supported when the relevant calibration files will become available (e.g. SAX, Astro-D, Jetx, etc.) A graphical user interface is currently being developed and will be available in the spring/summer 1992. Future versions of XIMAGE will access calibration files written in standard FITS format.

Additional information about XIMAGE can be obtained from P. Giommi (ESIS::GIOMMI, tel +39 6 94180266, fax +39 6 94180361).

REFERENCES

EXOSAT Observatory Team 1991, The EXOSAT Database System: BROWSE Version 3.0 User's Guide, ESA TM-11

EXOSAT Observatory Team 1991b, The EXOSAT Database System: On-line User's Guide, ESA TM-12

Giommi, P., et. al. 1992, this volume

Pearson, T. 1988, PGPLOT Graphics Subroutine Library manual, California Institute of Technology

Shafer R., et al. 1991, XSPEC User's Guide, ESA TM-9

Stella, L., & Angelini, L. 1992, this volume

Sternberg, J. 1985, The EXOSAT Express no. 18, p. 25

VanHilst, M. 1991, SAOimage user's manual, SAO

XRONOS: A TIMING ANALYSIS SOFTWARE PACKAGE

LUIGI STELLA [1]
Osservatorio Astronomico di Brera, Via Brera 28, 20121 Milano, Italy.

LORELLA ANGELINI [1]
HEASARC/USRA, NASA GSFC, Code 668, Greenbelt, MD 20771, USA

INTRODUCTION

XRONOS is a general purpose timing analysis package. Although it was designed primarily for X-ray astronomy, XRONOS is basically detector and wavelength-independent. It has been used to analyse data from the Einstein Observatory, EXOSAT and Ginga, as well as optical photometry and helioseismology data. XRONOS applications can be run within the EXOSAT Database System (EXOSAT Observatory Team 1991a,b). Algorithms and features are described in detail in the XRONOS user's guide (Stella & Angelini 1991). XRONOS is currently available for remote installation on VAX/VMS machines.

OVERVIEW

XRONOS consists of a set of programs which are accessed by the control program `xronos`. There are 3 different user interfaces which can be used within XRONOS:

QUESTION/ANSWER user interface, in which the user specifies input files and analysis parameters by answering individual questions;
PARTIAL QUESTION/ANSWER user interface, in which input filenames are specified only once and analysis parameters are entered by replying to questions;
COMMAND-DRIVEN user interface, in which applications are run by entering a command string followed by qualifiers and arguments.

XRONOS is designed to read in time series from many (up to 50) consecutive input files. Time series from up to 3 different energy ranges can be handled simultaneously within XRONOS. Additional flexibility is provided by "Input File Options", which are used, e.g. to perform algebraic operations on individual input files, shift times, apply barycentric corrections etc.

Currently, two different kinds of input files are supported: "Rate Buffers", which are binary files (4 different formats are available), and "QDP files", which are ASCII files written according to the QDP/PLT standard (Tennant 1991). XRONOS input files can either contain binned data or unbinned data (i.e. input files containing the arrival times of individual photons).

---

[1] Affiliated to ICRA

APPLICATIONS

XRONOS applications are the subset of XRONOS commands which are used to carry out different types of analyses. In the currently available version (No. 2.0), XRONOS includes the following applications:

`lda` - List DAta in input (header and points)
`asc` - convert binary file to ASCII
`lc1` - plot Light Curve for 1 time series
`lc2` - plot Light Curve and ratio for 2 time series
`lc3` - plot Light Curve and ratios for 3 series (and colour-colour)
`sta` - STatistical Analysis
`win` - create/read/change a WINdow file
`ef1` - Epoch Folding for 1 time series
`ef2` - Epoch Folding and ratio for 2 time series
`ef3` - Epoch Folding and ratios for 3 series (and colour-colour)
`efs` - Epoch Folding Search for periodicities
`qdp` - plot a QDP file
`psd` - Power Spectrum Density analysis with FFT
`pss` - Power Spectrum density analysis with Slow direct FT
`acf` - Auto-Correlation Function with fast method
`acs` - Auto-Correlation function with Slow method
`ccf` - Cross-Correlation Function with fast method
`ccs` - Cross-Correlation function with Slow method
`tss` - Time Skewness function with Slow method

Note that most applications include utilities to rebin the results of the analysis either by using a constant or a logarithmic step and to remove (or analyse) various types of trends from the input series.

MAIN FEATURES

Within XRONOS the user has full control of the binning to be used in the analysis, the way of subdividing a time series and/or grouping the results of the analysis. To this aim, 4 basic entities are used in most XRONOS applications:

BINS : these are the time bins of the input time series which is/are being analysed. Note that there can be more than one bin duration.
NEWBINS : these correspond to the time resolution with which the analysis is carried out.
INTERVALS : an interval is defined by the number of newbins over which the analysis is carried out (e.g. the user might decide to calculate a power spectrum over each interval of 256 newbins).
FRAMES : a frame consists of the average of the results of the analysis of one or more contiguous intervals. Each XRONOS output (QDP) file and therefore each plot produced by XRONOS correspond to a frame.

There are 4 different types of windows which can be used within XRONOS: time, phase, intensity and exposure windows. In applications requiring more than one input series (e.g. a cross-correlation analysis) intensity and exposure windows are specified separately for each series. Intensity and exposure windows can be specified separately for bins, newbins and intervals. Many XRONOS applications use some default exposure windows, which are designed to avoid analysing data sets which are too inhomogeneous with respect to their statistical properties.

Additional flexibility is achieved by using a number of global parameters controlling the XRONOS session. XRONOS global parameters are designed to: (a) control the chattiness of XRONOS applications; (b) write to a log file all the interaction with XRONOS applications and specify the log file chattiness; (c) change the default output filename root and i/o status; (d) replace data gaps with a moving average of the light curve; (e) force a given start time in the analysis; (f) calculate error bars for the results of various types of analysis by propagating theoretical error bars or by direct evaluation of the scatter around the mean values in a frame; (g) analyse window functions; (h) normalise the results in various ways; (i) force strict simultaneousness in time series from two or three energy bands, e.g. for cross-correlation analysis; (j) specify special windows for burst analysis; (k) specify a rescaling factor and an additive constant for the results.

FUTURE

A portable version of XRONOS (VAX/VMS and SUN and HP UNIX) is in an advanced state of development and will be available by early 1992. XRONOS will also be interfaced to FITS tables.

Future applications will include: (i) automatic searches for coherent periodicities both with power spectrum and epoch folding techniques; (ii) cross-spectrum analysis; (iii) display and analysis of "time images" (e.g. a large number of colour-coded power spectra displayed versus frequency and time); (iv) pulse phase determination for light curves with a coherent signal; (v) alternative power spectrum and correlation techniques in the presence of sparse and/or non-equispaced data; (vi) fractal analysis.

Further information about XRONOS can be obtained from Luigi Stella (tel: +39-2-72023751/9, fax: +39-2-72001600, SPAN: ASTMIB::STELLA) or from Lorella Angelini (tel: +1-301-286-3607, fax: +1-301-286-9240, SPAN: HEASRC::ANGELINI).

REFERENCES

EXOSAT Observatory Team 1991a, The EXOSAT Database System: BROWSE Version 3.0 User's Guide, ESA TM-11

EXOSAT Observatory Team 1991b, The EXOSAT Database System: On-line User's Guide, ESA TM-12

Stella, L., & Angelini, L. 1991, XRONOS: a Timing Analysis Software Package User's Guide Version 2.0, ESA TM, in press

Tennant, A. 1991, The QDP/PLT User's Guide, NASA Technical Memorandum 4301

ASTRO D SOFTWARE DEVELOPMENT

R.H.D. CORBET, C. LARKIN AND J.A. NOUSEK
Department of Astronomy and Astrophysics, Pennsylvania State University, 525 Davey Lab., University Park, PA 16802

ABSTRACT Astro D will be the fourth Japanese X-ray astronomy satellite and has major US participation. Penn State is writing software for the analysis of data from Astro D with emphasis on the Solid-state Imaging Spectrometers (SIS). Work is currently concentrated on producing a converter which will take satellite telemetry files and convert these to FITS files. The FITS output will be suitable for input to PROS and other analysis systems. This program is known as FRFREAD and is a stand-alone system independent of any other package. In addition to the conversion process, simple quick look analysis tasks are also supported. The program runs in two forms: one GUI (OpenLook) driven and another suitable for running in a pipe-line.

Other related software being developed at Penn State includes Robot, a general purpose plotting and data analysis program and PIXI, an experimental GUI for IRAF. Both Robot are PIXI are available by anonymous ftp.

INTRODUCTION

Astro D will be the 4th Japanese X-ray astronomy satellite (Tanaka 1990) and it follows on from the highly successful Ginga mission. Astro D is scheduled for launch in February 1993 and no slips are thought to be likely. Astro D will carry four sets of mirrors, two will be equipped with with CCD's (SIS - Burke et al. 1991) and two with imaging gas scintillation proportional counters (GIS - Ohashi et al. 1991). This equipment combines moderate imaging capability (arc mins) with good (2 to 8 per cent at 6.7 keV) energy resolution and a response extending beyond iron K_α (6.7 keV). Better energy resolution comes from the SIS but the GIS provides a wider field of view and greater sensitivity at higher energies. There is a large US involvement in this project: mirrors are being provided by GSFC, the SIS effort is led by MIT and software is being written at PSU and GSFC. In return for this contribution a significant fraction of observation time will be given to US or joint US/Japan observations.

ASTRO D SOFTWARE - FRFREAD

The software development for Astro D at PSU is presently concentrated on conversion routines. These will take telemetry data and convert them into FITS

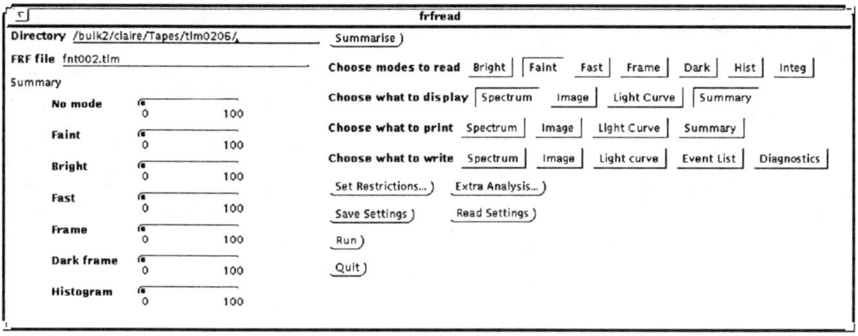

FIGURE I The main control panel for Frfread

binary table format. These FITS files will be similar to those used for ROSAT and it is intended that the Astro D data files will be analyzed using major existing software packages such as IRAF/PROS and Xanadu. The conversion routine, known as Frfread, is a portable stand-alone program, however, which does not rely on other packages within IRAF or Xanadu. It is intended to form part of the data analysis pipeline to be run at GSFC. FRF "first reduction file" files are close to raw telemetry data, but input from attitude and orbital files is also required. We intend, however, that some processing will be possible even if the auxiliary files are absent. As part of the processing Frfread will convert detector x and y positions to celestial coordinates. In addition to the conversion tasks simple quick look analysis and display is possible such as plotting summed spectra, summed images and an observation mode history. We use Saoimage as our display tool. Frfread will be run in two ways, with a GUI for quick look and other interactive use and without a GUI as part of the GSFC pipeline.

SIS - ANALYSIS ALGORITHMS

The SIS will be the first X-ray CCD on a satellite, the only previous experience of using CCD's in space as X-rays detectors comes from two short rocket flights (Burrows et al. 1989; Berthiaume et al. 1990). New algorithms for analysis of data from CCD's will be required because of a number of factors: The spectral resolution is much better than proportional counters and we will have spatial resolution (unlike BBXRT). In addition, the spectral resolution is spatially dependent because of charge transfer efficiency (CTE) effects. The CTE is also time dependent due to the, so far uncertain, effects of radiation damage which will cause a gradual decline in CTE. There will also be complicated data modes used. For example, a 3x3 pixel array is combined to give a single PHA value. Depending on the splitting of charge between pixels in this array the energy resolution can differ for the same incident photon energy. The combination of pixels will be done either on the ground or in space depending on the data mode employed.

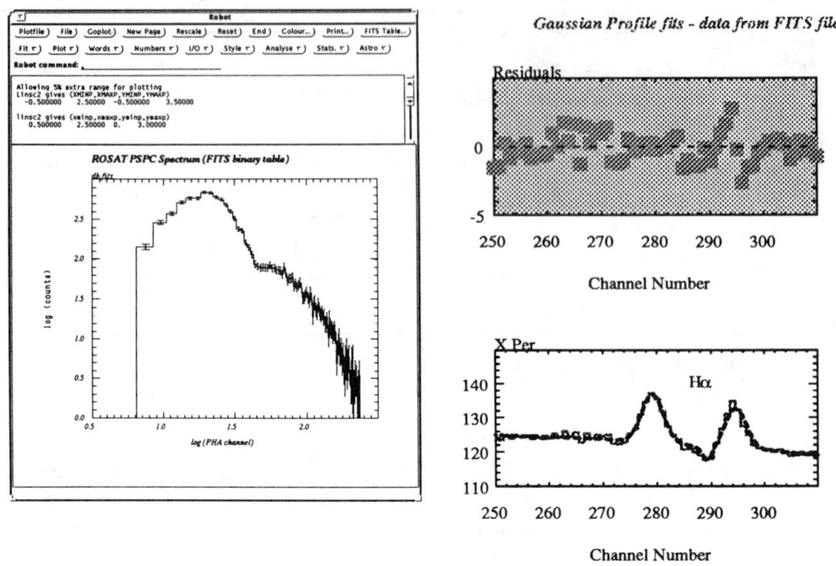

FIGURE II Robot and an example of output

ROBOT

Robot is a general purpose plotting and analysis program. It reads data as: ASCII, FITS binary table or cross sections of a FITS standard image. It is controlled through any of an OpenLook GUI, command line interface or file driven input. By reading from files it is possible to create simple subroutine-like structures. Robot will run under most major "flavors" of the X window system but has so far only been implemented on Sun workstations. PostScript color output is produced in either portrait or landscape mode. Nonlinear fitting can be done using a modified CURFIT routine which includes functions such as Gaussians, Lorentzians, polynomials, etc. Additionally, a user supplied FORTRAN routine may also used. There are a number of plotting modes such as lines, dashed lines, error bars, symbols, histogram etc. Simple operations (arithmetic and functions) on data can also be performed. It is written in C and FORTRAN, and uses the XView toolkit. We are considering using Robot for some stages of analysis of ground calibration data such as measuring changes in CTE and energy resolution with position on the CCD's. Robot may be obtained by anonymous ftp from astrod.astro.psu.edu (128.118.147.28) in the directory pub/astrod.

PIXI

PIXI stands for "Prototype IRAF X Interface" and was written as an experimental GUI for IRAF. This was motivated by the multi-national nature of Astro D. The figure shows the main control panel of PIXI. For each package loaded into IRAF a button and associated pull-down menu containing each IRAF task in that package are produced. It has its own GUI editor known as "PIE" or

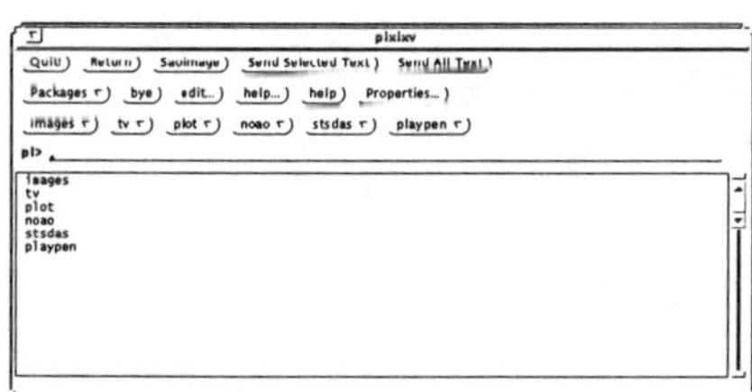

/home/astrod/corbet/uparm/xdofits2p.par

Quit Save Reset

fits

qpoe tempqp

naxes 0

axlen1 0

axlen2 0

events EVENTS

clobber Yes No

oldqpoename Yes No

display 1 0 5

fits_cards xdataio$fits.cards

qpoe_cards xdataio$qpoe.cards

ext_cards xdataio$ext.cards

wcs_cards xdataio$wcs.cards

qp_internals Yes No

qp_pagesize 1024

qp_bucketlen 2048

qp_blockfactor 1

qp_mkindex Yes No

qp_key

qp_debug 0

mode ql

FIGURE III Pixi and its associated editor Pie

the user can specify another editor such as vi or emacs. PIXI pipes its output into IRAF but doesn't directly receive any information back. No further work is presently planned on PIXI but copies of the source code are also available by anonymous ftp.

REFERENCES

Berthiaume, G. D., Burrows, D. N., Garmire, G. P., & Nousek, J. A. 1990, BAAS, 22, 1252

Burke, B., Mountain, R. W., Harrison, D. C., Bautz, M. W., Doty, J. P., Ricker, G. R., & Daniels, P. J., 1991, IEEE Trans. on Electron Devices, 38, 1069

Burrows, D. N., Nousek, J. A., Berthiaume, G. D., & Garmire, G. P. 1989, ApJ, 347, 1114

Ohashi, T., Makishima, K., Ishida, M., Tsuru, T., Tashiro, M., Mihara, T., Kohmura, Y., & Inoue, H. 1991, SPIE Proc., 1549-02

Tanaka, Y. 1990, Adv.Space Res., 10-2, 255

THE EUVE GUEST OBSERVER ANALYSIS SOFTWARE

E.C. OLSON
Center for EUV Astrophysics, Space Science Laboratory, University of California, Berkeley, CA 94720

C.A. CHRISTIAN
Center for EUV Astrophysics, Space Science Laboratory, University of California, Berkeley, CA 94720

ABSTRACT The Extreme Ultraviolet Explorer will provide EUV spectral data to Guest Observers in 1992. The EUVE Guest Observer Analysis Software, developed at the Center for EUV Astrophysics, is the system that delivers Guest Observer data. These data will be provided in FITS format. The system consists of an exportable IRAF layered package, a CEA host system, and an observation database. The system provides facilities for standard and specialized data reduction of EUVE spectral data. Guest Observers will be able to carry out their data reduction at CEA or at their home institutions.

INTRODUCTION

The Extreme Ultraviolet Explorer (EUVE) Guest Observer (GO) Center was created in the Spring 1990 at the Center for EUV Astrophysics (CEA). The center was created to provide dedicated resources to the GO phase of EUVE.

The principal concerns of the EGO Center are the delivery of EUVE spectrometer data and corresponding analysis software to Guest Observers. The EGO Center also supports operations planning of EUVE, and distribution of EUV related catalogs through the Astronomical Data System (ADS).

BACKGROUND

The EUVE mission has two components after the initial Inorbit Check-Out: a 6 month full-sky survey, and the pointed-mode Guest Observer phase. All data collected from EUVE will be archived at CEA.

Data from the EUVE all-sky survey will be processed from raw telemetry into skymaps and catalogs at CEA. At the same time, the science operations of EUVE will be conducted from CEA, including the health and safety monitoring of the science payload, instrument commanding, and mission science planning.

Data processing for the pointed spectroscopy observations is handled by the EGO Center. EUVE has seven photon counting detectors. Three of the detectors are used for spectral data. The band passes are 70 – 190 Å, 140

– 380 Å, 280 – 760 Å for the short, medium, and long wavelength detectors, respectively. Each photon event contains a detector position and a time stamp.

EUVE generates approximately 100Gb of telemetry per year. With a typical observation lasting 40,000 seconds, about 150 GO observations will be completed each year. Each observation corresponds to approximately 70Mb of night time data. It is expected that handling each observation will require several hundred megabytes of disk space and a workstation computer.

DESIGN GOALS

The main design goals of the EGO Center software are to provide data to Guest Observers shortly after the completion of their observation, provide the data in a "raw" format, provide a nominal reduction of their data, and to distribute the necessary calibration and analysis software.

The EUVE telemetry format is not portable and is too complex for typical reduction software. Instead, we represent the telemetry in a time-stamped tabular format. Each row in the table contains a column with a time-stamp. This format is flexible enough to represent all of the EUVE telemetry science and engineering data, as well as supplementary engineering data. FITS provides two suitable representations for tabular data: ASCII tables, and BINARY tables. Either representation can be used for EUVE tables.

Several points were considered when evaluating how to distribute the software. Portability was considered most important. It is expected that Guest Observers will have a variety of hardware platforms possibly running under different operating systems. Familiarity within the astronomical community with the software package was also considered. Software restrictions such as licenses and fees limit the distribution and availability of the software to astronomy community.

We selected IRAF/STSDAS since it provides portability, is freely available, and has an established user community. IRAF/STSDAS provides a binary table format and a set of tools for manipulating them. This includes a tool for reading and writing FITS tables. IRAF also provides QPOE, an image format, for event oriented image data. In addition, it has a large suite of analysis tools with proven performance on optical data sets.

SOFTWARE DESIGN

The EUVE software environment must change frequently in response to new calibration data and software refinements. This requires a flexible data reduction system that can incorporate new calibration data into the processing routines.

Since we decided to provide all of the spectral data reduction software in IRAF, we also have to provide access to all the calibration data from within IRAF. Instrument calibration data at CEA was not designed with portability in mind. Thus, the calibration data must also be converted to IRAF compatible formats. These IRAF data formats can also be converted into FITS files.

Accessing EUVE telemetry and calibration data requires CEA specific software. In order to minimize code errors, we tightly integrated the telemetry data and calibration data selection software into the existing CEA software system. This provides a stable interface from the CEA host system to the IRAF environment, even when the host system undergoes significant changes.

The reduction software must be flexible enough to handle changes in the calibration data. This is the main motivation of the event pipeline. The event pipeline also simplifies the prototyping of new calibration modules. The pipeline allows various event processing elements to be assembled into reduction algorithm. As changes in the reduction process occur, the pipeline processing elements are re-assembled.

As a first step in designing the EUVE software, the domain of the observing conditions was restricted to pointed mode spectrometer data with a known point source position. It is expected that the first round of observations will have these characteristics. Further, we decided that it was most important to model on-boresight line sources. Subsequent development will focus on off-boresight sources in both the reduction and modelling software.

Data quality usually requires significant amounts of analysis to be accurate, and can vary from task to task. We decided not to put data quality flags on the photon events themselves. Instead we provide a mechanism to select photons events based on engineering and health information. Thus, the Guest Observer is free to decide what is quality data for the particular task at hand. In addition, we provide a default data quality selection criteria that will work for typical observations. Guest Observer can tailor these selections to their own needs.

The engineering and health information used in data selection is not restricted to telemetry data. For example, data selection can be based on distance from EUVE to the South Atlantic Anomaly, or whether the moon is in the field of view. This type of information is incorporated into the data selection by adding supplemental time-stamped "history" data to a table. Many of these tables will be the result of detailed analysis of the observation. The most common supplements will be provided with the observation data.

SOFTWARE IMPLEMENTATION

The software, as designed by the EGO Center team is periodically reviewed and tested by an internal review team. In addition, feedback on the software design and implementation is obtained by external review boards including the EUVE Guest Observer Working Group. Currently, software releases occur about twice a year. It is planned that the software will be released less frequently after the first round of GO observations.

The Source Code Control System (SCCS) is used for revision control. CEASCS is an in-house configuration control system built on SCCS and is used to manage the entire EUVE GO program software package.

The GO analysis software is implemented as an IRAF layered package. Currently, the package contains four components of the EUVE analysis: data selection, data quality selection, data reduction, and instrument modeling. As with the CEA host system, code re-use is a high priority. Within IRAF this means that our tasks are closely integrated to IRAF system resources like the world coordinate systems, binary tables, position order event, and image interfaces.

Data Selection

The data selection task communicates with the CEA host software system. It is responsible of accessing the telemetry archive, formatting the telemetry into tables, and acquiring supplemental history tables. This task had to be completed first, since it is the only source of EUVE telemetry in IRAF. This task has been kept as simple as possible in order to give the Guest Observers data that is as similar as possible to the telemetry. This task is nearing completion. It has consistently performed well on internal and external tests.

Data Quality Selection

The data quality selection task converts engineering and instrument health data into time intervals. These time intervals are then used to filter out low quality photon events. We currently have a preliminary version of this tool. Although it is functional, a more sophisticated user interface is required.

Data Reduction

The data reduction task is encompassed by the event pipeline. This task assembles several modules including: time and region filters, simple arithmetic computations, detector distortion correction, wavelength mapping, and aspect correction. It reads input from binary tables or QPOE images. The processing time is dominated by the sorting algorithm used to convert the time ordered telemetry data into the position ordered format that is required in IRAF. Although this task has performed well on simulations, it needs some revisions. The most important change will be in its configuration file. Currently, this is difficult to modify.

The standard aperture extraction package in IRAF has difficulty with the low signal-to-noise, Poisson nature of EUV data. Additional tools are needed to compared observed data sets with modeled data sets.

Instrument Modeling

The instrument modeling task convolves a source spectrum by a model of the spectrometer. The resulting modeled spectrum can be compared to the observed spectrum. The preliminary version of this task convolves a line spectrum. Future development will expand it to handle continuum sources. Additional work will focus on modeling the instrument using observation specific information.

SOFTWARE TESTING

The GO program software is tested internally and externally. Internal tests focus on software functionality and integration. Emphasis is also placed on stress conditions. Externals test focus on scientific validation and robustness.

The GO program also participates in instrument simulations. These instrument simulations model EUVE over a period of a week. The results of these simulations are our closest model to what will actually be observed.

Currently, additional tests are being added for handling real telemetry. This telemetry is generated from actual transmission between CEA and the remote instrument location. These transmissions are characterize by line noise that are similar to those expected after launch.

ACKNOWLEDGMENTS

This work was supported by NASA contracts NAS 5-29298 and NAS 5-30180.

REFERENCES

Malina, R. F., & Marshall, H. L. 1988, JBIS, 41, 357–362

Christian, C. A., & Olson, E. C. 1990, Bull. Am. Astron. Soc., 22, 922

THE ESO–MIDAS SYSTEM

REIN H. WARMELS
European Southern Observatory,
Karl-Schwarzschild-Straße 2,
D-8046 Garching bei München

ABSTRACT This poster paper briefly discusses the concept, design, and functionalities of the Munich Image Data Analysis System (ESO–MIDAS), the astronomical data analysis system developed and maintain by the European Southern Observatory.

INTRODUCTION

ESO–MIDAS is the acronym for the European Southern Observatory – Munich Image Data Analysis System which is developed and maintained by the European Southern Observatory.

The MIDAS system provides general tools for image processing and data reduction with emphasis on astronomical applications including special reduction packages for ESO instrumentation at La Silla.

The initial design of MIDAS was made on a DEC/VMS system in the early 80's. However, with the acceptance of UNIX as a standard operating system and the introduction of workstations, the system was largely redesigned and now runs on a wide variety of computers, with either DEC/VMS or one of the various flavours of UNIX as the operating system.

OBJECTIVES

In the design of MIDAS a number of basic requirements were taken into account in order to ensure the system can involve in the future. These main objectives were:

- Modular Design, which makes it easy to adapt to different environments;
- Portability, to ensure MIDAS can run on different, and migrate to different computers.
- Use of Standards, like the programming languages Fortran and C, and the X Window system. Use of standard programming languages enables the use of CASE tools, and makes the shift to object oriented coding easier.
- Easy to Program by providing simple interface routines to access the data, and by implementing a flexible control language.

- Open Design of the system, which allows easy addition of contributed software from other institutes.

DESIGN

The basic system was designed to work on a single computer with appropriate peripherals. It consists of three parts:

- Monitor — which includes user interface, and administration of tasks and local variables. It controls the execution of tasks and interacts with the user.
- Applications — which perform the actual operations on the data. They may be written in standard Fortran77, C, or the MIDAS control language. All communication must go through the monitor and is done via keywords and frame descriptors.
- Interfaces — which bind the applications to the monitor, and define the possible interaction with application tasks and monitor. An additional level is used to interface the MIDAS system itself to the host operating system. This lowest interface level must not be used be application programs.

THE MONITOR

MIDAS is command driven. All the user interaction and the scheduling of processes to execute the commands is done via the monitor. In concrete, the monitor has the following functions:

- display on-line help with different levels of detail;
- keep a log of all operations during a session;
- serve as a command interpreter;
- preprocess input strings to translate user defined symbols and to facilitate command abbreviation, command numbering and buffering, etc.;
- execute applications in a subprocess.

MIDAS is a multiprocess system. The monitor executes in the context of the process active when starting up the session and the applications run in either of two subprocesses, the foreground or the background subprocess. A third option is to submit applications as a job to a queue.

THE MIDAS CONTROL LANGUAGE

The MIDAS command language, MCL, provides the tools to construct complex command procedures from existing commands. Since all application programs get their parameters via a set of standard interfaces, it is very simple to string commands together, each using results from the previous command. In fact, except for the system commands, all MIDAS commands are MCL procedures themselves.

In general, MCL provides the following basic features of a programming language:

- definition of parameters;
- looping;
- conditional statements;
- branching;
- global/local variables;
- procedure calls (also recurrent) with parameters;
- built-in functions.

MCL is an interpreted language which means that the definition of local variables may be done anywhere in a procedure and no compilation and linking steps are required before executing a MCL procedure.

To identify syntax and other obvious errors in long and complicated procedures the MIDAS Command Language Debugger is provided. It enables debugging of procedures as well as application code (F77 and C).

DATA OBJECTS

Data items in the MIDAS system can be divided into several groups:

- Frames — data sets which have a uniform sampling, e.g. images or spectra. Up to 16 dimensions can be supported; however, most applications are limited to 2.
- Tables — collections of heterogeneous data organised in columns and rows.
- Descriptors — variables related to the general database, e.g. frames and tables. They are used to describe the data in detail.
- Keywords — variables relating to processes or sessions. They are similar to descriptors and mainly used to transfer information between tasks, and the control of them.
- Catalogues — contain lists of either images, tables, or fit files for the purpose of grouping data together.

APPLICATIONS

The functions of the applications in MIDAS are split in several levels of importance, the top level being the primary or core applications without which image processing is hardly possible. The core applications take care of:

- Image Display — is comprised of all the usual functions expected from a mature image processing system, such as displaying and retrieving data in full or split screen mode, zoom and scroll, getting cursor values, modifying lookup tables, blinking, etc. The kernel of this package is the Image Display Interfaces library (IDI).
- Graphics Display — provides the functionality needed for data presentation in graphical form, as well as interactive data reduction. The graphics package is device independent, and based of the Astronet Graphic Library (AGL).
- General Image Processing — which include typical arithmetic operations like filtering, resampling, interpolation, rotation, extraction/insertion, FFT, etc.
- Table File System — provides a complete set of functions to process tabular data. These include reading, writing, editing, searching, sorting, regression, etc.
- Fitting Package — that offers necessary tools to fit non-linear functions, and to model data distributions both in table and image format.
- Data I/O — to transfer the data from and to tape or disk.

Besides the core applications mentioned before, MIDAS contains many additional application programs as well as complete packages. Packages are combined inside the system as so-called MIDAS contexts. These contexts contains a coherent set of commands suited for the reduction of data coming from ESO instrumentation (e.g. echelle reduction, long-slit spectroscopy, reduction of polarimetric data, etc.), as well as various areas in data analysis (e.g. stellar photometry, generation of test images, modeling interstellar or intergalactic absorption clouds, multivariate analysis).

SUPPORT

A number of services are provided to help the MIDAS users:

- MIDAS Hot–Line service, to obtain help quickly. The Hot–Line service can be contacted by phone, FAX, telex, and the main computer networks.
- Storage of all MIDAS problems and their solutions into the MIDAS Problem Database, which can be queried through STARCAT.
- The ESO–MIDAS Courier, the biannual newsletter for (potential) MIDAS users. Copies of the Courier are available at this conference.

- MIDAS courses, that are organised at regular intervals and covering various aspects of MIDAS.
- The MIDAS Users Meeting, organised yearly at ESO Headquarters in Garching, Germany.

INFORMATION

The main source of information about MIDAS is the MIDAS Users Guide, Volumes A and B (ESO–IPG 1991a).

The MIDAS system has been described in various papers. General overviews can be found in Banse et al. (1983) and in Grosbøl & Ponz (1990). MIDAS as a development environment is discussed in the document "MIDAS Environment" (ESO–IPG 1991b) and by Banse et al. (1992).

The performance of MIDAS on different platform is described by Grosbøl et al. (1988). The implementation of the table file system is described by Grosbøl & Ponz (1985). The first issue of the ESO–MIDAS Courier (ESO–IPG 1991c) contains information about the three dimensional extension of the table file system.

Further information about MIDAS can be obtained from the European Southern Observatory.

REFERENCES

Banse, K., Grosbøl, P. J., & Baade, D. 1992, this volume

Banse, K., Crane, Ph., Ounnas, Ch., & Ponz, D. 1983, "MIDAS" in Proc. of DECUS, Zürich, p.87

European Southern Observatory - Image Processing Group 1991a, MIDAS Users Guide, Volume A and B, ESO, Garching

European Southern Observatory - Image Processing Group 1991b, MIDAS Environment, Version 1.0, ESO, Garching

European Southern Observatory - Image Processing Group 1991c, ESO–MIDAS Courier, Volume 1, Number 1, 1991, ESO, Garching

Grosbøl, P. J., Banse, K., Guirao, C., Ponz, J. D., & Warmels, R. H. 1988, "MIDAS Benchmarks of Workstations" in ESO Messenger, ESO, Garching, 54, 59

Grosbøl, P. J., & Ponz, J. D. 1985, Mem. S.A.It., 56, 429

Grosbøl, P. J., & Ponz, J. D. 1990, "The MIDAS System" in Acquisition, Processing and Archiving of Astronomical Images, G. Longo & G. Sedmak (eds.), OAC and FORMEZ, 1990, p.109

MIDAS AS A DEVELOPMENT ENVIRONMENT

K. BANSE, P. GROSBØL, D. BAADE
European Southern Observatory,
Karl-Schwarzschild-Straße 2, W-8046 Garching, Germany

ABSTRACT MIDAS (Munich Image Data Analysis System) is the portable image processing system developed at ESO and used for off-line data reduction at ESO and many astronomical institutes all over Europe.

Many tools are provided to customize MIDAS to the individual needs of the user and set up a productive, personalized working environment. Rapid prototyping is supported via an interpreted macro language. In order to facilitate application development by the user, an advanced programming environment based on standard languages is integrated in MIDAS.

INTRODUCTION

MIDAS has been developed by the Image Processing Group at ESO since 1980 and provides a portable data analysis system as well as a programming environment. MIDAS is used at ESO Headquarters in Garching, Germany, and the La Silla observatory in Chile, as well as in many astronomical institutes mainly in Europe; but a sizable user community exists also outside Europe.

The main design objective for MIDAS was to create a system including a software development environment which enables the users to adapt MIDAS to their specific needs and to integrate their own software in an easy and straightforward manner.

Originally, MIDAS was built in a VAX/VMS environment but has been modified substantially in the years 1988/1989 to become a portable system supporting VMS and all flavours of Unix.

CUSTOMIZING MIDAS

The possibility to streamline and customize an individual working environment does not offer any additional developing tools *per se*. However, it is an essential component to make a user feel comfortable and at ease in a given environment which leads to more productive work. Therefore, MIDAS provides quite a few commands and built-in functions which support the setting up of a personalized user environment (ESO-IPG 1990).

In an interactive MIDAS session the commands are numbered and the most recently used commands are kept in a command buffer. All commands in that buffer are saved in a file upon termination of a MIDAS session and read back automatically when the session is resumed later on.

The command buffer serves as a convenient basis for extensive and flexible command recalling and editing, using either the command no. or a substring of the command line. Before being re-executed these commands may be edited in a way similar to the VMS command line editor. Command suspension lets the user interrupt the command input and continue at a later stage.

All MIDAS commands use default values for their parameters whenever it is sensible. The user may override these and enforce specific defaults for any MIDAS command. New commands may be added any time and can be either just aliases for existing commands or new functions implemented through a user written MIDAS procedure. These commands can be placed in the file `login.prg`, which is executed (if existing) at the beginning of each MIDAS session.

Several MIDAS sessions may be started in parallel and can communicate with each other. Thus, it is possible in a window environment to have two MIDAS sessions in different windows and use MIDAS in one window just for the display of the on-line help. The other MIDAS runs in the usual interactive way but can also be used to send commands to the "help-MIDAS".

THE MIDAS COMMAND LANGUAGE

Predefined functions and commands to adjust an environment to specific needs form just the first step towards a comfortable working environment. In our view, the next higher level should be provided with a `Macro` language, which is both simple and powerful enough to enable the user to build complex applications by combining the basic system functions, similar to UNIX shell scripts or DCL command files.

Programming constructs

In the case of MIDAS, the MIDAS Command Language (MCL) fulfills that role. The MCL operates on a sequence of commands stored in an ASCII file, called a *procedure*. Besides all interactively usable MIDAS commands it contains a set of basic commands to provide the necessary functionality for writing "programs", comprising

- global/local variables
- looping
- conditional statements
- multi way branching
- procedure calls (also recursive) with parameter passing
- built-in mathematical, logical and character-string functions
- direct access to the MIDAS data structures (images, tables, etc.)

The MCL is an interpreted language, *not* a full blown programming language with a rigid grammar, e.g. the definition of local variables may be done anywhere in a MIDAS procedure. Therefore, no compilation and linking step is required before executing a procedure.

Putting it all together

All MIDAS application programs get their input via global parameters, called *keywords*, through standard interfaces in a controlled manner. Also, all relevant output is written back to keywords, e.g. names of newly created images or tables. Therefore, it is simple to string separate commands together, each command using the results from the previous one. One merely needs to know which keywords are read and written by a specific application in order to integrate it together with other MIDAS commands in a MIDAS procedure. The on-line help for each application provides that information.

A MIDAS procedure can call another procedure (or itself) to build complex "programs" from simple procedures. Keywords defined inside a procedure serve as local variables, only known in the context of that procedure. The MIDAS keywords provide the "glue" to tie commands together which have been written individually without explicit planning for later pipelining.

This principle is exploited by the MIDAS system itself as well: except for the approximately 50 basic "system" commands, all MIDAS commands are procedures themselves. As a consequence we could keep the MIDAS monitor, the module which directly interacts with the user, very compact.

In the simplest mode the user just combines several MIDAS commands in one macro for later reuse. As a next step one can employ all the "language" constructs of the MCL to write programs and do rapid prototyping. The user can quickly try out different algorithms, modify parameters and change input data, all by just editing a MIDAS procedure and executing it right away. This aspect of fast and flexible modification is very important and often worth the slower execution time of an interpreted procedure when compared to the effort of writing a program in FORTRAN or C.

Debugging

The more complex a procedure gets, the more difficult it is to debug it. MIDAS has the usual means to echo each command line of a procedure during its execution and also to show the replacements of variables with their actual values. These tools are sufficient for short procedures but inadequate for more complicated ones. Therefore, we included a MCL debugger containing the basic functions necessary to walk through the procedure code:

- stepwise execution of procedures
- listing of the procedure code
- setting of breakpoints
- examination of the contents of any variable (keyword)

For a detailed description of the MIDAS Command Language see the MIDAS Users Guide (ESO-IPG 1990).

SOFTWARE DEVELOPMENT WITHIN THE MIDAS ENVIRONMENT

With the MIDAS Command Language described above it is already possible to create quite sophisticated and complex new MIDAS commands and also to integrate them smoothly into the system. However, a macro language is most useful for manipulations of complete data structures. For operations at the pixel level, programming languages necessarily offer superior solutions.

It would be naive to assume that within one software group, even at big observatories, there is enough expertise to build an astronomical reduction system that provides the "best" algorithm for every application. We think, the best one can do, is to offer basic and proven algorithms in as many fields as possible. To this stock of general algorithms, specialized software can be added for instruments for which in-depth knowledge and experience are locally available.

For all other topics one should acknowledge that the user knows his/her data best and most probably has a fine tuned reduction package of his/her own. Therefore, it is of highest importance to provide an easy migration path for that specialized software into the reduction system.

MIDAS programming language interfaces

For these reasons, we support also interfaces for standard programming languages to enable efficient access to all MIDAS data structures (ESO-IPG 1991). Within these interfaces we tried to strike a balance in keeping their number well limited but still offer complete control over the data structures.
The `ST`(andard) interfaces include routines to

- connect to and disconnect from the MIDAS environment
- access to the MIDAS data structures, e.g. creation, deletion, opening, closing of images (or tables, see below) and manipulation of their associated descriptive information
- manipulate the *keywords*, e.g. read, write, copy, prompt functions
- support basic error processing, e.g. signaling and identification of errors
- access MIDAS *catalogs*, which are lists of images, tables, etc.

Besides the images themselves, the most important data structures in MIDAS are the tables (Grosbøl & Ponz 1985). They are most effectively used for character or unequally sampled numerical data. Therefore, another set of routines, the `TB` interfaces, is devoted entirely to the access and manipulation of MIDAS tables, providing

- operations on table rows, columns and single elements
- creation, labelling, formatting, searching and sorting of table columns
- handling of the special selection column and null entries

We omitted interfaces to provide general utilities of operating systems like e.g. `'ls'` or `'cd'` (in the case of Unix), because host system commands may be executed inside MIDAS directly. Neither are there special routines for handling ASCII files nor interfaces for fancy terminal I/O.

Almost all capabilities of the interactive MIDAS are available also to the application programmer through the MIDAS interfaces in conjunction with the already existing MIDAS commands. The philosophy is that any application only manipulates its data. For example, code for displaying images or plotting table columns is normally not included in application programs. Instead, these functions are provided by the relevant commands existing in the core MIDAS system. This strict functional modularity enforces a consistent appearance to the user.

In our experience the `ST` and `TB` interfaces cover most of the needs of a programmer for integrating applications into MIDAS. For instance, the DAOPHOT-2 package of P. Stetson (Stetson & IPG 1991) was implemented in MIDAS in less than a week. Only the routines accessing the data had to be exchanged, all visualization of the data is performed by the relevant MIDAS commands.

Another example is EXSAS, the Extended Scientific Analysis System (MPE-EXSAS group 1991). The package by the ROSAT project team at the MPE (Max Planck Institute for Extraterrestrial Physics in Garching, Germany) is used for the reduction of the X-ray data from the ROSAT satellite. The group at MPE wrote all their ROSAT specific algorithms within the MIDAS environment. Especially important to them were the MIDAS table interfaces which support the construction even of unusually large data structures (EXSAS tables are generally larger than 100 Mbytes). For general image processing, display and graphics requirements EXSAS utilizes the standard MIDAS commands.

We recognize the fact that there are indeed also applications which do need direct control of image displays and graphics devices. Therefore, MIDAS supports the `IDI` (Image Display Interfaces) routines which grew out of a joint collaboration of major American and European astronomical institutes and provide a portable interface to image displays (Terrett et al. 1988). For plotting applications, MIDAS supports `AGL` (ASTRONET Graphic Library), the graphics package used by ASTRONET in Italy (Fini 1985). Also, an `os` library exists in MIDAS for portable access to the underlying host system.

Standard languages

The MIDAS interfaces are supported in both FORTRAN 77 and C. We chose FORTRAN 77 because it is an ANSI standard and widely used by scientists. The same applies to the C programming language. Furthermore, via the C-interfaces we can make MIDAS available to the C++ community.

We stress the importance of using well-known, standard and universally taught programming languages as the basis for a portable software system:

- Programmers can write code in the language they know well and run their application also outside an interactive MIDAS session. Familiar debugging and analyzing tools exist and speed up production of software.
- Users can easily adapt an existing algorithm within MIDAS, e.g. by adding new options.
- The code of MIDAS applications is transparent to the user. If something looks fishy during the data reduction, one gets to the relevant source

quickly and can check out the code. No user should have to trust an application blindly!

- The computer industry is evolving at a rapid pace. New technologies make their way into products and, e.g. new workstations come to the market at always shorter intervals. With standard languages as the basis it is much simpler and faster to port code to new machines which usually support these.

- No system will live forever! The more standardized the code, the easier it will be to salvage applications from an outlived system.

OUTLOOK TO THE FUTURE

In the coming year MIDAS will be used also for fast and versatile on-line analysis at the telescopes on La Silla (Banse et al. 1991). Furthermore, MIDAS will be the environment for the VLT instrumentation software effort. These developments will give MIDAS the benefits from extensive interaction with a much larger and more diverse community comprising also engineers and technicians.

We are currently working on a graphical user interface to the MIDAS commands, a facility which we shall make available also to the application programmers (the same goes for audio capabilities). Also, we plan to support programming interfaces which provide a higher level functionality, e.g. loading an image, to facilitate the execution of MIDAS applications outside an interactive MIDAS session.

REFERENCES

K. Banse, P. Crane, C. Ounnas, & D. Ponz 1983, in Proc. of the Digital Equipment Computer Users Society, 87

P. Grosbøl, & D. Ponz 1985, in Mem. Soc. astron. Italiana, 56, 429

L. Fini 1985, in Astronet Documentation, AUM2

D. L. Terrett, P. M. B. Shames, & R. J. Hanisch 1988, Astronomy and Astrophysics Suppl., 76, 263

European Southern Observatory - Image Processing Group 1990, MIDAS Users Guide - ESO Operating Manual No. 1, Garching

European Southern Observatory - Image Processing Group 1991, The MIDAS Environment, Garching

K. Banse, D. Baade, P. Biereichel, & P. Grosbøl 1991, in Data Analysis in Astronomy IV, Di Gesu et al.: Plenum Press, in press

Max Planck Institute for Extraterrestrial Physics - EXSAS Development Group 1991, EXSAS Users Guide, Garching

P. B. Stetson and Image Processing Group - ESO 1991, DAOPHOT II - Users Manual, Garching

THE ADAM SOFTWARE ENVIRONMENT

P. M. ALLAN
Rutherford Appleton Laboratory, Chilton, Didcot, Oxon OX11 0QX

ABSTRACT This paper describes the ADAM software environment and some of the facilities that it offers users and programmers. After a brief overview of ADAM, a more detailed description of the hierarchical data system is given.

INTRODUCTION

The ADAM software environment provides an integrated environment for writing all forms of astronomical software from telescope control and data acquisition systems, through data reduction, to complex data analysis packages. Although the environment provides many tools, it does not attempt to hide the underlying operating system from the user.

ADAM AND STARLINK

The ADAM kernel is maintained and developed by the ADAM support group, which is part of the Starlink central team at RAL. Most of the ADAM application programs for data analysis are written by Starlink contractor at universities.

Although ADAM is Starlink's preferred software environment, we also distribute software written to use other environments such as IRAF or AIPS.

WHAT IS ADAM?

The term 'ADAM' is used in different situations to describe differing amounts of software. ADAM provides a set of tools to several groups of users, which are conveniently described in a hierarchical manner.

ADAM provides:

- to users, a set of application programs to analyse astronomical data
- to writers of applications programs, a set of subroutine libraries
- to writers of subroutine libraries, a set of standards and guidelines

ADAM can be taken to mean the entire set of applications that use the ADAM tools, or just the toolkit of subroutine libraries.

ADAM APPLICATION PROGRAMS

ADAM contains application packages for analysing data from several wavelength ranges. Some packages are general purpose ones that can handle data from any wavelength whereas others are more specific. Packages are currently available in the areas of X-ray data processing; CCD reduction; infrared camera reduction; general image processing; CCD aperture photometry; CCD stellar photometry; position, intensity and shape analysis; catalogue searching; time series polarimetry; general spectroscopy; two dimensional spectroscopy and N dimensional data. More applications are being added all the time.

ADAM programs can be run in one of three ways. Firstly they can be run directly from the command prompt of the computer. Secondly they can be run from the interactive command language (ICL), a simple interpretive language written by Jeremy Bailey specifically for controlling ADAM programs. Finally they can be run from SMS, the screen management system. This last option provides a simple menu-driven graphical interface to a suite of ADAM programs. This has found most use in the telescope control environment, where a simple series of operations may be performed many times.

TOOLS FOR THE ADAM PROGRAMMER

ADAM does not impose a rigid structure on a programmer. You are free to use whatever language you like and can mix and match the subroutine libraries as you please. However, a completely unfettered approach would lead to chaos and there are certain guidelines that a good ADAM program should follow. These are described in the ADAM programmer's guide. In practice, most ADAM programs are written in Fortran 77, although C is a growing force, and some of the telescope control programs at La Palma are written in Pascal.

The use of standard programming languages means that all of the CASE tools that are available from the manufacturer of the host computer, such as a language sensitive editor and a source level debugger, are available to the programmer. Our experience with CASE tools is that astronomers do not make great use of them due to the learning curve associated with any new piece of software. However, they have greatly improved the productivity of our professional programmers.

ADAM has many subroutine libraries to offer the programmer, whether they are writing programs for data analysis or for telescope control. Most of the subroutine libraries come in two versions; a stand alone version for use in non-ADAM programs and in fully integrated ADAM versions. The calling sequences of the routines do not differ between the two versions, but the ADAM libraries often contain a few extra routines to integrate the library with the parameter system.

Some of the functions provided by the current set of libraries are a parameter system, graphics, sequential and direct file access, hierarchical data system, magnetic tape access, inter-process message system, inter-process notice board system, and networking.

THE HIERARCHICAL DATA SYSTEM

One of the acknowledged strengths of ADAM is its hierarchical data system. The concept of HDS dates back ten years and HDS has been in serious use for about eight of those years. HDS allows the programmer to construct arbitrary hierarchical data structures within a single data file. There are essentially no restrictions on what can be built as an HDS structure, although some structures are more efficiently implemented than others. HDS supports all of the standard FORTRAN data types except COMPLEX, as well as BYTE and WORD (16 bit integers) for when it is important to store the data as compactly as possible. Using these basic types, complex structures can be built up. This is both good and bad and some of the restrictions that it is sensible to adhere to are described in the next section. Almost the only restriction that is placed on the contents of an HDS file is that they must fit on a single disk. If this is not possible, then HDS components can be placed on separate disks using the REF library. Despite its very general nature, HDS is very efficient, to the extent that it can be faster to use HDS than to handle your own Fortran files.

An example of the contents of an HDS file is given in Figure I. This shows a file containing a data array and its variance, a title to be used in any plots of the data and an extension structure containing FITS header records read off the tape containing the data.

FIGURE I The structure of a simple HDS file

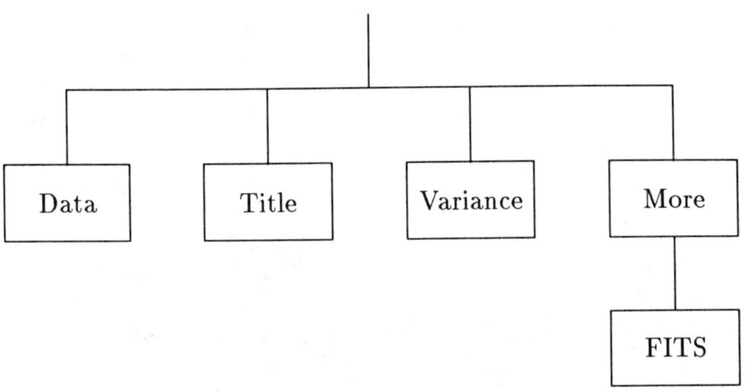

One of the major advantages of HDS over other data storage systems is that the data is portable from one machine architecture to another. For example, a file that was written on a VAX/VMS system can be read on a Sun. All data conversion processes are transparent to the user; you do not need to know what machine wrote the data in the first place.

THE EXTENSIBLE N-DIMENSIONAL DATA FORMAT

As mentioned above, the very flexibility of HDS can be a potential source of problems. It is vital that data written by one program be readable by other programs, often those written by other programmers. If a program constructs a complex data structure, then that structure must be known to all the programs that wish to read that file. One can try to invent an all embracing data structure that can cope with all possible data types, but this never works. Rather than take the 'we've thought of everything' approach, which invariably fails the first time a new type of data is encountered, ADAM uses an extensible data format called NDF. This defines a standard set of HDS components that all programs are expected to understand, and a place to put extensions that might only have meaning in special circumstances.

The standard data components are:

- The data array
- The variance of the data array
- A title for the data array
- A label for the data array values
- The units that the data are stored in
- Are there any bad pixels in the data?
- The quality of the data array
- Axis values (per pixel), labels, units and errors
- A history of the processing that has been done on this structure
- A place to put extensions

This is a long list. Fortunately the only component that is required to be present is the data array itself. Furthermore, it is not necessary for every program to explicitly process each component. NDF routines that process these data structures can automatically propagate certain components from an input data file to an output data file. The title and axis information are good examples of components that are unlikely to be changed by many operations.

Two features that are worthy of further mention are bad pixels and quality arrays. The ability to flag certain pixels as containing bad data can be very useful when processing CCD frames with cosmic ray hits. Rather than replacing pixels that have suffered cosmic ray hits with interpolated data at an early stage in processing, the fact that a pixel is bad can be retained throughout the processing. Handling bad pixel values requires no extra storage, but can increase the processing time. Consequently all NDFs contains a flag indicating whether it might contains bad pixel values. If even even more control is required over the processing of dubious pixels, a quality array can be defined which gives further information about each pixel. For example, if a detector has been driven into a non-linear regime, then it may be worth flagging the affected pixels. The data is still useful, but it needs extra care in processing it.

To reiterate: an NDF is just an HDS file that follows certain conventions to ensure that any ADAM program can process it.

FURTHER READING

A few of the Starlink documents on ADAM that may be of general interest are listed in the references. Printed copies or the LaTeX source of these and other Starlink documents are available on request from the Starlink software librarian, username STAR, at IP address rlstar.bnsc.rl.ac.uk (130.246.32.1) or DECnet address RLVAD (19457).

REFERENCES

Bailey, J. A. & Chipperfield, A. J. 1991, ICL – The Interactive Command Language for ADAM, Starlink Guide 5

Lawden, M. D. 1989, ADAM – The Starlink Software Environment, Starlink Guide 4

Murray, M. J. 1991, Introduction to ADAM programming, Starlink User Note 101

Warren-Smith, R. F. 1991, ADAM Programmer's Documentation, Starlink Guide 6

Warren-Smith, R. F. 1990, NDF – A Subroutine Library for Accessing NDF Data Structures, Starlink User Note 33

Warren-Smith, R. F. & Lawden, M. D. 1991, HDS – Hierarchical Data System Programmer's Manual, Starlink User Note 92

THE GRONINGEN IMAGE PROCESSING SYSTEM, GIPSY

J.M. VAN DER HULST, J. P. TERLOUW, K. G. BEGEMAN,
W. ZWITSER, P.R. ROELFSEMA

Kapteyn Astronomical Institute, Postbus 800, 9700 AV Groningen,
The Netherlands

ABSTRACT This paper describes the Groningen Image Processing System, GIPSY. GIPSY was built during the late seventies based on earlier systems developed in Groningen since 1971 and has recently been redesigned to run on a variety of Unix architectures. The general design philosophy, data structure and user interfaces are described in detail and a brief overview of the wide range of applications is provided.

INTRODUCTION

With the advent of synthesis radio telescopes in the early seventies a need for interactive image processing arose. At the Kapteyn Institute this led to the development of such an interactive software package: GIPSY, the Groningen Image Processing SYstem (Allen et al. 1985). Over the years GIPSY has been built up further and a wide variety of new applications has been added by many staff members and Ph.D. students at the Institute who designed software in GIPSY to suit their own needs.

At present GIPSY is particularly suited for the processing and analysis of spectral line synthesis data and Infrared data from the IRAS satellite. Especially the spectral line analysis has received great attention, motivated by the extragalactic HI line work at the Kapteyn Institute. As a result GIPSY has spectral line analysis capabilities which are quite unique and have not yet been implemented in other image processing packages. The importance of visualization of truly three dimensional data sets was recognized early on and has been implemented in several ways.

The new developments in computer hardware, in particular the quick spreading of Unix workstations combining a graphical environment with fast (and affordable) CPU capabilities have led to a major port and code overhaul of GIPSY. Although all of the principal functionality has been kept, many new features were implemented and this paper will describe the system as it runs now under Unix and the X-Window environment. An additional advantage of the new implementation is its portability as the system does no longer depend on particular hardware for its display functions.

In this paper we will describe the general design philosophy, the data structure, the user interface and display software, and give a short summary of application software present in GIPSY.

GENERAL DESIGN CONSIDERATIONS

The main objectives for the design of the second generation GIPSY were to maintain the original ease of adding application programs, to have a flexible, self descriptive, multidimensional data structure, to make use of the de facto standard X-Window (X11) environment for graphics and image display and to code the system in ANSI C and the applications in either ANSI C or standard FORTRAN 77.

The growth of GIPSY through the years into a versatile image processing package with a variety of applications, in particular in the area of spectral line analysis has been possible because of its friendly and simple environment for programming. Therefore many astronomers have contributed to the system, adding applications which fulfilled both their own and general needs. A very crucial ingredient for the ease of programming has been the modular structure of the system and the use of interfaces which shield the application programs from other parts of the system such as the data, and which provide simple tools for communication with the user and communication with other tasks. This very important aspect of the system has been kept in the new design, because the possibility for astronomers and students to contribute to the system with ease is considered crucial.

The main design requirement for the data structure is the capability to handle multi-dimensional data. In addition the data structure must be self descriptive and support a large variety of coordinate systems. The latter becomes increasingly important as astronomers use and combine data from different telescopes in different spectral domains. Few software systems existing at present support the large variety of coordinate systems used in astronomy.

The increasing use of workstations dictates the requirement that the user should be able to make use of the workstation hardware to view his data. This requirement then implies the use of X11 for image and graphics display, as X11 has become the standard graphics environment on workstations of all vendors. The additional advantage is that the same functionality is available on X-terminals run from either a workstation or a Unix based mini-super computer such as the Alliant and Convex systems.

DATA STRUCTURE

The GIPSY data structure (GDS) is a multi-dimensional crystalline structure with the possibility of describing the data at **all** levels. For the description of data structures GDS makes use of FITS type keywords. These keywords can be stored at different levels within the data structure, thus allowing a hierarchical header structure for the description of the data.

Figure I shows an example of a 3-D GDS data cube with its hierarchical header structure. Examples of such three dimensional data cubes are: images at a range of frequencies (wavelengths) resulting from a spectral line synthesis observation or a processed Fabry-Perot interferometer observation. In these cases the three axes typically are Right Ascension, Declination and frequency

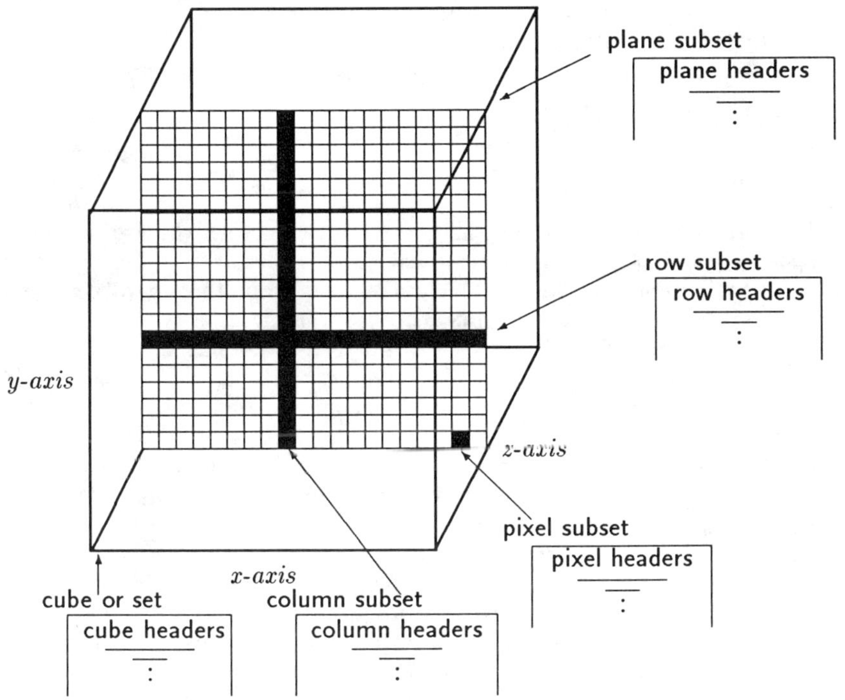

FIGURE I A three-dimensional GDS set

(wavelength) or velocity. Such a data-cube can be described at four levels: (i) the top level (here information about e.g. the pointing direction and central velocity, the three dimensional gridsize, bookkeeping information describing the observation is stored); (ii) the three intermediate levels corresponding to the three principal planes of the cube (here e.g. the velocity of each individual RA-DEC plane in the data cube can be stored. Information about whether a plane has been processed; image reconstruction, continuum subtraction, flat fielding etc., could also be stored at this level); (iii) the three 1-D levels and (iv) the pixel level. Note that Figure I for clarity only shows an $x - y$-plane, simultaneously headers exists for the $x - z$ and $y - z$ planes.

The hierarchical header structure appears to be quite unique to GDS and makes it suitable for storing many different kinds of data. Storing 2-D or 3-D image data in a multi-dimensional data structure such as GDS is relatively straightforward. GDS is, however, flexible enough to store much more complicated data such as sections of scans from the IRAS satellite covering a certain area of sky, including the time stamps and calibration information (Roelfsema & Kester, this volume).

The interface layer between GDS and the application programs has the necessary intelligence built in to address the data cube at all levels using simple subroutine calls. The interface can be requested to deliver a pixel, a one dimensional slice in any direction parallel to the primary axes, a two dimensional slice in any direction parallel to the primary axes, or a subcube, without having to worry which is the order in which the data is stored on disk as simple byte streams. Similarly the header access subroutines can obtain header information from any level within the data structure.

USER INTERFACE

As in the original design of GIPSY, all communication with the user is handled by a master control program Hermes, described in detail in Allen & Terlouw (1981). Its main functions are to shield the user from the operating system and to provide a standard environment for communication with the tasks in GIPSY. The main functionality of the original Hermes has been kept in the second generation GIPSY. This includes logging of the session in a file which can be inspected interactively, the use of a control language and the ability to construct macros, bookkeeping of inputs for tasks used in the processing including the possibility to rerun processes with the same parameters except for those specified explicitly, and program control using special escape sequences.

The present implementation of Hermes can run on any display terminal, including X11 terminal windows. The screen is divided into three areas. The top area displays the log, which can be inspected page by page, including a character string search mode. A five line area below displays the status of the running tasks (at present this can be up to four simultaneous applications) and the current page of the log file. A two line area at the bottom is used for user input.

Hermes provides in addition a keyword sensitive help facility which can be activated at any time and uses an overlay screen. Once called one obtains information about the keyword for which input is asked. At this point one can also page through the whole document describing the task one is running. Exit from the help page brings the terminal back to the original log/status/input Hermes display.

A new Hermes exploiting the many nice features of the X11 environment is under development. An experimental version does exist, but has limited capabilities such as the restriction that it can run only one task at a time (but more instances of X-Hermes can be active simultaneously). A feature which will be developed in addition is remote tasking, e.g. the possibility to start up a GIPSY task on another processor. This capability for distributed processing will be extremely useful in an environment where processors of different performance are networked together. One can do the simple operations on a small workstation and start heavy duty tasks on more powerful machines in the network without having to leave the GIPSY environment.

DISPLAY AND GRAPHICS

A special application has been developed for the display of images: GIDS, the GIPSY Image Display System. Up to 64 images can be loaded into GIDS, depending on the memory size of the workstation or X-terminal one is using. All display functions are contained within GIDS and are activated using using a mouse pointing device and pushbuttons. GIDS makes use of standard X11 calls and is approached from the application program level through a library of simple subroutine calls. The data transfer to (image loading) and from (cursor read out) GIDS is done via Unix or Internet sockets.

An important aspect of GIDS is that all display functions are integrated into the window, so that once an image or images are loaded one can control the display from within the GIDS window using the mouse. It is not necessary to type in display commands from the user input area of Hermes, other than the command for loading images.

The GIDS window appears default as a 512 × 512 display but can be resized. Seven fields at the top left give status information about which image plane is displayed, what the cursor position and pixel value at this position are and what the corresponding image name on disk is. The remaining fields at the left side of the image are push buttons to control the display.

The built in functions are: (i) display control, including a large variety of color schemes and the additional possibility to provide user defined color tables (contrast enhancement in this mode is performed by moving the mouse pointer over the screen where vertical motion controls the slope and horizontal motion controls the zero point of the transfer function); (ii) zoom and pan with the useful feature that one can zoom in x and y independently; (iii) a blinking and movie-loop mode where the rate is controlled by the x position of the mouse pointer; (iv) a blotting mode for defining polygons (which could be input to application programs e.g. for calculating integrated fluxes) using the mouse; and (v) a split screen mode. The speed of the movie and blink mode depends on the CPU performance of the workstation. A 512 × 512 movie on a SUN Sparc 1^+ approaches the standard 24 frames per second movie speed, provided there are no other users on the machine. On a HP 720 workstation the same speed is obtained with a 800 × 800 image size.

For graphics output such as contour plots, GIPSY uses the extensive PGPLOT library of routines. The plot files can be displayed in a Tektronix 4010 window. Graphics capabilities within GIDS are under development and will in addition allow graphics overlays on displayed images. Hard copies are made using a PostScript laser printer.

APPLICATIONS

The application programs in GIPSY now constitute some 100.000 lines of code, partly written in ANSI C, partly in FORTRAN 77. The main part of the system is radio synthesis image processing and includes many standard algorithms. Examples are conditional or unconditional arithmetic operations on single images or series of images, convolution, reprojection (including rotation) onto another coordinate system, pixel statistics in images, surface photometry routines, and CLEAN (Högbom 1974; Schwarz 1978; Clark 1980). An implementation of the Multi Resolution Clean (Wakker & Schwarz 1988) is also planned.

A strong emphasis in GIPSY has been on the processing and analysis of radio spectral line synthesis data and this has led to the coding of a unique variety of applications for handling three dimensional data (such as routines to construct maps of velocity versus an arbitrary direction through the image plane, an average of such maps etc.), for the analysis of velocity profiles using Gauss components, for the analysis of velocity fields and for modelling the gas motions in galaxies assuming realistic mass distributions.

In addition to general and spectral line synthesis applications, routines to extract Infra Red images from the IRAS raw data base have been developed in GIPSY, including destriping routines (Roelfsema & Kester, this volume).

AVAILABILITY

GIPSY now runs on the following platforms: Alliant FX 80, SUN 4, DEC station 3100, and HP 400 and 700 series workstations. GIPSY will become generally available in the spring of 1992 and can then be obtained via anonymous ftp. More information can be requested via E-mail from "terlouw@rug.nl" or "zwitser@rug.nl".

REFERENCES

Allen, R. J., & Terlouw, J. P. 1981, Proceedings of the Workshop on IUE Data Reduction, ed. W. Weis, (Vienna Observatory), 193

Allen, R. J., Ekers, R. D., & Terlouw, J. P. 1985, Data Analysis in Astronomy, ed. V. di Gesù, L. Scarsi, P. Crane, J. H. Friedman & S. Levialdi, (Plenum Press, N.Y.), 271

Clark, B. G. 1980, Astron. Astrophys., 89, 377

Högbom, J. A. 1974, Astron. Astrophys. Suppl., 15, 147

Schwarz, U. J. 1978, Astron. Astrophys, 64, 345

Wakker, B. P., & Schwarz, U. J. 1988, Astron. Astrophys., 200, 312

ASTRONOMICAL SOFTWARE DEVELOPMENT AT THE IUE REGIONAL DATA ANALYSIS FACILITIES

J. BONNELL, R. THOMPSON, P. LAWTON, N. OLIVERSEN, L. TAYLOR, J. GAGLIARDI, R. BRADLEY
CSC/IUE Observatory, Goddard Space Flight Center, Greenbelt, MD 20771

E. BRUGEL, G. ALLISON
CASA, University of Colorado, Boulder, CO 80309

ABSTRACT In 1981 the International Ultraviolet Explorer (IUE) project established two Regional Data Analysis Facilities (RDAFs) to aid in the analysis of observations made with the IUE satellite, one located at NASA/GSFC, and one at the University of Colorado. The RDAFs presently offer data analysis software packages for UNIX and VAX/VMS computers based on the Interactive Data Language (IDL). These packages were designed to support the post pipeline processing analysis of IUE data. The RDAFs also provide access to computer facilities for researchers, archival IUE data access, and access to the Astrophysics Data System (ADS). A brief description of these RDAF services and software is given.

INTRODUCTION

The International Ultraviolet Explorer satellite (IUE) was launched to geosynchronous orbit on January 26, 1978 and has been in continuous operation since. The satellite is operated in real time and has generated an archive which presently contains over 80,000 ultraviolet spectra of an extensive range of astronomical objects. The onboard scientific instruments are ultraviolet spectrographs which operate from 1150-3000 angstroms in two dispersion modes, low (6Å) and high (0.2Å). The detectors are SEC Vidicon Cameras with ultraviolet converters which produce a raw image of 768 x 768 pixels with 0-255 intensity levels.

Two Regional Data Analysis Facilities (RDAFs) were subsequently established by the IUE project, one at NASA/Goddard Space Flight Center (GSFC) and one at the University of Colorado, Boulder. The RDAFs provide computer hardware and software support for any researcher who wishes to analyze IUE data, with priority given to those individuals with accepted NASA research programs. The GSFC RDAF has seen the total number of users triple since 1987 and currently supports about 60 users/month. The majority of these users access the RDAF facilities remotely. Additionally, the GSFC RDAF supports users as one of the first operational nodes of the Astrophysics Data System.

RDAF SOFTWARE

The IUE RDAFs have developed a complete software package primarily designed for IUE data reduction. This package is a collection of about 250 staff developed and user contributed procedures coded in the Interactive Data Language (IDL). It was initially developed in IDL on a PDP-11 and was subsequently ported to the VAX/VMS computing environment using IDL Version 1. Recently the package has been ported to IDL Version 2 for UNIX-based workstations as well as VAX/VMS machines.

The RDAF software package has been in use, in some form, by the general astronomical community for a decade and its evolution has been substantially affected by input from this community. The evolution of the software has also been affected by changes in computer hardware technology over the decade of its existence. Consequently, the present package can be used on relatively "dumb" terminals which provide a tektronix graphics emulation but also contains many enhancements which take advantage of the windowing environment provided by workstations.

The raw IUE data are first processed using the IUE Spectral Image Processing System (IUESIPS). The RDAF IDL software libraries are focused on post processing analysis and offer a number of important data reduction capabilities which are specifically tailored to the data produced by IUESIPS and the IUE spectrographs. An overview of some of these IUE specific capabilities follows.

Data correction and calibration routines are provided which apply time dependent camera sensitivity degradation corrections, absolute flux calibrations, and assorted wavelength and velocity corrections specific to the IUE spectrographs and pipeline image processing history.

Customized extraction routines are also provided. A principle result of the IUE pipeline processing of low dispersion data is a geometrically and photometrically corrected two dimensional image file. With this file it is possible to perform customized extractions of spectral data that provide spatial as well as spectral information for extended sources and which improve the signal-to-noise ratio in the extracted spectra. RDAF software tools in this category perform Gaussian and weighted extractions, deconvolutions with the IUE point-spread function, and simple standard or customized "boxcar" extractions.

Routines which allow straightforward data quality determinations are also available. This crucial class of RDAF routines provides researchers with the ability to easily and correctly identify regions in the processed IUE data affected by telemetry dropouts, camera fiducial marks, microphonics noise, and camera blemishes such as those caused by charged particle "hits". Software aids for identifying camera artifacts and for flagging data that may be affected by pipeline image processing errors and calibration problems are also included.

On-line documentation is available in the form of descriptive procedure prologs accessed by the IDL 'doc_library' function. An example-oriented users tutorial and other descriptive documents are listed in the reference section. Distribution of the software is provided via anonymous ftp from iuesn1.gsfc.nasa.gov and cetus.colorado.edu. Magnetic tape formats compatible with UNIX and VMS platforms are also available on request.

RDAF SERVICES

RDAF user accounts are available on the GSFC IUE VAX 8350. These accounts are offered to any researcher interested in preparing for a visit, searching the IUE merged log, retrieving archival IUE data, or continuing IUE data analysis. Remote access to the IUE VAX 8350 is available via NSI DECnet, Internet and dialup modems. Prospective users can contact R. Thompson at GSFC for further information (on span, IUE::RTHOMPSON).

Archived IUE data is easily available through the IUE RDAF at GSFC. from an online mass-store device. The mass-store archive is supported by IUE project, NSSDC, and GSFC LASP personnel. The use of the IUE archives continues to increase rapidly. Since July 1982, approximately 188,000 images have been distributed in the U.S., implying that astronomers have requested archival data at an average rate of more than 20,000 images per year. The request rate for the first 9 months of 1991 has been over 3,000 images per month.

User accounts for use in conjunction with the ADS software are also available through the GSFC RDAF utilizing a SUN workstation. The ADS service is broad in scope and allows access to a wide range of astronomical databases. For IUE, the ADS allows users access to the IUE merged log and bibliography, along with general IUE project documentation. Interested parties can contact P. Lawton at GSFC for further information (on span, IUE::LAWTON).

FUTURE PLANS

The IUE project will implement new pipeline processing software (NEWSIPS) in 1992 (Nichols-Bohlin 1990). NEWSIPS will be used to produce the IUE Final Archive, a homogeneous reprocessing of the entire IUE archive. The NEWSIPS software will contain significant enhancements to IUESIPS software and will produce a number of similar, although improved data products. NEWSIPS will also produce additional classes of data products not represented in the current IUESIPS output, particularly for high dispersion data. The RDAFs will produce software to support the analysis of all NEWSIPS data products.

An important feature of the NEWSIPS data products is that they will be available to the astronomical community in FITS format. Image data will be stored as a FITS primary array with associated byte encoded data quality flags stored in a generalized table extension, 'IUEIMAGE'. Extracted spectral files will be stored as a binary table extension.

REFERENCES

Nichols-Bohlin, J. 1990 , "IUE Final Archive", Proc. Int. Symp. 'Evolution in Astrophysics', Toulouse, France, ESA SP-310

RDAF staff 1990, IUE RDAF Users Tutorial

Thompson, R. 1990, "Remote Users Guide", NASA IUE Newsletter, No. 41

Bonnell, J., et al. 1990, "An IUE RDAF Software Package for UNIX Workstations", NASA IUE Newsletter, No. 41

IMAGE RECOVERY AND PATTERN RECOGNITION APPLICATIONS IN A VISUAL LANGUAGE ENVIRONMENT

DR. TIMOTHY SCHNEEBERGER
Logicon, RDA, 2600 Randolph S.E., Albuquerque, NM 87106

ROBERT PIERSON AND DAVE DAYTON
Applied Technology Associates, Inc., 1900 Randolph S.E., Albuq., NM 87106

JOHN D. GONGLEWSKI
PL/LIII, Kirtland Air Force Base, NM 87117-6008

ABSTRACT Khoros is a software system which implements a visual programming language. It uses graphical icons and pipeline data flow to simplify the development and testing of new signal and image processing algorithms. We briefly describe two different applications developed under Khoros: one in pattern recognition, the other in astronomical imaging.

INTRODUCTION

The evaluation of new astronomical image recovery or pattern recognition algorithms requires rapid prototyping of new ideas as well as extensive testing against simulated and real image data. In these fields, there is a core of common software modules (e.g., Fourier transforms, filters) which often recur in applications. An ideal working environment allows the software developer to rapidly recombine these building blocks into new algorithms. In addition, the ideal environment allows the inclusion of new software modules developed by the analyst. These concepts of rapid prototyping and user introduced functions formed the basis of our past image processing codes (Schneeberger et al. 1987).

A few years ago, we sought a new simulation and image processing environment based on the following requirements:

Portability: minimum machine dependence; adherence to software standards
Efficiency: rapid prototyping; automatic code generation; easy incorporation of new functions; broad base of standard algorithms; a high level visual user interface
Flexibility: suitability for a variety of image processing applications

Based on these criteria, we selected the Khoros software system (Rasure & Williams 1991) developed at the University of New Mexico Vision Laboratory. The following pages describe some of our work within the Khoros software environment. We have chosen two very different applications to illustrate the flexibility and efficiency of the Khoros visual user interface.

KHOROS

Khoros is a visual programming environment. It provides a consistent interface across applications. Interaction with the environment is primarily mouse driven through pull down menus, buttons, and icons. Khoros incorporates a pipeline data flow paradigm particularly well suited to signal and image processing. In addition, the system provides over 260 image and signal processing functions, corresponding utility libraries, a suite of one- and two-dimensional display utilities, and full support for a wide variety of data types and image file formats. The FITS format is only one of many supported standards. Khoros is written in C, based on the X-windows X11R4 standard, and is implemented under the UNIX operating system for a wide variety of computer platforms.

Under Khoros, application codes are developed in a graphical "workspace" which may be configured to suit the particular application. Functions are represented as boxes ("glyphs" or "icons") within the workspace screen, and connections between functions indicate data flow. The user creates an algorithm by selecting appropriate icons and attaching their data input and output connections. Each function has an associated information form which allows access to function parameters and help files. Default values for processing parameters and intermediate data file names are generated automatically by Khoros. New functions may be added by the user, and these new functions behave identically to the standard functions supplied with Khoros. Khoros provides code generation and documentation generation utilities to expedite new function development.

A finished workspace becomes the equivalent of an executable program. The contents of the workspace may be saved to disk and reloaded. In addition, a workspace may contain another workspace or a set of nested workspaces. For example, an entire algorithm may be saved to disk and then reinvoked as a single "macro" glyph from within a new workspace.

For a further description of Khoros, refer to Rots & Herreid (this volume). The remainder of this paper is devoted to examples of Khoros programming for two disparate applications.

PATTERN RECOGNITION APPLICATION

Integration of kinematic information with target imagery improves the performance of tracking systems (Kendrick, Maybeck & Reid 1981). If we know what we are tracking and where it is pointed, we can better predict its future behavior. Our implementation of this data fusion problem requires the extraction of three-dimensional target orientation from two-dimensional target images. In addition, the target itself is identified by feature-vector classification. The extracted orientation is temporally synchronized and spatially aligned with a set of radar tracking data. The combined data set is passed to an extended state Kalman filter for track estimation and prediction. This is a multi-disciplinary sensor fustion problem involving image processing, pattern recognition, track estimation, and performance assessment.

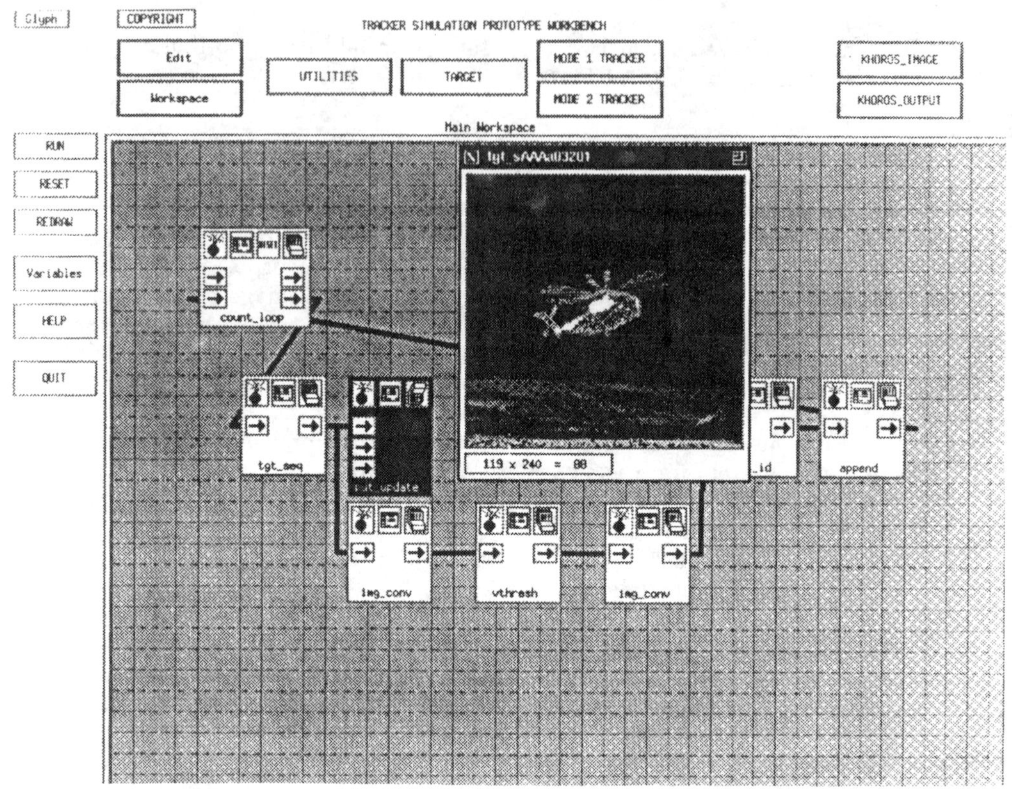

FIGURE I This Khoros workspace runs a simulation and display loop for pattern recognition.

We implemented the tracker simulation under Khoros on a SUN-4 SparcServer. Figure I illustrates how individual functions are placed as boxes in the Khoros workspace screen and joined by lines representing data connections. This particular workspace contains a simple loop which processes and displays target images. Note that one glyph is darkened to indicate that it is running; this glyph is controlling the image display at the center of the screen.

The workspace shown in Figure I is only a fragment of the entire simulation. The full tracker algorithm is a complex web of function glyphs and interconnections. However, despite the complexity, the advantage of the Khoros system remains the direct interaction with any component of the system. Any module may be accessed, queried, executed, removed, or refined from within the workspace. The use of standard forms associated with each function allow easy selection of different estimator and predictor algorithms.

The flexibility of Khoros is demonstrated by our ability to incorporate large independently developed programs as functions within our processing pipeline. Specifically, we incorporated two programs: the VISIX image processing code and extended state tracker routines. Developed at Cornell University, the VISIX code includes image moments and Fourier descriptor calculations. The extended state tracker routines were developed at Purdue University. Neither of these codes was developed for the Khoros environment, yet we integrated both into Khoros with little difficulty. They appear under Khoros as "glyphs" resembling any other standard Khoros function.

Although Khoros is primarily a visual environment for algorithm prototyping, it does not restrict users to this mode of operation. We implemented the tracker algorithm as both a batch simulation for handling pre-processed image files as well as an interactive session.

ASTRONOMICAL IMAGE RECOVERY APPLICATION

Self-Referenced Speckle Holography is a high resolution astronomical imaging technique for compensation of atmospherically induced aberrations (Gonglewski et al. 1990). This technique uses a narrow-band focal plane camera and broadband wavefront sensor to simultaneously record both an atmospherically distorted image and the distorting wavefront. Post-processing allows correction of the distorted image by removal of the distorting wavefront. For sufficiently bright objects, this method produces a correction from a single short exposure snapshot.

The digital field system consists of an image acquisition VME bus, and a post-processing system controlled by a SUN SparcServer. The SUN SparcServer runs Khoros, and all analysis takes place from within the Khoros environment. To accelerate centroid and deconvolution calculations, the system includes accessory boards based on the i860 chip and digital signal processors. These boards are accessed transparently from our Khoros functions. Access to raw data stored on the real-time disk is also provided by custom Khoros functions.

The data reduction code is too complex to be easily programmed as a single Khoros workspace. Instead, the processing is divided into several stages and implemented as a nested set of workspaces or "macros" within the Khoros environment. In Figure II, the top level workspace is seen in the background; it consists of a set of unconnected glyphs representing other workspaces. One of these workspaces has been opened revealing the interconnected processing glyphs shown at the bottom of the screen. These glyphs run darkmap subtraction and wavefront sensor calibration. The two windows at the right-hand side of the screen illustrate plot and image data presentations.

Previous versions of self-referenced speckle holography software had been developed as FORTRAN programs under VAX/VMS on a microVAX. These FORTRAN modules were easily converted to new Khoros functions using the Khoros code generation utilities; only a small shell of C code had to be written for each function. In addition, the coding of new functions was accelerated by the large library of image processing modules provided with Khoros.

SUMMARY

Khoros has proven flexible and powerful enough for a variety of image processing applications. The visual interface and the user expandable workbox of data processing tools allow rapid prototyping of practical applications such as astronomical image recovery.

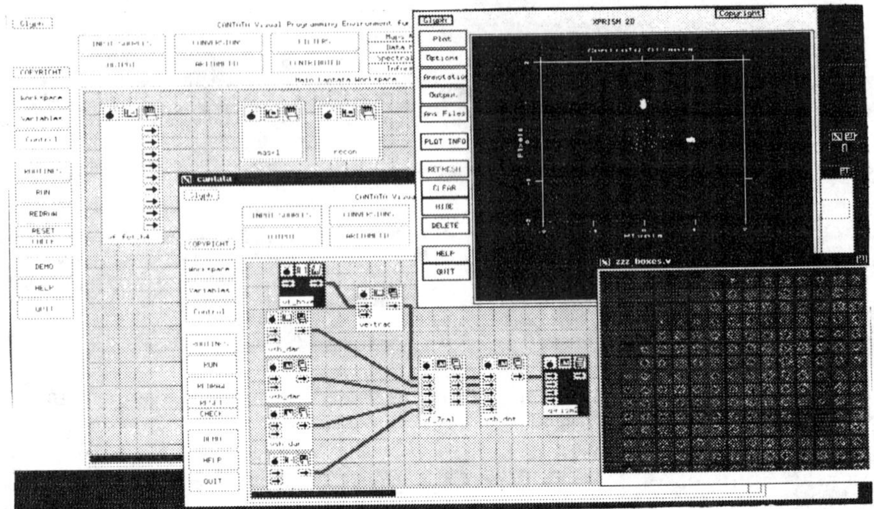

FIGURE II A set of nested Khoros workspaces performed data reduction and analysis for astronomical image recovery.

ACKNOWLEDGMENTS

We thank Dr. John Rasure and his team at the University of New Mexico, as well as Mr. Hal McIntire and Mr. Brian Kelly of Applied Technology Associates, Inc. Dr. Dominick Andrisani and Timothy Tao of Purdue developed the tracker modules, and Dr. Tony Reeves of Cornell provided the VISIX routines for the pattern recognition application. The image recovery work is sponsored by the Phillips Laboratory and the pattern recognition work by Picatinney Arsenal.

REFERENCES

Gonglewski, J. D., Voelz, D. G., Fender, J. S., Dayton, D. C., Spielbusch, B. K., & Pierson, R. E. 1990, Applied Optics, 29, 4527

Kendrick, J. D., Maybeck, P. S., & Reid, J. G. 1981, IEEE Trans. Aerospace and Electronic Systems, AES-17, 254

Rasure, J., & Williams, C. 1991, Journal of Visual Languages and Computing, 2, 1

Schneeberger, T., Dayton, D., Hill, R., & Tenorio, R. 1987, SPIE, 829, 339

PROTO-TYPING ASTRONOMICAL SOFTWARE IN *KHOROS*

A. H. ROTS
Universities Space Research Association, Goddard Space Flight Center, Code 666, Greenbelt, MD, 20771

L. A. HERREID
Hughes-STX, Goddard Space Flight Center, Code 664, Greenbelt, MD, 20771

ABSTRACT We consider the requirements for software proto-typing tools for the XTE Science Operations Center. We conclude that *Khoros* meets these requirements and describe its characteristics.

INTRODUCTION

The X-ray Timing Explorer (XTE) is a 100% guest investigator mission that is slated for launch in April 1996. It will carry three instruments, two of which are pointed instruments with a spatial resolution of 1° and an instantaneous energy range coverage of 2-200 keV, at a time resolution of the order of 1 μs; the third instrument provides an image of about 70% of the sky in the energy band 2-10 keV at moderate resolution every 90-100 minutes.

The data will be processed in the XTE Science Operations Center (SOC) at Goddard Space Flight Center, as soon as the telemetry is received. The software to be developed will have to satisfy two basic requirements: that it serves the full spectrum of the diverse tasks to be performed at the SOC; and that it be adequate (*i.e.*, not obsolete) during the full life of the mission. Consequently, we are not interested in the software that is in use today, but in the software that will be used five years from now. In order to investigate the software requirements for such a future package we decided to first mount a modest proto-typing effort. This project is intended to aid in our system analysis and formulation of the final design specification. Special emphasis is put on an "open" system design (easy expandibility), easy application programming, easy maintenance, and a thoroughly modern user interface.

REQUIREMENTS FOR A PROTO-TYPING ENVIRONMENT

A proto-typing environment for (astronomical) software should offer support in the following areas:

- Easy to add code
- Include a user interface

- Contain a utility toolkit
- Flexible in usage
- "Open" system design

In practical terms, one would like to be able to create new applications by just having to code the application algorithm and without having to worry about such trivial matters as how to translate a simple graphical user interface into lines of code. At the same time one would like to have a sufficient repertoire of tools (e.g., display utilities, format translators) available to be able to test and use the new application, and to offer the user an environment that allows considerable flexibility in setting up data reduction streams. In regard to this last point, one would like to avoid a collection of rigid monoliths. An open system means to us that non-native modules can be used in conjunction with the package as well, without modification.

It goes without saying that we want to be able to run this on Unix workstations using X-windows.

KHOROS

We have chosen to use *Khoros* for our proto-type development. *Khoros* is a product of the Vision Lab research group directed by John Rasure at the Department of Electrical and Computer Engineering of the University of New Mexico. It provides a combination of a software development environment, a visual programming environment, and a data analysis package. *Khoros* is described by Rasure et al. (1990) and in Schneeberger et al.'s contribution elsewhere in this volume. It may be obtained by anonymous *ftp* and is supported in part by a user consortium.

Khoros appears to offer most required characteristics for a rapid prototyping environment. From the *user's* point of view it includes:

- A "visual programming" tool that offers the user full flexibility in designing and implementing data reduction streams; it includes a macro capability and allows for canned processing pipelines as well as customized data reduction and analysis.
- A graphical user interface with interactive multi-page form fill-out.
- A command line interface with a variety of modes: direct specification, prompting, or prepared specification of input parameters.
- A play-back mechanism that is especially useful for demonstrations.
- An integrated Help facility.

The visual programming environment allows the user to select application or utility modules from a menu and control the flow of data between modules, thus creating data reduction pipelines. Looping and conditional constructs are provided as well. Modules may be fired individually or by an automatic scheduler, either sequentially or (as far as the flow logic permits) simultaneously. The data transfer between modules is effected through permanent files, temporary files, or shared memory.

From the *software developer's* point of view *Khoros* provides:

- A proto-typing tool for user interface specification.
- Code generators for graphical user interface, command line interface, and system interface.
- Maintenance and installation tools.
- Tools for generating and maintaining documentation.
- An extensive subroutine library (including FITS).
- Configuration management that allows for heterogeneous networks.
- A reasonable selection of ready-to-use applications, tools, and utilities. *Khoros* was developed in the context of remote sensing and digital signal processing. Its current application toolkit is therefore not optimal for astronomical purposes, but quite adequate for rapid proto-typing.

COMMENTS

Khoros fulfills essentially all requirements. Experience so far indicates that the system is very robust and easy to use. The task of actually implementing one's first, simple *Khoros* application, including the GUI, does not take more than a day's work. *Khoros* can be used in conjunction with other software development tools like debuggers. There are a number of draw-backs, however. In particular, we would have preferred a system that fully supports C++ and one that supports inter-proces communication. *Khoros* supports neither. Also, as noted before, the included applications and utilities are not optimal for our use. We feel, though, that these shortcomings are acceptable for a proto-typing environment; and it is quite conceivable that the support may eventually be added.

REFERENCES

Rasure, J., Argiro, D., Sauer, T., & Williams, C. 1990, Intern. J. of Imaging Systems and Technology, 2, 183

THE GRO/BATSE DATA ANALYSIS SYSTEM

SETHANNE HOWARD
Universities Space Research Association/NASA/MSFC, ES 62,
NASA/Marshall Space Flight Center, Huntsville, AL 35813

G. FISHMAN, C. MEEGAN, R. WILSON
NASA/Marshall Space Flight Center

W. PACIESAS
University of Alabama, Huntsville

The successful launch of the Gamma Ray Observatory in April, 1991 from the shuttle Atlantis began a several year mission to investigate the gamma ray sky. This second of NASA's Great Observatories is now called the Compton Gamma Ray Observatory in honor of Arthur Compton, discoverer of the Compton Effect. There are four experiments aboard. One experiment is called BATSE (Burst And Transient Source Experiment). It consists of eight modules - each placed at a corner of the spacecraft. The eight modules form an octohedron of NaI detectors that can monitor the entire sky all the time. Each module has two detectors: a large area detector (2000 cm^2) and a spectroscopic detector (100 cm^2) which are sensitive to energies from tens of keV to several hundred MeV. The scientific objectives of BATSE are:

- to observe gamma ray bursts
- to detect transient sources
- to observe pulsars
- to measure the high energy background

Since launch BATSE has met and exceeded its objectives. The data are of excellent quality and the spacecraft continues to perform in an exemplary fashion. Telemetry data arrive daily from the data link from Goddard Space Flight Center to Marshall Space Flight Center. The raw data are archived daily onto optical disk at MSFC. There are several types of data gathered by the instrument. Each data type is also archived to optical disk. There are data types for pulsars, data types for individual bursts, and data types for continuous coverage of the background, for each of the eight BATSE detectors in varying time and energy resolutions. All data are available to Guest Investigators.
The hardware used at MSFC to analyse these data are:

- two microvaxen which can receive telemetry data. One is a back-up system identical to the other. A two slot erasable optical disk jukebox is attached to this system.

- six clustered Dec Workstations that share several disks.

The software system consists of Fortran, TAE, Mongo, IDL, Ingres all under configuration management operating in a VMS environment. TAE stands for Transportable Applications Executive, a software system available from GSFC. IDL stands for Interactive Data Language, a commercial software package. Ingres is a relational database management system, also a commercial software package. The daily operations team produces a first look at the data products through a system of Fortran, TAE and NCARgraphics. This system runs with little user interaction and has been fully operational since launch. The analysis team can access either on-line data (kept on-line for three to four days after receipt) or archived data. The methods used to access these data are similar for all users — both science team members and Guest Investigators. The data fetch from the on-line data or the archive data is user friendly, driven by an Ingres menu system. This system does the following:

- provides on-line help
- prompts the user for information on the desired data. Minimal information is needed to find the data. For example, the Julian date and the burst number are sufficient to set up the system to retrieve burst data files.
- searches the database for the location of the data file requested.
- loads the correct optical disk
- retrieves the data into the user's directory, filtering the data as it copies. The filters are chosen by the user from a standard list. They permit the user to omit data during undesirable times (e.g. when the spacecraft high voltage is off).
- unloads the optical disk

Future enhancements to this data fetch include processing the data into FITS and into IDL for individual analysis. MSFC will provide the user with the data structure of each data type along with the actual file so decoding the file is straightforward. The system is working and is in use both locally and remotely. Guest Investigators can obtain their data in a routine fashion. Please contact either the GRO Science Support Center/GSFC/Greenbelt, MD or Dr. M. Finger/NASA/MSFC for GI information.

The BATSE Science Team has a gamma ray burst location algorithm which is run on each cosmic burst. It folds in the sky sensitivity as well.

The pulsar analysis is especially well developed. The analysis system links Fortran, Ingres into IDL where the final period fit to the data is displayed. Please contact R. Wilson, ES 62, MSFC, Huntsville, AL for further information.

In addition there is a large body of software to handle spectral analysis: BSAS (BATSE Spectral Analysis Software), developed at GSFC and installed there, at MSFC and UCSD. It will process BATSE gamma ray burst data by folding the detector response functions into the burst count data to produce photon spectra. Both user specified model parameters and non-specific models can be used. This software links Fortran, Ingres and and Mongo in a VMS environment. Information on BSAS is available from Brad Schaefer/USRA/NASA/GSFC.

The earth occultation data is similarly well developed. Since BATSE is an all sky monitor there are times in each orbit that the earth occults a large fraction of the sky for several detectors. Over twenty-five known sources are followed as they disappear and re-emerge from behind the earth. Please contact Alan Harmon, NASA/MSFC for more information.

BASS: BASIC SCIENTIFIC SOFTWARE FOR SAX

M.C. MACCARONE, B. SACCO
IFCAI/CNR, Via M. Stabile 172, 90139 Palermo, Italy

L. CHIAPPETTI
IFCTR/CNR, Via L. Bassini 15, 20133 Milano, Italy

D. DAL FIUME, M. TRIFOGLIO
ITESRE/CNR, Via de' Castagnoli 1, 40126 Bologna, Italy

L. PIRO, P. UBERTINI
IAS/CNR, Via E. Fermi 21, 00044 Frascati, Italy

G.C. PEROLA
IOA, Univ.'La Sapienza', Via G.M. Lancisi 29, 00161 Roma, Italy

J. HEISE
SRON, Sorbonnelaan 2, 3584 CA Utrecht, Holland

A. PARMAR
SSD-ESA/ESTEC, Kepleerlaan 2, 2200 AG Noordwijk, Holland

INTRODUCTION

SAX (X-ray Astronomy Satellite) is a programme jointly developed by the Italian Space Agency (ASI) and the Netherland Agency for Aerospace Programs (NIVR) devoted to systematic, integrated, and comprehensive studies of galactic and extragalactic sources in the energy band 0.1-200 KeV.

Scientific objectives are: imaging (with moderate angular resolution of 1 arcmin) and broad band spectroscopy over the energy range from 0.1 to 10 KeV; spectral measurements, spectroscopy and timing on sources from 3 to 200 KeV; all sky monitoring (2-30 KeV) for the investigation of long time variability and localisation and study of transients.

The spacecraft has a total mass of 1200 Kg, is three axis stabilised, and it will be placed into a circular equatorial orbit at 600 Km with an inclination of 2 degrees, by an Atlas G-Centaur.

The payload includes: a low energy (0.1-10 KeV) concentrator/spectrometer, LECS; a medium energy (1-10 KeV) concentrator/spectrometer, MECS, consisting of three units; a high pressure gas scintillation proportional counter, HPGSPC,(3-120 KeV); a phoswich detector system, PDS, (15-200 KeV), all of which have narrow field of view with coaligned optical axis; and two wide field cameras, WFC, (2-30 KeV), field of view 20x20 degrees, which point in diametrically opposed directions perpendicular to the narrow instrument axis.

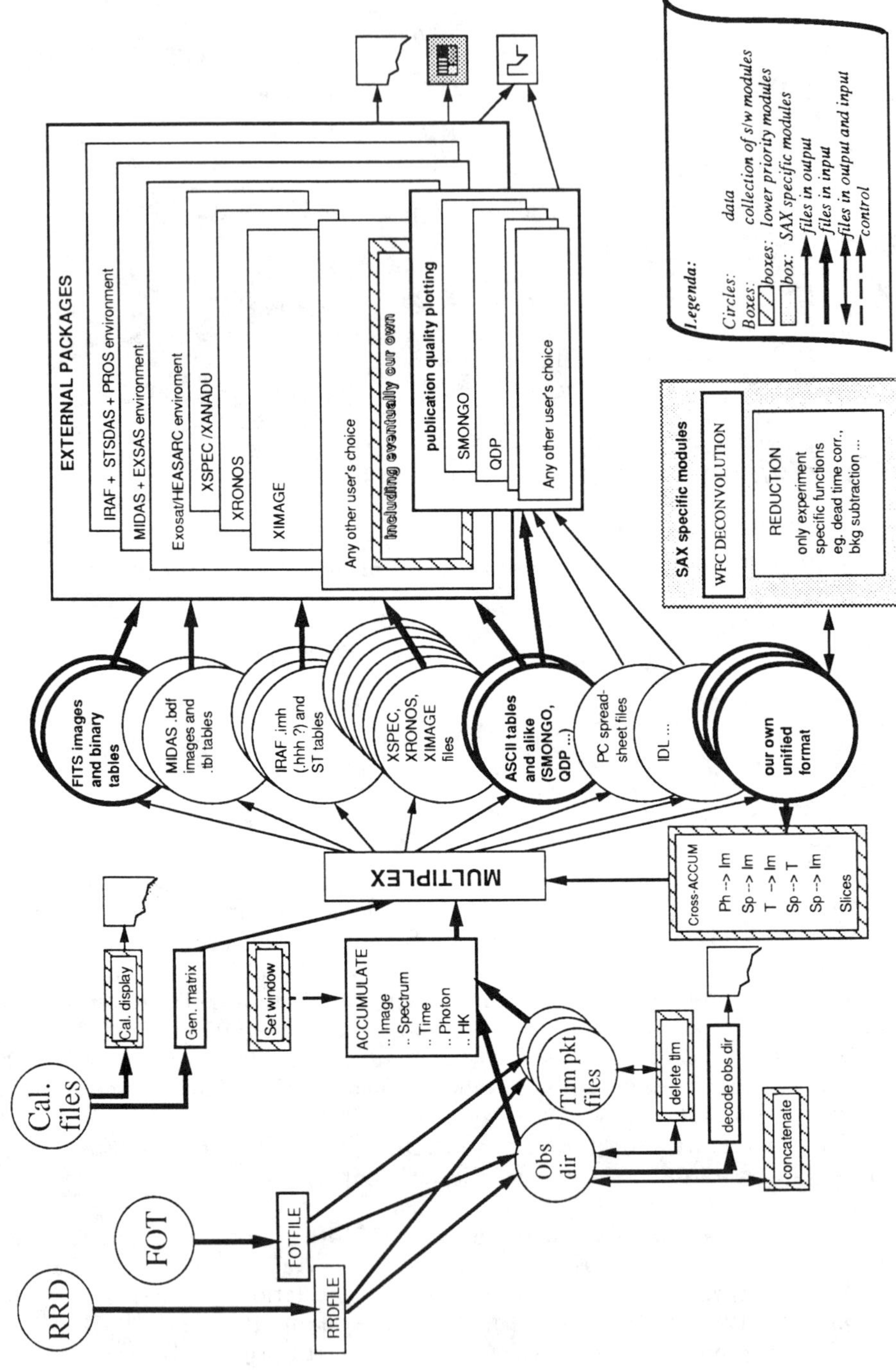
RRD
FOT
Cal. files
RRDFILE
FOTFILE
Obs dir
Tlm pkt files
delete tlm
decode obs dir
concatenate
Cal. display
Gen. matrix
Set window
ACCUMULATE
.. Image
.. Spectrum
.. Time
.. Photon
.. HK
MULTIPLEX
Cross-ACCUM
Ph --> Im
Sp --> Im
T --> Im
Sp --> T
Sp --> Im
Slices
FITS images and binary tables
MIDAS .bdf images and .tbl tables
IRAF .imh (.hhh ?) and ST tables
XSPEC, XRONOS, XIMAGE files
ASCII tables and alike (SMONGO, QDP ...)
PC spread-sheet files
IDL ...
our own unified format
EXTERNAL PACKAGES
IRAF + STSDAS + PROS environment
MIDAS + EXSAS environment
Exosat/HEASARC enviroment
XSPEC /XANADU
XRONOS
XIMAGE
Any other user's choice
Including eventually our own
publication quality plotting
SMONGO
QDP
Any other user's choice
SAX specific modules
WFC DECONVOLUTION
REDUCTION
only experiment specific functions
eg. dead time corr.,
bkg subtraction ...
Legenda:
Circles: data
Boxes: collection of s/w modules
boxes: lower priority modules
box: SAX specific modules
files in output
files in input
files in output and input
control

SAX, to be launched at the end of 1993, will have a minimum mission life of two years, extendable up to four years.

THE BASIC SCIENTIFIC SOFTWARE FOR SAX

SAX-Data Analysis Working Group, DAWG, is one of the groups defined in the framework of the SAX mission. SAX-DAWG has been charged by the SAX Consortium to provide:

- definition of the software specifications for the SAX data scientific analysis,
- development and realization of the operating scientific software,
- participation and support to simulations and calibrations, essentially from the software point of view,
- support to the Observation Program Working Group for what concerns simulations and information on calibrations,
- contribution to the definition of the functional structure and activities concerning the SAX-Scientific Data Center, SDC.

Several data analysis software systems (IRAF, MIDAS, XSPEC, XIMAGE, ...) are today available to analyse high-energy astronomy data. The choice of one or more among these systems is depending on several elements, such as their availability and documentation, and/or the personal expertise and taste of the observer. In any case, DAWG considers as mandatory to have a basic scientific software, BASS, which, starting from the Final Observation Tapes, FOT, permits the observer to reduce data in spectra, images, and time profiles, to be analysed with the chosen external package. This basic scientific software is strictly specific of the mission and it can be developed only within the hardware institues which have the necessary knowledge about the scientific performances of the payload. In this sense, the following tasks will be covered by BASS: read FOT, access to the calibration data, select data, perform background subtraction, accumulate spectra, sky images, and time series, provide specific modules (such as a WFC deconvolution).

As sketched in the figure, concerning the SAX data flow, it is to be noted that: i) data filing is from FOT or RRD to disk resident telemetry files and observation directories; ii) accumulation step is to decode observation directories, accumulate images, spectra, time profiles (including HK), and photon lists from telemetry files; and response matrices from calibration files; iii) cross-accumulation are possible (e.g. images from photon lists, or pseudo-images with counts as function of time and energy, etc.); iv) a dedicated format is likely required to support the small number of SAX specific reduction modules (grouped in the figure inside a dotted box); v) non-specific reduction, analysis and display may occur through a variety of existing external analysis and plotting packages. Therefore BASS will provide a variety of output files in order to be immediately compatible with the main external packages.

As a baseline, BASS will support both VMS and Ultrix platforms. BASS will be well-documented (programmers manuals, data file structure, user's guide) and it will be checked by an 'Acceptance' service before its official release.

DEVELOPMENT OF A DATA ANALYSIS SYSTEM FOR THE SAX X-RAY WIDE FIELD CAMERAS

R.JAGER, J.J.SCHUURMANS, J.J.M.IN 'T ZAND AND J.HEISE
SRON - Laboratory for Space Research Utrecht, Netherlands

ABSTRACT This paper describes the principles of the image reconstruction for the SAX Wide Field Cameras. Also requirements for a data analysis system are generated from the expected amount of data and the effort needed for a standard all-data analysis.

INTRODUCTION

SAX is an Italian/Dutch X-ray satellite that will be launched december 1993. SAX will consist of, among other instruments, two X-ray Wide Field Cameras built by the SRON Laboratory for Space Research at Utrecht. These WFC's are based on the coded mask principle, which implies that the actual image reconstruction can only be done on ground by means of correlating the detected photon positions with the coding mask. One of the functions of the WFC's will be monitoring of all the sources in the Field Of View on timescales from milliseconds to months or years. The FOV is 40 degrees square (FWZR) and the angular resolution is 5 arcmin, the effective area at 6 keV is 200 cm^2. The energyband is from 1.8 to 30 keV.

Here we will describe the principles of the image reconstruction and the outline of the data analysis system that is needed to deconvolve, analyse and store routinely the expected amount of 150 MByte raw data per day during four years. A more complete description of the instrument is given by Jager et al. (1989).

PRINCIPLES OF THE IMAGE RECONSTRUCTION

Imaging with a coded mask camera is basically a multi step process. Photons from X-ray sources in the observed part of the sky are spatially modulated by a coded mask and detected with a two dimensional detector (Multi Wire Proportional Counter). The output of the instrument consists of an eventlist where each detected photon is labelled with X and Y positions on the detector, the energy E of the photon and the time T of detection. On ground a detector image can be formed from the eventlist by applying appropriate filters around X, Y, E or T. This image will be binned to 512 x 512 pixels but occasionally it is foreseen to use the full detector resolution of 768 x 768 pixels.

The reconstruction of the sky image can then be performed by a correlation of the detector image with the mask pattern. In order to apply corrections for local mask transparency and shadowing of the detector window support structure, the correlation is done separately for the open and closed maskelements. This procedure results in the sky image and the accompanying error image. See Figure I for a pictorial outline.

In order to reach optimum sensitivity and accuracy it may be necessary to 'clean' the image. This is done by a method called Iterative Removal Of Sources (IROS, Hammersley 1986). The method is as follows: find the strongest sources in the reconstructed sky image, construct a simulated detector image by folding these sources through a model of the instrument, subtract the simulated image from the observed detector image and do the reconstruction of the skyimage again. This way the coding noise, which is caused by incomplete shadowgrams of the mask, can be eliminated. Only the poissonian noise will remain.

After the image reconstruction the final source positions and fluxes have to be extracted from the image. To do so, a rather complex model Point Spread Function is needed, i.e. a PSF that is a function of source position and energy.

This procedure must be repeated for all time and spectral bins one wants the data to be resolved in, resulting in lightcurves and spectra of all sources simultaneously.

If a more detailed analysis is needed for just a few sources, a (faster) method of direct correlation can be applied. Hereby the eventlist is scanned and each photon is directly correlated with the relevant part of the mask (determined by the source position).

REQUIREMENTS FOR A STANDARD DATA ANALYSIS SYSTEM

SAX will be placed in a circular, nearly equatorial orbit at an altitude of 600 km. The two WFC's onboard SAX are watching opposite directions of the sky, each perpendicular to the optical axis of the other narrow field instruments. This implies that there is usually always one WFC active, while the other one is pointed to Earth.

The expected countrate due to the detector and the diffuse X-ray sky background is about 250 c/s. With some moderate strong sources in the field of view we expect an overall mean countrate of about 500 c/s. (cf. the Crab nebula is 400 c/s). Each event is labelled with 28 bits (including overhead), this amounts to an average data rate of 14 Kb/s (or 1 GByte/week).

In order to keep up with this data stream and to have time available to do dedicated analysis work, the standard analysis of all data is required to be accomplished in about 10% of the real time. Assuming a lightcurve time resolution of one minute and an energy resolution into 16 channels, this calls for a floating point processing capacity of about 15 million operations per second. To minimise the data transfer (keeping the mask and correction matrices in memory), at least 32 MByte RAM is needed.

The results of the standard analysis system consist mainly of positions, fluxes, lightcurves and spectra for each detected X-ray source or upper limits for non detections. This amounts to roughly 150 KByte/source/day. Furthermore it is foreseen to store detector images in four energy ranges per observation. This amounts to 1 Mbyte/day. The result data base will be written on Rewriteable Optical Disks.

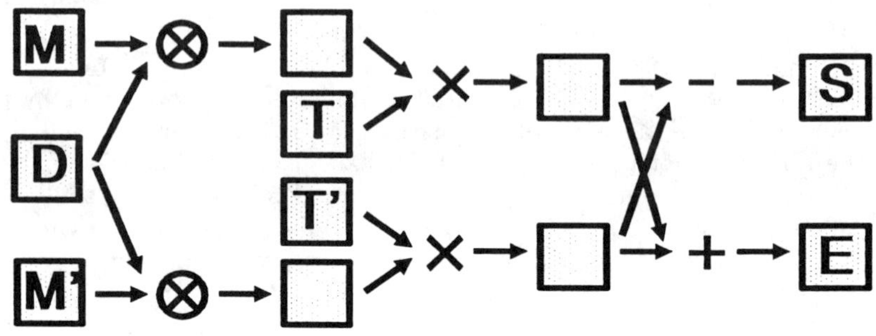

FIGURE I The detector image D is convolved not only with the mask pattern M (where transparent mask elements have value 1 and opaque elements 0) but also with the complementary mask M' (with values of respectively 0 and 1). After multiplication with correction factor matrices T respectively T' (separate for the open and closed mask elements) the two resulting matrices subtracted give the reconstructed skyimage S, added together they give the errors in the skyimage E. These images are standard 1024 x 1024 pixels large, with maximum detector resolution they can be as large as 1536 x 1536 pixels.

For reasons of compatibility with the software that is to be developed for the other instruments in SAX, Fortran is chosen to be the main programming language. For the astronomical image processing and database, the possibilities of XANADU, IDL, IRAF and MIDAS are being considered.

REFERENCES

Jager, R., in 't Zand, J. J. M., Schuurmans, J. J., Heise, J., Mels, W. A., & Brinkman, A. C. 1989, Proc. SPIE, 1159, 2

Hammersley, A. P. 1986, Ph.D. Thesis, Univ. of Birmingham (UK)

SEARCH FOR A COMMON DATA MODEL FOR ASTRONOMICAL DATA ANALYSIS SYSTEMS

A. FARRIS
Space Telescope Science Institute, Baltimore, MD 21218

R. J. ALLEN
Space Telescope Science Institute, Baltimore, MD 21218

THE BROADENING PERSPECTIVE OF MODERN ASTROPHYSICS

The goal of astrophysics is to develop comprehensive physical models and the theoretical framework to explain observations obtained over broad regions of the electromagnetic spectrum. Consequently, such a comprehensive theory of astrophysics must be based on measurements from radically different instruments, requiring fundamentally different observational techniques and methods of calibration, with different methods for analyzing and storing data from traditionally distinct domains. The expanded domains of modern astronomy, such as optical, radio, infrared, X-ray, gamma-ray, and neutrino, have produced a variety of types of data structures: 2-dimensional images, spectra, u-v data, maps, time-tagged lists, photon event counts. Modern astronomy is also characterized by increasingly complex instrumentation: single-dish telescopes, synthesis-arrays, ground-based and space-based telescopes, space-VLBI. Many data analysis systems have been developed to cope with this increased complexity: AIPS, MIDAS, IRAF/STSDAS/PROS, EXSAS, COMPASS, GIPSY. Modern astronomy faces a major problem in gaining access to data, because it exists in so many different locations, media, forms, and structures.

There are many problems with current approaches to data storage and analysis in astronomy. Rapid development of technology has caused areas of modern astronomy to develop in relative isolation from one another with a consequent sense of provincialism. Observational techniques, methods of calibration, and data storage methodologies are frequently instrument dependent or mission dependent. Real-time hardware processors and data collection devices often dictate data structures. Data analysis systems originated within these same specialized domains and frequently reflect the same local dependencies. To achieve the goal of astrophysics a much more unified approach to the development of astronomical databases and data analysis systems is needed.

PROJECT GOAL: CAN A UNIFIED DATA MODEL BE DEFINED?

Can an astronomical data model be defined in such a way that it is at once both practical and general? 'Practical' means capable of being implemented with reasonable efficiency given the current state of computer technology. 'General' means applicable to a broad range of astronomical problems, research areas, and observational techniques. Such a model should be independent of specific computer hardware, operating systems, or data analysis packages.

This paper describes a project to determine if it is possible to define such an astronomical data model. Its goal is theoretical. Its goal is not to construct such a model; but to determine if it is possible to define such a model and, if so, to answer the question, "Can it be constructed?" The answers will be of great importance to future software developments in astronomy. If such a model exists, it will help break the barriers to traditionally distinct domains, shed light on the most logical methods for organizing large bodies of data, guide the development of distributed data management and analysis systems, and guide the development of common user interfaces to data storage and analysis systems.

WHAT IS A DATA MODEL?

A model, in this context, is a collection of meaningful astronomical concepts, such as an 'image' or 'astronomical position' that are used to describe astronomical observations, together with the logical relationships that hold among these concepts. This set of concepts maps onto a set of computer-supported, abstract data structures and functions that are associated with these astronomical concepts. Any statement or operation that can be made using these conceptual terms can also be made or carried out using the computer system. A model is a mapping between a conceptual space and a computer-supported space. The difference between a 'two-dimensional array of floating point numbers' and an 'image' is an astronomical interpretation. In other words,

Abstract data structures
+ Access functions
+ An astronomical interpretation
= An astronomical data model

The purpose of a data model is to promote conceptual clarity and consistency. Common data modeling approaches used in various areas in computer science are enterprise modeling, entity-relationship modeling, object-oriented approaches, and knowledge representation. A data model serves as a blueprint for system developers. It also serves as a high-level guide for end-users of a system. A data model serves as a bridge between users and system developers. The end-user employs the conceptual terms of the model to guide his or her interactions with the computer system. The system developers see the model as a goal and a guide to implementation. This role the data model plays is analogous to the role a "Principles of Operation" plays between hardware developers and software developers. The "Principles of Operation" are the ultimate goal of the hardware designer, but are base-level concepts to the software designer.

BASIC CONSIDERATIONS

Doesn't astronomy already have a standard data model in FITS? No, because FITS is not a data model. FITS is a standard to promote the interchange of data contained in certain types of data structures. There is an underlying model of an astronomical image, but this is not part of the FITS standard. FITS could just as easily be used to transport bit-mapped images of Greek text. Nor do the recent additions of tables to the FITS standard make it a data model. Tables are data structures, together with conventions for interchanging data that conform to these structures. The concepts employed in FITS to promote self-defining, transportable data are useful in implementing a data model but do not in themselves constitute a data model.

A possible model may be constructed in the following manner. Exploit the physical properties of the measured data: position, time, intensity, frequency, polarization of incoming electromagnetic radiation. Generalize to an N-dimensional data space in which the axes correspond to measurable physical quantities. Any physical measurement can be represented as a region in such an N-dimensional space. Higher level concepts, such as 'source name' or 'map', can be related to these basic physical properties. However, corresponding to each measurement is a complex body of information about the attributes and state of the measuring instruments and about the calibration and data reduction steps performed to deduce the measured quantities from the raw data. The N-dimensional data space model does not represent this complex body of information, sometimes called the meta-data. **It is this meta-data that represents the most serious challenge.** Is there sufficient commonality to form a unified model of the complex processes that must be represented?

METHODOLOGY: ANALYZE EXISTING DATA ANALYSIS SYSTEMS

The set of existing analysis systems may be regarded as forming a minimum set of requirements for a more unified analysis system. Each such analysis system represents the cumulative experience of years of astronomical and software expertise. It is in these systems that the knowledge of how to do data analysis in astronomy is buried. The following four steps will generate a unified data model. 1. Select major data analysis systems representing each division of astronomy. 2. Identify in detail the data model implicitly assumed by that system. 3. Form the union of all such data models from the individual systems. 4. Use high-level astronomical considerations to remove inconsistencies and redundancies and then refine the model into a coherent structure.

This methodology is a blend of bottom-up and top-down approaches. Generating the data models from existing analysis systems is a bottom-up approach. Refining the model, and possibly making additions to it, is a top-down approach. If such a model is defined and complete, it can be verified in two ways. First, there should be a mapping between the model and the data handling portion of any existing analysis system (bottom-up). Second, any proposed astronomical experiment should be able to represent its data handling using the model (top-down).

Currently, we are engaged in determining the major divisions of astronomy to consider and establishing criteria for selecting data analysis systems for study, e.g., general purpose systems within some domain of applicability, used by a broad class of users within astronomy, or with interesting or unique features for representing data within its domain. Then, possible data analysis systems, such as AIPS, IRAF/STSDAS/PROS, MIDAS, EXSAS, COMPASS, GIPSY, and IDL, will be selected for study.

SECTION 4

DECONVOLUTION AND IMAGE RESTORATION

DECONVOLUTION FOR REAL AND SYNTHETIC APERTURES

T.J. CORNWELL
NRAO, PO Box O, Socorro, NM 87801

ABSTRACT Deconvolution of observations is now commonplace for synthetic apertures such as occur in radio astronomy and is performed by nearly every VLA observer as a standard part of the data reduction. It is less common in real aperture observing, especially at optical/IR wavelengths, though the well known spherical aberration of the Hubble Space Telescope is causing a resurgence of interest. Although the two problems are substantially different, much can be learnt from a comparison. I will attempt such a comparison and will also give some advice as to how to go about devising new processing algorithms.

INTRODUCTION

Many astronomical instruments don't make images in the way one thinks of a camera making an image. Interferometric arrays collect their primary data in the Fourier plane, and even ostensibly imaging telescopes such as the HST collect data which are not straightforwardly interpreted as a slightly smoothed version of the true sky brightness (in the case of the HST this distinction is due not only to the well-known spherical aberration but also to effects such as the severe undersampling which afflicts the WFPC at some observing wavelengths). So, although the title of my talk includes the word "deconvolution", this is actually an inappropriate and old-fashioned term since in many circumstances there is no straightforward convolution equation between object and image to be solved. Instead it is probably better to think of the central problem as one of *Inferring the sky brightness distribution from observations and prior knowledge.* I intend to talk about procedures for performing such an inference and, in particular, whether the methods now used at radio wavelengths can be applied at optical and IR wavelengths. In many respects there is a considerable convergence between the problems in these different wavelength ranges and it is therefore tempting to transfer methods wholesale from the radio (where presumably they are well-understood) to the optical (where they are sorely needed). I want to offer some cautionary remarks concerning these assumptions and, in a slightly more positive vein, to offer some advice on how imaging problems can fruitfully be addressed without necessarily adopting radio wavelength algorithms as black boxes.

Another caution about my title is in order: although the dichotomy specified is between real and synthetic apertures, there are many other splits which are important and so a more complete list would be:

- Real/Synthetic Apertures
- Filled/Unfilled Apertures
- Photon/Gaussian noise
- Fourier plane Extrapolation/Interpolation
- Radio/Optical Wavelengths

Imaging procedures tend to split along one or more of these lines and therefore in moving an algorithm from one application to another one must be very careful not to move it into a regime whether it is inapplicable or simply does not work.

THE IMAGING PROBLEM

Establishing some terminology, let $\underline{f}$ be the unknown sky, $\underline{g}$ be the measured data (represented in the image plane), H be the imaging kernel (*i.e.* point spread function), and $\underline{n}$ be the noise. Then the imaging equations for real and for synthetic apertures are:

$$\underline{g} = H\underline{f} + \underline{n} \tag{1}$$
$$\underline{g} = H(\underline{f} + \underline{n}) \tag{2}$$

So now we ask what do we usually know about the various elements of these equations?

First consider the *noise*, $\underline{n}$. The physics expressed in these equations is that the noise is usually independent in the space where the data were taken, and so for a synthetic aperture, the noise has the same "shape" as the PSF. We also usually know the statistics of the noise quite well. Hence we can develop algorithms which know a lot about the noise, but with the disadvantage that any such sophisticated algorithm may be difficult to transfer to another application in which the noise has different characteristics. Furthermore, a well-tuned algorithm will fail when the noise is not what you thought it was.

Second, the *imaging kernel* or Point Spread Function can be split into known, $H_{\text{theoretical}}$, and unknown parts, $H_{\text{aberrations}}$. Both components can be spatially varying. In this regard, there is a remarkable, if dismaying, convergence between the HST PSF and the VLA PSF at wavelengths of 1m and longer. We usually know quite a lot about $H_{\text{aberrations}}$: for both real and synthetic apertures it is formed from sums of the autocorrelation of various apertures. As described in Cornwell (1989), this insight leads directly to the family of phase closure/speckle methods (though it must be admitted that historically this insight was hard won).

Third, since we build finite diameter telescopes, the *theoretical imaging kernel* or PSF extends more or less forever in the image plane. Examples which support this are the HST PSF, and the VLA PSF (which never drops below 0.5% to 1% even in the best cases). Computing is then tedious and all points in the final image tend to be quite strongly coupled.

Fourth, we know a fair amount about the "unknown" *sky brightness.* Of course, we know that the sky brightness must be positive (and that the fractional polarization must be less than one *i.e.* in terms of the Stokes parameters, $I^2 - Q^2 - U^2 - V^2 \geq 0$). Also, the sky may have been imaged at lower resolution or at a different wavelength and we may be able to classify it into one of the following categories: point-like, limited in support but complex, or fractal.

Finally, the context of these equations is the same over all wavelengths. Per unit time, the cost of a reasonably capable computer is now much less than either that of a telescope or an astronomer. Also better software practices and open systems are increasing the transportability and thus exchangeability of computing solutions (anyone who doubts this should be introduced to the world of anonymous ftp).

SYNTHETIC APERTURE DECONVOLUTION ALGORITHMS

The principal "deconvolution" algorithms used in radio-interferometric imaging are the CLEAN algorithm (Hogbom 1974) and the Maximum Entropy Method (*e.g.* Narayan & Nityananda 1986).

CLEAN

CLEAN is a misleadingly simple algorithm with a wealth of strange features which arise as one delves deeper and deeper. Although Schwarz (1978, 1979) has published an analysis of this method, it says little about the solutions produced in the underdetermined case where the *a priori* constraints such as limited support are insufficient to constrain the solution to a single image. Marsh and Richardson (1987) wrote a tantalizing paper which suggests that CLEAN produces a minimum L1-norm solution. That this cannot be always be true is evident from the investigation by Tan (1986) which shows amongst other things that CLEAN shows chaotic behavior as a function of loop gain. Other attributes of CLEAN are also less than satisfactory: for example, in this proceedings, Briggs and Cornwell describe the exceedingly poor convergence sometimes seen in CLEAN.

Despite these drawbacks, CLEAN is widely used in radio-interferometry and has spread to other fields with some success. The basic idea of decomposing an object into point sources works in many applications and even in cases where an object is somewhat extended. It is fast for simple sources and can be easily coded. Variants which promise some improvements in speed have been describe by many, the most notable being Clark (1980) and Steer, Dewdney & Ito (1984). Also, an unpublished CLEAN algorithm developed by Schwab and Cotton of NRAO is one of the most used tasks in the AIPS package.

Schwarz (1978) showed that convergence is guaranteed if:

1. H is positive semi-definite. This means that the eigenvalues must be non-negative.

2. $\underline{g}$ must be in the range of H. This means that there are no spatial frequencies in $\underline{g}$ which are not also in H. The most straightforward interpretation of this is that the noise must be "PSF-shaped".

Convergence here means that the maximum error always decreases with increasingly iteration number.

In practice, CLEAN often works even if these conditions are violated. This probably reflects on the poor convergence of CLEAN. It is hardly ever run to convergence at the machine precision, instead often it is stopped when the maximum residual is comparable to the noise level. Running deeper into the noise distorts the noise histogram, though there has never been a good demonstration of this defect.

All in all, CLEAN prospers despite these defects. Probably the key strikes against it are pragmatic: it is too slow for very extended objects (even in the Steer-Dewdney-Ito variant), it is prone to striping instabilities (Cornwell 1983) and is too hard to adapt to different types of measurement. For these reasons, radio interferometrists turn to the other popular deconvolution algorithm, the Maximum Entropy Method. However, before describing MEM, I want to first talk about the use of CLEAN in processing data from real apertures.

The criteria for convergence described above immediately rule out most types of data from real apertures. First, in many cases, the PSF H will not be positive semi-definite. For example, a simple motion blur has negative eigenvalues as does the HST PSF in some circumstances. Second, the noise will not be PSF-shaped. For example, it is well known that photon noise has spatial frequency content beyond that of the PSF. So, given these defects, why does CLEAN work when applied to optical data (as it does: see *e.g.* Keel 1991)? The answer again probably lies in the fact that CLEAN is hardly ever run even close to convergence. In the event that these violations of the convergence criteria become important and limit running of CLEAN, they can be alleviated to some degree by pre-filtering of the data to remove out of band noise and to correct negative eigenvalues in the PSF.

The Maximum Entropy Method

MEM is in some senses the opposite of CLEAN: it seems rather complex but is actually quite simple and well-understood. Unlike CLEAN, we can calculate properties of the final solution and therefore can show when it is satisfactory and when it fails. We also know basically how to calculate error bars (Gull & Skilling 1989). What has hurt MEM over the years has been a continuing dispute over the rationale for MEM (see Narayan & Nityananda 1986, for a clear and, in my opinion, unbiased summary). Now most people in astronomy use the following form of entropy:

$$\mathcal{H} = -\underline{f}^T \left(\ln \underline{f} - \ln \underline{m}\right) \tag{3}$$

where $\underline{m}$ is a default image representing the image reconstructed from no data.

The central idea of maximizing the image entropy $\mathcal{H}$ subject to data constraints leads to a smooth, stable, positive image, which can be calculated reasonably quickly using any of a number of algorithms. The MemSys5 routine is by far the most robust and capable algorithm around but others such as that of Cornwell & Evans (1985) may be substantially quicker in easy cases.

MEM has many known defects: biased residuals (see *e.g.* Cohen 1991), SNR-dependent resolution, poor reconstruction of point sources on top of extended emission (*e.g.* Narayan & Nityananda 1986), and dependence on accurate knowledge of the noise level. That these are documented and understood is attributable to the fact that the form of the MEM solution is known and can be manipulated to derive certain properties of the solution. Fixes are now in hand for most of these problems, some based upon re-posing of the basic question being asked in image reconstruction (see Gull & Skilling 1989), and some based upon heuristics (see *e.g.* Narayan & Nityananda 1986, Cornwell & Braun 1989).

There have been some interesting developments in the theoretical interpretation of MEM. The fully Bayesian analysis of imaging by Gull (1989) leads to the startling but convincing conclusion that image entropy should have relatively little weight in reconstructing an image. Simply put, the full and consistent application of Bayes theorem destroys the power of entropy in smoothing images. Gull then proposed the use of intrinsic correlation functions (ICFs) to re-introduce some smoothing. This is undoubtedly an ingenious solution to a problem generated by a clever argument. Weir (this volume) has developed the ICFs further but it is probably fair to say that the end of this line of development remains some time away.

One of the most important attributes of MEM is that it is easily adaptable to new types of data. For example, in my work on the radio-interferometric imaging of fields covering more than one antenna field of view, it was trivial to adapt MEM to provide one image satisfying many complementary datasets (Cornwell 1988). MEM has also handled the peculiarities of data from the HST easily (see *e.g.* Weir, this volume), allowing correction for cosmic ray hits, the spatially variant PSF, and the undersampling of the image plane.

HOW TO SOLVE IMAGING PROBLEMS

So we have seen from the last section that it pays to think carefully about imaging problems: you have to understand the measurement equation, the *a priori* information you may have and also how various deconvolution methods work. It is best to take a systematic approach to these points to ensure that all are covered adequately. With such an approach, it is really quite easy to come up with good ways of handling imaging problems. My recommended course of action is as follows:

Start with physics : write down the imaging equation in as complete a form as you think is appropriate for your application. For example, for imaging with radio interferometers we have that the visibility function is the Fourier transform of the sky brightness tapered by the antenna power reception pattern.

$$V(\mathbf{u},\mathbf{x}_p) = \int A(\mathbf{x}-\mathbf{x}_p)\, I(\mathbf{x})\, e^{j2\pi\mathbf{u}.\mathbf{x}}\, d\mathbf{x} + \epsilon \tag{4}$$

Repeat this exercise for all relevant data. In this example, another special case is a beam-switched single dish:

$$V(\mathbf{u}=0,\mathbf{x}_p) = \int [A(\mathbf{x}-\mathbf{x}_p) - A(\mathbf{x}-\mathbf{x}_p-\Delta\mathbf{x})\,]\, I(\mathbf{x})\, d\mathbf{x} \tag{5}$$

Choose type of solution : **Either** maximize entropy subject to all data constraints, **or** formulate via Bayes' theorem with other prior constraints on the sky brightness (see *e.g.* Gull 1989). The measurement equations then enter in via the probability distribution of the noise ϵ.

Worry about algorithms, convergence, uniqueness, robustness, etc : If you can afford it, the MemSys5 package is probably the way to go for an MEM algorithm. Failing that, there are a number of simple alternatives published (see *e.g.* Narayan & Nityananda 1986). For hard problems such as when the phase is unknown, additional heuristics may be needed to help even the best algorithm.

Simulate and test : This is tremendously important and if you discover anything worthwhile scientifically, you will have to do it so it is best to just accept it and do it from the very start.

Apply to real data with cross checks : simulations are only partly convincing as to the effectiveness of an algorithm. Ultimately, you will want to compare results from your new method to results from a well-understood method operating on better data.

Interpret algorithm physically : ask just what piece of information a method is really trying to uncover. For example, in the imaging of large fields, one is really just trying to synthesize the effect of having an array made of smaller antennas spaced closer together (see *e.g.* Cornwell, Holdaway & Uson 1992). This insight then aids in both understanding the method and in developing new variants.

The key ingredient in this list is not the numerical methods, which can be purchased or experts consulted, but instead the physical understanding of an instrument and of the sky. These are the insights all of us can bring to imaging problems.

REFERENCES

Clark, B. G. 1980, A&A, 83, 377

Cohen, J. 1991, AJ, 101, 734

Cornwell, T. J. 1983, A&A, 121, 281

Cornwell, T. J. & Evans, K. F. 1985, A&A, 143, 77

Cornwell, T. J. 1988, A&A, 202, 316

Cornwell, T. J. 1989, Science, 245, 263

Cornwell, T. J. & Braun, R. 1989, in Proc. NRAO Third Synthesis Imaging Summer School, Socorro 1988, R. A. Perley, F. R. Schwab, & A. H. Bridle, Astronomical Society of Pacific, 167

Cornwell, T. J., Holdaway, M., & Uson, J. M., 1992, A&A, (submitted)

Gull, S. G. 1989, in Maximum Entropy and Bayesian Methods, Cambridge 1988, J. Skilling, Dordrecht: Kluwer, 53

Gull, S. G. & Skilling, J. 1989, MemSys5 Quantified Maximum Entropy Users' Manual

Högbom, J. 1974, A&AS, 15, 417

Keel, W. C. 1991, PASP, 723

Marsh, K. A., & Richardson, J. M. 1987, A&A, 182, 174

Narayan, R., & Nityananda, R. 1986, ARA&A, 24, 127

Schwarz, U. J. 1978, A&A, 65, 345

Schwarz, U. J. 1979, in Proc. IAU Col. 49, Image Formation from Coherence Functions in Astronomy, C. van Schooneveld, D. Reidel, 261

Steer D. G., Dewdney, P. E., & Ito, M. R. 1984, A&A, 137, 159

Tan, S. 1986, MNRAS, 220, 971

AN ALTERNATIVE INTERPRETATION FOR THE PHYSICAL BASIS OF CLEAN

D.S. BRIGGS AND T.J.CORNWELL
National Radio Astronomy Observatory, Socorro, NM, 87801

INTRODUCTION

The original Högbom iterative sidelobe removal algorithm, CLEAN, is one of the oldest image processing algorithms still extant today. There are many variations of the algorithm, designed both to remedy certain known defects in its operation, and also to reduce the computation time required for its use. The essential nature of the algorithm, however, is unchanged. CLEAN is an integral step in more sophisticated calibration schemes. Unfortunately, while CLEAN has the advantage of great conceptual simplicity, it is not completely rooted in a sound theoretical basis.

The original CLEAN was presented as a pattern matching algorithm in the image plane. It solves the convolution equation $I_D = I * B$, where I_D is the measured "dirty" image, the "beam" B is some instrumental blurring function, and I is the true image to be determined. The correlation between the dirty image and the beam is calculated, and the point of maximum correlation is presumed to have some true intensity. (If the beam is sharply peaked, the calculation of the correlation is often omitted.) A delta function is added to the image model at that position, with an an amplitude equal to some fraction of the dirty image intensity at that point. The scaled delta function convolved with the beam is subtracted from the dirty image. The scaled delta function is a clean component, the fraction used is the loop gain. The process is iterated, building up an image model composed of delta functions. There is no completely satisfactory termination criterion, and this is usually estimated by hand. The clean components are convolved with some function, (the "clean beam"), to reduce the resolution of the clean image to approximately that of the dirty image, and the remaining residual dirty image is added back into the output clean image.

CLEAN COMPONENTS AS A VISIBILITY REPRESENTATION

A central assumption of the CLEAN algorithm is that the image is mostly empty, and can be represented by a modest number of clean components. Even in the continuous case, the precise meaning of these assumptions is not clear, and the situation is more complicated when one carefully considers the issue of discretization. In the case of interferometry in particular, the continuous equations are discretized

$$I_D(\ell_i) = (I_C * B)(\ell_i) = \int_{-\infty}^{\infty} I_C(\ell')B(\ell_i - \ell')\, d\ell' \qquad (i = 1, \ldots, M)$$

This is equivalent to the criterion that the transform of I_C match the visibility data. In general, these are ill posed integral equations, and very difficult to solve. To progress further, we must assume a form for $I_C(\ell')$. It can be any function that satisfies the convolution equation. If we choose a form for I_C whose transform V_C is complete in the visibility plane, then any possible measured visibility function may be represented in this form. There is no guarantee, however, that I may be representable in the form of I_C. It is easy to find cases where it cannot be so represented. This is the basis of the statement, *clean components model the visibility data, not the image data.* While other choices are possible, the usual basis function selected for I_C is delta functions spaced periodically on the sky. The transform of such a shifted delta function is a complex exponential. Clearly, the transform of

$$\sum_{j=-\infty}^{\infty} C_j \delta(\ell - j\Delta\ell) \qquad \text{is} \qquad \sum_{j=-\infty}^{\infty} C_j e^{-i2\pi j\Delta\ell}$$

We can simply recognize this as a Fourier sum. In the case where we have complete sampling in the visibility plane, and the clean components are defined as those matching the visibility data, the clean components can be found in closed form by the standard integrals for Fourier coefficients. For the particular case of critically sampled pixels, the CLEAN algorithm takes a particularly simple form, and it can be verified directly that the clean components produced by the iterative process are in fact the classical Fourier coefficients of the visibility. In the more general (and usual) case of incomplete sampling, U.J. Schwarz has shown the CLEAN algorithm to be a least squares fit of complex exponentials to the measured visibility data. If we take this result over to the completely sampled case, a well known theorem in Fourier analysis shows that minimizing the squared deviation of the visibility and fitted exponentials leads to the standard form for the Fourier coefficients.

Clearly, clean components are intimately related to Fourier coefficients in situations where we have complete data. Normal CLEAN sitations often approach this as a limit, so it is reasonable to use results of Fourier theory as an illustrative case in seeking understanding of CLEAN. In particular, note that the visibility function of a shifted delta function will have sharp discontinuities at the edge of the uv plane. The Fourier representation of a discontinuous function requires an infinite number of terms, and the coefficients drop only as $1/n$. The series representation will be the worst in regions of the discontinuity, and may well require negative values of the coefficients. While in practice the visibilities are not constrained at the edge of the uv plane, and hence the discontinuity is not sharp, we should not be surprised to see modest amounts of such "non physical behavior" in the actual clean components. Thus:

- Clean components do not necessarily obey physical constraints such as positivity or finite support.

- In the case of complete sampling, clean components are the complex Fourier series coefficients in the expansion of the visibility data on the interval $(-1/2\Delta\ell, 1/2\Delta\ell)$.

CONVERGENCE RATES IN SIMULATIONS

Simulations were run of an ideal VLA observation in 'C' configuration. The data was noiseless, free of calibration error, and complete over the full uv tracks. The simulation frequency was 1.5 GHz, and the declination of the sources were 60°. Uniform weighting led to a nearly circular synthesized beam with FWHM of 11.8″, and natural weighting led to a beam of size 19.4″ by 16.4″. Standard maps had pixel size of 5″ in both axes, and were of size 512 by 512. The model extended source was a cell centered pillbox convolved with a Gaussian of the same half width as the pillbox radius.

Results for single point source simulations:

- When the point source is located at a pixel center, CLEAN operates to a dynamic range comparable to the machine precision.
- When the point source is located halfway between pixels, and the cell size only somewhat larger than the Nyquist criterion, CLEAN is limited to a dynamic range of less than 1000 to 1.
- Significant oversampling of the clean image will alleviate this effect somewhat, but the dominant effect is still the placement of the point source. Five times Nyquist sampling on a shifted point source still resulted in a dynamic range of only 2000 to 1.
- The error pattern is a high frequency ringing, centered around the point source, and qualitatively similar to the error pattern of a calibration error.
- The residuals in the uv plane are the worst near the edges, and are much larger than conventional wisdom would suggest.
- Högbom and Clark CLEANs reach a common solution, though they differ slightly in how they approach it.

Results for extended source simulations:

- As with the point sources, the residuals are worst near the edges of the uv plane.
- The smaller disk tests converged at a rate consistent with that expected by merely counting pixels with significant emission.
- In the case of the larger disks the rate of convergence was prohibitively slow. The solutions were stopped far from convergence, and yet the final images were already showing clean stripes indicative of numerical instability.

REFERENCES

Högbom, J. A. 1974, A&AS, 15, 417–426

Schwarz, U. J. 1978, A&A, 65, 345–356

APPLICATIONS OF *CLEAN* TO DECONVOLUTION OF DIRECT IMAGES

WILLIAM C. KEEL
Department of Physics and Astronomy, Box 870324, University of Alabama, Tuscaloosa, AL 35487

THE CLEAN ALGORITHM.

The use of CLEAN has a long and only slightly checkered pedigree in aperture-synthesis mapping, for which it can be very powerful in removing spurious structures caused by inadequate sampling of the Fourier plane. I present here modifications for efficient work on direct images, where the noise behavior is quite different than for measurements made in the Fourier domain. Noise in the image domain contains all spatial frequencies up to pixel-to-pixel scales, which may produce features sharper than any real structure could be. Two schemes have been used to suppress the effect of such noise peaks - smoothing between iterations, and incorporating a noise model to assess the statistical significance of features in the observed data.

CLEAN is described by Högbom (1974), Clark (1980), and Cornwell's treatment in this volume. At each iteration, the peak (in absolute value) of the current image is found. A point source of some fraction (the *loop gain* ϵ) of the associated flux is subtracted from this image at the peak location, and added into a list of point sources (in the limit, one point source per pixel gives a continuous image). This process continues until some stopping criterion is satisfied, set by number of iterations, remaining peak flux, or some convergence behavior exists. The image of point sources is convolved with a *clean beam* whose dimensions set the final resolution, and added to the *residual image* left over after all these components have been subtracted from the input image.

The normal CLEAN implementation just described does not work well for direct images, because once the first spurious (noise) peak is found, the iterations lock up trying to fit this nonphysical peak with increasingly strong pairs of positive and negative point sources. No further information can be obtained on the image. In aperture-synthesis maps, there can be no such fine structure, since the noise is in the Fourier domain.

Two strategies have been useful in circumventing this "noise locking". For ground-based data, the core of the point-spread function (PSF) is nearly flat, so the resolution gain is modest (factor of 2) even for data of high quality. Any real peak extends over several pixels, and for appropriately well-sampled data, the image may be smoothed over a scale comparable to the FWHM before finding each peak, reducing the amplitude of a noise peak by the linear factor over which smoothing is done (Keel 1988, 1990). Experience suggests that a minimum sampling requirement is three pixels per FWHM (especially since sharpening can go only to about two pixels per FWHM unless subpixellation has been applied).

HST images more nearly resemble those produced by radio interferometers, with a sharp core and extended wings. These respond well to a version of CLEAN with a noise model, in which the *most significant*, rather than largest, peak is used at each iteration. The code was described, and a basic version reproduced by Keel (1991). This problem is mostly concerned with removing extended wings of the PSF (sidelobes in aperture-synthesis parlance) rather than achieving super-resolution.

A series of tests (presented by Keel 1991) on simulated data demonstrates that the results of such a "σ-CLEAN" deconvolution are reliable both as to structure and photometric scale. For this, an image of the luminous spiral NGC 309 was resampled to match its appearance with the HST WFC at $z = 0.6$. Poisson noise was added appropriate to the WFC at various exposures and background levels, and the results were deconvolved with this algorithm. The sensitivity of the results naturally varies with signal-to-noise ratio, but no spurious features are introduced, and surface photometry is still reliable.

Results on WFC images are in accord with these simulations. Inner spiral features were recovered on a 17th-magnitude spiral in a Science Assessment Observation (King et al. 1991). Tidal structures and companions have been found in radio galaxies at $V \approx 22$ in a program with R. Windhorst et al.

BELLS AND WHISTLES.

Several enhancements to CLEAN are possible to improve execution speed or final noise distribution.
(1) Hybrid iterations. The σ-CLEAN scheme can leave a residual map that contains a ghost of the original image, since that is how the noise map was produced. It is sometimes useful to apply additional passes of the standard CLEAN before terminating, to flatten the noise distribution.
(2) Preprocessing the image with a global smoothing, to remove large-scale smooth features so that CLEAN doesn't have to move mountains with a teaspoon. This is analogous to using (for example) the Lucy-Richardson method for a single pass, since this accounts for most of the flux at small spatial frequencies. Some tests have been done using a median filter much larger than the PSF peak, convolved with the PSF and subtracted before applying CLEAN. This saves an important fraction ($> 50\%$) of the CPU time for a similarly deep deconvolution.
(3) Adaptive loop gain. At low flux levels, the loop gain $\epsilon \approx 0.01$ that are needed for accurate reconstruction of bright features will force the routine to spend a lot of time making tiny corrections that are by themselves physically insignificant; 1% of a 2.5σ peak is too small to worry about in one iteration. Another 25% or so of the CPU time may be saved by increasing the loop gain once a specified significance threshold has been passed.

WHY USE CLEAN?

With several other deconvolution procedures on the market (Lucy-Ruchardson, maximum-entropy) what are CLEAN's advantages, and are there special applications for which it is especially suitable? Unique attributes are:

(1) The code is quite simple and transparent, so that application-specific features may be added easily. These might include restrictions on where flux can be found, or rejection of discrepant pixels such as cosmic-ray hits.
(2) CLEAN operates only locally on the image. A spatially-variable PSF may be incorporated as long as its variation is well understood, at the expense of added memory use (but not execution time).
(3) The residual image is carried into the result, so that photon noise has familiar behavior and may be used to assess the significance of features.
(4) The technique is photon-conserving, so that meaningful absolute photometry and surface photometry can be performed after deconvolution.

ACKNOWLEDGMENTS

This work was partially supported by Cray Research and EPSCoR grant RII-8996152. CRAY time for some of these computations was provided by the Alabama Supercomputer Network. David Block provided the NGC 309 test image. Rogier Windhorst and Tony Ferro suggested adaptive loop gain and hybrid iterations.

REFERENCES

Högbom, J. 1974, A&AS, 15, 417

Clark, B. G. 1980, A&A, 83, 377

King, I. R., et al. 1991, AJ, 102, 1553

Keel, W. C. 1988, ApJ, 329, 532

Keel, W. C. 1990, AJ, 100, 356

Keel, W. C. 1991, PASP, 103, 723

RESTORATION OF IMAGES AND SPECTRA FROM THE HUBBLE SPACE TELESCOPE

RICHARD L. WHITE
Space Telescope Science Institute, Baltimore, MD, and Joint Institute for Laboratory Astrophysics, University of Colorado, Boulder CO

ABSTRACT The spherical aberration of the primary mirror of the Hubble Space Telescope has seriously degraded the resolution of images and spectra from the instruments onboard the satellite. However, computer restoration techniques make it possible to restore some of the lost spatial and spectral resolution. This paper describes various aspects of the Hubble image restoration problem, shows results for some real data, and discusses outstanding problems.

INTRODUCTION

The very first image taken a few weeks after the launch of the Hubble Space Telescope (HST) showed evidence for spherical aberration in the HST optics. The results of the next month of testing proved conclusively that the HST primary mirror has about 1/2 wave RMS of spherical aberration ($\lambda = 5000$ Å). Burrows et al. (1991) give a detailed description of the HST spherical aberration.

It was quickly recognized that image restoration methods should allow the removal of much of the blurring. The point-spread function (PSF) consists of a sharply focussed core that contains about 15% of the light surrounded by a very extended halo. Even though relatively little of the light is well-focussed, the surface brightness of the core is much greater than that of the halo so that the core is readily visible in typical astronomical images containing clusters of point sources (stars). There is consequently considerable high-frequency information still present in the blurry images. Experiments have shown that restored images can be computed successfully using many different methods even for images having relatively low signal-to-noise (S/N) ratios. The proceedings of a workshop held at the Space Telescope Science Institute in 1991 August (White & Allen 1991) include the results of early experiments in image restoration as well as suggestions for future work.

This paper describes HST's spherical aberration and the resulting HST point-spread function. The characteristics of the HST instruments that are relevant to image restoration are discussed. A variety of image restoration methods are discussed, and results are presented for some typical HST data. Finally, some of the outstanding problems in HST image restoration are discussed.

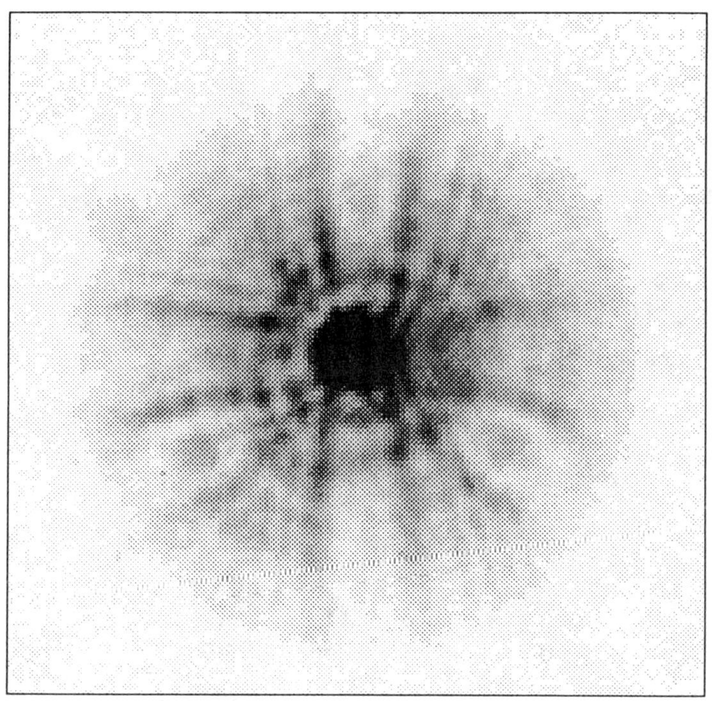

FIGURE I Grey scale representation of a bright star obtained on 15 July 1990 with the Planetary Camera. The field of view of this 128 pixels square sub-image is 5.5 × 5.5 arcsec.

THE PROBLEM

The HST primary mirror is too flat. The difference Δ between the designed mirror surface and the actual surface varies as $\Delta = 2.3\mu\text{m}(r/R)^4$, where r is the radial distance from the center of the mirror and $R = 1.2$ m is the radius of the mirror. This error leads to an optical path length error twice as large; the minimum resulting wavefront RMS error is 0.5 waves at $\lambda = 5000$ Å. This error has been determined independently from measurements in orbit and from the flawed ground test equipment that was used to figure the mirror. The measurements currently differ by 10%, but the difference seems to be due to aberrations within the Wide Field/Planetary Camera used to make the measurments, so there is relatively little uncertainty about the nature of problem.

As a consequence of this aberration, light reflected from the center of the HST primary mirror ("paraxial" rays) does not focus at the same point as light reflected from the edge of the mirror ("marginal" rays). The marginal focus is about 4 cm beyond the paraxial focus. There is no single best focus for this optical system: as the focus is changed between the paraxial and marginal positions, there is always a annular zone on the mirror for which the rays are focused, but rays inside or outside that zone are unfocused. The resulting images

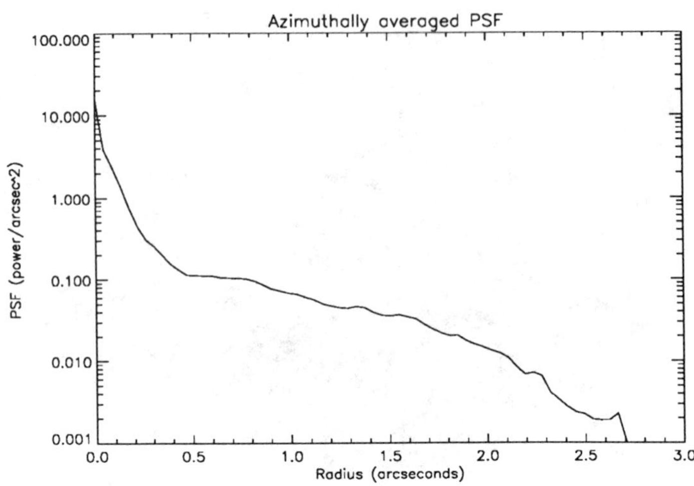

FIGURE II The intensity profile of the star in Figure I.

have a small, bright core surrounded by a halo of unfocused light. The adopted focus position was chosen to maximize the fraction of light in the core of the images; this leads to a focus position that is about 1 cm from the paraxial focus.

THE HST POINT-SPREAD FUNCTION

Figure I shows a stellar image obtained close to the best focus setting with the Planetary Camera through a filter with a central wavelength of 5470 Å. The size of the image core is close to the diffraction limit expected for HST (about 0.1″ radius), but contains only about 15% of the energy. The core is surrounded by a halo that contains the remaining 85% of the energy and extends out to a diameter of about 4″ from geometrical optics, with light scattered beyond 4″ by diffraction effects. The halo is composed primarily of unfocused marginal rays; shadows of obscurations in the pupil such as the four secondary mirror support arms and the three support pads for the primary mirrors are visible in the halo. The fine structure in the halo results from diffraction and varies with wavelength. The point-spread function (PSF) is strongly focus-dependent, and it varies with field position in the Wide Field/Planetary Camera (WFPC) because of the camera's internal vignetting.

Figures II and III show intensity and encircled-energy profile plots for the stellar image in Figure I. As can be seen, 70% of the energy is contained within a radius of about 1.25″. (For comparison, the design requirements for the HST optical system called for 70% encircled energy within 0.10″ radius.)

It may be useful to think of HST images as being the sum of *two* images of the same astronomical object: a sharp, well-focused image containing 15% of the light, and a blurry image containing 85% of the light. The sharp image may sometimes be difficult to see against the bright background of the blurry image, but the sharp image is always detectable if the signal-to-noise ratio of the data

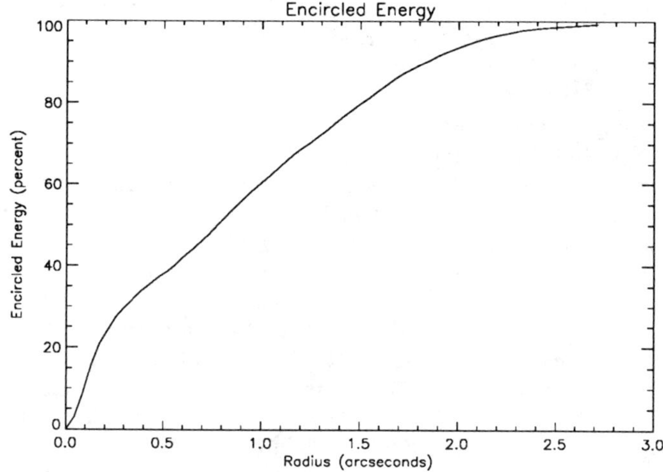

FIGURE III The encircled energy profile of the star in Figure I.

is sufficient.

The fundamental loss of HST imaging science as a result of the spherical aberration is not a loss of resolution; rather, it is a loss of the ability to detect faint objects, especially in crowded fields. The detection of the faintest sources in astronomical images is limited by the background light from the sky. HST was to have a tremendous advantage over the ground for faint sources because its small images meant that only a very small amount of the sky light need be included when measuring the brightness of faint stars. With the current images, even isolated stars are harder to detect because only 15% of the light is in the PSF core; a naive attempt to sum the extended light in the halo to help detect faint objects actually *reduces* the S/N because much more sky light is also included. The situation is much worse in crowded fields, where the overlapping halos of many sources create a much brighter background against which faint objects are even harder to detect.

The end result of all this is that with current HST images the faintest detectable objects are 6 to 10 times brighter than with a perfect HST. No image processing technique can possibly restore these faint objects because the S/N is just not sufficient.

The S/N for brighter stars will obviously also be reduced as a result of the aberration. However, for brighter objects it is possible to increase the exposure time by a factor of a few to recover the lost S/N. For the faintest objects, though, the exposure times were already as long as is practical, so increasing the exposure time is not an option.

HST CAMERAS AND SPECTROGRAPHS

An ideal detector for HST would have high efficiency from the ultraviolet to the infrared, perfect linearity of response over a large dynamic range of illumination levels, and no dark noise or readout noise. In most astronomical observations we are starved for light, so even a perfect detector will be limited by the Poisson noise that comes from counting individual photons. Of course perfection is unattainable, but the instruments on HST in many cases do produce data that is ultimately limited by the photon-counting statistics.

The Wide Field/Planetary Camera (WFPC) CCDs have very good quantum efficiency, especially in the red, but do have significant readout noise (RMS $\simeq$ 15 counts/pixel). Consequently, the WFPC is limited by counting statistics only for exposure levels $>$ 250 counts/pixel. The photometric response of the WFPC is quite linear until its A/D converter saturates at about 30,000 detected photons per pixel. The flat-field response of the WFPC CCDs can be calibrated by observing roughly uniform sources such as the sun-illuminated Earth but will nonetheless be uncertain at about the 1–2% level. This will limit the accuracy of WFPC measurements for brighter sources.

The WFPC has several peculiarities that must be taken into account in restoration attempts. First, in its Wide Field configuration it has pixels that are almost twice as big as the core of the PSF, so the image is undersampled. This leads to some nasty properties such as stars that change rather dramatically in appearance when they are shifted by a fraction of a pixel. Second, the PSF in the WFPC changes with position in the field of view of the camera due to variable obscurations in the camera's internal optics. Finally, the CCDs are subject to hits by high energy particles (cosmic rays) that can generate bright pixels or trails in the image. These cosmic rays must be removed from the image either before or during the restoration process. Despite these various shortcomings, the WFPC represents an excellent compromise between the conflicting requirements for a camera and will be the most heavily used camera on HST.

The Faint Object Camera (FOC) has true photon-counting detectors with low dark count rates and no readout noise. The FOC detectors are considerably less sensitive than the WFPC CCDs in the red, but are superior in the ultraviolet. The main drawback to the FOC is its very limited dynamic range. The FOC detector response becomes non-linear at count rates of only $\sim$0.5 counts/sec/pixel. This means that even faint astronomical objects will often have nearby "spoiler" stars that are bright enough to be saturated in the FOC. This greatly complicates the image restoration problem for FOC images.

Another consequence of the limited dynamic range of the FOC is that it will be difficult to collect high signal-to-noise data. Exposures longer than 1 hour will be rare with HST because the HST orbital period is only 90 minutes. In a 60 minute exposure, the signal-to-noise ratio for a source that is just saturated (0.5 counts/sec) will be about 40:1; typical observations will have a lower S/N ratio. It is hard to measure the flat-field response of the FOC accurately because of the time required for high S/N observations.

The small field of view of the FOC also complicates image restoration. The size of the FOC field-of-view ranges from 44 arcseconds square to 7.3 arcseconds square depending on the instrument configuration. Since the PSF has significant light over a diameter of at least 10 arcsec, the PSF can be comparable in size to or even bigger than the field of view. Careful treatment of both edge effects and contamination by objects just off the edge of the field is of great importance.

Another difficulty with the FOC is that it suffers from significant geometrical distortion. Although the FOC does not (in most configurations) have any internal obscuration and so does not have a spatially variant PSF, the geometrical distortion effectively makes the PSF vary with field position anyway.

The HST spectrographs (the Faint Object Spectrograph and the High Resolution Spectrograph) are similar in design. They both have quiet, photon-counting detectors. For most purposes they can both be treated as essentially ideal detectors that are entirely limited by the statistics of photon counting.

Several different focal plane apertures are available for the spectrographs. The astronomer has a choice between using a small aperture that passes little of the light but produces spectra with resolution undegraded by the spherical aberration, or using a large aperture that passes most of the light but leads to degraded spectral resolution. In the latter case one can attempt to restore higher resolution through deconvolution methods. For modest resolution this approach can be successful; however, if one needs the highest resolution possible it is better to use the small apertures and pay the penalty of lost light (Gilliland 1991).

The spectrographs provide an excellent proving ground for image processing methods. Not only are numerical experiments on one-dimensional spectra much more computationally tractable than experiments on large images, but it is also possible to determine the "ground truth" for blurry, large-aperture spectra by taking spectra of the same objects through the small apertures.

RESTORATION METHODS AND RESULTS

There are a number of important features of HST data that need to be incorporated into image deconvolution methods adapted to HST images and spectra:

(1) The statistics of the data are dominated by the Poisson nature of photon detection. The most important effect is that the variance of the data depends on the data, $\sigma^2 \propto N$; the non-Gaussian distribution of the data is of secondary importance for the WFPC because the WFPC readout noise dominates the counting noise for low count levels, but the skewed Poisson distribution can be important in modelling the sky background for the FOC.

(2) HST data is generally subjected to stringent quantitative analysis in order to extract the scientific information from an observation. Photometric fidelity for both point sources and extended sources is consequently a highly desirable property for restoration methods.

(3) For the WFPC, restoration methods should allow for the spatially varying PSF, for undersampling of the image, and for contamination of the image by cosmic ray hits.

(4) For the FOC, the methods should allow for the presence of saturated pixels and for edge effects due to the large size of PSFs compared to the FOC field of view.

This is not, of course, an exhaustive list; however, an image restoration method that takes proper account of the above points will be suitable for the great majority of the images and spectra from HST.

Many methods have been applied to HST data, and one of the first results is that even the simplest techniques give quite good results. Hanisch (1991) discusses the use of simple linear techniques (Fourier quotient, Wiener filter) on HST data and shows that while these methods do generate some ringing artifacts around stars, they give remarkably good results. The major advantage of the linear methods is that they are very fast — they require only a few seconds of compute time on a modern high-speed workstation even for large images. These techniques also have good photometric characteristics: experiments have shown that good photometry can be done on the deconvolved images.

The disadvantages of the Fourier quotient and Wiener filter methods lie in the inflexibility that results from their simplicity. One cannot readily allow for Poisson statistics; the PSF cannot vary with position; there is no obvious way to allow for the undersampling in the WFPC; and pixels hit by cosmic rays and saturated pixels must have their values editted to avoid objectionable artifacts. To overcome these limitations it is necessary to go to a more sophisticated, iterative method.

The Richardson-Lucy (R-L) method (Lucy 1974) has been shown to converge to the maximum likelihood solution when the data has Poisson statistics. Snyder (1991) has developed simple modifications of the method that also allow the WFPC readout noise to be included. The one major disadvantage of the R-L method compared with the simple Fourier techniques is that R-L requires 10–100 times as much compute time to reach a solution. However, R-L has numerous advantages: it incorporates a much more complete noise model, it enforces positivity in the image, it permits a spatially varying PSF, and it can handle incomplete data (*i.e.*, bad pixels can be flagged as such — it is not necessary to create values for pixels hit by cosmic rays.) The R-L method also has the great advantage that it draws a clean distinction between object space and data space. For example, it is possible to construct the image on a finer grid than the data (White 1991), which is a promising approach for handling the undersampling in the WFPC. Experiments have shown that the R-L method can, like the Fourier methods, produce restorations with quite linear photometric properties.

Two examples of HST data processed with the R-L method (modified as suggested by Snyder [1991] for the WFPC readout noise) are shown in Figures IV and V. Figure IV(a) is a 90×90 pixel ($9 \times 9''$) image of a dense star cluster taken with the Wide Field Camera. It is displayed with a logarithmic gray scale to emphasize the fainter structures. Figure IV(b) shows the R-L result using 2×2 oversampling. Figure V(a) is a 256×256 pixel image of Saturn, and Figure V(b) is the R-L deconvolution. In both cases the R-L method does a good job of de-blurring the images without creating some of the objectionable artifacts generated by the Fourier techniques.

The major disadvantage of the R-L method is that it does not incorporate any regularization other than that generated by starting with a smooth initial guess. If one tries to go "too far" with this method, the noise in the data can be amplified and the images become very rough. If, on the other hand, one halts the iteration before it converges (either using some objective criterion or simply because the image "looks good"), the image does not fit the data uniformly and the restoration loses its linear photometric properties. Snyder (1991) and Llacer & Nunéz (1991) have discussed some approaches for regularizing the Richardson-Lucy iteration, but both of their methods have some drawbacks. Regularization will be discussed further below.

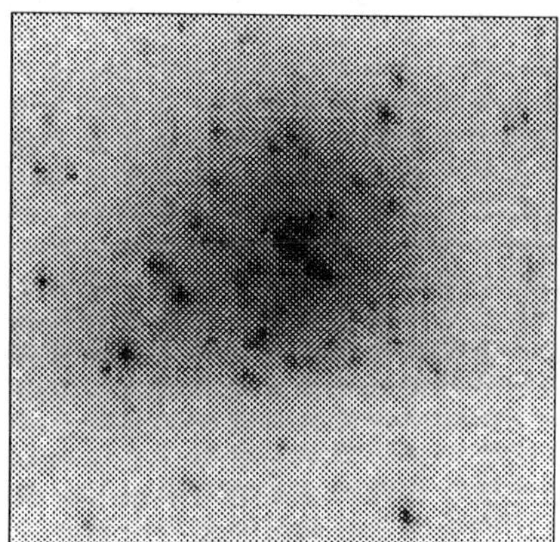

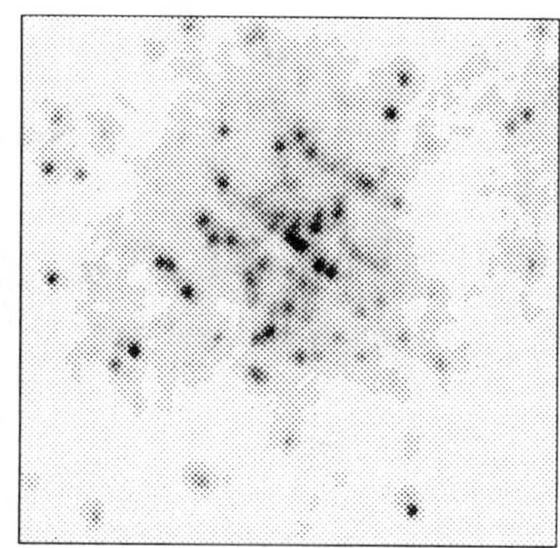

FIGURE IV Raw Wide Field Camera image of the R136 cluster in the Large Magellanic Cloud, and image restored with Richardson-Lucy method using 2×2 oversampling.

FIGURE V Raw Wide Field Camera image of Saturn, and image restored with Richardson-Lucy method.

The maximum entropy method (MEM) has also been applied with success to HST data (Horne 1991; Weir & Djorgovski 1991). MEM is in some ways the ultimate deconvolution method: in addition to sharing most of the advantages of the R-L method, it includes complete regularization. MEM gives visually impressive results, but is so highly non-linear in its photometric characteristics that MEM images are not obviously of great use for quantitative analysis. What is needed is a version of MEM that adaptively removes the entropy regularizing condition where there is evidence for real structure in the image so that one can derive unbiased estimates of brightnesses (Weir, private communication). MEM and similar methods are highly computationally intensive, however, so it is doubtful that such methods can be applied on a routine basis to HST data unless much faster approaches can be discovered.

OUTSTANDING PROBLEMS

The first obstacle to be crossed when attempting to restore an HST image is where to get a good PSF for the observation. The variation of the PSF with time, wavelength, and position makes it impossible to maintain a library of observed PSFs for all filters, positions, etc. Furthermore, observed PSFs have noise and sampling problems that may make them unsuitable anyway. However, at present getting good PSFs from optical models is also nontrivial, especially as one goes into the ultraviolet. The derivation of a decent PSF for any observations is at present the biggest practical problem for HST image restoration. Possibly a "blind deconvolution" approach, in which the PSF is derived along with the model image, will be fruitful for HST images if one can incorporate an optical model to insure that the derived PSF is consistent with the known optical properties of HST.

Once the PSF is known, the biggest roadblock to accurate restorations of HST images (and other images with Poisson noise) is the need for *efficient adaptive regularization*. We have in hand methods that incorporate good models of the Poisson noise and that can handle the peculiar characteristics of HST data (variable PSF, bad pixels, undersampling, etc.) The Richardson-Lucy method can generate deconvolved images with good photometric fidelity if one uses the method of sieves for regularization (Snyder, Miller, Thomas & Politte 1987), but considerable smoothing is required to control the ringing and roughness of the unregularized R-L method. On the other hand, the maximum entropy method and its variants can generate beautiful, smooth images that unfortunately are grossly biased in the vicinity of both faint and bright stars.

What is wanted is an adaptive regularization method that applies the minimum amount of smoothing needed to control the amplification of the Poisson noise. The regularization should also adapt to features in the image so that it does not smooth across edges and in the vicinity of stars.

I am aware of some promising approaches to the adaptive regularization problem. For example, Jeng & Woods (1990) use simulated annealing to find edges in the image across which no smoothing constraint should be applied. Their method is very intriguing and appears to have the desirable features we are looking for; however, simulated annealing is notoriously slow in convergence for most optimization problems and may not be practical for general use.

The computational efficiency of the restoration methods is an important issue for HST data because the images are large and arrive from the spacecraft in copious quantities. Methods with compute times that increase much faster than the number of pixels are simply not practical for most purposes.

One approach I am pursuing is the use of wavelet transforms for adaptive filtering. Wavelet transforms combine computational efficiency with great flexibility, so they seem an obvious candidate for efficient adaptive regularization.

There are other outstanding problems: Is there a computationally efficient way to incorporate color and position-dependent PSFs? What are the limits to crowded field photometry with these variable PSFs? Regardless of the questions still to be answered about the restoration of HST images, it is clear that image restoration will play an important role in the analysis of HST images.

ACKNOWLEDGMENTS

Thanks are due to Hans-Martin Adorf, Ron Gilliland, Bob Hanisch, John Holtzman, Keith Horne, Ivan King, Tod Lauer, Don Snyder, and Nick Weir for helpful discussions during my work on this problem.

REFERENCES

Burrows, C. J., Holtzman, J. A., Faber, S. M., Bely, P. Y., Hasan, H., Lynds, C. R., & Schroeder, D. 1991, ApJ, in press

Gilliland, R. 1991, in White and Allen (1991)

Hanisch, R. 1991, in White and Allen (1991)

Horne, K. 1991, in White and Allen (1991)

Jeng, F.-C., & Woods, J. W. 1990, IEEE Trans. Information Theory, 36, 94

Llacer, J., & Nunéz, J. 1991, in White and Allen (1991)

Lucy, L. B. 1974, AJ, **79**, 745

Nisenson, P. 1991, in White and Allen (1991)

Snyder, D. L. 1991, in White and Allen (1991)

Snyder, D. L., Miller, M. I., Thomas, L. J., Jr., & Politte, D. G. 1987, IEEE Trans. Medical Imaging, MI-6, 228

Weir, N., & Djorgovski, S. 1991, in White and Allen (1991)

White, R. L., & Allen, R. J., eds. 1991, Restoration of HST Images and Spectra, STScI Workshop Proceedings

White, R. L. 1991, in White and Allen (1991)

A MULTI-CHANNEL METHOD OF MAXIMUM ENTROPY IMAGE RESTORATION

NICHOLAS WEIR
Department of Astronomy, Caltech 105-24, Pasadena, CA 91125

ABSTRACT We describe an extension to the maximum entropy (MaxEnt) method of image restoration which may improve results for images with multiple scales of pixel-to-pixel correlation. The technique is a straightforward generalization of modern MaxEnt methods, whereby one solves not just for a single restored image, but for multiple images with varying degrees of spatial correlation.

INTRODUCTION

An outstanding problem in the field of image restoration is how to properly restore images containing both sharp and extended structure (e.g., point sources coincident with extended galaxies). Most methods to date demand a frustrating trade-off between high-resolution reconstruction of the sharp features and lack of artifacts in the extended structure. Even when opting for higher resolution, however, the results are always less than optimal if the structure lies atop a non-zero background. In such instances, a non-negativity constraint no longer prevents ringing in the image. Ringing is not only a troublesome restoration artifact, but significantly reduces resolution.

THE MAXIMUM ENTROPY FRAMEWORK

The present work is based upon the MaxEnt framework developed by Gull (1989) and Skilling (1989) and more fully described in Gull & Skilling (1989) and Weir (1991). In these references, the derivation of the method is presented in terms of Bayesian probability theory, and we refer the reader to these sources for detailed treatment of the theory and mathematics.

The basic imaging equation is

$$d_j = (i * p)_j + n_j,$$

where "$*$" is the convolution operator, d is the observed image (data), i is the original (restored) image, p is the point spread function, and n_j is the noise associated with pixel j. According to Bayes' theorem, we know

$$\Pr(i|d,m) = \frac{\Pr(d|i,m)\Pr(i|m)}{\Pr(d|m)}, \qquad (1)$$

where m represents the image model parameters, $\Pr(i|d,m)$ is the posterior probability of the restored image given the data and the model, $\Pr(d|i,m)$ is the likelihood function, $\Pr(i|m)$ is the prior probability of the restored image given the model, and $\Pr(d|m)$ is a normalizing term denoted the Evidence for reasons that will become apparent below.

The characteristic feature of MaxEnt methods of image restoration is the use of a prior of the form $\Pr(i|m) \propto \exp(\alpha\mathcal{S}(i))$, where $\mathcal{S}(i)$ is an entropic function of i and α, the so-called regularization constant, is a free model parameter. The entropy expression we employ is (Skilling 1989)

$$\mathcal{S}(i) = \sum_{j=1}^{L}(i_j - \mu_j - i_j \log(i_j/\mu_j)),$$

where L is the number of pixels in i and and μ_j is the measure assigned to pixel j. The measure, sometimes referred to as the image model or default image in the MaxEnt literature, is generally set to some constant value.

Approximating the errors as Gaussian distributed, the likelihood function is of the form $\Pr(d|i,m) \propto \exp(-\mathcal{L}(i))$, where $\mathcal{L}(i) = \chi^2(i)/2$. Therefore, by combining the likelihood and prior terms on the right side of equation 1, we find that the image with the maximum *a posteriori* probability (the MAP solution) is that which maximizes the objective function

$$\mathcal{F}(i) = \alpha\mathcal{S}(i) - \mathcal{L}(i). \tag{2}$$

The MAP i is generally referred to as the MaxEnt solution.

The traditional MaxEnt method simply involves maximizing $\mathcal{F}$ subject to a choice of α. Historically, the chosen value has been that which makes the χ^2 misfit of the MAP solution approximately equal to the number of pixels, N, in the data image. While this is an admittedly *ad hoc* criterion, the method generally provides reasonable solutions, meaning moderately enhanced resolution and a fair amount of noise smoothing. These results are achieved, however, at the expense of very biased residuals and less than optimal resolution. Figure *1c* depicts a traditional MaxEnt restoration of the simulated data set shown in Figures *1a* and *1b*. We will use this simple 1-D example to illustrate the differences between the various MaxEnt approaches we describe.

An alternative method of choosing α, one which is fully consistent with the probabilistic foundations of MaxEnt, was proposed by Skilling (1989) and Gull (1989). As α is a part of the model m, they proposed applying Bayes' rule to find the most probable value of this regularization constant, *ie.* to find that α which maximizes

$$\Pr(m|d) \propto \Pr(d|m)\Pr(m).$$

Assuming equal prior probability of the models, we see that maximizing the term on the left side is equivalent to maximizing $\Pr(d|m)$, previously named the Evidence, which appears in the denominator of equation 1. One may estimate the value of this normalizing term for any given choice of m (or α), providing a consistent means of comparing models with different numbers and types of free parameters.

The Bayesian choice for α will always be less than or equal to the historic $\chi^2 = N$ choice, lending less weight to the regularizing term in the objective function, $\mathcal{F}$. The likelihood term is of consequently more importance, forcing a better fit to the data and less biased residuals (see Figure *Id*). The unfortunate trade-off is an unacceptable amount of high-frequency artifacts in the restoration where extended structure should appear. This new choice for α highlights a severe deficiency in the MaxEnt image model: it assumes no preference for pixel to pixel correlation. The occasional "smoothness" of historic MaxEnt reconstructions is achieved only indirectly, by over-regularizing the solution and systematically underfitting the data.

Gull (1989) and Charter (1990) formally incorporated preference for spatial correlation in the MaxEnt image model with the intrinsic correlation function (ICF). In this model, the restored image is defined as the convolution of some "hidden" image, h, with the ICF, C,

$$i = h * C,$$

where C is some blurring function, such as a Gaussian. The MaxEnt solution is derived by maximizing the objective function with respect to h, and h is substituted for i in the entropy expression. Figure *Ie* represents the MaxEnt solution obtained by using an ICF of Gaussian form with a standard deviation of one pixel.

The ICF model limits the set of all possible restorations to those with some minimal degree of spatial correlation. By applying different ICFs, one can obtain any desired degree of uniform smoothing in the restoration. Unfortunately, there are generally features with vastly different degrees of intrinsic spatial correlation in an image, and no single choice for C will be appropriate to them all.

THE MULTI-CHANNEL MODEL

A possible extension to the ICF model is to allow for multiple hidden image/ICF combination, or channels. In this framework, the restored image i is represented by

$$i = \sum_{k=1}^{M} i_k = \sum_{k=1}^{M} h_k * C_k,$$

where M is the total number of channels. The final restored image is simply the sum of a set of restored image channels, i_k, each of which has an associated spatial scale defined by its ICF, C_k. Initial results from an implementation of this multiple channel model appear in Weir (1991).

The multi-channel MaxEnt solution is still derived by maximizing an objective function similar to equation 2, except the optimization is with respect to the h_k's, and one uses the revised entropy expression,

$$\mathcal{S} = \sum_{k=1}^{M} \sum_{j=1}^{L} (s_{k,j} - \mu_{k,j} - s_{k,j} \log(s_{k,j}/\mu_{k,j})),$$

where the sum is over all M channels and L image pixels. The s_k terms in the expression are weighted versions of the hidden images, $s_k = h_k w_k$, where w_k is an "entropic weight" assigned to channel k. By varying this weight, one varies the degree to which each channel is weighted in the regularizing term, $\mathcal{S}$. To explicitly incorporate a preference for smoothness in the image, one can assign the w_k's so as to bias flux into channels with larger degrees of intrinsic spatial correlation. Figure *1f* illustrates how a relatively high-resolution restoration, with only moderate ringing, was achieved using this method. We particularly note the residuals and how much more accurately this restoration fits the data than the historic MaxEnt solution.

An unaddressed problem is how to appropriately set all of the newly introduced parameters: M, C_k, w_k, and μ_k. Thus far we have achieved rather encouraging results by applying simple heuristic guidelines. Most of these suggestions appear in Weir (1991). In particular, we have experimented with different means of scaling the w_k's, but have achieved the most satisfactory results with $w_k \propto \sigma_k^{-n}$, where σ_k is the linear scale of C_k and $n = 1$ or 2. With future research, we hope to be able to objectively determine the most probable number and value of the these parameters for any given data set by maximizing the Evidence. At this time, however, we have only been able to successfully estimate it for simple 1-D examples, such as the one presented here.

This multi-channel technique is currently implemented in three image reconstruction systems utilizing the MemSys5 package of subroutines by Maximum Entropy Data Consultants, Ltd.: MEM, for deconvolution of direct image data (Weir 1991); and independent systems by Weir (1992) and Bontekoe & Koper (1992) for producing high-resolution images from IRAS scan data. While the results from none of these implementations indicate that this model is the final answer to properly reconstructing images with multiple correlation scales, they do suggest that the method has significant potential in a variety of applications.

ACKNOWLEDGMENTS

I am grateful to D. MacKay and S. Djorgovski for useful discussions and assistance. This material is based upon work supported under a National Science Foundation Graduate Fellowship and NASA HST grants to S. Djorgovski.

REFERENCES

Bontekoe, T. R., & Koper, E. 1992, this volume

Charter, M. K. 1990, in Maximum Entropy and Bayesian Methods, Dartmouth College 1989, P. F. Fougère, Dordrecht: Kluwer, 325

Gull, S. G. 1989, in Maximum Entropy and Bayesian Methods, Cambridge 1988, J. Skilling, Dordrecht: Kluwer, 53

Gull, S. G., & Skilling, J. 1989, MemSys5 Quantified Maximum Entropy Users' Manual

Skilling, J. 1989, in Maximum Entropy and Bayesian Methods, Cambridge 1988, J. Skilling, Dordrecht: Kluwer, 45

Weir, N. 1991, in Proceedings of the ESO/ST-ECF Data Analysis Workshop, Garching 1991, P. Grosbol, in press

Weir, N. 1992, in preparation

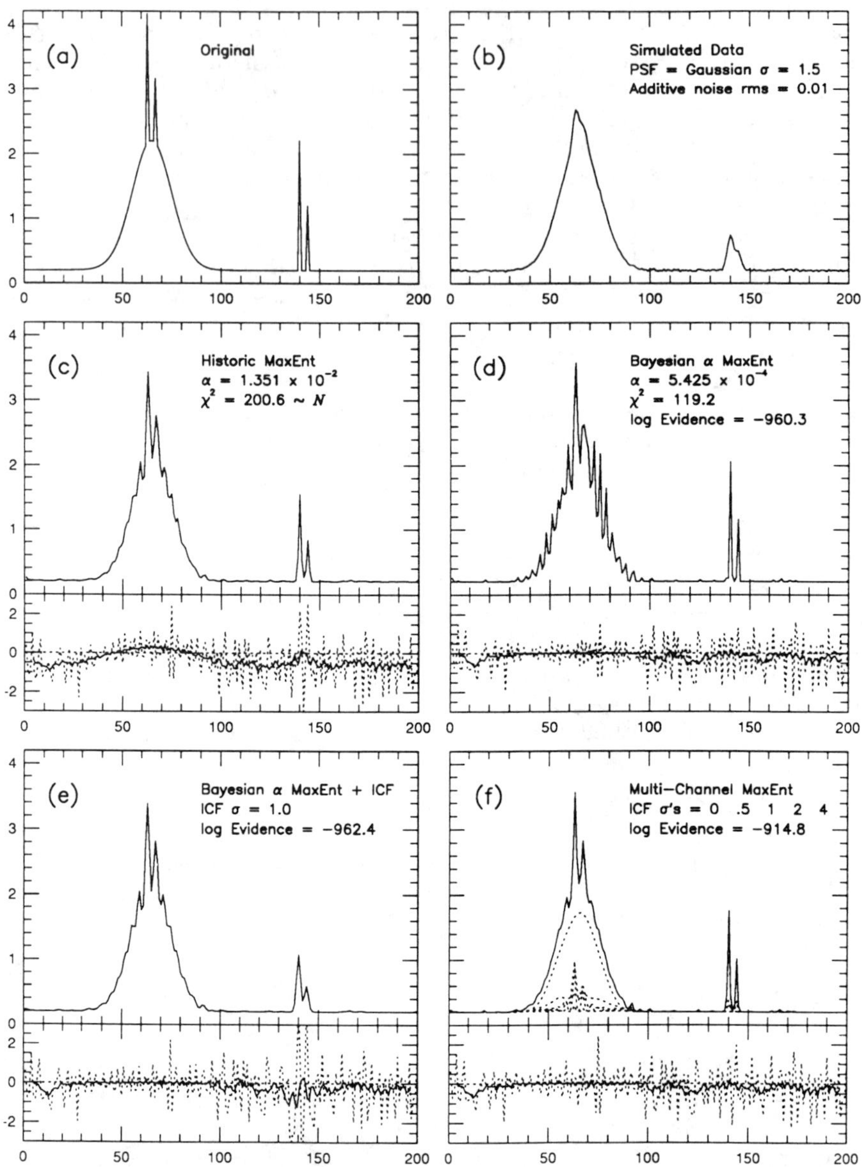

FIGURE I (*a*) The original, uncorrupted 1-D image; (*b*) a noisy, blurred version of (*a*) used for the following restorations; (*c*) MaxEnt solution and normalized residuals using the historic, $\chi^2 = N$ choice of α, with a solid line indicating a running mean of the residuals; (*d*) MaxEnt solution using the Bayesian choice of α; (*e*) MaxEnt solution using a single Gaussian ICF of $\sigma = 1$ pixel; (*f*) Multi-Channel MaxEnt solution using five channels and entropic weights scaled as $w_k \propto \sigma^{-1}$, with the dotted lines indicating the contributions from individual image channels.

PITFALLS IN IMAGE RECONSTRUCTION

TJ. ROMKE BONTEKOE
European Space Agency, ESTEC-SA, Postbus 299, 2200 AG Noordwijk, The Netherlands

ENRICO KOPER
Sterrewacht Leiden, Postbus 9513, 2300 RA Leiden, The Netherlands

ABSTRACT We report on advanced use of MemSys5, including multi-channel reconstruction, random samples from the maximum entropy probability distribution, and their use in constructing high-resolution maps, error maps, and signal-to-noise maps. We also report on problems we encountered in this process, specifically, the treatment of background subtraction or flat fielding. Since similar problems have been encountered in other disciplines, they seem to be of a general nature. Careful data massaging and parameter control nevertheless yield an astronomically credible image of the galaxy M101, showing spiral structure at 60 μm, and its corresponding error and signal-to-noise maps which quantify this spiral structure as at least a 2–3 σ detection.

INTRODUCTION

The InfraRed Astronomical Satellite (IRAS) was the first cryogenically cooled satellite using liquid helium, which flew in 1983. Only recently it has become possible to reconstruct high-resolution images from the survey data. In an earlier paper a high-resolution map of M51 at 60 μm has been obtained by use of MemSys3 (Bontekoe et al. 1991).

A difference between MemSys3 and MemSys5 is the introduction of spatial correlations in images, through an Intrinsic Correlation Function (ICF), which usually is a two-dimensional gaussian giving the desired spatial correlation length scale. The image reconstruction problem is then defined on three different vector spaces: hidden space, visible space, and data space (see Gull 1989; Skilling 1989; Charter 1990; and Weir in this volume).

A requirement of modern maximum entropy techniques is that the entropy of an image must be computed with reference to a default image in hidden space. Usually a flat default suffices, but Gull (1989) has investigated various advanced default models.

The most important improvement in MemSys5 is the possibility to draw random samples from its posterior probability density, each sample being a whole image. Statistics then can be applied: the average forms a high resolution map, the standard deviation an error map, etc. This opens a whole new field for

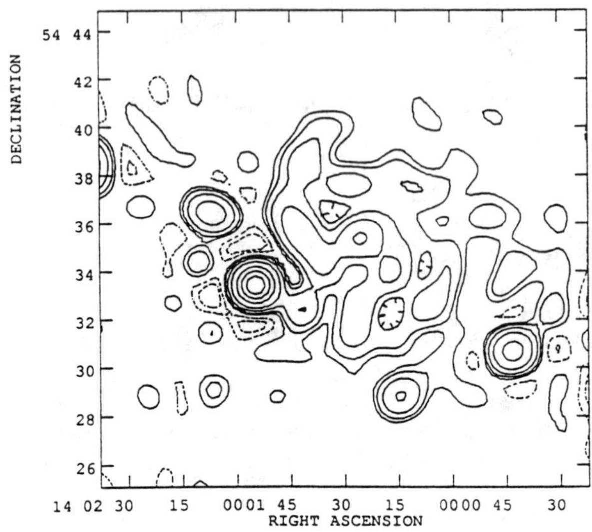

FIGURE I Visible space image M101 60μm. Contours: 4 (dashed), 6 (dashed), 10, 12, 16, 24, 40, 56, 72 MJy/Sr. Dashed contours indicate levels below the background of 8 MJy/Sr.

quantifying final results in a systematic way.

PITFALLS

A straightforward application of MemSys5 to IRAS survey data covering the galaxy M101 is shown in Figure I. Here the width of the gaussian ICF is equal to 1.1 times the diffraction limit of the telescope at 60 μm. Many features in this map are genuine: the HII regions NGC 5447, 5455, 5461, 5462, 5471 (edge of map), as well as the galaxy centre are clearly resolved. Around strong sources, however, artifacts can be discerned. The reconstruction can not restore a constant background around those, but seems to over-compensate by forming depressions in the background. Since the extended structure in the galaxy is at relatively low levels it could have been affected as well. The background is artificial and originates from the zodiacal foreground emission.

What causes this ringing and what can we do about it? Straightforward subtraction of a constant from the input data removed the ringing, but it was felt artificial at the time: how much should one subtract?

A first attempt was to increase the ICF width gradually, even to unrealistically large values. The images became smoother, but the ringing phenomenon remained.

Since we are dealing with real data some of it may be corrupt. It is possible to flag such data points after convergence of MemSys5 and effectively remove them (Bontekoe & Kester 1992; Weir 1992). Although it converged was much faster the second time, the result was virtually the same as in Figure I.

The ringing seems associated with the strong point sources. Therefore it

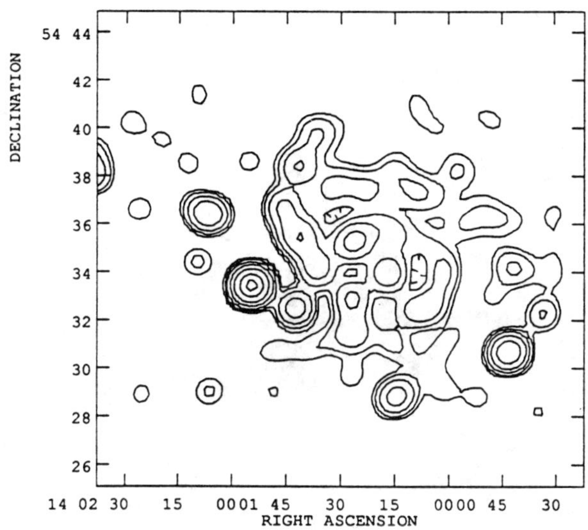

FIGURE II M101 high-resolution map, 7 MJy/Sr background subtracted. Contour levels 2, 4, 8, 16, 32, 64, and 96 MJy/Sr.

could be possible that the data do not sufficiently constrain the solution to single sources. Since 20% of the total entropy stems from NGC 5461, the next try was to improve the default map to a flat image plus a delta function. The position and flux were taken from the IRAS Point Source Catalog (IRAS 1988). To our surprise, again the result was essentially the same as in Figure I.

Perhaps the 'continuity' of the image should be more encouraged. Therefore, the default map was changed into a low-resolution co-added map, however, without any improvement.

The next step was to implement a multiple channel approach (Weir in this volume). The underlying rationale is to separate hidden and visible space into two or more channels, each with their own ICF. The different channels are sensitive only for spatial structure larger than their ICF width. One generally encourages broad over fine structure by applying different weights to the various channels. In a number of cases this has worked quite successfully in separating background and image (Charter 1990; Weir 1992).

In our case, however, this was unsuccessful. We have experimented with the number of channels (2 – 7), various ICF widths, different weights per channel, but were unable to find an optimum set of parameters. Summarizing, usually only 2–3 channels contain all structure, leaving the other ones virtually empty, but a small change of weights alters this into an entirely different set of channels. As it turned out, the problem also depends on the assignment of the defaults in each channel. At this point we concluded that the elegant way of separating background from image through multiple channels did not work our case.

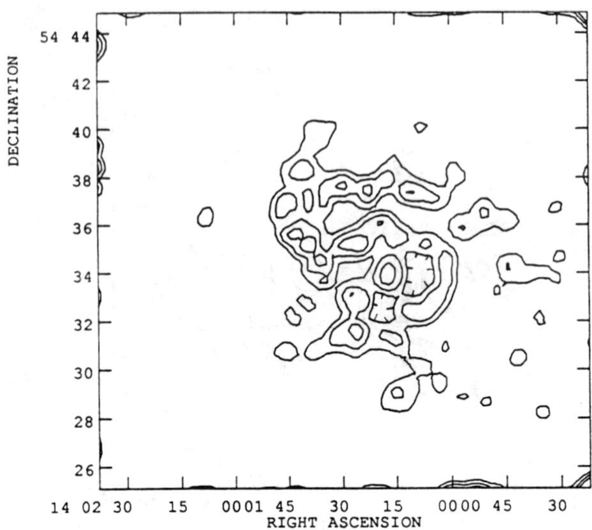

FIGURE III Error map. Contour levels 3, 4, and 5 MJy/Sr.

RESULTS

There seems to be no panacea for subtracting a constant background, in our case as well as in others (Skilling, private communication). The amount to subtract somewhat depends on personal taste. Subtracting too much leaves the background of the map completely empty, which looks unnaturally. We like to set the amount roughly one noise standard deviation less than the background level in Figure I, which leaves some noisy features in the map.

Also we retain two channels, with ICF widths of 0.8 and 2.0 times the diffraction limit, having weights 0.25 and 0.75 respectively. Figure II is the sum of both visible space channels. Spiral structure at the East side of the galactic disk as well as the spiral arm along the North side bending around towards NGC 5447, coincide with spiral arms in the visual. This map is actually the average of 100 random samples/images using the MOVIE5 option in MemSys5.

Figure III is a map of the standard deviations of the 100 samples and can be regarded as the error map of Figure II. The background error level is about 2 MJy/Sr. Note the 'absence' of the point sources, which apparently are not difficult to reconstruct from the data. The errors are clearly correlated with extended structure of the galaxy.

The ratio of Figures II and III is shown in Figure IV. The Figures II and IV quantify the results in a well defined manner, and can be used sensibly to make inferences about the astronomical interpretation of these results, such as the spiral structure at a 2–3 σ level. In a subsequent paper we will discuss the results in view of their astronomical importance.

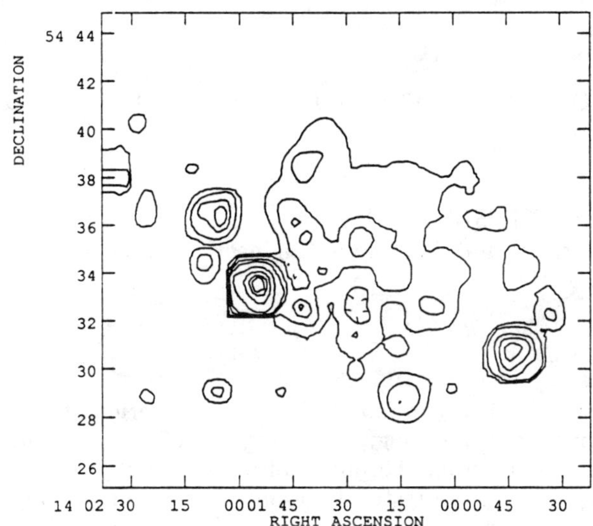

FIGURE IV Signal-to-noise map. Contour levels 1.5, 3, 6, 9, 15, 24, 36, and 48σ. The straight edges near NGC 5461 are artifacts due to the cut-off of the gaussian ICF.

ACKNOWLEDGMENTS

We thank the Laboratory for Space Research in Groningen for using their IRAS data, and specifically Do Kester for his excellent destriper. We also acknowledge many helpful suggestions by Mark Charter and John Skilling (Cambridge).

REFERENCES

Bontekoe, Tj. R., Kester, D. J. M., Price, S. D., de Jonge, A. R. W., & Wesselius, P. R. 1991, A&A, 248, 328

Bontekoe, Tj. R., & Kester, D. J. M. 1992, in Maximum Entropy and Bayesian Methods, G. J. Erickson, P. O. Neudorfer, & C. R. Smith, Dordrecht: Kluwer, in press

Charter, M. K. 1990, in Maximum Entropy and Bayesian Methods, P. F. Fougere, Dordrecht: Kluwer, 1990, 325

Gull, S. F. 1989, in Maximum Entropy and Bayesian Methods, J. Skilling, Dordrecht: Kluwer, 1989, 53

IRAS 1988, IRAS Catalogs and Atlases, Beichman, C. A., Neugebauer, G., Habing, H., Clegg, P. E., Chester, T. J., Washington DC: US GPO

Skilling, J. 1989, in Maximum Entropy and Bayesian Methods, J. Skilling, Dordrecht: Kluwer, 1989, 45

Weir, N. 1992, in Proceedings of the ST-ECF/ESO Data Analysis Workshop, P. Grosbol, Garching, ESO, in press

DESCRIPTION OF A POISSON IMAGERY SUPER RESOLUTION ALGORITHM

B.R. HUNT AND P. SEMENTILLI
Dept. of Electrical and Computer Engineering, University of Arizona, Tucson, AZ 85721

INTRODUCTION

Image restoration is a problem commonly described in terms of correcting for the defects of an imaging system, e.g., removing the effects of a specific optical transfer function (OTF) from the image of a specific object. As such, the conventional techniques of image restoration only correct for the object spatial frequency degradation below the spatial frequency limit imposed by diffraction (Andrews & Hunt 1976). An extension of the problem of image restoration is the problem of *super-resolution*, first presented in an algorithmic form by Gerchberg (1974), which seeks to recover object information from *above* the spatial-frequency limit of diffraction. There was substantial debate and theoretical justification for the prospect of super-resolution (e.g., see the discussion in Goodman 1968), prior to the algorithm of Gerchberg. A critical distinction in the nature of super-resolution vs. image restoration algorithms is non-linearity in computations. Conventional image restoration algorithms, e.g., Wiener filter, constrained-least-squares, singular value decomposition, etc. (Andrews & Hunt 1976), are linear in the processing computations. However, any algorithm capable of recovering information above diffraction must be nonlinear, since nonlinear operations on a signal generate high-order harmonic frequencies, a simple fact demonstrable in elementary Fourier analysis. However, an arbitrary nonlinear process can not be considered to qualify as a super-resolution algorithm. The issue is not constructing spatial frequency information above the diffraction limit, but constructing spatial frequencies with a proper relationship to the original object.

To insure that an algorithm constructs meaningful spatial frequencies above the diffraction limit, additional knowledge must be brought into the problem. Gerchberg's algorithm incorporates additional information on constraints in the object space and spatial frequency domains. The Maximum Entropy Method of Frieden (1972) uses an explicit model of image/object statistics to constrain solutions and incorporate additional knowledge. These are examples of model-based super-resolution algorithms, and we present another model-based algorithm herein.

THE POISSON MAP MODEL

Our motivation is the method of *maximum-a-posteriori* estimation. Our point of departure is Bayes' law, i.e.,

$$p(f|g) = \frac{p(g|f)p(f)}{p(g)} \tag{1}$$

where f is the original object and g is the image. To estimate the object we find the maximum probability of f given the image g, i.e., we maximize $p(f|g)$, which means we manipulate so that the right hand side of equation (1) is a maximum on f given the data g. This *maximum-a-posteriori* (MAP) estimate is obviously governed by the assumed form for the probability density functions $p(g|f)$ and $p(f)$. It is the specific forms of these quantities that represents the models of object and image relations that we use in deriving the algorithm.

We have chosen Poisson models for the statistics of image and object. We further adopt a discrete sampling of both object and image. Assume the image g is sampled by detector elements (e.g., CCD's) which are T units apart, and that T is chosen to satisfy the Nyquist criterion imposed by the diffraction spatial frequency limit of the OTF. Then we have samples $g(jT)$ for $j = 1, 2, 3, \ldots, M$. Likewise, we assume the object is to be computed (estimated) on a grid of samples. However, super-resolution implies that the sampling in the object be with samples *more closely spaced* than the samples in the image. The reason for this requirement is both physical insight and mathematical rigor. Physically, super-resolved object detail can be meaningful only if the sampling of the object is smaller than the image. Mathematically, the Nyquist theorem specifies the need to raise the sampling rate, if an algorithm extends the spatial frequency spectrum, in order to prevent artifacts from the phenomenon of *aliasing* (Oppenheim & Schafer 1991). Thus, we assume the object will be samples $f(jT/\alpha)$, where $\alpha > 1$ measures the extent of super-resolution being sought by the processing.

The final aspect of modelling object and image relations is the imaging equation. We assume that the point-spread-function is known as the quantity h. This function must also be sampled.

Since we must convolve h with f to establish g, the sampling of h must be on a grid of spacing T/α, so it is compatible with f. This convolution is evaluated or sampled on a grid of spacing T, however, giving the equation:

$$g(jT) = \sum_{k=1}^{\alpha N} h(jT - kT/\alpha) f(kT/\alpha), \quad \text{for } j = 1, 2, 3, \ldots M, \tag{2}$$

where N is the number of samples in f when sampled at a spacing T. We now draw together all these relevant model assumptions. For the object we assume a series of Poisson (photon) emitters, each characterized by a mean that varies with spatial position.

Thus, for the probability density function (*pdf*) of f we have:

$$p(f) = \prod_{k=1}^{\alpha N} \frac{(\overline{f(kT/\alpha)})^{f(kT/\alpha)} \exp(-\overline{f(kT/\alpha)})}{f(kT/\alpha)!}, \tag{3}$$

where the overbar is the mean that characterizes the emitters as a function of position. For the *pdf* of the image, given the object we have,

$$p(g|f) = \prod_{j=1}^{M} \frac{\left(\sum_{K=1}^{\alpha N} h(jT - kT/\alpha) f(kT/\alpha)\right)^{g(jT)} \exp\left(-\sum_{k=1}^{\alpha N} h(jT - kT/\alpha) f(kT/\alpha)\right)}{g(jT)!}. \tag{4}$$

The estimate of the object, given the above *pdf's* is derived in a straightforward (although algebraically tedious) way. The steps are as follows. We substitute from equations (3) and (4) into equation (1). The maximum on f of equation (1) is not affected by $g(j)$, so it can be considered as an unimportant constant. We now take the log of equation (1) after substitution, and the monotonicity of the log does not affect the maximum on f while simplifying the products and exponents of equations (3) and (4). Next, we use Stirling's approximation to simplify the factorial on f. Finally, we take the derivative of the resulting expression with respect to $f(kT/\alpha)$ for all $k = 1, 2, 3, \ldots, \alpha N$. The resulting equations are set equal to zero and rearranged. To construct a working algorithm we use the iteration of Picard's form, i.e.,

$$f^{n+1} = \Phi(f^n), \tag{5}$$

and rearrange into the final form for the algorithm:

$$f^{n+1}(kT/\alpha) = \overline{f(kT/\alpha)} \exp\left(\sum_{j=1}^{M}\left(\frac{g(jT)}{\sum_{k=1}^{\alpha N} h(jT - kT/\alpha) f(kT/\alpha)} - 1\right) h(jT - kT/\alpha)\right).$$

$$\text{for } k = 1, 2, 3, \ldots, \alpha N, \quad \text{for } n = 1, 2, 3, \ldots \tag{6}$$

In practice we have found that, lacking knowledge of the actual mean $\overline{f(kT/\alpha)}$, we can produce convergent behavior in the algorithm by assuming on the first iteration that g is the best estimate of $\overline{f}$. On successive estimates, we assume that $f^n(kT/\alpha)$ is the current best estimate of $\overline{f}$. This leads to the algorithmic form:

$$f^{n+1}(kT/\alpha) = f^n(kT/\alpha) \exp\left(\sum_{j=1}^{M}\left(\frac{g(jT)}{\sum_{k=1}^{\alpha N} h(jT - kT/\alpha) f^n(kT/\alpha)} - 1\right) h(jT - kT/\alpha)\right), \tag{7}$$

where $f^0 = g$ to start the processing.

EXPERIENCE WITH THE ALGORITHM.

We have gained extensive experience with the algorithm by using it in computer simulations and with real images. Typical experience is summarized in the following comments:

1. Convergence to final solution is quite slow, and hundreds or thousands of iterations of equation (7) may be required. Indeed, what constitutes the final solution and stopping-point at convergence is problematic, as is true with similar algorithms, such as Gerchberg's (1974).

2. The extent of super-resolution produced by the algorithm, i.e., the amount of bandwidth extension, is dependent on the object, the point-spread-function, and the SNR. For simple point objects it is possible in simulations to observe the bandwidth extend by a factor of 4 to 6. For more complex objects, such as scenes of the earth from aerial perceptive, the expansion of the bandwidth is much less, e.g., effective visible quality increase equivalent to extension beyond the diffraction limit of 25%.

3. The choice of α, the super-resolution factor, is best for practical implementation when α is an integer. In such cases the point-spread-function is over-sampled by an integer factor, and the correspondence between g and f in the convolution of equation (2) is simple subsampling. For example, $\alpha = 2$ implies that g in equation (2) arises from selecting every-other pixel produced by the convolution. Conversely, the convolution on the index j, seen embedded within equation (7), implies the data within the innermost parentheses must be up sampled before the outermost operation with h. Inspection of the relationship between the number of samples in the innermost parentheses and the number of samples in the psf reveals that a simple upsampling is to use pixel replication of the data from the innermost parentheses. For example, $\alpha = 2$ implies the data from the innermost parentheses is upsampled by duplicating every sample prior to convolution.

REFERENCES

Andrews, H. C., & Hunt, B. R. 1976, Digital Image Restoration, Prentice-Hall, Englewood Cliffs, NJ

Gerchberg, R. 1974, "Super-resolution through error-energy reduction," Optica Acta, 21, 709-720

Goodman, J. W. 1968, Introduction to Fourier Optics, McGraw-Hill, New York

Frieden, B. R. 1972, "Restoring with maximum likelihood and maximum entropy," J. Opt. Soc. Am., 62, No. 4, 511-518

Oppenheim, A. V., & Schafer, R. 1991, Discrete-Time Signal Processing, Prentice-Hall, Englewood Cliffs, NJ, ($2^{\underline{nd}}$ Ed.)

SUPER-RESOLUTION OF IMAGERY USING THE MAXIMUM ENTROPY METHOD

J. E. DORBAND AND J. M. HOLLIS
Space Data and Computing Division, Code 930, NASA/Goddard Space Flight Center, Greenbelt, MD 20771

ABSTRACT The theoretical basis for the ability to increase the resolution of an image beyond the diffraction limit of the optical system used to form that image has been addressed by many investigators during the past twenty years (e.g., Gerchberg 1974; Holmes 1988; and references therein). Here we address the practical aspects of achieving such super resolution of images by using the maximum entropy method.

INTRODUCTION

The Maximum Entropy Method (MEM) is a technique of statistical inference that can restore astronomical imagery which is plagued by incomplete or missing data (e.g., detector saturation). MEM infers the values of the invalid data from the redundant information contained in the valid data. This redundant information manifests itself in the point spread function (PSF) which extends outside the region of the invalid data. As an extension of MEM, we demonstrate how to potentially effect an increase in the resolution of an image by statistically inferring information in gaps that we have interspersed between rows and columns of valid data. A key prerequisite in the methodology is to first obtain a PSF that is sampled at the desired super resolution. Clearly, a PSF with such spatial characteristics must ideally be obtained from some modelling process, and, in fact, such noise-free PSFs have been successfully simulated by Lyon, Miller & Grusczak (1991) for the Hubble Space Telescope (HST). However, to demonstrate the validity of our method, we will simply use an observed PSF taken with HST Faint Object Camera (FOC) in the f/96 mode as our super sampled PSF. As a result of our PSF choice, we are defining an instrument with half the resolution of HST with super-resolution results at the resolution of the present HST-FOC combination to simply illustrate our methodology. In any case, we then form a true object image of known spatial characteristics by superimposing this PSF in a regular pattern on a field. In this way, one can synthesize an extended image since, in the limit, an extended, unresolvable object could be approximated by an infinitude of superimposed points. Using this artificially constructed image as a point of departure, we then deconvolve test cases with our MEM application which attempts to super-resolve features of the simulated "known" true object image. Results suggest that our MEM application could aid in the design of more cost effective instrumentation or aid in the increased resolution of imagery taken with a poorly designed instrument.

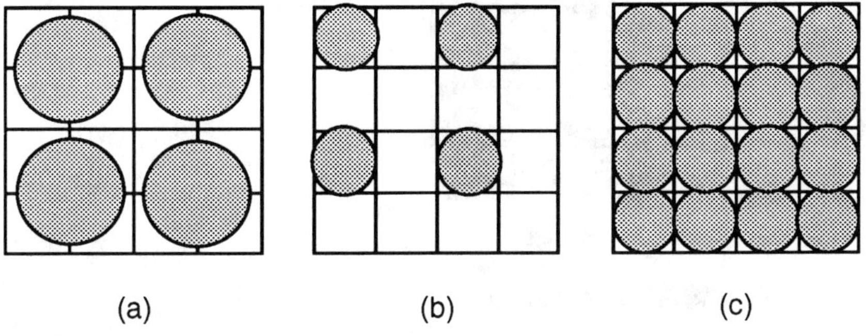

FIGURE I Sensor (or pixel) spacing

FORMATION OF THE TRUE OBJECT IMAGE OF KNOWN SPATIAL CHARACTERISTICS

One would like to obtain the true image of an object (true object image) which is unaffected by the optical system used in the observation. In other words, we want to know what the object looks like *before* it passes through the optical system whose characteristics are manifested in its point-spread-function (PSF). The PSF is simply how the optical system responds to a true point source of light which has no spatial extent. Once an object passes through the optical system, the image detected at the output sensor array is a convolution of the true object image and the PSF (i.e., the output image detected is blurred by the PSF).

To test various methods of "true object image" restoration, we must first fabricate a true object test image. Image 1 is an arbitrary true object image comprised of a collection of equal intensity points which are distributed on a grid that has the spatial resolution of the HST FOC operating in the f/96 normal mode (hereafter denoted HST/FOC). The points in Image 1 should be viewed as a collection of equal intensity delta functions since they have no spatial extent. Note that the distance between any two points in Image 1 is always greater than or equal to the square root of two times the pixel size.

The convolution of Image 1 with a real PSF from the HST/FOC yields blurred Image 2. Image 2 would be the image formed at the output of an instrument with the resolution of the HST/FOC. Image 2 provides a useful point of departure for the process of super-resolving imagery where we must treat the data points of a given image as points of data of an image of higher resolution that are separated by a grid of data points whose values are unknown. These data points could have been obtained by an array of equally spaced sensors that either cover the entire field of view (Figure Ia) or have gaps between them (Figure Ib). In either case the objective is to resolve an image with a greater number of effective data points (Figure Ic).

FORMATION OF THE DECONVOLVED IMAGE

We degrade the resolution of Image 2 by a factor of 2 to show what can be retrieved from such a severely blurred image. This degradation is effected by sim-

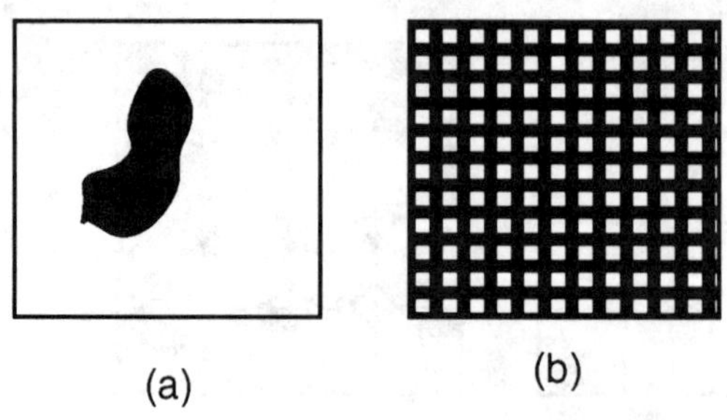

FIGURE II Regions of restoration

ply summing every 2x2 pixels (see Figure Ic) into one pixel with the same areal size (see Figure Ia); this results in blurred Image 3. Application of the MEM technique (see Hollis, Dorband & Yusef-Zadeh 1992 for algorithmic details) using Image 3 and the PSF yields the deconvolved Image 4 which represents what such an instrument (i.e., one with half the resolving power of the HST/FOC) should have been able to produce in the absence of spherical aberration. In other words, Image 4 is at the design resolution of our new instrument with half the resolving power of the HST/FOC.

CASE 1: FORMATION OF A SUPER-RESOLVED IMAGE

To prepare a "high resolution" blurred input image for deconvolution, we can place the sum of the values of each 2x2 group of pixels (see Figure Ic) in Image 2 into (say) the top left corner pixel (see Figure Ib) of each group and set the three remaining pixel values to zero. The result is Image 5 which can also equivalently be formed from Image 3 by placing rows and columns of zeros between the existing rows and columns. In principle, the technique for restoring the missing data in Image 5 is described in Hollis, Dorband & Yusef-Zadeh (1992). However, rather than restoring one large region of invalid data as these authors did (Figure IIa), we are restoring regions that are distributed throughout the image (Figure IIb). The maximum entropy method is used to solve a set of linear equations which is under-specified, and, in our case, an image for which we know only a quarter of the data points.

Subsequent deconvolution of Image 5 yields the super-resolved Image 6 which is a decided improvement over the deconvolved Image 4 in representing the true object image (Image 1). Carefully note that the resolution of the PSF used to super-resolve Image 5 to Image 6 must be 2 times that of the PSF used to resolve Image 3 to Image 4. Image 6 shows more detail of the true object image structure but is by no means perfect. We will address this point later in our concluding remarks.

CASE 2: FORMATION OF A SUPER-RESOLVED IMAGE WITH A NEW SENSOR ARRAY WITH GAPS

Unlike Case 1 where we used all the data in Image 2 (or equivalently Image 3), we now prepare a new blurred input image by subsampling Image 2. One pixel out of each 2x2 group of pixels (see Figure Ic) is selected and the remaining three pixel values are set to zero (see Figure Ib). That is, we are eliminating 75% of the data contained in Image 2. The result is Image 7 which cannot be formed from Image 3, since Image 7 represents a 4-fold increase in pixels. What we have essentially done is place the same number of smaller sensors within the same array area so that there are intervening gaps (rows and columns) between sensors, thereby throwing away 75% of the light falling on the sensor array. One reason to do this would be to cut down on the light from bright sources so that extended structure could be more effectively detected.

Subsequent deconvolution of Image 7 yields the super-resolved Image 8 which is a further improvement over the HST super-resolved Image 6 in representing the true object image (Image 1). Again, as with case 1, the resolution of the PSF used to super-resolve Image 7 to Image 8 must be 2 times that of the PSF used to resolve Image 3 to Image 4.

CONCLUDING REMARKS

The design resolution of our instrument with half the resolving power of the HST/FOC can produce a deconvolved image (Image 4) which resolves some individual points if they are far enough apart, as expected. Image 6 (super-resolution case 1) resolves the individual points of the true object image much better but not completely, which is also not unexpected. Image 8 (super-resolution case 2) is almost completely resolved with nearly imperceptible glitches. This can be explained by looking at how the images of case 1 and 2 were generated. In case 1, each group of 2x2 pixels are summed into a single data value, which increases the ambiguity as to the location of any point source within such a group. Whereas in case 2, only one pixel of the 2x2 pixels of each group is represented in the data used to reconstruct the image. The one pixel value is always chosen as the same relative position with in each group. Hence, we know something about the value of the pixels within each group by knowing the specific value of the pixel that was chose as the representative of the group. Note, however, that image recovery or increasing of resolution is only possible if the PSF is sufficiently extended to cover many valid data points and is accurately known to the resolution of the "super-resolved" image. This would imply that higher resolution can be obtained by sensors that sense over small areas with space between rather than crowding the sensors together. The amount of space that can be placed between the sensors depends on how well the PSF is sampled in the gaps between the sensors. Nevertheless, as demonstrated by case 1, more detail can be brought out even in an instrument not designed with gaps between the sensors if a sufficiently high resolution PSF can be determined.

REFERENCES

Gerchberg, R. 1974, Optica Acta, 21, 709

Hollis, J. M., Dorband, J. E., & Yusef-Zadeh, F. 1992, ApJ, Feb. 10 issue

Holmes, T. J. 1988, J. Opt. Soc. Am. A, 5, 666

Lyon, R. G., Miller, P. E., & Grusczak, A. 1991, SPIE, 1567, in press

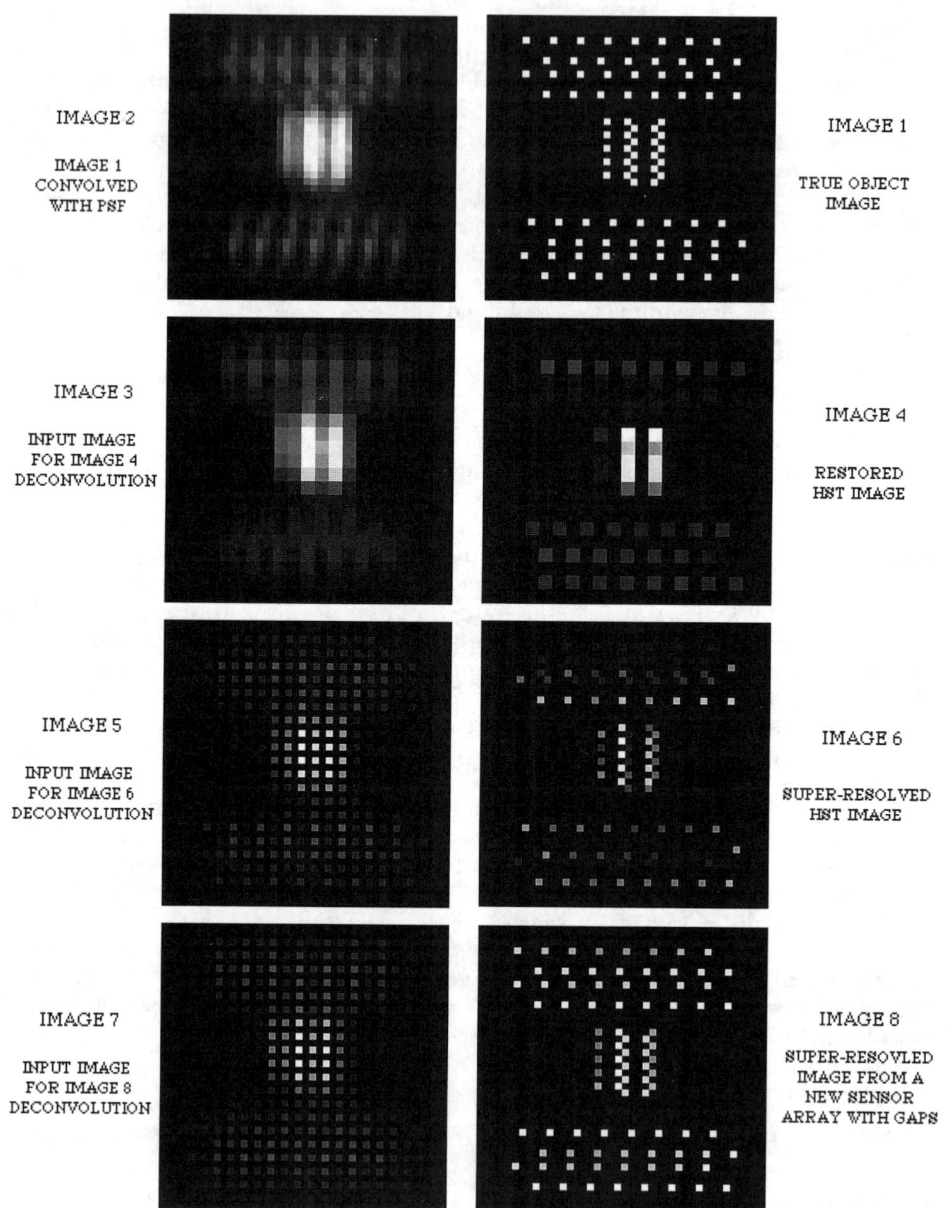
IMAGE 2
IMAGE 1
CONVOLVED
WITH PSF
IMAGE 1
TRUE OBJECT
IMAGE
IMAGE 3
INPUT IMAGE
FOR IMAGE 4
DECONVOLUTION
IMAGE 4
RESTORED
HST IMAGE
IMAGE 5
INPUT IMAGE
FOR IMAGE 6
DECONVOLUTION
IMAGE 6
SUPER-RESOLVED
HST IMAGE
IMAGE 7
INPUT IMAGE
FOR IMAGE 8
DECONVOLUTION
IMAGE 8
SUPER-RESOVLED
IMAGE FROM A
NEW SENSOR
ARRAY WITH GAPS

APPLICATION OF SUPERRESOLUTION IN DIFFRACTION-LIMITED SYSTEMS

TERESA A. BIPPERT-PLYMATE
Steward Observatory, University of Arizona, Tucson, Az. 85721

ANDREW E. PAUL
Physics Department, University of Arizona, Tucson, Az. 85721

GEORGE H. RIEKE
Steward Observatory, University of Arizona, Tucson, Az. 85721

ABSTRACT Two geometric parameters can be controlled in gathering data for superresolution (SR) processing. First, the pixel scale relative to the image can be adjusted either by a physical adjustment of the size of the pixels or by modifying the focal length of the optical system that feeds the pixels. Second, the telescope can be moved to place the image at differing positions on the detector array. We have investigated the optimization of these parameters for SR with a diffraction-limited system. Simulations allow us to explore the influence of projected pixel size, overlapping exposures, noise, and telescope tracking errors on the fidelity of the resolution-boosted image. We find that overlapping exposures on centers of half the pixel width can significantly improve the restorations. The optimum pixel size is signal-to-noise dependent. Pixels of about $\lambda/2D$ (where λ is the wavelength and D is the telescope aperture) used with overlapping exposures are well-suited for observations at signal-to-noise ratios of 100:1.

A REVIEW OF THE IMAGING SIMULATION PROCESS

The images used in the simulations were created artificially so that there are no ambiguities about them. A Fortran program was written to generate random "sources", with the program weighting towards low-luminosity sources using an integral power law distribution, $N(>S) = S^{-1}$. Eight hundred sources were generated in a 640×640 pixel image. The sources were convolved with an Airy pattern corresponding to $\lambda/D = 32$ pixels. A random Gaussian distributed noise map was generated with another program, and added to the master image using IRAF. The "master" represents an ideal, diffraction- limited image. The noise has an rms amplitude of 1% of the maximum signal in the master image.

A copy of this master image with noise was fed into the program "pixsum". Pixsum generates sub-maps with different geometric parameters in two ways: 1) changing the pixel scale relative to the image and; 2) microstepping, simulating movement of the telescope. Pixsum changes pixel scale by summing (averaging) blocks of pixels of predetermined size in the master image into a single pixel; it simulates telescope movement by sampling across and down the sub-map at 3 movement intervals. The movement intervals, or "steps", are called whole, half, and quarter. A whole step means that the sampling interval is the size of an entire pixel block (no overlap). A half step means that the sampling interval is half the size of a pixel block (overlap by half a pixel width). Likewise, a quarter step is a sampling interval one-quarter the size of a pixel block. Pixsum takes into account the sampling interval it will be dealing with when it is summing the master into pixel blocks so that the edge effects are dealt with in a realistic way, as if the array is overfilled with sky.

The stepped and summed sub-maps were written to a FITS file and imported into AIPS for MEM processing. The master map and the sub-maps were processed until the best (determined by fidelity number) deconvolved examples of each were generated. Resolution boosts can be seen in areas of high luminosity; areas that were previously smooth can now be seen to have features.

One set of simulations was run without the addition of the noise map for comparison purposes. Three separate sets of masters with the same noise map added were used in the simulations, and their data averaged after MEM processing. Each of these sets were summed into pixel blocks of 8, 10, 12, 16, 20, and 24 pixels, and microstepped in whole, half, and quarter steps. Fidelity calculations were performed for each sub-map to use as a means of mathematical comparison. The fidelity calculation is the mean of the square of the difference between the deconvolved master and sub-maps. The lower the fidelity number is, the better the restoration.

One of the master sets with noise was used in a telescope tracking error test. To simulate tracking error, a copy of the 640×640 master with noise was cut to 512×512, but the cut was shifted by a certain number of pixels in the x-axis relative to the master 512×512. This shifted image was added to the unshifted 512×512 to make a blurred image. Blurred images were made for pixel shifts of 2, 4, 6, 8, 10, and 12 pixels. These blurred images were processed with MEM, and fidelities calculated. The fidelity numbers were plotted versus the number of pixels shifted.

A simulation was set up independently and deconvolved by another individual to investigate the effects of noise on SR. A separate, smaller master was made. Noise was added to copies of the master image at levels of zero noise, 3% noise, and 8% noise. These masters were summed into $\lambda/2D$ pixels, with microstepping at whole, half, and quarter steps. Unlike the above simulations, the fidelity was calculated using a deconvolved master without noise.

RESULTS

The results of the noiseless simulation are depicted in Figure 1. The data show that in all cases the fidelity becomes larger as the blocksize is increased. There is a major gain in resolution by microstepping in half or quarter steps instead of whole steps, with a very slight increase in resolution by stepping with quarter steps over half steps below the $\lambda/2D$ level (16 pixel block). The half/quarter pixels at $\lambda/2D$ have nearly the same fidelity as the much smaller $\lambda/4D$ pixel (8 pixel block) that was sampled at whole step intervals. Likewise, the half/quarter stepped pixels at λ/D are nearly as good as the $\lambda/2D$ pixel sampled at whole steps.

Figure 2 depicts the data with the 1% noise added, with the half and quarter step data averaged as these points are very close in value for pixels at the $\lambda/2D$ level and smaller. The results from the masters with noise are not as smooth as the noiseless case, but point to the same conclusions. The dotted line drawn through the first few half/quarter step points shows that after blocksizes of about 16 pixels the fidelity quickly becomes much worse. The "jiggling" in the lines appears to be due to noise effects, especially the very low fidelity numbers for a blocksize of 16 pixels, since the noiseless master does not show the same large dip at that point. These simulations suggest that for the best image restoration at a signal-to-noise ratio of 100:1, blocksizes should be kept below 16 pixels ($\lambda/2D$ level), and microstepping by half pixels be utilized.

Figures 3 and 4 plot the results of the tracking error simulation. Figure 3 shows that the effects of tracking errors tend to be minor up to errors of 4 pixels, and then begin to increase more rapidly with increasing error. Figure 4 shows that microstepping by half steps brings a large improvement in fidelity, with only a very minor further improvement from quarter stepping. A plausible argument is that tracking errors below 4 pixels ($\lambda/8D$ level, peak-to- peak) have little effect on the quality of the image restoration, and slightly larger errors are only a little worse.

The effect of noise on SR is plotted in Figure 5. These simulations concur with the previous independent simulations, in that there is a large advantage to using half steps over whole steps at all noise levels, and the half and quarter steps are very close. There is a very small advantage to using a $\lambda/4D$ pixel over a $\lambda/2D$ pixel at the 8% noise level. The results indicate that half steps should be used to achieve a higher-quality restoration.

CONCLUSIONS

These simulations suggest that for the best results in SR imaging, the pixels used at a S/N of 100:1 should be at the $\lambda/2D$ level or smaller, and that the telescope should sample at intervals of half the pixel width. Sampling at quarter-pixel intervals yields only marginally better restorations, so may not be worth the extra effort. Larger pixels can be used with fine telescope stepping to arrive at fidelities similar to those of smaller pixels with coarse microsteps. At other S/N levels, the half-step sampling scheme also seems to increase the quality of the restoration greatly. The optimal size of the pixel seems to be S/N dependent. The telescope should have a maximum tracking error of $\lambda/8D$.

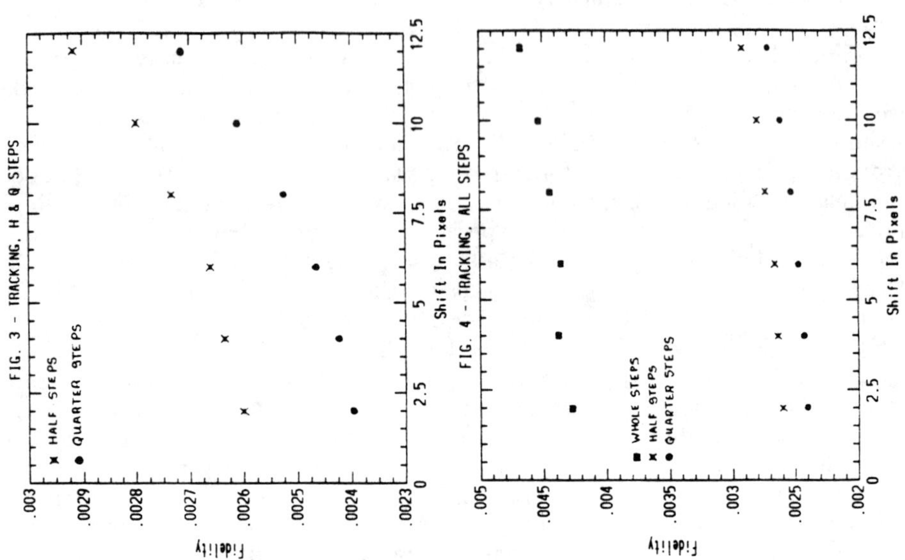

FIG. 3 - TRACKING, H & Q STEPS
HALF STEPS
QUARTER STEPS
Fidelity
.003
.0029
.0028
.0027
.0026
.0025
.0024
.0023
0
2.5
5
7.5
10
12.5
Shift In Pixels
FIG. 4 - TRACKING, ALL STEPS
WHOLE STEPS
HALF STEPS
QUARTER STEPS
Fidelity
.005
.0045
.004
.0035
.003
.0025
.002
0
2.5
5
7.5
10
12.5
Shift In Pixels

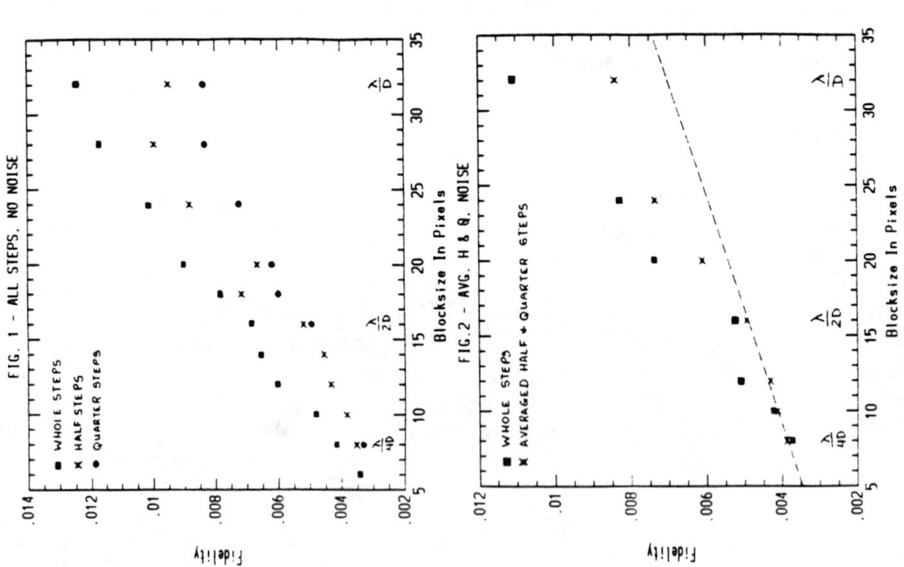

FIG. 1 - ALL STEPS, NO NOISE
WHOLE STEPS
HALF STEPS
QUARTER STEPS
Fidelity
.014
.012
.01
.008
.006
.004
.002
5
10
15
20
25
30
35
λ/4D
λ/2D
λ/D
Blocksize In Pixels
FIG. 2 - AVG. H & Q. NOISE
WHOLE STEPS
AVERAGED HALF + QUARTER STEPS
Fidelity
.012
.01
.008
.006
.004
.002
5
10
15
20
25
30
35
λ/4D
λ/2D
λ/D
Blocksize In Pixels

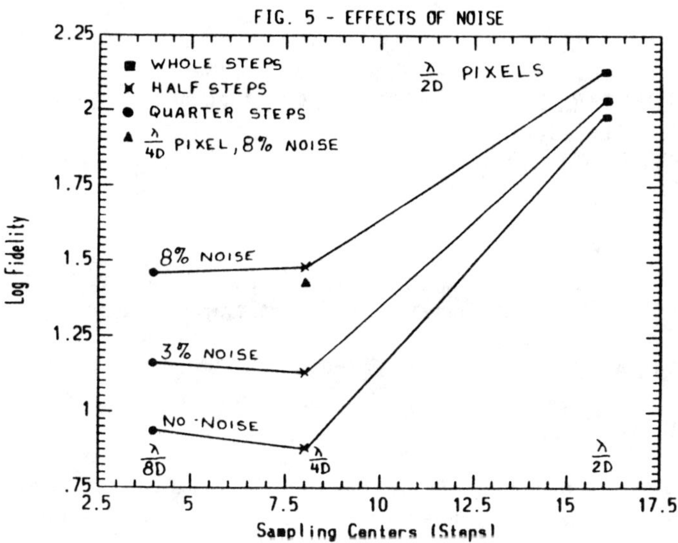

ACKNOWLEDGMENTS

The authors would like to thank Irene Barg and Tom Milner for all their programming help.

REFERENCES

Cornwell, T. 1983, VLA Scientific Memorandum, 149

MAXIMUM LIKELIHOOD ESTIMATOR AND BAYESIAN RECONSTRUCTION ALGORITHMS WITH LIKELIHOOD CROSS-VALIDATION

JORGE NUNEZ
Departament d'Astronomia, Universitat de Barcelona, Av. Diagonal 647, E-08028 Barcelona and Observatorio Fabra, Barcelona, Spain

JORGE LLACER
Engineering Division, Lawrence Berkeley Laboratory, University of California, Berkeley, CA 94720

ABSTRACT In this paper we describe the cross-validation method as applied to astronomical image reconstruction and present results of reconstructing images from the Hubble Space Telescope Faint Object Camera. Both Maximum Likelihood Estimator (MLE) and Bayesian reconstruction methods with Entropy prior benefit from the approach: in the MLE case, cross-validation provides a stopping rule which results in optimized images; in Bayesian reconstruction with Entropy prior, the cross- validation allows specifying de hyperparameter that controls the relative weight of Likelihood vs. Entropy in the reconstruction to obtain the best Maximum a Posteriori image.

INTRODUCTION

During the last few years, we have been developing statistically based algorithms for Image Reconstruction in Emission Tomography (Llacer et al. 1989; Nunez & Llacer 1990) and in Astronomy (Llacer & Nunez 1990; Nunez & Llacer 1991). In particular, our group has focused its attention on the Maximum Likelihood Estimator Method (MLE), and in Bayesian Reconstructions for Poisson data (the extension to Gaussian is straightforward). The MLE seeks the maximization of the Likelihood function, resulting in an image for which the measurements have the highest prôbability. However, the image reconstruction problem in presence of noise is an ill-posed problem making the Maximum Likelihood solution very sensitive to the noise. To avoid this problem, it is necessary to stop the process of the algorithm before reaching the maximum.

Another approach is to use Bayesian probability theory to define a different target function to be maximized in order to obtain the image of highest probability given the data. The Bayesian target function and the Likelihood are related through Bayes Rule, which includes the probability distribution of the image, also known as image prior. We use Entropy to define this prior probability. We have developed an iterative algorithm called FMAPE that maximizes the Bayesian function.

ALGORITHMS

Let $p_j, j = 1, \cdots D$ be the projection data (measured data); $a_i, i = 1, \cdots B$ the emission density in the image (parameters to be estimated); f_{ji} the Point Spread Function (PSF) or the probability that an emission in pixel i in the source be detected at detector j; C_j the detector gain correction at detector j; $f'_{ji} = \frac{f_{ji}}{C_j}$ the corrected PSF; $q_i = \sum_{j=1}^{D} f'_{ji}$ the total probability of an emission from pixel i being detected at any detector; $h_j = \sum_{i=1}^{B} f'_{ji} a_i$ the forward projection or blurring operation and Q_i the prior distribution density or default image (if any). In this notation, our Bayesian Maximum a Posteriori Algorithm for Poisson data with Entropy Prior (FMAPE) leads to the following iterative formula:

$$a_i^{k+1} = K a_i^k \left[\frac{1}{q_i} \sum_{j=1}^{D} \frac{f_{ji} p_j}{\sum_{i=1}^{B} f_{ji} a_i^k} - \frac{1}{q_i \Delta a} \left(log \frac{a_i^k}{Q_i} + 1 \right) + C \right]^n$$

where k is the index of the iteration, K is a constant to preserve the energy in the form $\sum_{i=1}^{B} q_i a_i = \sum_{j=1}^{D} p_j$, computed at the end of each iteration; C is an arbitrary constant (usually $C = 1$) to secure positivity in the solution; n is a constant to accelerate convergence up to approximately three times ($n = 3$). Constants C and n do not affect the point to which the algorithm converges. The parameter Δa controls the point of convergence of the solution between the trivial solution and the Maximum Likelihood solution. Δa is thus the balancing parameter of the Bayesian approach and is the inverse of the weight of the entropy versus Likelihood.

The algorithm has a number of desirable characteristics: It is iterative, easy to implement, always positive, stable, its main loop (projection and back-projection) is identical to the Expectation Maximization algorithm, up to three times faster than the Richardson-Lucy algorithm (Lucy 1974), does not compute exponentials and include gain (flatfield) corrections.

The parameter Δa can be made "global" for all parts of the image or "local" allowing different degrees of contrast control (sharpening) in areas of high and low signal/noise ratio. If Δa is very small the trivial solution $a_i = Q_i$ is obtained. If Δa is very high the FMAPE algorithm becomes:

$$a_i^{k+1} = a_i^k \left[\frac{1}{q_i} \sum_{j=1}^{D} \frac{f_{ji} p_j}{\sum_{i=1}^{B} f_{ji} a_i^k} \right]^n$$

that is our Maximum Likelihood algorithm (MLE) which for $n = 1$ is identical to the Richardson-Lucy algorithm but it includes the detector corrections.

CROSS-VALIDATION

The questions are now: a) How far to iterate towards maximizing the Likelihood in the Maximum Likelihood method, and b) How to balance the terms of prior information (Entropy) and Likelihood in the Bayesian approach. Several answers have been proposed (including a pure personal decision). We have proposed answers based in feasibility criteria in the following way:

We define an image as strongly feasible if the data is a Poisson (or Gauss) sample with means given by the projection of the image. We define that an image is weakly feasible of order n if the first n moments of the distribution are consistent with the Poisson/Gauss hypotesis. We have implemented tests based on the feasibility criteria that give a stopping rule for the MLE and a value for the balancing parameter Δa of the Bayesian approach. However, to work well, two conditions should be satisfied: a) the hypotesis made about the statistics of the data (Poisson/Gauss) should be satisfied by the true statistics of the data, and b) a perfect knowledge of the Point Spread Function is needed.

The feasibility tests, as well as the other proposed tests, work well with computer generated data in which we can control the statistics and we know the PSF exactly. The tests work also in some real cases but they often fail with real data because the nature of the statistics in not sufficiently well known. Disturbances caused by readout noise, corrections for geometric distorsion, flatfield corrections, detector failures, reseaux marks, etc. contribute to that uncertainty. The PSF is also not known exactly because it is usually the image of a field star and it has its own noise, background, different color, etc.

We present here the astronomical application of a new approach based on Cross-validation (Coakley 1991) that answers the questions posed above in a more robust way. We have applied and tested successfully the cross-validation technique in medical emission tomography.

Consider the data split into two halves, A and B. This splitting can be obtained in two ways: a) as a result of two independent but successive exposures (p.e. if two exposures have been made in the Hubble Space Telescope to detect cosmic-ray hits), and b) as a result of a thinning process if only one data set is available. The thinning process results in two new Poisson/Gauss distributed data sets with means that are one half of the means of the original data set. We have used as a thinning the following process: for each individual count (photon) in each pixel of the data, generate a random number between 0 and 1, if the number is less than 0.5 the count is assigned to the corresponding pixel of data set A and else it is assigned to data set B.

The method consists of the following steps: Reconstruct the data set A and compute the following two quantities at the end of each iteration (for Maximum Likelihood), or at convergence (for Bayesian reconstruction with a given Δa): a) the log Likelihood of data set A with respect to the reconstructed image A (Direct Likelihood) and b) the log likelihood of data set B with respect to the reconstructed image A (Cross-Likelihood).

While the reconstruction process (Maximum Likelihood during the iterative process or Bayesian at convergence for a given Δa) is working on features that are common to both data sets (the signal), both Direct Likelihood and Cross-Likelihood will increase because both data sets "recognize" the features of the image and the data and the projection of the image are approaching each other.

When the process begins to reconstruct artifacts that are due mainly to the noise in data set A, the Direct Likelihood will continue increasing because the reconstruction continues being consistent with the data set from which it is coming. However, the Cross-Likelihood will reach a maximum decreasing afterwards because the noise in data set B is different from the noise in data set A and the image is now acquiring features that are specific to the statistical random fluctuations of data set A.

The maximum of the Cross-Likelihood curve gives us a stopping point for the Maximum Likelihood algorithm (MLE) or a method to compute the balancing parameter Δa in the Bayesian (FMAPE) reconstruction. The process is now repeated but reconstructing the data set B and obtaining the Cross-Likelihood respect to data set A. After obtaining the two stopping points or the balancing parameters in the Bayesian case for the two halves, both reconstructions are added to obtain the final result. A slight filtering with a Gaussian kernel corresponding to the theoretical unaberrated diffraction limit of the telescope (or a fraction of it) is advised.

The cross-validation method is robust for the following reasons: a) both data sets have the same signal; b) the nature of the noise is identical in both cases; c) the noise is independent in the two data sets (they have independent random values) and d) they have the same Point Spread Function. The cross-validation method is qualitatively similar to the feasibility test requiring that the residuals do not become too small, i.e., that the reconstructed image does not follow the statistical noise in the data too closely. However, the cross-validation method is more robust than the feasibility test because a perfect knowledge of the statistical nature of the data or of the PSF are not needed.

RESULTS AND CONCLUSIONS

We have applied the Cross-validation method to real images of the Hubble Space Telescope Faint Object Camera (FOC) to obtain stopping points for the MLE reconstructions and values for the balancing parameters in the Bayesian FMAPE algorithm. The raw images consisted of only one frame, which we have split in two halves (A and B) by the thinning process described.

Figure I shows the ratio of the change in log Cross-Likelihood divided by the change in log Direct Likelihood as a function of the Δa parameter (divided by 10) for FMAPE and as a function of the iteration number for a MLE reconstruction of the data set A of the image of the Supernova 1987A taken with the HST-FOC working at f/96. The point at which the curves cross the zero value corresponds to the maximum of Cross-Likelihood curves. The curves of data set B look almost identical, giving the same stopping point and Δa. The near linear relationship displayed in the figure has been obtained not only in astronomical data but also in more than one hundred reconstructions in medical Tomography with real and simulated data.

After adding the reconstruction of the two halves and filtering slightly with a Gaussian kernel of 0.033 arc seconds (1.5 pixels), both methods result in high quality images. Of course, it is possible to choose a point other than the zero crossing of the ratio of Figure I as the point to determine the stopping in MLE and the Δa in FMAPE. In any case, once a value different from zero has been chosen (0.3, for example), the general characteristics (contrast, sharpness, bias, etc.) of the images reconstructed with such criterion will be similar.

We can conclude that the Likelihood Cross-validation is a robust technique in Astronomical Image Reconstruction to balance the Entropy and the Like-

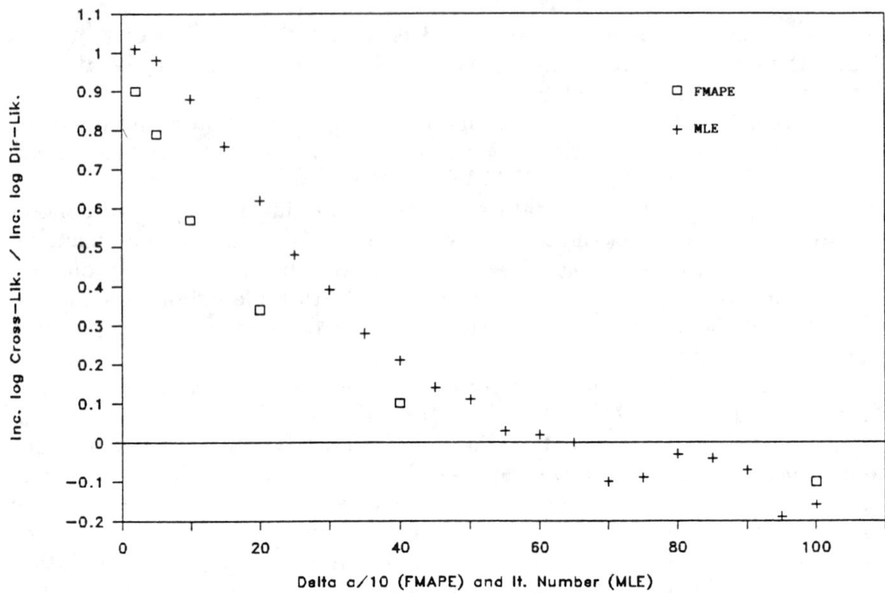

FIGURE I Cross-validation tests for FMAPE and MLE methods

lihood in the Bayesian approach and to determine the stopping point for the Maximum Likelihood method. The technique is more robust than the other methods because it is less demanding in terms of the knowledge of the exact characteristics of the data and of the knowledge of the PSF.

ACKNOWLEDGMENTS

We are particularly grateful to E. Veklerov of Lawrence Berkeley Laboratory for many discussions and suggestions. This work has been supported partially by the DGICYT Ministerio de Educacion y Ciencia (Spain) under grant n. BP90-0478 and by the DGU Generalitat de Catalunya. J. Llacer's work was been supported by the U.S. Department of Energy under Contract No. DE-AC03-76SF00098.

REFERENCES

Coakley, K. J. 1991, IEEE Trans. Nuclear Sci., NS-38, 1, 9

Llacer, J., & Nunez, J. 1990, in Proc. Workshop "The restoration of HST Images and Spectra", ST-ScI, Baltimore, Aug 1990, 62

Llacer, J., Veklerov, E., & Nunez, J. 1989, International Journal of Image Systems and Technology, 1, 132

Lucy, L. B. 1974, AJ, 79, 745

Nunez, J., & Llacer, J. 1990, IEEE Trans. Medical Imaging, 9, 159

Nunez, J., & Llacer, J. 1991, ST-ECF Newslett., 15, 6

PARALLELISM FOR HST IMAGE RESTORATION – SOME EXPERIENCES AND OPTIONS

H.-M. ADORF
Space Telescope – European Coordinating Facility, European Southern Observatory, Karl-Schwarzschild-Str. 2, D-8046 Garching b. München, F.R. Germany, Internet: adorf@eso.org – SPAN: ESO::ADORF

M.J. OLDFIELD
Mullard Radio Astronomy Observatory, Cavendish Laboratory, Madingley Road, Cambridge CB3 0HE, England, Internet: mjo11@cus.cam.ac.uk

ABSTRACT We report on recent experience gained in a project set up to explore the power of parallel computing for the restoration of Hubble Space Telescope images. Using our vector-processor implementation of the standard SIPSF Richardson-Lucy algorithm, we achieve a full 40-iteration restoration of a 512x512 image frame in less than 90 seconds. With this considerable computing power we carry out Monte Carlo simulations which allow the quantification of the probable statistical errors in a restoration. We compare the relative merits of our approach, which utilizes fine-grained parallelism, with several alternatives including a low-cost one transforming a workstation cluster into a coarse-grained "virtual supercomputer".

BACKGROUND

Restoring the almost diffraction limited images from the Hubble Space Telescope's two on-board cameras is a compute-intensive undertaking which is compounded by the spherical aberration of the present optics (Burrows et al. 1991). Any attempt to restrict image processing to smaller regions of interest is unfeasible due to (i) the large size of the aberrated point-spread function (PSF) coupled with its near-critical sampling, (ii) the high dynamic range of the data and its severe spill-over effects and (iii) the wrap-around effects caused by discrete Fourier transforms (DFTs). While the initial discussion of the HST-restoration problem centred mainly around the question of which methods to use (White & Allen 1991; White 1991; Adorf 1990, 1991b, c) – an issue largely resolved in the meantime – it is now appropriate to investigate further the options offered by contemporary hardware and software technology for optimally processing larger quantities of HST image data.

The need to increase the performance of image processing (IP) operations on HST image data sets has long been recognized. Efforts are underway e.g. at the ST-ECF to accelerate the original Richardson-Lucy (RL) algorithm (Richardson

1972; Lucy 1974) by modifying its fixed-point iteration (cf. Holmes & Liu 1991). Cobb et al. (1991) report on an early attempt to use the Connection Machine-2, a fine-grained parallel processor, for the difficult task of restoring HST Wide Field & Planetary Camera images with their space-variant PSFs (SVPSFs).

Generally speaking, the state of the art in parallel computing has substantially advanced over that found in the mid-seventies (Adorf 1991a), when radio-astronomers intensively investigated vector-processing options. Apart from a considerable increase in raw computing speed (and an accompanying decrease in hardware cost), important progress has been made on the software side. Vectorizing Fortran-, C- and Lisp-compilers are available now, which permit a simple port of existing programs from scalar architectures to vector-processors with substantial speed-up gains (particularly when double-precision calculations are involved), even when the code is left unmodified (Lasenby 1991, *pers. comm.*).

A single vector-processor (VP), namely Intel's i860^{TM} 64 bit RISC-microprocessor, stands out from its digital-signal processor competitors. The i860^{TM} – actually a computer in its own right – exploits parallel instruction execution and other supercomputer architectural concepts to achieve a peak performance of 80 MFlops for single-precision floating-point (FP) calculations. The sustained performance for IP-applications lies in the 40–45 MFlop range (Gull 1991, *pers. comm.*). The i860^{TM} microprocessor has made its way into several workstations as a graphics accelerator. Although tasks such as surface rendering require massive amounts of FP-computations, none of the main workstation vendors seems to target i860^{TM}-based accelerator boards specifically at image processing.

Despite its importance for the scientific sector, the IP-area appears to be dominated by a number of smaller companies. Besides the major players in this field, i.e. Sky Computers, CSPI and Computer Systems with their respective Skybolt/Skystation, SuperCard and Mercury boards, Maximum Entropy Data Consultants (MEDC) is now offering a low-priced i860^{TM}-based accelerator board (sometimes called the "Flyer"), which we have used to implement the standard SIPSF RL-algorithm.

IMPLEMENTING THE RICHARDSON-LUCY ALGORITHM ON THE I860

MEDC's *MemSys accelerator* i860^{TM}-board plugs into an S-bus based Sun workstation or clone, although other configurations, e.g. for PCs, also exist. Being limited by the physical space available inside the Sun pizza-box, the "Flyer" currently carries a maximum of 8 Mbyte of fast dynamic memory, with the implication that the number of image frames simultaneously required by the algorithm has to be minimized. The MEDC driver software currently achieves a (Sun host) memory to (i860^{TM}-board) memory transfer rate of 5–8 Mbyte/s, which is sufficient to allow intermediate results from the i860^{TM}-board to be read into host memory for visualization and assessment.

While in principle any existing standard C/Fortran-77 program could be compiled with one of several existing vectorizing compilers for the i860^{TM}, this approach was not taken, because we had no efficient, stand-alone C/Fortran-implementation of the SIPSF RL-algorithm to begin with. Instead, we started afresh using MEDC's *MemSys Hostmode Library,* a special purpose single-precision math-library comprising those VP-operators needed to efficiently implement the MemSys5 package on the i860^{TM}. All code development for a hostmode program takes place on the host machine. (Consequently the end-user needs no new compiler.) The compiled master program executes on the workstation controlling program flow and data streams as well as a slave "executive" on the i860^{TM}. The standard RL-algorithm for space-invariant PSFs (SIPSFs), implemented in this environment, can succinctly be summarized in the recursion (cf. Adorf 1990)

$$rest^{(1)} = data \star psf \quad \& \quad rest^{(k)} = rest^{(k-1)} \left[\left(\frac{data}{rest^{(k-1)} \otimes psf} \right) \star psf \right]$$

where $\otimes$ and $\star$ denote the convolution and the (unnormalized) correlation operators, respectively. The second equation forms the core of our implementation of the RL-algorithm. When coded in Fortran plus MEDC's Hostmode Library, it becomes

```
call apmac(lucyh)                              ; begin macro definition
   call convolve_f(p_rest, p_psf_f, p_work)    ; calculate prediction
   call divide(p_data, p_work, p work)         ; derive correction factor
   call correlate_f(p_work, p_psf_f, p_work)   ; map correction into model space
   call multiply(p_rest, p_work p_rest)        ; apply correction to restoration
call apendm(lucyh)                             ; end macro definition
```

In the RL-algorithm, the majority of the compute time is incurred during the fast Fourier transforms (FFTs). Each of these takes about 450 milliseconds for a standard 512x512 frame, which is about 7–10 times faster than our best FFT on a Sun Sparcstation-2. With four real FFTs per iteration, a 40-cycle RL-restoration takes less than 90 seconds to complete. This is commensurate with the time it takes to transfer the data across a loaded local area network (LAN) and thus *large local disks* are required to realize the full benefit of the VP.

A MAP OF THE PROBABLE STATISTICAL ERROR IN A RESTORATION

Given the substantial compute-power of a VP, we can address one of the outstanding problems of HST image restoration, namely how to quantify those errors that are introduced by the unavoidable statistical fluctuations in the science data or, as long as empirical PSFs are being used, in the PSF. In order to trace the propagation of statistical errors from the input data to the restoration, we used the Monte Carlo technique to generate seven variants of a (simulated) FOC science observation and seven variants of a (simulated) PSF observation. Each of the 49 possible dataset-PSF pairings was subsequently restored and used to estimate a map of the probable statistical errors in the restoration. We also generated a movie, similarly to that implemented in MemSys5 (Skilling et al. 1991), which dynamically displays these statistical errors by showing the series of restorations in quick succession. This method of assessing the sensitivity of the restoration to statistical errors in the input data is completely general and thus equally applicable to other restoration algorithms.

DISCUSSION

Our experience has shown that it is not unduly difficult to implement specialized IP-tasks on a VP. Implementations are likely to be facilitated in the future by the emergence of optimized libraries such as NAGlib (Brankin 1991, *pers. comm.*), common to both VPs and traditional workstations. Code using these libraries should capture most of the VP's power, yet remain completely portable. The key benefit of the VP is, of course, its greater FP-capability leading to a faster turn-around time for a single restoration. Backed by a VP, the RL-method can be considered an interactive technique, obviating the need to resort to e.g. Wiener-filtering for real-time data analysis. A memory based IP-package such as IDL seems a logical route to transparently further harness the VP's power and no fundamental problems are envisaged in calling VP-routines interactively (Stern 1991, *pers. comm.*).

The "fine-grain" parallelism on the instruction level described above is not the only form of parallelism with potential for image restoration. Often, the same restoration task has to be run repeatedly with different parameters. Such a job quite naturally splits into several unrelated IP-tasks, which, given appropriate tools, could be processed in parallel by otherwise idling workstations on a LAN. Indeed, according to a recent survey (Carver 1991) in a Unix environment several options exist for such a "virtual supercomputer", including the NASA funded distributed queue manager NQS. Network Linda (Bjornson et al. 1991), a low-cost commercial product, apparently offers great flexibility in distributing tasks across a heterogeneous workstation network with only marginal code modifications. Existing applications are reported to show an almost linear speed increase for small ($\approx$ 10 node) networks. Further investigations in this direction should be rewarding.

In conclusion, we find both fine-grained and coarse-grained parallelism complement each other. The latter technique allows more efficient utilization of existing workstation resources, whereas the former permits individual workstations to tackle larger problems and opens new avenues in interactive image processing.

ACKNOWLEDGMENTS

We appreciate support of and/or discussions with Gerry Gilmore, Derek Richardson, David Schade, Bob Thomson and Tim Wilkins (IoA, Cambridge), Steve Gull and Anthony Lasenby (MRAO, Cambridge), Don Wells (NRAO, Charlottesville), David Stern (RSI, Boulder) and Richard Brankin (NAG, Oxford).

REFERENCES

Adorf, H.-M. 1990, ST-ECF Newslett., 14, 8–12

Adorf, H.-M. 1991a, ST-ECF Newslett., 15, 8–11

Adorf, H.-M. 1991b, in 4th Int. Workshop "Data Analysis in Astronomy", Apr. 1991, Erice, Sicily, V. di Gesù et al. (eds.), Plenum Press, New York, USA, (in press)

Adorf, H.-M. 1991c, in 3rd ESO/ST-ECF Data Analysis Workshop, 22–24 Apr. 1991, P. Grosbøl et al. (eds.), ESO, Garching b. München, FRG, (in press)

Bjornson, R., Kolb, C., & Sherman, A. 1991, SIAM News, 24, No. 1, (reprint by Sci. Computing Associates)

Burrows, C. J., Holtzman, J. A., Faber, S. M., Bely, P. Y., Hasan, H., Lynds, C. R., & Schroeder, D. 1991, ApJ, 369, L21–L26

Carver, G. 1991, message to sun-managers@nsfnet-relay.ac.uk from glenn@atm. ch.cam.ac.uk, 27. Oct. 91

Cobb, M. L., Hertz, P. L., Whaley, R. O., & Hoffman, E. A. 1991, NRL, Washington, DC, USA, (preprint)

Holmes, T. J., & Liu, Y.-H. 1991, J. Opt. Soc. America A, 8, 893–907

Lucy, L. B. 1974, AJ, 79, 745–754

Richardson, B. H. 1972, J. Opt. Soc. America, 62, 55–59

Skilling, J., Robinson, D. R. T., & Gull, S. F. 1991, in "Maximum Entropy and Bayesian Methods 1990", Laramie, J. W. T. Grandy (ed.), Kluwer, Dordrecht

White, R. L. 1991, in 25th Conf. "Computer Science and Information Processing", Baltimore, March 1991, (in press)

White, R. L., & Allen, R. J. (ed.) 1991, "The Restoration of HST Images and Spectra", Proc. of the Workshop held at the Space Telescope Science Institute, Baltimore, MD, 20–21 Aug. 1990

WIENER RESTORATION OF HST IMAGES: SIGNAL MODELS AND PHOTOMETRIC BEHAVIOR

I. BUSKO
Astrophysics Division, National Space Research Institute, C.P. 515, CEP 12201, São José dos Campos, SP, BRAZIL

THE PROBLEM

Fourier noniterative image restoration techniques (inverse and pseudoinverse filters) were already explored for Hubble Space Telescope (HST) image restoration (Hanisch 1990). Their main advantages over other restoration methods are speed and linearity.

The minimum mean-square error (MMSE) linear deconvolution operator (Wiener filter) implicitly assumes that some *a priori* knowledge about the undegraded original image, here called *signal*, is independently available. In other words, a *signal model* must be adopted prior to applying the filter to a particular image. The Wiener filter Fourier implementation is possible only because some constraints are imposed on the imaging problem, which otherwise would be intractable. One of these important constraints is *stationarity* of signal and noise processes. In other words, the adopted signal model will be implictly assumed to be stationary (see Andrews & Hunt 1977 for a detailed discussion).

In this work, we study the role played by several assumed signal models in the restoration of HST images. We also evaluate the photometric properties of images restored by Wiener filtering, and compare them with images restored by the Richardson–Lucy algorithm. Software is implemented under the STSDAS/IRAF environment, and as such will be made available to all users of these systems.

ANALYSIS

For the Wiener filter implementation, it is enough to specify the correlation matrix of the original signal. This translates, in Fourier space, to the signal *power spectrum.* Our implementation allows to choose between several different power spectrum models: (i) estimate it from the input degraded image; (ii) estimate it from another available undegraded image with similar statistics as the one being deconvolved; (iii) assume that the object autocorrelation is first–order Markov, with correlation parameter equated to the *average* correlation in the input degraded image (Pratt 1978); (iv) in the case of scenes dominated by stellar images, assume a spatially Gaussian–shaped signal. Notice that this model violates the stationarity hypothesis; the result will be good reconstruction of star images an poor reconstruction of other structures (e.g. background).

A comparative evaluation of signal models was performed in a variety of images, both real and simulated. A qualitative assessment may be obtained from one example in Figure I. Quantitative analysis performed in this image is shown in Table I. Two stars were measured, a bright and a faint one, as well as the elliptical ring and a blank sky patch.

TABLE I

Signal model	Sky stdev.	Signal/noise ring	bright ⋆	faint ⋆	Centering error bright ⋆	faint ⋆
original image	0.66	11.4	194	19	0.31	1.45
Gaussian	6.98	5.1	202	19	0.07	0.78
input	3.09	9.7	252	23	0.11	0.56
Markov	1.40	17.7	271	22	0.15	0.51
lucy	1.24	25.2	460	40	0.18	0.70

Star field simulations were designed to roughly approximate HST FOC $f/96$ mode. A synthetic PSF image, very similar to the one shown by Alighieri & Walsh (1990), was used.

Figure II show the basic results of aperture photometry performed on the original and restored images.

CONCLUSIONS

Use of a specific signal model tuned for specific images produces measurably better restorations, both qualitatively and quantitatively, than restorations using a simple pseudoinverse filter. This filter is able to produce results similar to the Wiener filter, but has the disadvantage that the user does not have any idea on which spatial features the filter is restoring optimally.

Stellar aperture photometry, impossible to be carried out in a HST–degraded image, can be done in the restored image. Linearity is conserved, and crowding effects are minimized. Errors will be larger than in comparable photometry in non–degraded images.

Signal–to–noise ratio in a Wiener–restored stellar field image tends to be larger than in the original, HST–degraded one. This, however, is limited to brighter stars. For stars near the detection limit, no improvement (but also no degradation) in S/N is attained by the restoration process.

REFERENCES

Alighieri, S. S., & Walsh, J. 1990, ST-ECF Newsletter, 14, 4

Andrews, H. C., & Hunt, B. R. 1977, Digital Image Restoration, Prentice-Hall, New Jersey

Hanisch, R. J. 1990, in The Restoration of HST Images and Spectra, ed. White, R. L., & Allen, R. J., Space Telescope Science Institute

Pratt, W. K. 1978, Digital Image Processing, John Wiley, New York

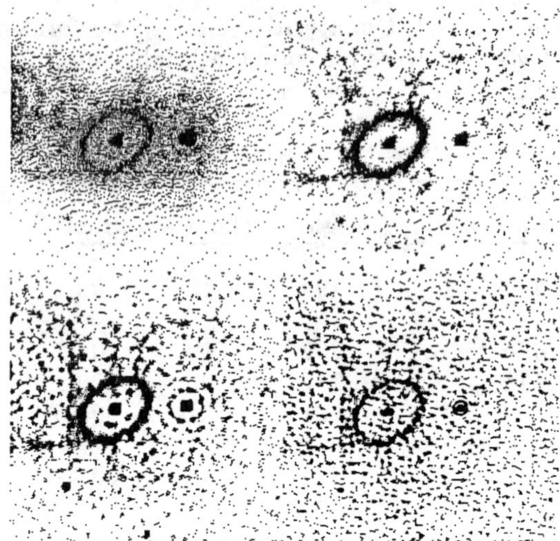

FIGURE I FOC f/96 image of SN1987a. Upper left: original image. Upper right: Lucy restoration. Lower left: Wiener restoration, Markov signal model. Lower right: Wiener restoration, Gaussian signal model. Lucy background is smoother, but stellar *shape* is not good. Gaussian restoration optimizes star image reconstruction.

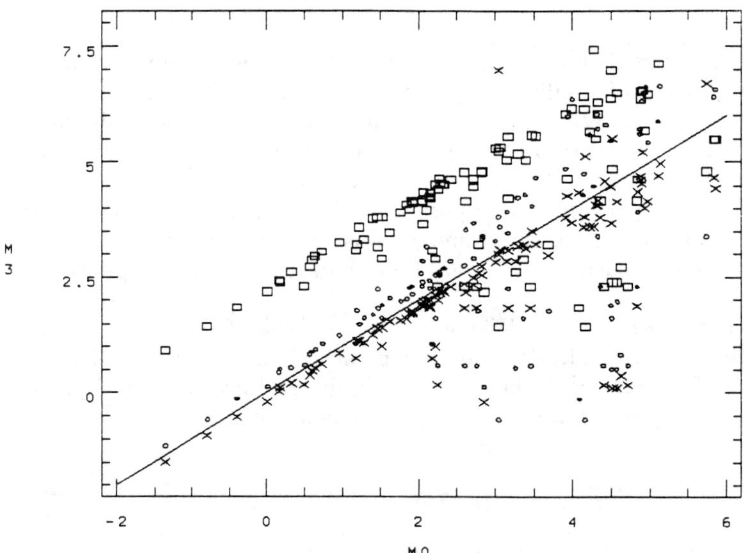

FIGURE II Aperture photometry in simulated HST (FOC f/96) crowded field. Ordinate is measured magnitude in 3 pixel diaphragm. Abcissa is "catalog" magnitude. Squares: original image. Crosses: Wiener image, Gaussian model. Points: Lucy image. Wiener restoration is linear and with less crowding effects. Lucy is clearly non–linear and with more crowding.

TELESCOPE IMAGE MODELLING AT STSCI

JOHN E. KRIST, HASHIMA HASAN, AND CHRISTOPHER J. BURROWS
Space Telescope Science Institute, Baltimore, MD 21218

INTRODUCTION

At the Space Telescope Science Institute (STScI) there has been a continuing effort to model Hubble Space Telescope point spread functions. PSF simulations played a critical role in determining that spherical aberration was causing the distortion visible in HST data, and since the discovery of the problem, work has been underway to more accurately simulate PSFs in order to provide a proper prescription for the corrective optics in COSTAR and WFPC II, as well as for use in deconvolving images and spectra.

THE HST POINT SPREAD FUNCTION

The general properties of the HST PSFs have been described in detail elsewhere (Burrows 1991), so small scale characteristics important for simulating PSFs are discussed here.

Variations of as little as 0.02 waves (referenced at 547 nm) of path error caused by aberrations such as defocus, coma, and astigmatism can visibly alter the PSF. The large amount of structure caused by diffraction effects from the mirror pads, secondary mirror obscurations, and spiders make these changes evident. Coma, for instance, may make the distance between one pad and the core less than that for the other two. Astigmatism causes a slight elongation of a PSF, as well as a cross shaped pattern in the core. With such high sensitivity, structure caused by mirror zonal errors is also evident. Figure I illustrates the effects of these errors, which become more obvious at shorter wavelengths. The diffraction patterns are smoothed by jitter and wavelength dependent variations over the filter bandwidth.

The Wide Field and Planetary Camera (WFPC) repeater optics introduce another set of secondary mirror and spider obscurations to the light path. Because of the spherical aberration, the field dependent shifts of these obscurations relative to the telescope's are obvious, causing a "shadow" of the repeater secondary and distortion of the core near the detector edges. In addition, the entrance baffles to the repeater optics (von Lossberg 1991) cause significant vignetting at the edges of the field. Accurate models require that the positions of these obscurations be known to within about twenty pixels.

STSCI PSF MODELLING SOFTWARE

The primary PSF simulation program at STScI is TIM (Telescope Image Modelling software). In addition, routines written in IDL (Interactive Data Language) provide rapid generation of moderate quality PSFs, primarily for prototyping and phase retrieval fitting.

The TIM package consists of five main stages. The first computes the amplitude distribution function for a given pupil function and a set of aberration and wavelength values. This forms the input to the second stage, which generates a critically sampled PSF for each desired wavelength, and can include mirror microroughness, zonal errors, and jitter. The third stage integrates the PSF onto detector pixels using sinc interpolation with subsampling to ensure accuracy. Since the detector size is user selectable, an oversampled PSF can be produced for deconvolution. The first three stages generate a set of data files for each wavelength, and the fourth combines these via a weighted average determined from instrument and filter throughputs to produce a polychromatic PSF. If necessary, the fifth stage can simulate detector defects, such as saturation, cosmic rays, and bad pixels.

For WFPC, TIM can generate a pupil function for a given detector position.

The TIM software currently executes on VAX VMS systems and is being ported to the IRAF environment by the STSDAS group at STScI. The executables are available via anonymous ftp on STEIS (the Space Telescope Electronic Information System) at stsci.edu. TIM accesses the calibration database, which is also on STEIS, for instrument and filter throughputs. For more information contact the User Support Branch at STScI.

FUTURE DEVELOPMENTS

A great deal of work is being done at STScI to determine the aberration values to a high degree of accuracy, largely through the use of phase retrieval methods. In addition, studies of the effects caused by mirror scattering and jitter, as well as other lower magnitude aberrations, are underway.

In response to the need by scientists outside of the Institute and without VAX systems, a portable PSF generator is being written by Krist which will contain the basic features needed to provide high quality models for deconvolution. The first release of this system is planned for February, 1992.

REFERENCES

Burrows, C. J., Holtzman, J. A., Faber, S. M., Bely, P. Y., Hasan, H., Lynds, C. R., & Schroeder, D. 1991, ApJ, 369, L21

Burrows, C., & Hasan, H. 1991, Telescope Image Modelling User Manual (STScI)

von Lossberg, Bryan "Vignetting at Entrance Baffle of WFC (f/12.9) Relay Optics Housing," JPL interoffice memo dated February 12, 1991

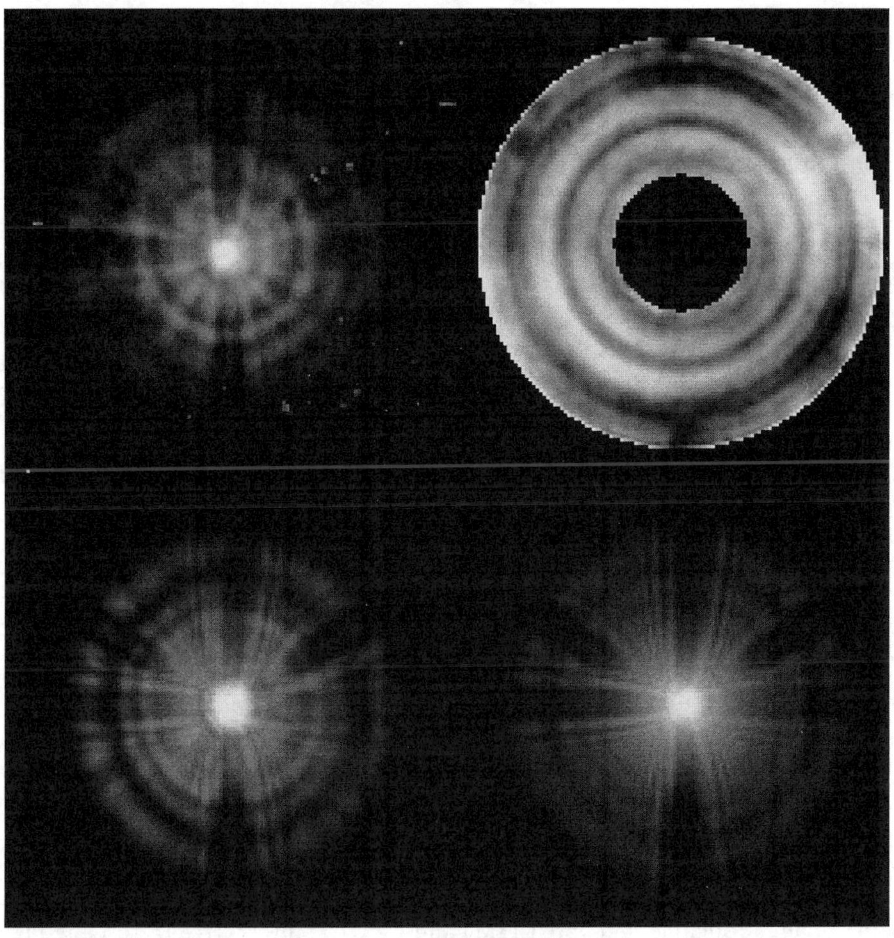

FIGURE I Observed and simulated point spread functions for a star observed through a wide band filter centered at 230 nm (F230W) using the Planetary Camera (chip 6). On the top left is the observed star, with cosmic rays visible. To the right is the mirror zonal error map. The maximum peak-to-valley error is about 0.02 waves (at 547 nm). In the lower right is a TIM model which does not incorporate the mirror map, and to the left is one that does. Both models are polychromatic and were generated with the same aberration settings, which were not fully optimised for this PSF.

DECONVOLUTION OF HUBBLE SPACE TELESCOPE IMAGES USING SIMULATED POINT SPREAD FUNCTIONS

JOHN E. KRIST
Space Telescope Science Institute, Baltimore, MD 21218

INTRODUCTION

Most deconvolution methods used on Hubble Space Telescope (HST) images are limited by the properties of the point spread function (PSF) kernels. Because of a combination of diffraction effects and the presence of spherical aberration (Burrows et al. 1991), the HST PSF is highly dependent on wavelength, field position, and telescope settings. Observed star images are typically used as PSF kernels, but they introduce additional noise and artifacts. Also, it is often difficult or impossible to obtain a high quality observed star for a specific instrument, filter, and position.

One possible solution is to use computer generated PSF models, which can be produced for nearly any condition as long as the necessary optical parameters are accurately known. Models also contain no noise and can be created for any detector pixel spacing, which allows for improved deconvolution results via oversampling (Lucy & Baade 1989, and White 1990) for instruments, such as the Wide Field / Planetary Camera (WFPC), which undersample the PSF.

Presented here are the results of deconvolutions which utilized simulated point spread functions.

GENERATING PSF MODELS FOR DECONVOLUTION

The point spread functions used in this study were generated by the STScI Telescope Image Modelling software (TIM) and by IDL (Interactive Data Language) programs based on routines written by Chris Burrows at STScI. The input parameters were determined by careful examination of observed star images taken at the same settings as the data to be deconvolved. Besides spherical aberration, focus and coma terms were significant, as well as effects caused by the WFPC obscuration shifts.

The PSF models generated by both software packages utilized maps of mirror zonal errors obtained during pre-launch testing. The IDL software was used when time was a limiting factor. The TIM package produces higher quality PSFs because it can generate polychromatic models to account for filter bandwidth and object colors and uses sinc interpolation with subsampling to integrate the critically sampled PSF onto detector pixels. TIM can also simulate jitter to some degree, as well as mirror scattering.

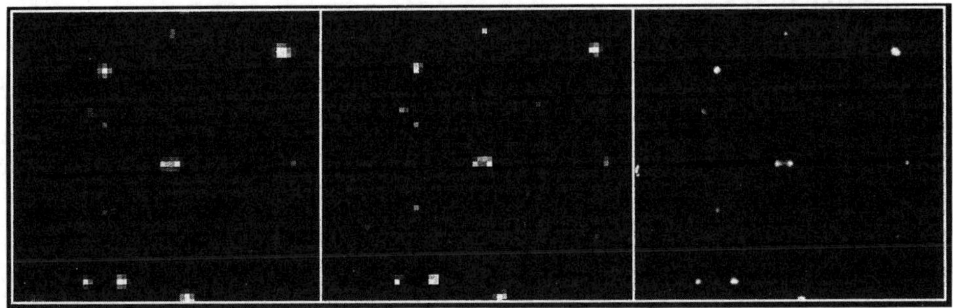

FIGURE I On the left is a 6.4 by 6.4 arcminute (256 by 256 pixels) Wide Field Camera (chip 2) image of a portion of NGC 1850. Two PSFs, one normally sampled and the other oversampled by four in each direction, were generated for the coordinates of the double star. The middle image is the result of Lucy deconvolution using normal sampling. The right image is the result of using the oversampled PSF and block replicating the image by the same factor. The separation of the double star is 0.4 arcseconds.

DECONVOLUTION PROCEDURES

For this project, the STSDAS Wiener routine and a Lucy-Richardson program written in IDL by Rick White (STScI) were used to deconvolve the images. The Wiener task provided quick results at the expense of additional high frequency artifacts and negative flux values. Recent improvements in the STSDAS routine have been implemented which reduce these effects. The Lucy-Richardson algorithm was used without the amplification of noise and additional features associated with the Wiener routine. However, it is an iterative, non-linear process which required, in this study, at least forty to eighty iterations for reasonable results. A significant advantage of the Lucy-Richardson scheme is that it can utilize oversampling, which was implemented using the method described by White (White 1990).

DECONVOLUTION RESULTS

A selection of Wide Field Camera (WFC) and Planetary Camera (PC) images were deconvolved with simulated PSFs. Because the aberration values for the Faint Object Camera were not well determined, images from that instrument were not used.

Figure I illustrates the result of using an oversampled PSF. The original image was block replicated by a factor of four in each direction, and an oversampled PSF was generated.

CONCLUSIONS

From the images presented here and the results of other experiments, it is evident that simulated PSFs can be used successfully for deconvolution of HST data. However, the model PSF must be generated with reasonable estimates of the aberrations and obscuration positions, and work is being done at STScI and other institutions to obtain this information.

The Wiener routine is very sensitive to errors in the PSF, but the Lucy-Richardson method is much less so, since it uses the PSF as a smoothing kernel rather than using it directly as a deconvolution kernel. As the number of iterations increases, Lucy-Richardson tends to adapt itself to the PSF. However, it is a nonlinear process, and enhances some objects more than others, sometimes completely removing fainter stars.

The ability to utilize subsampling is a distinct advantage of simulated PSFs, since trying to obtain a series of slightly displaced observed PSFs is impractical given the various constraints of HST. More HST users, specifically those with Wide Field Camera images, should consider using oversampling methods.

REFERENCES

Burrows, C. J., & Hasan, H. 1991, Telescope Image Modelling User Manual (STScI)

Burrows, C. J., Holtzman, J. A., Faber, S. M., Bely, P. Y., Hasan, H., Lynds, C. R., & Schroeder, D. 1991, ApJ, 369, L21

Lucy, L. B., & Baade, D. 1989, in Proc. 1st ESO/ST-ECF Data Analysis Workshop, Garching, April 1989, eds. P. Grosbol, R. H. Warmels, & F. Murtagh, 219

White, R. 1990, in The Restoration of HST Images and Spectra, August 1990, eds. R. L. White & R. J. Allen, 139

A DIRECT METHOD FOR SPECTRAL AND IMAGE RESTORATION

TIPEI LI AND MEI WU
High Energy Astrophysics Lab., Inst. of High Energy Physics, Chinese Academy of Sciences, Beijing, China

ITERATIVE RESTORATION WITH CONSTRAINTS

The equations connecting the true spectrum $f(i)$ with the observed data $d(i)$ are as follows

$$\Sigma p(i,i')f(i') = d(i) \quad (i = 1,\ldots,N) \tag{1}$$

where $p(i,i')$ is the point-spread function (PSF) of measurement system.

We can use an iterative procedure to derive an approximate solution of eqs.(1) for the true spectra $f(i)$.With the aid of the cross-correlation analysis the positions and intensities of apparent line features can be found and subtracted from the data. Smoothing the rest we obtain the background data $d_b(i)(i = 1,\ldots,N)$. Using d_b instead of d in eqs.(1) we can derive a background distribution $f_b(i)(i = 1,\ldots,N)$ by iteratively solving the equations. In the iterations the following continuous condition is applied to suppress the sudden change of intensity

$$|f_b(i+1) - f_b(i)| \leq \xi d_m/N \tag{2}$$

where $d_m = max(d_b) - min(d_b), \xi \approx 1$. On the basis of continuum f_b, the spectrum f can be derived from the observed data $d(i)(i = 1,\ldots,N)$ by iteratively solving eqs.(1) with the constraint

$$f(i) \geq f_b(i) \quad (i = 1,\ldots,N) \tag{3}$$

MULTI-FITTING SPECTRAL DECONVOLUTION

The direct method of restoration by solving eqs.(1) directly has fairly good sensitivity and resolution capability to show discrete lines existing in a true spectrum. But in the case of poor statistics and worse instrument resolution a real line may be split into two or more discrete lines and the fluctuation in data may easily cause the emergence of pseudo lines in the resultant spectrum. For the purpose of improving restoration quality, we modified the spectral formation equations by the aid of the curve fitting technique.

We imaginarily divided observation period into N_s sets which are equal in length, correspondently the observed spectrum $d(i)(i - 1,\ldots,N)$ could be divided into N_s spectra $d(i;s)(i = 1,\ldots,N; s = 1,\ldots,N_s), \Sigma d(i;s) = d(i)$. For the N_s spectra the spectral formation equations are

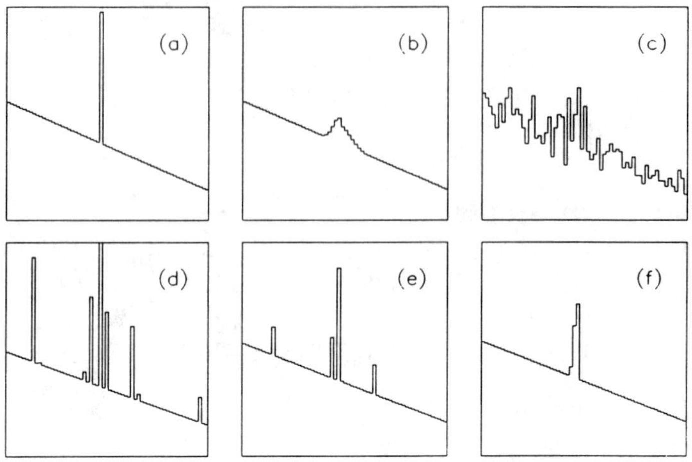

FIGURE I Direct restoration. (a) The Spectrum of object. (b) The theoretical observed spectrum obtained by convolving (a) with PSF. (c) A Monte-Carlo sample from (b). (d)-(f) restorations from (c) by iteratively solving eqs.(7) with M=0, 1, 2 respectively.

$$\Sigma p(i,i')f(i')/N_s = d(i;s) \quad (i=1,\ldots N; s=1,\ldots,N_s) \tag{4}$$

Solving eqs.(4) by least-squares we derived the normal equations

$$\Sigma p_1(i,i')f(i') = \Sigma p(k,i)d(k) \quad (i=1,\ldots,N) \tag{5}$$

where

$$p_1(i,i') = \Sigma p(k,i)p(k,i') \tag{6}$$

The equations (5) and (6) can be used as a new equation system instead of the eqs.(1) to be solved directly.

The above procedure can be repeatedly performed. The equation set derived through M-fold fitting procedures is

$$\Sigma p_M(i,i')f(i') = d_M(i) \quad (i=1,\ldots,N) \tag{7}$$

where

$$\begin{aligned} P_0(i,i') &= p(i,i') \\ d_0(i) &= d(i) \\ P_M(i,i') &= \Sigma p_{M-1}(k,i)p_{M-1}(k,i') \\ d_M(i) &= \Sigma p_{M-1}(k,i)d_{M-1}(k) \end{aligned} \tag{8}$$

In Figure I are presented results of multi-fitting direct spectral restoration for a simulated data.

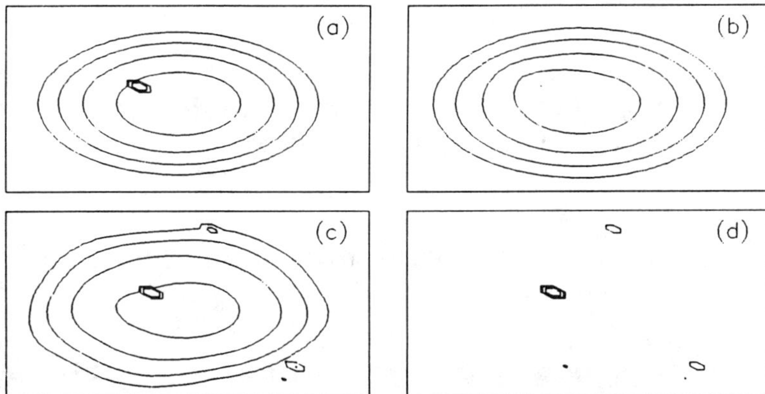

FIGURE II Direct imaging. (a) The true intensity distribution. (b) Theoretical observed image obtained by convolving (a) with PSF. (c) The restoration with M=2 for a Monte-Carlo sample from (b). (d) is obtained by subtracting the background distribution from (c). The total number of photons from the point source is 120 and that from extended background is 1500. Contours are spaced logarithmically and equal for all images.

MULTI-FITTING DIRECT IMAGING

If the intensity distribution of object is $f(i,j)$ and observed distribution $d(i,j)$, for the image region $i = 1,\ldots,N_x$ and $j = 1,\ldots,N_y$, the image formation equations are

$$\Sigma p(i,j;i',j')f(i',j') = d(i,j) \quad (i = 1,\ldots,N_x; j = 1,\ldots,N_y) \tag{9}$$

The procedure discussed above for spectral deconvolution can be naturally extended to the two-dimensional case for performing direct image restoration.

Figure II shows the direct image restoration for a Monte-Carlo sample of a point source lying on top of an extended background. Figure III illustrates the result for data containing two point sources.

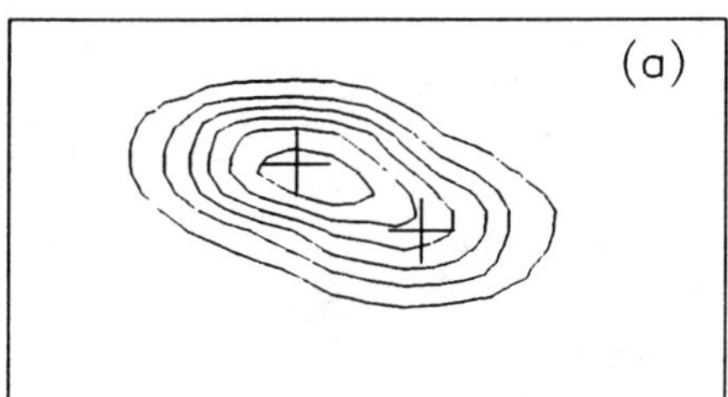

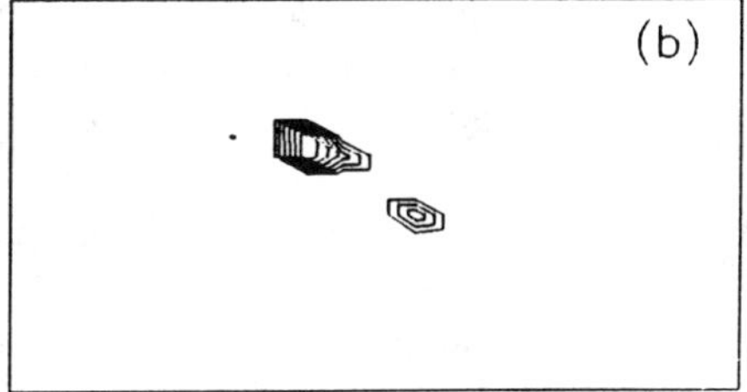

FIGURE III Smoothed simulated data (a) and restoration (b) of two point sources of intensity ratio 2:1 placed at the positions marked by crosses with total 300 photon events.

REAL-TIME MAXIMUM ENTROPY DECONVOLUTION IMPLEMENTED ON 386/486-BASED PCS

AJAI SEHGAL
Canadian Department of National Defense, NDHQ Ottawa, 101 Colonel By Drive, Ottawa, Canada K1A OK2

L. ROBERT MORRIS
Systems Engineering, Carleton University, Ottawa, Canada K1S 5B6

INTRODUCTION

Radio astronomers have pioneered the use of computerised image enhancement techniques for many years, possibly due to the nature of radio astronomy where the computer itself forms the "image". Optical astronomers, on the other hand, have traditionally seen their images directly or have captured them on photographic plates. The development of charge-coupled devices (CCDs) was initially perceived by astronomers as being a simple digitisation of the photographic process, with the associated computer being little more than an instrument controller. Any processing of the optical image was considered to be "doctoring" the data, with a fear that the processing might generate artifacts which would corrupt the data. In a letter to the Astrophysical Journal (Sharp 1989), Nigel Sharp of the National Optical Astronomical Observatories (NOAO) observed that optical astronomers were not making better use of computerised image processing. He demonstrated that considerable image improvement is possible and illustrated clearly that there is no creation of spurious, misleading structure. The challenge implicit in Sharp's letter was the principal motivation for this project.

In the two years since Sharp's letter was published, the use of computerised optical image enhancement by some professionals has grown dramatically, spurred on by events such as the faulty manufacturing of the Hubble Space Telescope and the requirement to "correct" its blurred images. The MAXEN_386 software we have developed enables use of the maximum entropy technique to reach a wider circle of the professional astronomy community, and to be applied, for the first time, in amateur applications.

OBJECTIVES, LIMITATIONS, AND PRECONCEPTIONS

The project was intended to explore the feasibility of providing advanced real-time image processing capabilities on low-cost ($\ll$ \$5000) hardware platforms. The definition of "real-time" adopted was the CCD image acquisition time for deep-sky objects, which is on the order of 5 minutes. The target architecture was the Intel i386/i486 32-bit family. Historically, Intel CPU-based systems have been universally available at a comparatively low cost. However, there has been a general belief that the implementation of sophisticated image processing techniques such as maximum entropy image deconvolution was not possible on such PC's. This preconception was based on at least two limitations traditionally associated with such systems.

First, until very recently, PC-based compilers supported arrays limited to a maximum of only 64-kbytes each. This is incompatible with efficient processing of images larger than 128x128 pixels. Second, without application of non-standard memory manipulation techniques, which are both clumsy and inefficient, total program/data memory space is limited to 640-kbytes. It was not until early 1990 that the first PC-compilers capable of transparently dealing with large arrays and programs in a manner similar to mainframe compilers appeared. These new compilers yield a large, flat address space, with programs and data limited only by physical memory. Also, there has been a misconception with respect to the intrinsic speed capability of i386/i486- based machines. It is not fully appreciated that most compilers on PC's have traditionally emphasized speed of compilation and elegance of programming environment, rather than run-time-efficiency of code. The incredibly sophisticated optimisation capabilities available in most mainframe compilers in the 1970s have simply failed to appear on PCs. And, duc in part to the demands imposed by the pseudo-stack structure of the 80387 register file, even the better PC-based compilers have generally lacked the capability for true optimisation of floating-point code. Thus, few implementations of any floating-point intensive algorithms on PC's have demonstrated the full processing power of these machines.

OPTIMISATION STRATEGIES

Most advanced image processing techniques are Fast Fourier Transform (FFT) bound. We have used techniques (Morris 1977) which employ algorithm-to-processor architecture matching strategies to bring FFT performance on 33Mhz 80486's to within a factor of 5 of contemporary digital signal processing micro-processors.

The first step in the optimisation of maximum entropy deconvolution was in fact the selection of an appropriate FFT algorithm. Examination of the i386/i486 and 80387 architectures suggests that a radix-4 FFT is "optimal", since the radix-4 requirements for 4 data pointers and 8 floating-point accumulators is fully satisfied by these processors. Thus, the major benefits of radix-4 over radix-2 – namely 1/2 the data traffic and 3/4 of the multiplications – are fully realized.

Structures internal to the radix-4 butterflies were also optimised. For example, one can carry out a 3-multiply/3-add "complex multiply" rather than the usual 4-multiply/2-add "complex multiply" (Morris 1977). Besides substitution of a faster add instruction for a slower multiplication, the "3+3" in fact yields less 80387 code.

The FFT was further optimised by other measures: array driven bit reversal; pre-generated trig coefficient arrays; minimisation of the number of operations involved by using multiple butterfly kernels; use of intrinsic i386 processor string manipulation instructions where appropriate; and, the elimination of disk I/O by efficient use of memory available in standard i386/i486 configurations.

Other time-saving techniques applied to the non-DSP portions of the system software include program analysis and optimisation techniques, and program profiling – to find and resolve bottlenecks.

RESULTS AND CONCLUSIONS

The above measures result in a 256-point complex FFT on a 33Mhz 80486 being completed in 2ms. This fast kernel is the core of the maximum entropy software.

Execution times for MAXEN_386 are comparable to those on powerful mainframes: for example, VAXes costing several hundred thousand dollars. Typically, ten iterations in MAXEN_386 require 90 seconds for a 128x128 pixel image and about 5 minutes for a 256x256 pixel image, both on 33MHz i486-based PCs. 1Kx1K pixel images can also be processed with the same time proportionality constant.

The combination of low-cost CCD cameras, relatively straightforward image processing techniques (such as unsharp masking), and MAXEN_386 provides a cost-effective and robust technique known to be capable of consistently yielding gains in resolution of 2-3x (Sharp 1989; Sharp 1991). Our tests have shown that instruments of small aperture (8-16 inch mirror) can consistently yield detailed images, many of which are comparable to published, film-based results on 200 inch telescopes.

Thus, the capability of using advanced image processing techniques, previously afforded only to big-budget professional institutions, is now more cost-effective for all professionals. Also, amateur astronomers – using widely available, low-cost CCDs – can now reliably discern structure in deep space objects (such as NGC 4565) and can resolve significant, changing surface detail on large planets (e.g., Jupiter), even with less-than-optimal seeing and imperfect optics.

REFERENCES

Morris, L. R. 1977, IEEE Trans. Acoust. Speech Sig. Proc, ASSP-26, No. 2, 141-150

Sharp, N. 1989, ApJ, 345, L37-L38

Sharp, N. 1991, Sky & Telescope, May, 571-572

Skilling, J., & Bryan, R. 1984, Mon. Not. R. Astr. Soc., 211, 111-124

SECTION 5

ERROR ANALYSIS AND STATISTICS

IMPROVING THE STATISTICAL METHODOLOGY OF ASTRONOMICAL DATA ANALYSIS

ERIC D. FEIGELSON
Dept. of Astronomy & Astrophysics, Penn State University, University Park PA 16802

GUTTI JOGESH BABU
Dept. of Statistics, Penn State University, University Park PA 16802

ABSTRACT Contemporary observational astronomers are generally unfamiliar with the extensive advances made in mathematical and applied statistics during the past several decades. Astronomical problems can often be addressed by methods developed in statistical fields such as spatial point processes, density estimation, Bayesian statistics, and sampling theory. The common problem of bivariate linear regression illustrates the need for sophisticated methods. Astronomical problems often require combinations of ordinary least-squares lines, double-weighted and errors-in-variables models, censored and truncated regressions, each with its own error analysis procedure. The recent conference *Statistical Challenges in Modern Astronomy* highlighted issues of mutual interest to statisticians and astronomers including clustering of point processes and time series analysis. We conclude with advice on how the astronomical community can advance its statistical methodology with improvements in education of astrophysicists, collaboration and consultation with professional statisticians, and acquisition of new software.

ADVANCED STATISTICS AND OBSERVATIONAL ASTRONOMY

Modern physical scientists have had little exposure to, and typically express little interest in, statistical methodology. This may be due to an underlying approach to empirical science in which the experiment is perfected until the results can be convincingly demonstrated without much statistical treatment of the data. A number of valuable books designed to help physical scientists with statistical analysis are available (e.g. Bevington 1969; Martin 1971; Eadie et al. 1971; Box et al. 1978; Press et al. 1986). But they generally cover only a limited range of methods and are not all widely read. Observational astronomers are thus typically exposed only to a few simple methods during their training.

Astronomical data analysis and interpretation often require quite complex and specialized statistical techniques which are not usually associated with physical experimentation. The objects astronomers observe can not be manipulated (as a chemist might purify a compound), and the experimental conditions can not be adjusted to optimize the results. The astronomer observes only a few salient characteristics of very complex objects. In this sense, the astronomical enterprise resembles the effort the paleontologist studying ancient animals from a few bones, or the mythical blind men inferring the structure of an elephant from sparse and possibly conflicting data. The astronomer also often seeks to understand the underlying properties of huge populations of objects (say, dM stars or starburst galaxies) from study of small, and often not randomly chosen, samples. This situation resembles the dilemma of social scientists and pollsters, who seek to understand human behavior from limited samplings. Finally, the astronomer is often investigating the spatial distribution of objects in N-space, such as galaxy locations in supergalactic coordinates, photons recorded in a CCD array, or objects in parameter space. This problem resembles the efforts of ecologists and geographers in studying spatial patterns.

Specialized statistical sub-fields have developed for all of these purposes. Thus, the astronomer may greatly benefit from study of volumes with unlikely names like *Numerical Ecology* (Legendre & Legendre 1983), *Survival Analysis* (Miller 1981), *Numerical Petrology* (Le Maitre 1982), *Statistical Analysis of Spatial Point Patterns* (Diggle 1983), *Analysis of Messy Data* (Milliken & Johnson 1984-1989), *Incomplete Data in Sample Surveys* (Madow, Nisselson & Olkin 1983, Vol. 2), *Advanced Econometrics* (Amemiya 1985), and the extremely helpful *Encyclopedia of Statistical Sciences* (Kotz & Johnson 1982-90).

LINEAR REGRESSION: A SURPRISINGLY COMPLEX PROBLEM

Consider the common situation where the scientist seeks the slope and intercept, and the uncertainties of these coefficients, of a bivariate dataset (x_i, y_i). Most astronomers would formulate the problem as $Y = \alpha + \beta X$, and compute the slope $\beta = \sum(x_i - \bar{x})(y_i - \bar{y})/\sum(x_i - \bar{x})^2$ and corresponsing intercept $\alpha = \bar{y} - \beta\bar{x}$ by minimizing the sum of the square of the Y residuals from the line. This is the ordinary least-squares (OLS) solution which appears in all elementary texts. However, many astronomers are not aware that this solution is mathematically valid only under quite restrictive conditions, including: the property measured by Y actually depends on X (rather than the converse, or both X and Y depending on other variables); the x_i values have no scatter at all (either intrinsic to the objects or arising from the measurement process); all of the y_i values have the same weight and the scatter does not change with X; the sample is drawn from the underlying population in an unbiased fashion. If any of these conditions do not hold, then the standard OLS(Y|X) line is not a mathematically correct solution.

The astronomer often does not know the causal relations between variables. Consider, for example, the Cepheid period-luminosity relation where a detailed astrophysical theory based on ionization zones in red giant interiors exists. It is not clear that luminosities depend on period and not *vice versa*. Also, almost all observational quantities in astronomy, including the x_i variable in a regression, have scatter due to **both** the measurement process and to intrinsic differences between astronomical objects. Since the measurement errors often depend on many poorly controlled observational conditions (e.g. different telescope exposures, detector signal-to-noise, interstellar absorption, target distances, and so forth), the y_i values often do not have the same weight. Finally, due to cloudy nights, satellite scheduling constraints, and especially flux limits at the telescope, the objects under study are often biased samples of the underlying populations. Thus, it is not difficult to encounter a real-life astronomical regression situation that violates **all** of the assumptions underlying the standard OLS solution.

Some astronomers have been aware of these deficiencies, and have suggested alternatives to the standard OLS line. For unweighted (x_i, y_i) datasets, astronomers since 1940 have plotted the standard OLS(Y|X) and the inverse OLS(X|Y) lines, have used the geometric or arithmetic mean of the two OLS slopes, have minimized the orthogonal distances from the points to the line, and have constructed the angular-bisector of the two OLS lines. The last four of these lines are invariant to the reversal of the X and Y variables, and thus avoid the choice of causal dependency that is so hard to determine in many cases. What many astronomers have failed to realize is that all six of these lines are totally independent functionals of the population from which the data (x_i, y_i) are obtained. They have no more mathematical relation to each other than a linear fit gives the same fit as a quadratic or exponential fit. Furthermore, the uncertainties of the regression coefficients are not the same, and can not be determined analytically for small or poorly correlated samples. Some important studies involved in Hubble constant estimates using galaxy clusters have also used a slope estimate based on the mean of the ratios y_i/x_i. This solution is valid only in the specific case where the scatter in Y increases linearly with X, which generally does not apply.

We have attempted to clarify some of these issues and provide error analyses in our recent papers. Computer codes are available upon request. Schmitt (1985) and Isobe et al. (1986) treat the regression problem where some points are undetected, giving rise to censored values in X and/or Y. This problem has been extensively addressed in the mathematics of epidemiology, industrial reliability and econometrics. Isobe et al. (1990) and Babu & Feigelson (1991) present the unweighted (x_i, y_i) problem discussed above; the debate over OLS lines has waged for over a century in statistics and biometry. Feigelson & Babu (1992a) discuss various other linear regression problems, including truncated regression (which occurs when the sample is biased by a flux limit in one variable), measurement error models (when the scatter is due mainly to known measurement uncertainties rather than to the objects themselves), and problems associated with applying the linear fit of a well-studied calibration sample to a new sample. The truncated regression problem is widely encountered in econometrics, where a maximum likelihood solution is known assuming both variables have underlying Gaussian distributions. Generalizations to non-Gaussian cases are still under investigation by statisticians. The measurement error models are well-established, though many astronomers are not aware of the solutions. In the astronomical literature, we recommend the long-neglected review by Deeming (1968) for the case where the uncertainties are the same for all points, and the review by Murtagh (1990) for the case where the measurement uncertainties differ from point to point.

MORE POTENTIAL APPLICATIONS OF ADVANCED STATISTICS

We recently hosted a conference called *Statistical Challenges in Modern Astronomy*, where astronomers and mathematical statisticians had a rare opportunity to describe problems and potential solutions arising in observational astronomy. The conference covered many topics ranging from observational cosmology (e.g galaxy clustering, Lyα cloud statistics) to flux-limited problems (luminosity functions, nondetections), Bayesian approaches (image restoration, low signal-to-noise and other data analysis problems), multivariate and classification techniques, and more. We mention here two of the principal topics aired at the conference, concentrating on methods that may be unfamiliar to astronomers yet with potential for solving problems encountered in observational data analysis. Details can be found in the conference proceedings (Feigelson & Babu 1992b).

Spatial point processes.

Many problems in astronomy involve analysis of complicated distributions of point in N-space. When is a distribution not random, clustered or sheets-plus-voids, smooth or multimodal, elongated or isotropic? T. Beers discussed several methods. The Lee (1979) statistic gives a shape-invariant measure of substructure by projecting the distribution onto lines with various orientations. Beers is distributing a code implementing an adaptive kernel smoothing algorithm (Silverman 1986) has proved useful in finding structure within Abell clusters. Partitioning and fuzzy clustering algorithms described by Kaufman & Rousseeuw (1990, with software) may also be useful. J. Hartigan recommended tests based on bimodality and second-largest clusters on the minimum spanning tree to determine the presence of significant substructure (Hartigan 1988). F. Bookstein reviewed approaches to characterization of voids and shapes using contact distribution functions (Stoyan et al. 1987). He also described methods that will uncover shape- and size-invariant structures, based on measures of curvature and second-derivatives of the distribution (Romeny 1991).

Time series analysis.

T. Subba Rao recommended the monograph by Priestley (1981) regarding detection of periodicities, and suggested astronomers investigate the second order spectral density function (or bispectrum) of complex systems. It can detect periodicities in the presence of red noise, and can show whether a system is chaotic (Subba Rao & Gabr 1984; Subba Rao 1991). To characterize low-frequency noise and quasi- periodic oscillations seen in X-ray binaries, W. Liggett suggests minimum entropy methods of Donoho (1981). T. Loredo recommends Bayesian spectral methods as improvement over standard Fourier methods in many cases (Bretthorst 1988). Finally, J. Scargle and T. Speed reviewed the extensive new literature on the presence of 'chaos' in time series, and their application to determining when chaos may be present in astronomical time series.

HOW TO IMPROVE ASTRONOMICAL STATISTICAL METHODOLOGY

The situation described above suggests that there is far more complexity – and far more capability – to statistical analysis than astronomers are generally aware. The most frequent comment we received from the astronomers attending the *Statistical Challenges in Modern Astronomy* conference was: 'I did not know there was so much that I did not know'. We further suggest that astronomers should generally not attempt to derive new statistical procedures (a glance at Appendices of recent *Astrophysical Journal* articles will indicate this is quite common) until they have studied in detail, or consulted professionals concerning, the existing statistical literature on the subject of interest. Astronomers should learn the state-of-the-art in relevant branches of applied statistics, adopt what is useful and proceed to develop new techniques (if possible, collaborating with statisticians) only where necessary. The product of such interactions will be rapid acquisition of advanced and appropriate methodology, the elimination of unnecessary rederivations, and the concentration on important methodological problems that truly need development to further astronomical research.

At present, such interactions between astronomers and statisticians are infrequent. Historically, this has not always been the case. In fact, one could justifiably argue that modern statistics was founded by astronomers (Stigler 1986; Hald 1990)! Some modern 'robust' procedures have their seeds in the work of Galileo. Laplace proposed the least-squares procedure, and Gauss developed the normal distribution, in treatises on celestial mechanics. Royal Astronomer George Airy wrote a monograph on error analysis. However, during the 20th century when astrophysics rose to prominence and statisticians looked more to the biological or social sciences for applications, the connections between the two disciplines waned. A resurgence of communication and mutual enrichment is quite possible. The *Statistical Challenges in Modern Astronomy* conference drew an impressive group of distinguished scientists and journal editors from the statistical community to hear about astronomical problems.

On a practical level, we have a number of recommendations for the improvement of statistical methodology in observational astronomy. In many cases, investment of funds as well as effort is needed to achieve the potential improvements.

1. *Education.* The level of statistical expertise among all astronomers should be raised somewhat with the incorporation of statistics into the curriculum. A 2-credit course of this type has been taught to Astronomy graduate students at Penn State by John Nousek. Statisticians should be encouraged to develop with astronomers courses and readings oriented to the physical sciences.

2. *Collaboration and Consulting.* A significant number of statisticians are eager to apply their expertise to exciting problems in cosmology and space astronomy. Several mechanisms can be envisioned. First, close collaboration could be established, resulting in joint publications. The National Science Foundation has expressed considerable interest in funding such cross-disciplinary research efforts, and some funding is available through the NASA Astrophysics Data Program. Second, university astronomers

should allocate some grant funds to hire consultants through existing statistical consulting centers, used primarily by social scientists. Third, national observatories, NASA institutes and satellite mission groups, and major software development teams should hire a few applied statisticians as staff scientists.

3. *Software.* First, an enormous range of statistical methods are implemented in commercial software packages. The *Statistical Analysis System* or *SAS* is perhaps the most comprehensive. These packages are already available on university computer systems (though rarely used by astronomers). Second, a smaller but significant amount of statistical software is available in public domain software repositories (Feigelson & Murtagh 1992). Third, when specialized software is needed, astronomers should produce and distribute portable user-friendly packages (e.g. Jeffreys et al. 1990; LaValley et al. 1992; Feigelson & Babu 1992). Fourth, we recommend that managers of the large distributed astronomical software systems, like IRAF/STSDAS/PROS, contract for the inclusion of a large array of standard statistical methods into the astronomical packages.

ACKNOWLEDGMENTS

This work is supported by NASA grants NAGW-1917 and NAGW-2120, and NSF grant DMS-9007717.

REFERENCES

Amemiya, T. 1985, Advanced Econometrics, Cambridge:Harvard Univ. Press

Babu, G. J. & Feigelson, E. D. 1991, Commun. Statist. Sim. and Comput., in press

Bartlett, M. S. 1991, J. Roy. Statist. Soc. A, 153, 321

Bevington, P. R. 1969, Data Reduction and Error Analysis for the Physical Sciences, N.Y.:McGraw-Hill

Box, G. E., Hunter, W. G. & Hunter J. S. 1978, Statistics for Experimenters: An Introduction to Design, Data Analysis and Model Building, N.Y.:Wiley

Bretthorst, G. L. 1988, Bayesian Spectrum Analysis and Parameter Estimation, N.Y.:Springer-Verlag

Deeming, T. J. 1968, Vistas in As., 10, 125

Diggle, P. J. 1983, Statistical Analysis of Spatial Point Patterns, London:Academic Press

Donoho, D. L. 1981, in Applied Time Series Analysis II, D. F. Findley, N.Y.:Academic Press

Eadie, W. T., et al. 1971, Statistical Methods in Experimental Physics, Amsterdam: North-Holland

Feigelson, E. D. & Babu, G. J. 1992a, submitted for publication

Feigelson, E. D. & Babu, G. J. 1992b, Statistical Challenges in Modern Astronomy, N.Y.:Springer-Verlag

Hald, A. 1990, A History of Probability and Statistics and Their Applications Before 1750, N.Y.:Wiley

Hartigan, J. A. 1988, in Classification and Related Methods of Data Analysis, H. H. Bock, Amsterdam:North-Holland

Isobe, T., Feigelson, E. D. & Nelson, P. I. 1986, ApJ, 306, 490

Isobe, T., Feigelson, E. D., Akritas, M. & Babu, G. J. 1990, ApJ, 364, 104

Jeffreys, W., et al. 1990, Bull. Am. As. Soc. 22, 927

Kaufman, L. & Rousseeuw, P. J. 1990, Finding Groups in Data, N.Y.:Wiley

Kotz, S. & Johnson, N. L. 1982-90, Encyclopedia of Statistical Sciences, vols. 1-10, N.Y.:Wiley

LaValley, M., Isobe, T. & Feigelson, E. D. 1992, this volume

Lee, K. L. 1976, J. Amer. Stat. Assoc., 74, 708

Legendre, L. & Legendre, P. 1983, Numerical Ecology, Amsterdam:Elsevier

Le Maitre, R. W., Numerical Petrology: Statistical Interpretation of Geochemical Data, Amsterdam:Elsevier

Madow, W. G., Nisselson, H. & Olkin, I. 1983, Incomplete Data in Sample Surveys: Theory and Bibliographies, 2, N.Y.:Academic Press

Martin, B. 1971, Statistics for Physicists, London:Acad. Pr.

Miller, R. G. Jr. 1981, Survival Analysis, N.Y.:Wiley

Milliken, G. A. & Johnson, D. E. 1989, Analysis of Messy Data: Nonreplicated Experiments, 2, N.Y.:Van Nostrand Reinhold

Murtagh, F. 1990, in Errors, Bias and Uncertainties in Astronomy, C. Jaschek & F. Murtagh, Cambridge:Camb. Univ. Press, p. 385

Murtagh, F. & Feigelson, E. D. 1992, this volume

Press, W. H., et al. 1986, Numerical Recipes:The Art of Scientific Computing, Cambridge:Camb. Univ. Press

Priestley, M. B. 1981, Spectral Analysis and Time Series, N.Y.:Academic Press

Romeny, B., et al. 1991, in Information Processing in Medical Imaging, A. Colchester & D. Hawkes, Berlin:Springer-Verlag

Schmitt, J. H. 1985, ApJ, 293, 178

Silverman B. 1986, Density Estimation for Statistics and Data Analysis, London:Chapman & Hall

Stigler, S. M. 1986, The History of Statistics: The Measurement of Uncertainty before 1900, Cambridge:Harvard Univ. Press

Stoyan, D., Kendall, W. & Mecke, J. 1987, Stochastic Geometry and Its Applications, Chicester:Wiley

Subba Rao, T. 1991, in Nonlinear Modelling and Forecasting, M. Casdagli & S. Eubank, N.Y.:Addison-Wesley

Subba Rao, T. & Gabr, M. M. 1984, An Introduction to Bispectral Analysis and Bilinear Models, Lecture Notes in Statistics, 24, Berlin:Springer-Verlag

Tong, H. 1990, Non-linear Time Series:A Dynamical System Approach, Oxford:Oxford Univ. Press

ASURV: ASTRONOMY SURVIVAL ANALYSIS PACKAGE

M. LAVALLEY
Statistics Department, Penn State University, University Park PA 16802

T. ISOBE
Center for Space Research, Massachusetts Institute of Technology, Cambridge MA 02139

ERIC FEIGELSON
Department of Astronomy and Astrophysics, Penn State University, University Park PA 16802

BACKGROUND

Observational astronomers frequently encounter the situation where they observe a particular property (*e.g.* far-IR emission in spiral galaxies, X-ray emission in young stars, CO emission in starburst galaxies) of a previously defined sample of objects, but fail to detect all of the objects. The data set then contains nondetections as well as detections, preventing the use of simple and familiar statistical techniques in the analysis.

A number of astronomers have recently recognized the existence of statistical methods, or have derived similar methods, to deal with these problems. The methods are collectively called 'survival analysis' and nondetections are called 'censored' data points. These methods recover important information implicit in the failure to detect some objects under reasonable mathematical assumptions. ASURV is a menu-driven stand-alone computer package designed to assist astronomers in using methods from survival analysis. Rev. 1.0 of ASURV provides all of the functions described in Schmitt (1985), Feigelson & Nelson (1985) and Isobe, Feigelson & Nelson (1986), plus some additional calculations.

METHODS AVAILABLE IN ASURV

The statistical methods for dealing with censored data might be divided into a 2x2 grid: parametric *vs.* nonparametric, and univariate *vs.* bivariate. We have chosen to concentrate on nonparametric models, since the underlying distribution of astronomical populations is usually unknown.

Univariate Methods

The Kaplan-Meier estimator gives the distribution function of a randomly censored sample. First derived in 1958, it is the unique, self-consistent, generalized maximum-likelihood estimator for the population from which the sample was drawn. In its cumulative form it has analytic asymptotic (for large N) error bars, but is otherwise identical to the differential 'redistribute-to-the-right' algorithm derived by Avni et al. (1980) and others.

Two-Sample tests measure whether two censored samples are drawn from the same parent population. Several tests for censored data are provided: the logrank, Gehan, Peto-Peto, Peto-Prentice tests. They are generalizations of standard tests for uncensored data (*e.g.* Wilcoxon and logrank two-sample tests). They differ in how the censored data are 'scored', and thus vary in sensitivity to differences at the two ends of the distribution.

Bivariate Methods

Correlation tests quantify the probability that two variables are correlated in a censored sample. ASURV provides the Cox hazard model and generalizations of Kendall's τ and Spearman's ρ. The latter methods allow censoring in both variables.

Linear regressions give slope and intercept estimates (and their standard deviations) for a censored correlated dataset. Lines are calculated assuming normal residuals (the EM algorithm), Kaplan-Meier residuals (the Buckley-James method), and using the binned 2-dimensional Kaplan-Meier (Schmitt's method). The latter permits censoring in both variables.

REV 1.0 MODIFICATIONS AND ADDITIONS

All areas of ASURV have been updated in going from Rev 0.0 to Rev 1.0. Below are some of the most important changes.

- In Rev 0.0, the Kaplan-Meier estimator for left-censored (upper limits) data started at zero and increased to one, inconsistent with the statistical definition of the survival function. In Rev 1.0, this estimate has been changed so it decreases from zero to one for both right and left-censored data.

- A differential, or binned Kaplan-Meier estimator has been added to the package. This allows the user to find the number of points falling into specified bins along the X-axis according to the Kaplan-Meier estimated survival curve. However, no error analysis is available.

- The two-sample tests in Rev. 0.0 used a permutation variance which assumes that both groups being considered are subject to the same censoring pattern. Rev. 1.0 now uses a hypergeometric variance for most of the tests, which have recently been shown to be more 'robust' against differences in the censoring patterns. Also, the Peto-Prentice test, which uses a very effective asymptotic variance, has been added to the package.

- A new correlation coefficient has been added, a generalized Spearman's ρ developed by Akritas (1990). It is less computer-intensive than the Kendall's τ coefficient in Rev. 0.0.

- For Schmitt's linear regression, Rev. 1.0 now calculates standard error estimates for the slope and intercept based upon the bootstrap technique.

- Coding standards have been improved, with updating from FORTRAN 66 to 77. Portability to Sun UNIX, Vax VMS, IBM CMS, IBM-PC MS-DOS and Macintosh has been demonstrated.
- The interface and output has been streamlined and improved. Warnings are now printed when there is reason to suspect the results of a test are unreliable.

OBTAINING ASURV

Rev. 1.0 is available as a stand-alone package to any member of the astronomical community without charge. We provide the FORTRAN source code (59 subroutines, about 15,000 lines or 0.5 MBy), a 30-page manual, and test files. No external libraries are needed. We prefer to distribute the package by electronic mail. Please send your request to:

`code@stat.psu.edu`

specifying 'ASURV Rev. 1.0' and providing both email and regular mail addresses. The code can also be mailed, if necessary, on a 3 1/2-inch double-sided high-density MS-DOS diskette, or a 1/2 inch 9-track tape (1600 BPI, written on a UNIX machine). In addition to the stand-alone version, Rev. 0.0 has also been implemented in IRAF/STSDAS.

ACKNOWLEDGMENTS

The production and distribution of ASURV Rev 1.0 is funded by a grant from the Center for Excellence in Space Data and Information Sciences, NASA grants NAGW-1917 and NAGW-2120, and NSF grant DMS-9007717. We are grateful to Prof. Michael Akritas (Dept. of Statistics, Penn State), Drs. Mark Wardle (Dept. of Physics and Astronomy, Northwestern University), Paul Eskridge (Havard-Smithsonian Center for Astrophysics), and Eric Smith (Laboratory for Astronomy and Solar Physics, NASA-Goddard Space Flight Center) for advice and assistance.

REFERENCES

Avni, Y., Soltan, A., Tananbaum, H., & Zamorani, G. 1980, ApJ, 235, 694

Feigelson, E. D. 1990, in Errors, Bias and Uncertainties in Astronomy, C. Jaschek & F. Murtagh, Cambridge U. Press: Cambridge, 213

Feigelson, E. D., & Nelson, P. I. 1985, ApJ, 293, 192

Isobe, T., Feigelson, E. D., & Nelson, P. I. 1986, ApJ, 306, 490

Schmitt, J. H. M. M. 1985, ApJ, 293, 178

A K-NEAREST NEIGHBORS METHOD TO FIND EMBEDDED CLUSTERS IN THE CEPHEUS-CASSIOPEA REGION

L. PÁSZTOR[1] AND L.V. TÓTH
Department of Astronomy of the Loránd Eötvös University, Budapest
Ludovika tér 2., H-1083, Hungary

L. G. BALÁZS
Konkoly Observatory, Budapest Box 67, H-1525, Hungary

INTRODUCTION

Our paper describes a non-parametric statistical pattern recognition procedure performed on a sample selected by far infrared properties to detect surface density anomalies which may indicate grouping in the sample set. The method itself is a nearest neighbors method but developed to provide a measurable significance indicating the reliability of the results.

The aim of the recent work is to obtain a more automatic way for finding groups and to show its applicability.

SELECTION OF POINT SOURCES

After some preliminary results (Pásztor & Tóth 1990), we decided to study the IRAS point sources detected in the galactic region $110^o < l < 130^o$ and $5^o < b < 25^o$. We set up the following selection criteria to obtain a convenient sample: (1) let the source have moderate or good flux qualities in both 25 and 60 μ bands and/or moderate or good flux qualities in 60 and 100 μ bands; (2) let the source have non-stellar-like IRAS colors, that is let $F(12) \leq F(25)$; (3) let the source have at least 90% point-like feature in longer wavelengths; (4) finally we excluded the sources associated with known galaxies. The sample obtained after the former selections contains 272 objects.

IDENTIFICATION OF GROUPS

Distribution of separations of nearest neighbors is a fairly good indicator of small scale structures of spatial point patterns (eg. Bahcall & Soneira 1981). Nevertheless, we need some reference data if we want to understand the distribution functions of separations derived from the sample. However except for simple cases, to give an adequate analytic model for the spatial structure is not trivial.

[1]Research Institute for Soil Science and Agrochemistry, Budapest Herman Ottó út 15., H-1025, Hungary

For this reason we turned to Monte-Carlo simulations to generate artificial samples with same marginal distributions as in the real case (but otherwise the x and y coordinates were independent) to eliminate the large-scale structures modulating the surface distribution of the points, since our aim was the study of small scale deviations.

Then we compared (χ^2, Kolmogorov-Smirnov homogenity tests) the distribution functions of separations of 1st to 10th nearest neighbors of the real sample with those of the generated ones. All the homogenity tests showed significant deviation between the real and generated samples at a level of more than 99%, so we believe that our following statements are statistically well-founded.

For the identification of possible group members we set up criteria for the separations based on differences of frequencies. Since the patchiness is located on small scales the deviations are significant at small separations. At the distance where these differences dropped below zero we set a limit for each separations, thus we gained criteria for the distance of the 1st to 10th nearest neighbors. Then we selected the point sources which fulfilled at least 3 of the 10 criteria. The final 70 objects are clearly situated at the most populated regions of the area and so they probably outline the central part of possible groups.

To determine the number of groups a generalized measure of information, namely the Akaike information criterion (AIC) may be proposed (Akaike 1973). The definition of AIC is:

$$AIC = -2ln(maximum\ likelihood) + 2(\#\ of\ parameters)\ .$$

The smaller is AIC the better is the classification of the points into groups. Its estimate can be computed by , $AIC(estimated) = n\ ln(R) + 2p$, where n is the number of objects to be grouped, p is the estimated number of groups and R is the residual sum of squares of deviation from the fitted model. In the present sample the optimal number of groups was 10.

After fixing the the number of groups the cluster centers may be determined by a common clustering routine based on nearest centroid sorting with constant number of clusters .

MAPPING THE GROUPS

Clearly, the objects gained by the selection criteria do not necessary total the whole groups, since the method prefers the central members to the outliers. Thus if we want to get complete picture of the groups we might keep nothing else but the cluster centers obtained in the former analysis.

We estimated the surface density function by star counting in radial direction around the identified cluster centers. The distance where the cluster density reaches the field density is considered as the radius of the given cluster and sources inside this radius are treated as its members. Our groups have a radius of about 0.5-1 degrees and some of them can be divided into two parts, a dense core and a sparse outer region.

RESULTS

With the aid of our group identification procedure we found 10 groups in the sample with 88 members totally. The main characteristics of the derived groups, and the associated galactic objects are listed in Table I. The associated objects are either young stellar groups (open clusters) and/or dark clouds where star forming processes are supposed to take place. We believe that the groups are extreme young star clusters in their state of birth, or youngest parts of open clusters.

TABLE I Short description of the groups.

Gr	l	b[1]	#[2]	R[3]	Density[4]	Associated objects
1	114.61	14.57	10	1	4.5	LDN1251, KHAV776
2	111.81	11.92	5(4)	1(0.5)	2.3(7.2)	LDN1217, LDN1219, OCL255
3	110.27	11.24	7	0.5	12.6	B175 OCL255
4	115.02	8.57	8(5)	1(0.5)	3.6(9.0)	KHAV788, OCL271
5	118.55	5.15	15(13)	1(0.5)	6.8(23.4)	LDN1266, LDN1267, LDN1268, LDN1269 LDN1275, OCL286, 2 HII REGIONS
6	117.56	6.15	9	0.5	16.2	LDN1267, OCL280, HII REGION
7	119.57	5.62	10	0.6	12.5	LDN1273, KHAV797, HII REGION
8	118.63	6.17	6	0.5	10.8	LDN1273, KHAV797, HII REGION
9	117.17	5.66	9(7)	1(0.5)	4.1(12.6)	HII REGION
10	129.17	6.00	9	1	4.1	LDN1336, LDN1328, KHAV030, VB060

[1]The group center's galactic coordinates [1950].

[2]Number of identified group members.

[3]Radius of group in degrees.

[4]Surface density relative to field density.

[5]Numbers in brackets refer to the dense cores.

Our method is applicable in the description of any kind of objects such as stars or galaxies. The digital catalogues like HST catalogues or the catalogues of the planned DENIS or the ISO satellite may be subject of studies using this method; combined with suitable user interface to select samples for our procedure.

ACKNOWLEDGMENTS

L. Pásztor was partially supported by a grant of MHB 'For the Hungarian Science' Fundation.

REFERENCES

Akaike, H. 1973, in 2nd International Symp. on Information Theory, B. N. Petrov, F. Csáki, 267

Bahcall, J. N., & Soneira, R. M. 1981, ApJ, 246, 122

Pásztor, L., & Tóth, L. V. 1990, in Star clusters and associations, B. A. Balázs, G. Szécsényi-Nagy, Eötvös Univ., Budapest, 91

MULTIVARIATE STATISTICS ON HI AND IRAS IMAGES

L. VIKTOR TÓTH
Department of Astronomy, Eötvös University Budapest, Ludovika tér 2., H-1083, Hungary

LAJOS G. BALÁZS AND PÉTER ÁBRAHÁM
Konkoly Observatory, Budapest, P.O. Box: 67., H-1525, Hungary

ABSTRACT A procedure based on multivariate statistical methods have been worked out, making possible the simultaneous study of multi-wavelength data with large number of variables.

Testing the method a data set of HI (21cm) spectra and IRAS surface brightnesses from 891 positions of a Cep-Cas region have been analysed. The 891 HI spectra were replaced by 891 × 3 factor coefficients: the new coordinates in the space of the first 3 factors (as orthogonal axes) resulted by a principal component analysis (PCA).

A cluster analysis have been performed in the 7 dimensional parameter space of 3 factor coefficients and 4 IRAS fluxes. Results have been compared to the previous knowledge on the area.

INTRODUCTION

Large number of observed quantities, especially those of spectral data with a high number of channels requires multidimensional statistics for their handling. The methods try to reduce the dimensionality of the problem, and to find structures in the parameter space, thus prepare data for evaluation.

Our aim was to provide a fast, automatic procedure for simultaneous analysis of radio and IRAS data. This procedure has been used in the study of interstellar matter in a nearby region.

INPUT DATA

HI spectra were taken from Heiles & Habing (1974) having a velocity resolution of 2.11 km/s, a beamwidth (HPBW) of 36' and an average spacing of 0.5°. Only 60 channels from the 9th to 68th channels (corresponding to -45.38 km/s and +16.89 km/s respectively) were used because we expected that the velocities of the Local Arm clouds and their parts lay into this range.

There are 891 spectra in the survey which fall on the $b \geq 10°$ part of the IRAS Sky Flux Map (SFM) PL2; with the approximate boundaries of $b_{min} = 10°$ $b_{max} = 22°$ galactic latitude and $l_{min} = 110°$, $l_{max} = 130°$ longitude.

The four bands of the PL2 IRAS SFM were convolved with a two dimensional Gaussian to match the resolution of the HI survey. The convolved images were sampled at the positions of HI spectra. The original SFM fluxes were used without any correction (for Zodiacal light or galactic background).

DATA ANALYSIS

Principal Components Analysis (PCA)

In order to reduce the number of parameters describing our data set, we performed principal components analysis assuming the individual spectra as variables and the channels as cases. Detailed description of the PCA was given by Murtagh & Heck (1987). To spare computing time, 40 from the original 891 spectra have been selected by random sampling, and carried out the analysis on this reduced sample.

The PCA resulted in three main eigenvectors (with eigenvalues > 1) describing 97% of the variance of variables included in the computation. The 3 significant principal component vectors are the so called factors. We defined the factor coefficients as the scalar product of the of the original spectra with factors, and computed them for all the 891 spectra in the data set.

Cluster Analysis on the Multiwavelength Data

Providing three new variables representing the HI data (see above), our procedure gives a possibility to analyze HI and IRAS data simultaneously. In order to localize regions with similar HI spectra and similar infrared emission, we performed a cluster analysis in the 7 dimensional (7D) parameter space of the 3 HI factors and the IRAS brightnesses at four wavelengths. We used a K-means clustering algorithm to find the possible groups in the 7D space.

The location of the 9 clusters (found by the analysis) is shown on Figure I reprojected onto the sky. The distribution of the points with similar HI and similar IRAS appearance is not random; several well defined regions can be recognized. The typical HI spectra of these regions are also plotted, all of which were computed by averaging the factor coefficients inside one region.

RESULTS

The Procedure:

Our method may be used in an almost fully automatic way, and requires a short computational time. It provides information on the characteristic spectral features, and possible clouds for further studies.

The Particular Case Show:

I. The two main components of the HI gas belong onto the nearby 0 km/s clouds (expanding shell of a local bubble), and the so called -12 km/s clouds, see Dame et al. (1987).

II. The Cep/Cas bubble found by CO measurements (Grenier et al. 1989) neighbouring the Cepheus Flare giant molecular cloud can be identified on Figure I; as we recognized from the HI and IRAS data set.

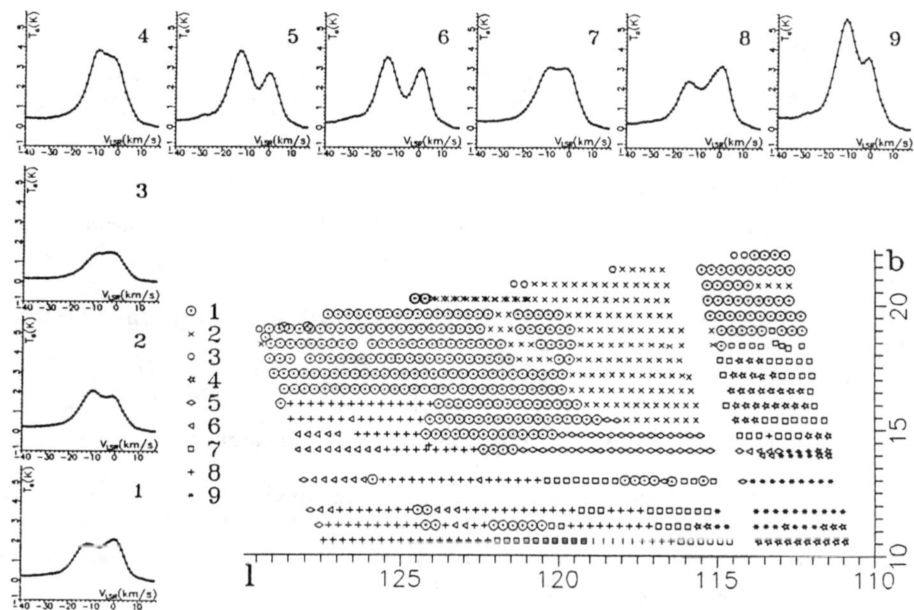

FIGURE I The 9 clusters of the 7D space of HI factor coefficients and 4 IRAS fluxes reprojected onto the sky. Typical HI spectra of the well defined regions are also presented.

ACKNOWLEDGMENTS

The authors thank the kind help (sending the machine readable tape version of the Heiles-Habing survey to Hungary) of the National Space Science Data Center.

REFERENCES

Dame, T. M., Ungerechts, H., Cohen, R. S., de Geus, E. J., Grenier, I. A., May, J., Murphy, D. C., Nyman, L.-Å., & Thaddeus, P. 1987, ApJ, 322, 706

Grenier, I. A., Lebrun, F., Arnaud, M., Dame, T. M., & Thaddeus, P. 1989, ApJ, 347, 231

Heiles, C., & Habing, H. J. 1974, A&AS, 14, 1

Murtagh, F., & Heck, A. 1987, Multivariate Data Analysis, Ap. Sp. Sc. Lib. Dordrecht Riedel Publ., Dordrecht, Holland, 13

FAINT OBJECT CLASSIFICATION USING NEURAL NETWORKS

M. SERRA-RICART
Instituto de Astrofísica de Canarias, E-38200 La Laguna, Tenerife, Spain

V. GAITAN
Laboratori de Física d'Altes Energies, Universitat Autònoma de Barcelona, E-08193 Bellaterra, Barcelona, Spain

S. DELGADO AND I. PÉREZ-FOURNON
Instituto de Astrofísica de Canarias, E-38200 La Laguna, Tenerife, Spain

ABSTRACT In this paper we propose a method to classify faint objects from digital astronomical images based on multilayer feed-forward neural networks which have been trained by the back-propagation procedure (Rumelhart 1986) from real data.

INTRODUCTION

The morphological classification of astronomical objects has usually been carried out by measuring some properties of their images which allow objective separation in different types. The variety of approaches to resolve the problem can be divided in two main groups: a) Bayesian methods (Sebok 1979; Valdes 1982a), in which the objects are classified maximizing the probability that the observed distributions of pixel intensities agree with given models of different object classes and, b) parametric methods (Jarvis & Tyson 1981) where the main parameters of the distribution of pixel intensities are measured and the classification is made according to the clustering of data points in the parameter space.

The classifier described in this paper combines both methods using a layered feed-forward neural network. The input of the network is a set of parameters measured from the image, as in a parametric method, but since the network output is equivalent to a Bayesian decision rule (Garrido & Gaitan 1991), the method could also be called Bayesian.

In a typical architecture for a layered feed-forward neural network the neurons are grouped into layers (Murtagh 1991). There is one input layer where the information comes in, one or several hidden layers where the information is processed, and one output layer which gives the output of the network. The input of neuron i in layer $l+1$ is given by

$$I_i^{l+1} = \sum_j w_{ij}^{l+1} S_j^l + B_i^{l+1}, \qquad (1)$$

where the sum is extended over the neurons of the preceding layer (l), S_j^l is the state of the neuron j, w_{ij}^{l+1} is the connection weight between the neuron j and the neuron i, and B_i^{l+1} is a bias input to neuron i. The state of a neuron is a function of its input, $S_j^l = f(I_j^l)$. In this work we take $f(I) = 1/(1 + exp(-I))$, the so-called 'sigmoid function', which is similar to the response curve of the biological neuron. The choice of this function constrains the state of the neuron to a value between 0 and 1. The activity of the neurons in the input layer is propagated forward to the output layer by equation (1). In general, the network must be trained. This is done by applying learning algorithms to the network, in our case the error back-propagation algorithm (Rumelhart 1986), to modify the weights and the bias term in such a way that the differences between the actual and the desired output is minimized. By interpreting the activity of the output units as a measure of the probability that the input pattern belongs to a certain category, the network can be used for classification.

SUPERVISED CLASSIFICATION

The objects classified with the neural network were contained in astronomical images obtained during the night of June 17th, 1990, at the 2.5 m Nordic Optical Telescope (La Palma). We used a TK512-011 512 × 512 chip at the Cassegrain focus. The scale was 0.2 *arcsec/pixel*, and the field of view was 102" × 102". In a total of 23 fields, 1273 objects were detected with the FOCAS package (Valdes 1982b) in the IRAF environment. These objects were classified in four categories by a human expert: a) uncertain objects, b) cosmic rays, c) stars, and d) galaxies. Our working sample of 478 objects included all those classified as cosmic rays, stars or galaxies. This sample was separated in two sub-samples, the learning sample and the test sample. To perform the classification we chose a multilayer feed-forward neural network with 10 units in the input layer, one hidden layer with 21 units, and 3 output units. The 10 input parameters were the second intensity weighted moments and unweighted moments of the object detection area, and normalized luminosities. They were calculated by the task CATLIST of the FOCAS package (Valdes 1982b).

RESULTS

The performance of the neural network classifier, after the learning procedure has been run on the testing sample, can be represented by two indices (Table I): a) the efficiency (e.g. number of galaxies found by the classifier divided by the total number of galaxies), and b) the purity (e.g. number of real galaxies divided by the total objects classified as galaxies).

The performance of the classification method using a multilayer feed-forward neural network is good, and the neural network presents the advantage that, once it has been trained for a particular problem, it is faster than other methods. This is particularly useful for the analysis of deep photometric data obtained with large-format CCDs or photographic plates.

TABLE I Neural network classifier.

Objects	efficiency [%]	purity [%]
cosmics	95 ± 2	98 ± 2
stars	85 ± 25	66 ± 21
galaxies	88 ± 8	81 ± 8

CONCLUSIONS

For the limited data and types of objects we have considered, we can conclude that the neural network performance is good, but further work is necessary in order to improve the previous results. The neural network classifier will be applied to images of larger field of view containing more types of objects, and the results compared with those from FOCAS. Another interesting research line is to attempt the same classification working directly on the data (Murtagh & Adorf 1991).

Hardware implementations could strengthen the potential of the neural network classifier.

ACKNOWLEDGMENTS

We are grateful to X. Calbet and A. Alonso for discussion and suggestions.

REFERENCES

Garrido, L., & Gaitan, V. 1991, International Journal of Neural Systems, in press

Jarvis, J. F., & Tyson, J. A. 1981, AJ, 86, 476

Murtagh, F. D. 1991, Neurocomputing, 183

Murtagh, F. D., & Adorf, H-.M. 1991, in Data Analysis in Astronomy IV, in press

Rumelhart, D. E., et al. 1986, Nature, 323, 533

Sebok, W. L. 1979, AJ, 84, 1526

Valdes, F. 1982a, in Inst. in Astronomy IV, S.P.I.E. proceedings, 331

Valdes, F. 1982b, FOCAS user's manual (KPNO:Tucson)

PROBABILITY PLOTS: A TOOL FOR ASTRONOMICAL DATA ANALYSIS

THOMAS E. LUTZ
Program in Astronomy, Washington State University, Pullman, WA 99164

ROBERT B. HANSON
Lick Observatory, University of California, Santa Cruz, CA 95064

INTRODUCTION

As astronomers make more and more use of computers to analyze and model their data, they find that graphical displays can be very powerful tools. One type of graphical display which has been used by statisticians for nearly a century is the probability plot. Statisticians and data analysts use probability plots to investigate whether or not data samples have the properties of one of the standard statistical distributions (usually the normal distribution). The method is extremely useful because it presents a visual means of assessing the normality of the data by comparing it with a straight line, something the eye is quite good at doing. Deviations from a straight line indicate deviations from normality, and the shape and size of the deviations from a straight line indicate the manner and amount of the deviations from normality. Among the advantages of probability plots are rapid and accurate assessment of normality, rapid identification of outliers, robust estimations of the mean and standard deviation, and, particularly in comparison with histograms, the lack of the need for any subjective decision regarding bin size.

HOW A PROBABILITY PLOT IS GENERATED

The cumulative distribution of a normal random variable varies from zero to one as the variable ranges from $-\infty$ to $+\infty$. Thus, a plot of the cumulative normal distribution appears as a smooth curve. If $F(i)$ is the cumulative probability up to the ith point then in general, for N points, a good approximation to $F(i)$ is $F(i) = (i - 1/2)/N$. We could then plot the values of $F(i)$ computed in this fashion against the sorted data values. The result would be a curved line, asymptotically approaching 0 for small i and 1 for large i.

We could make a better assessment of the fit of the points to a normal curve if $F(i)$ were linearized so that, if the points did fit a normal curve, they would lie along a straight line. To do this, we need to plot the data against the inverse of $F(i)$. For the normal distribution, the inverse of $F(i)$ is simply the deviation from the mean expected for the ith point in units of the standard deviation. Algorithms for computing $F^{-1}(i)$ for the normal distribution are readily available (Hamaker 1978; Beasley & Springer 1977). Alternatively, special graph paper (available from virtually all suppliers of graph paper and known simply as

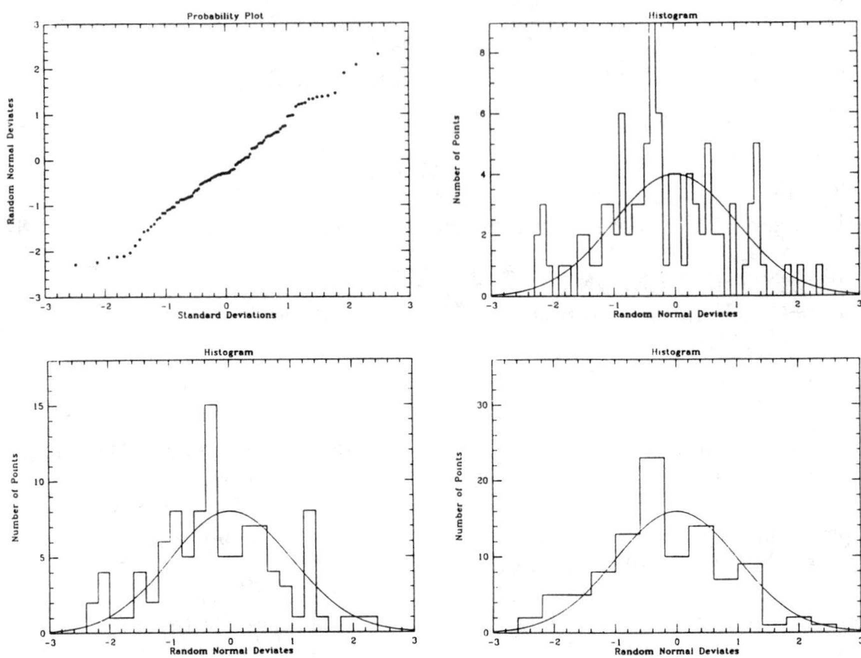

FIGURE I Probability Plot and Histograms of the Same Data

"probability" paper) may be used. Such paper is drawn so as to linearize $F(i)$. The reader is referred to the excellent book by Daniel & Wood (1971) for details and numerous examples of probability plots.

PROBABILITY PLOTS OF PSEUDO-RANDOM NORMAL DEVIATES

Figure I shows an example of a probability plot and 3 histograms of the same set of data. The data consist of 100 pseudo-random normal deviates. The three histograms differ only in their bin size. Note how easy it is to to see that the data in the probability plot lie close to a straight line with mean zero and slope one. This is not at all obvious for the histograms. Moreover, a careful look at the probability plot shows that all of the detail of the histograms is contained in the probability plot, but not *vice versa*.

APPLICATION TO DETERMINATION OF SUBDWARF MAIN SEQUENCE

Astronomers sometimes discuss subdwarf stars as if they have a unique main sequence, whereas at other times it is recognized that the location of the main sequence is a function of chemical abundance ([Fe/H]). Figure II shows a linear fit to the main sequence (of the form $M_V = a(B-V)+c$) for stars with accurate trigonometric parallaxes whose colors lie in the range $0.4 < (B-V) < 1.4$. The

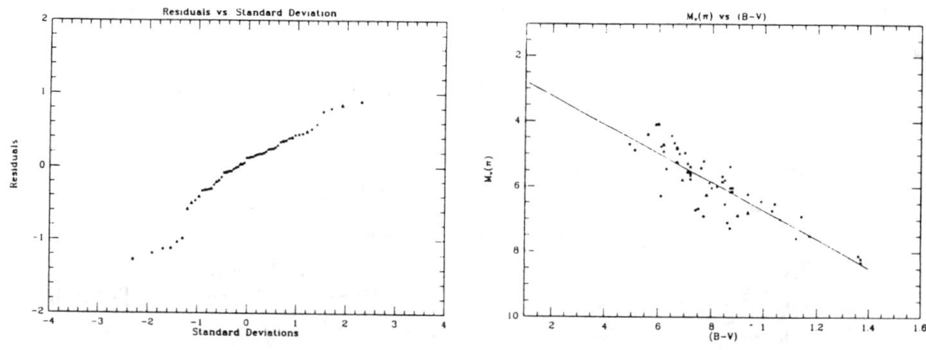

FIGURE II Probability Plot and Main Sequence (see Text)

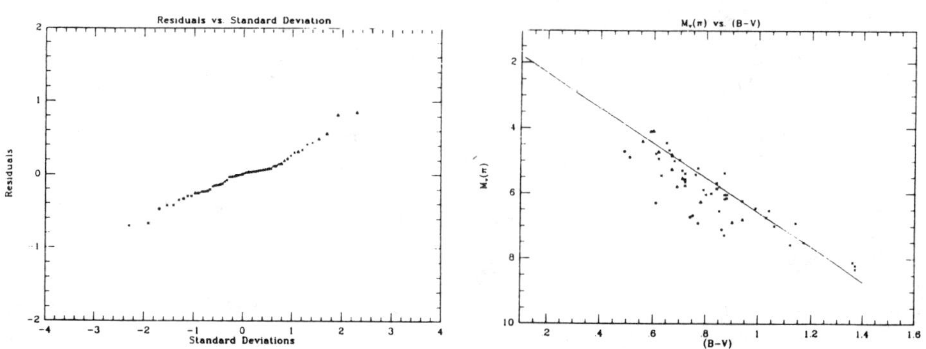

FIGURE III Probability Plot and Main Sequence (see Text)

symbols indicate stars with [Fe/H] < -1.1 ($*$), with $-1.1 <$ [Fe/H] < -0.4 ($\triangle$), and with [Fe/H] > -0.4 ($\times$). The probability plot shows that the residuals from this fit are not normal, and that the most under-abundant stars have the largest positive residuals. When a multivariate analysis is done with the same data (of the form $M_V = a(B-V) + b[\text{Fe/H}] + c$), the main sequence and probability plot of residuals are as seen in Figure III . Here, the line indicates the main sequence for stars with [Fe/H]=0.0. In the probability plot it can be seen that the distribution of residuals is closer to being normal and in addition, the distribution of residuals is no longer a function of [Fe/H].

REFERENCES

Beasley, J. D., & Springer, S. G. 1977, Appl. Stat., 26, 118

Daniel, C., & Wood, F. S. 1971, Fitting Equations to Data: Computer Analysis of Multifactor Data, Second Edition, New York: John Wiley

Hamaker, H. C. 1978, Appl. Stat., 27, 76

SOME NOISE CHARACTERISTICS OF DIRBE PHOTOMETRY

KENNETH J. MITCHELL (1), SALLY W. STEMWEDEL (2),
H. T. FREUDENREICH (3), M. G. HAUSER (4),
THOMAS KELSALL (4), MILA P. MITRA (1), NILS ODEGARD (1),
WILLIAM J. SPIESMAN (1), AND JANET L. WEILAND (1)
(1) General Sciences Corp., 6100 Chevy Chase Dr., Laurel, MD 20707
(2) Applied Research Corp., 8201 Corporate Dr., Landover, MD 20785
(3) Hughes STX Corp., 4400 Forbes Blvd., Lanham, MD 20706
(4) NASA/Goddard Space Flight Center, Greenbelt, MD 20771

ABSTRACT The sources of photometric fluctuations in DIRBE maps are reviewed. Measurements are presented showing the noise characteristics of DIRBE photometry at 2.2, 25, and 250 μm within a small patch of sky.

INTRODUCTION

The DIRBE instrument flown onboard NASA's COBE satellite has been reviewed by Hauser et al. (1991). DIRBE's main science objective is to search for the isotropic cosmic IR background radiation (CIBR) which could be attributed to remnant radiation from the epoch of galaxy formation. The task of isolating the CIBR within DIRBE's maps is rendered difficult by the extreme dominance of bright foreground sources of celestial IR emission (Hauser et al. 1991). A detailed understanding of the statistical nature of DIRBE photometry will be required to support the accurate removal of these bright foregrounds so that the underlying faint cosmological signal may be revealed.

The DIRBE data used for this study were obtained during four weeks in late December 1989 and early January 1990, and have not been absolutely calibrated.

SOURCES OF NOISE FLUCTUATIONS IN DIRBE PHOTOMETRY

One source of statistical fluctuations in DIRBE photometry is instrumental (or observation) noise. Figure I shows that there are both signal-independent and signal-dependent components to this noise. Being random and independent, the instrumental noise can be expected to integrate down as the square root of the number of DIRBE observations.

Another source of photometric fluctuations is confusion, which is caused by multiple discrete sources occurring within the 0.7°x 0.7°DIRBE beam. These fluctuations will follow a skewed confusion probability distribution (Condon 1974). Since confusion fluctuations are true sky signals, they will not integrate down by increasing the number of observations over the same region of sky.

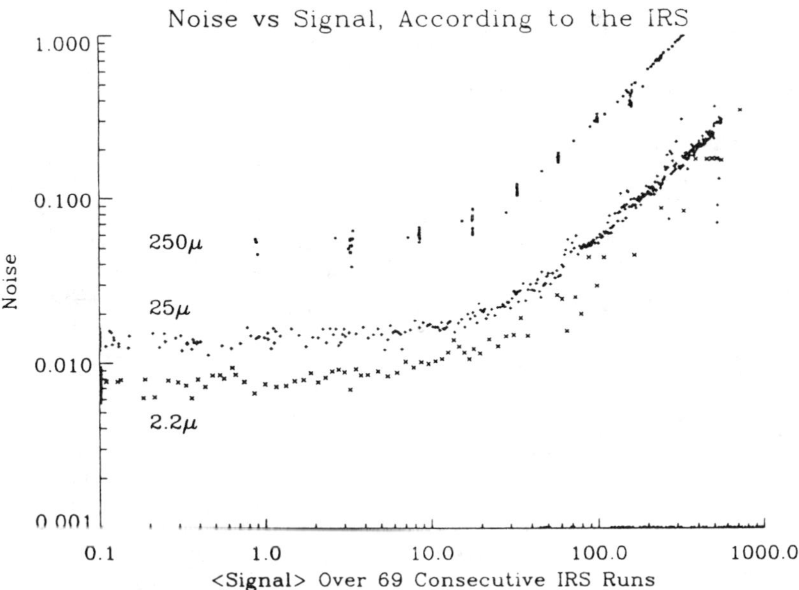

FIGURE I Plotted are measurements of rms instrumental noise vs signal for individual DIRBE observations, as obtained from multiple runnings of DIRBE's Internal Reference Source.

A third source of fluctuations in DIRBE photometry is an artifact of the sky-scanning nature of the instrument and the map pixelization, and occurs where structure/gradients are present in the map. It is observed that aperture photometry performed on DIRBE weekly maps is significantly noisier for point sources (large map gradients) than for nearby off-source regions (small map gradients). DIRBE photometry in the presence of sky gradients can be expected to stabilize as the fluctuation distributions within the individual map pixels become better sampled through an increased number of observations.

A MEASUREMENT OF THE NOISE IN DIRBE MAPS

DIRBE observations of a 10x10 pixel ($\sim$ 3°x 3°) patch of "dark" sky at high galactic and ecliptic latitudes were used to construct the flux-distribution histograms shown in Figure II. The histograms from left to right are constructed from DIRBE data averaged over progressively longer timescales. The widths of the distributions are a measure of the photometric fluctuations in the maps. Within this patch of "dark" sky, fluctuations due to the map pixelization are minimized, and the instrumental noise and confusion fluctuations will dominate.

At 2.2 μm, the skewed shapes of the distributions clearly show the effects of confusion due to stars, and the histogram of individual DIRBE observations has a width significantly larger than that expected from instrument noise alone. Inspection of the weekly and monthly histograms reveals that the confusion limit is essentially reached in weekly DIRBE maps at 2.2 μm.

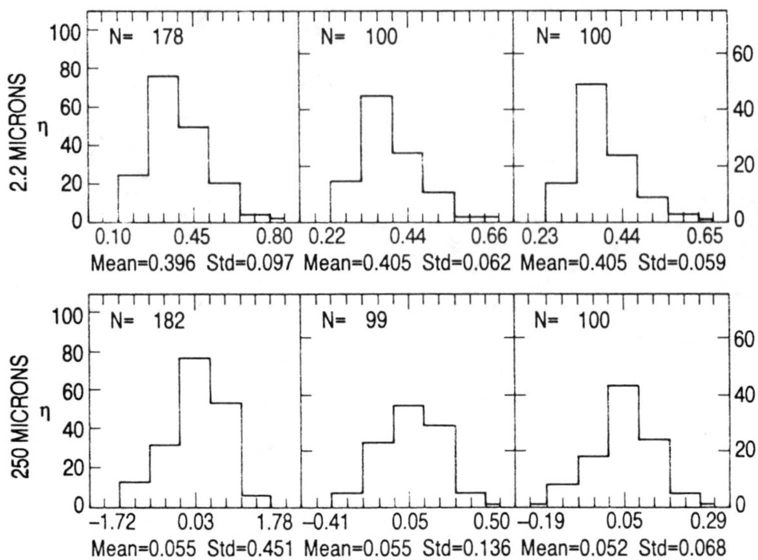

FIGURE II Flux-distribution histograms for DIRBE observations at 2.2 μm(top) and 250 μm(bottom) made within a 10x10 pixel-patch of "dark" sky. The histograms on the left are comprised of all the individual DIRBE observations falling within the pixel patch for a typical day. Those in the center are from typical weekly maps, where each pixel flux is a robust average of $\sim$ 13 individual DIRBE observations. The histograms on the right are from the monthly maps, where each pixel flux is a weighted average of four weekly map pixel fluxes.

At 250 μm, the distributions are more symmetric, and the histogram of individual DIRBE observations has a standard deviation consistent with the expected instrumental noise. Inspection of the weekly and monthly histograms reveals that instrumental noise integrates down as expected and continues to dominate the photometric fluctuations. Similar results are obtained for DIRBE maps at 25 μm, after subtraction of a crude zodiacal model.

ACKNOWLEDGMENTS

This research is supported by the Astrophysics Division of the Office of Space Science and Applications at NASA Headquarters.

REFERENCES

Condon, J. J. 1974, ApJ, 188, 279

Hauser, M. G., et al. 1991, in AIP Conf. Proc., No. 222, "After the First Three Minutes", 161

SECTION 6

IMAGE ANALYSIS METHODS

COSMIC RAY DISCRIMINATION ON HST WF/PC IMAGES: OBJECT RECOGNITION-BY-EXAMPLE

F.D. MURTAGH[1]
Space Telescope – European Coordinating Facility,
European Southern Observatory,
Karl-Schwarzschild-Str. 2,
D-8046 Garching, Germany

ABSTRACT A trainable classifier approach to cosmic ray hit detection on Hubble Space Telescope (HST) Wide Field and Planetary Camera (WF/PC) images is described. The method is applicable to the general problem of object recognition. We describe the implementation carried out, and the structure of the software which is now publicly available. We show experimental results, illustrating that the system can be trained to replicate human visual judgement. Finally, we discuss some future research directions.

THE PROBLEM: OBJECT RECOGNITION

The WF/PC's charge coupled device (CCD) detectors suffer from extensive cosmic ray hits (CRs). These are either (i) direct hits by high energy particles, or (ii) secondaries, originating in instrument housing. A number of adjacent pixels may be affected; CRs may be faint, and may be undetectable due to the noise background; and CRs may be streaks, if the inclination angle is oblique. An example is the image shown in Figure I. It is from the open cluster NGC 1850, taken with the planetary camera, and is of 10 seconds duration.

Image arithmetic on the basis of multiple, aligned frames is a recommended way to recognize CRs. It was felt in past months that HST's pointing difficulties would make this difficult: transforming images to make them alignable would be impossible in the case of undersampling. For this reason, our point of departure is to consider the single frame CR-recognition problem. Human, visual specification of CRs on the single frame is used to define a training set, to be discussed below. The availability of CRs reliably derived from multiple frames could supplant such human input in defining the training set.

Among approaches to CR recognition, we have: IRAF's tasks `cosmicrays` for single frames, and `combine` and `icombine` for multiple frames (see Shaw, this volume); a locally adaptive modified Haar transform approach due to Richter (1991) and Richter et al. (1991); a median filter approach, due to Meurs et al. (1991) and van Moorsel (1991); and image restoration, followed by flagging

[1]Affiliated to Astrophysics Division, Space Science Department, European Space Agency.

FIGURE I Part of a WF/PC image of the open cluster NGC1850: see text for details.

badly restored "objects" as bad pixels, and re-restoring (Weir 1991). To this last could be added point spread function (PSF) fitting, coupled with the flagging of poorly restored "sources" as bad pixels.

The foregoing approaches define CRs in terms of particular models, or procedurally. We, on the other hand, define CRs in the following way: a CR is whatever a user says one is.

Such an approach is feasible if the user provides sufficient information for a decision system to be constructed. A simple and well-established decision system is called a classifier. In the pattern recognition literature, this is also termed supervised classification, or learning with a teacher; in the statistics literature, the term discriminant analysis is used. We will now look at salient aspects of classifier systems.

A SOLUTION: TRAINABLE CLASSIFIERS

A trainable classifier in a first *training* phase uses the known information regarding class memberships, or labels, on a set of observations. In a second *test* phase, the correctness of the classifier is assessed relative to a set of observations which differ from the test set. Subsequently, one may proceed to full application. Insofar as the trained classifier is being used on new observations, one may speak of its *generalization* ability.

Testing of a classifier is usually based on two trade-off measurements. Such a pair of correctness parameters are estimates of the *rate of detection* and the *false alarm rate*. In the current case, the rate of detection is the percentage of CRs correctly classified by the trained classifier as CRs. The false alarm rate is the percentage of non-CRs (stars, say) which were incorrectly classified by the trained classifier as CRs. Clearly a trivial classifier can be defined which defines everything under consideration as a CR; this will achieve a 100% accurate rate of detection, but at the expense of a very unacceptable false alarm rate. In practice, therefore, there is a trade-off between these measurements. Together, they characterize how well the trained classifier performs. Further discussion of the completeness versus contamination issue is in Prusti et al. (1991), and Meurs & Adorf (1991).

The test set, on the basis of which these measurements are arrived at, is assumed to be representative of the full application set.

The mechanism used by the trainable classifier can be of various types. We have examined two families of methods, which were found to perform quite similarly. In both cases, the training set furnishes class exemplars (e.g. of the classes CR and non-CR).

K-nearest neighbors, a widely-used nonparametric discriminant analysis method works as follows. In testing or application, class assignment of a new object is decided on the basis of the majority assignment of K of the training set exemplars. We investigated various values of K, small integers, and finally used K = 5 since it offered a good balance between effectiveness and robustness (Murtagh & Adorf 1991a).

The multilayer perceptron is the most widely used neural network method. Here we provide a cursory (and simplified) description. The parametrized input objects are fed into the network. Weights on links to one or more hidden layers, and between such hidden layers and an output layer of units, are determined in the training phase. The desired output units in a 2-class case have values 0 and 1, so that an output of (0, 1) signals class 2. We found the often described backpropagation approach for weight determination to be outrageously time-demanding. As an alternative, we used conjugate gradient optimization in the training phase, which provides satisfactory results. Three-layer networks, with various numbers of units in the hidden layer, were investigated (Murtagh & Adorf 1991a).

We ultimately opted for the K-nearest neighbors method, on the grounds of implementation convenience. If it were the case that much greater training sets were being used, then the multilayer perceptron could offer considerable computational advantages.

Further discussion of these methods may be found in Murtagh & Heck (1987); Murtagh (1991a, 1991b, 1991c, 1991d, 1991e).

A DECISION: PIXEL-BASED OR OBJECT-BASED?

The objects under consideration are CRs, extended objects of various types, point sources, and detector faults of various types. Such objects are, of course, given by their discrete pixel values. A question arises as to whether to work at the most basic, and more assumption-free, pixel level; or alternatively to work at the level of already-derived objects.

In the pixel-based case, one could take every possible window or subimage of, say, 7×7 pixels. In the object-based case, we opted for a point spread function (PSF) fitting procedure to determine objects; followed by the use of a window of 5×5 pixels values, together with additional features, – roundness and sharpness – of the object within such a window.

We chose an object-based approach, on computational grounds. We believe that a pixel-based approach would necessitate specialized hardware to achieve reasonable speed of performance (for our purposes, a few minutes, at most, for processing 800×800 images on a machine of the Sun SPARCstation 2 class).

The next major issue to consider is how to define an object. Answers are provided by 2-dimensional photometry packages such as DAOPHOT, FOCAS, INVENTORY, ROMAFOT, etc. For convenience, we chose to work with a PSF-fitting routine motivated by DAOPHOT's FIND module. The routine used was initially implemented by W. Landsman in IDL. This routine smooths the image with a Gaussian of user-specified full width at half maximum (FWHM) in order to avoid spurious detections. Subsequently this FIND routine detects objects on the basis of a user-specified threshold.

In experiments carried out to date, the choice of background has not been problematic. A total number of in or around 1000 objects in the training and application sets were used, and the threshold may be selected to yield approximately this number of objects.

The choice of FWHM is more open. We used this choice to solve another problem which arises in the implementation of this method. The problem is that any single run of FIND outputs objects of a particular type, – let us term them "FIND-objects". Of necessity, such objects will be star- or CR-like, which for the present problem is satisfactory. However the "FIND-objects" may be an overly restrictive subset of stars and CRs. We carried out a number of runs of FIND, each with a different FWHM, in order to augment the set of objects which were output. In our experiments we used FWHM values of $1.0, 2.0, \ldots, 7.0$. The set of objects yielded by any one run of FIND were used, for training, testing or application. The results from the various runs of FIND were fused to produce an overall result.

In summary, some object-finding module is used, for an efficient implementation. One could envisage a set of cooperating and/or competing object-finding modules, each of which would concentrate on finding a different class of object (CR or star, and extended objects of various types). We used the somewhat blunt tool of a DOAPHOT-inspired FIND routine, – "blunt" because we sought to find a range of objects with what was initially construed as a routine for finding stellar point sources. In spite of the bluntness of this module for our purposes, we found it to perform well. The employment of various FWHMs considerably improved this performance. For the experiments carried out, this procedure provided a completely satisfactory outcome.

SOFTWARE FOR THE TRAINABLE CLASSIFIER

The system was implemented in IDL (image input/output and object finding); and Fortran spawned from the IDL session (trained classifier, calculation of performance measures, and assignment or classification).

User inputs required are (i) the image to be processed; and (ii) a boolean image of the same dimensions. In the latter, 1s indicate the training set objects of interest (CRs) and elsewhere are 0s. Part of this boolean image, only, may be used for training: the user may specify a rectangular subimage. Further user inputs are the FWHM and threshold for object-finding.

A routine, `assess`, structures the training set and outputs the performance measures, – the estimates of detection and false alarm rates. This is therefore an internal consistency study of the training set.

A routine, `classify`, is used to apply the trained classifier to the entire image.

An output boolean image is created. This may be augmented through a number of runs of FIND with a range of FWHM values as described in the previous section.

This software is available by anonymous ftp from the following address: `ns3.hq.eso.org` (134.171.11.4). `Cd` to directory:

```
pub/swlib/restoration/cosmic-rays
```

where brief documentation is also to be found.

EXAMPLE OF USE

An image of NGC 1850, provided by Jon Holtzman (Lowell Observatory) was used (Figure I). CRs were flagged. Figure II shows a boolean image following this flagging of CRs, and their removal. Further details of complementary experiments are in Murtagh & Adorf (1991a, 1991b). One experiment which may be noted was the internal consistency of the human visual capability, determined by re-flagging CRs on the same image with a few months delay from the first time this was done. It is a truism to say that the human visual facility is not infallible, and that the recognition of human faces (for example) is considerably easier than consistent recognition of CRs.

Figure III shows the procedure followed. The lower part of the image was used for training. Figures I, II, and IV relate to the upper right hand corner of the image (selected arbitrarily).

Seven different values of the FWHM in FIND were used in the producing of the results shown in Figure IV. Each of these may therefore be considered as a contribution to the overall analysis. Even though part of the analysis only, the detection rates on the training set varied between 66.38% and 75.32%. The false alarm rates on the training set varied between 2.98% and 4.84%. These figures provide an internal consistency check of the training data. We conclude from these measurements that the visually-derived training data was not entirely self-consistent for any given value of FWHM in FIND. From the comparison of Figures II and IV, taking multiple values of FWHM increases performance considerably.

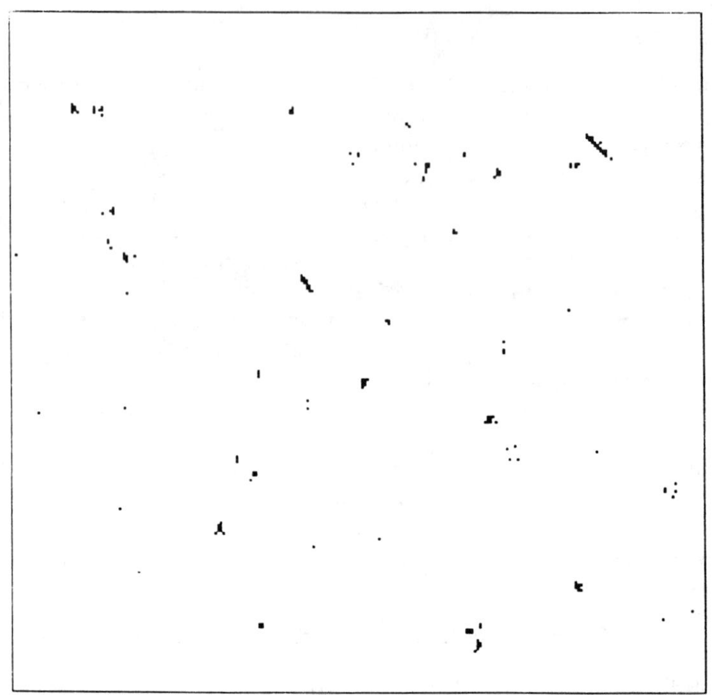

FIGURE II Cosmic ray hits (CRs) defined visually. Entire CR objects are flagged.

In Figure IV, the objects found by the classifier to be CRs are indicated. Note that this figure does not attempt to indicate the extent of these objects. The agreement between Figure IV (the classifier's result, using training data from elsewhere in the image) and Figure II (the human, visual result) is very high.

From human input, the classifier has produced a result which is in near-complete agreement with a human's judgement.

CONCLUSION: OBJECT "RECOGNITION-BY-EXAMPLE"

The generalization abilities of a classifier, replicating the visual process, have been illustrated using the example of CR detection.

It is clear that generalization to images, other than that used for training, could perform less well. However, a comprehensive selection of training data would alleviate this difficulty and perhaps solve the problem completely. Research has yet to be undertaken in this direction.

In the context of quick-look and browse systems, one implication of this work may be noted: it would be convenient to have sample subimages available for user perusal at all times.

Although using the CR problem, and selecting particular "tools" appro-

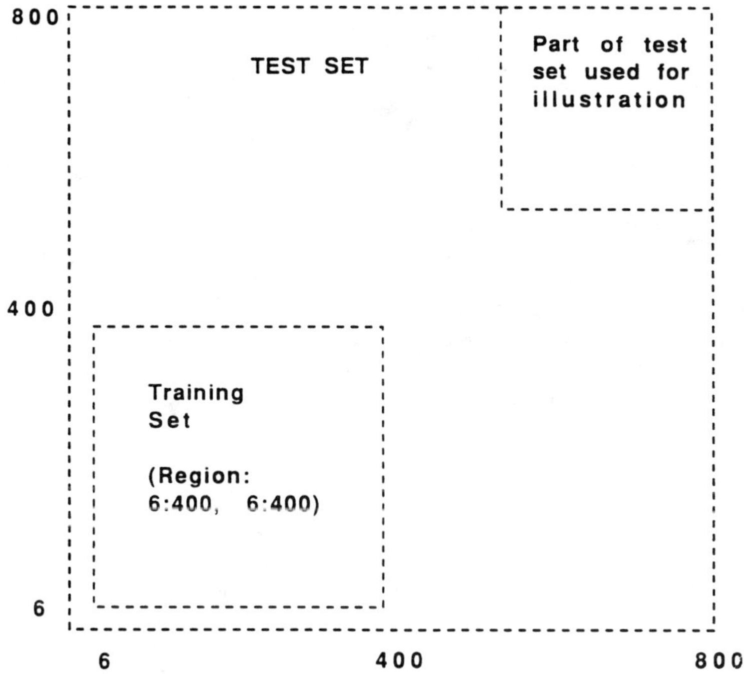

FIGURE III Illustration of aspects of the procedure followed. Results for the upper right hand corner (of the NGC 1850 image) are shown in other figures.

priate to this problem (especially in the object-finding aspect), the method described here is quite general. It could be termed "object recognition-by-example". The increasing number of image databases (Hubble Space Telescope among others) raise the following interesting question: would it not be user-friendly, and furthermore would it not allow for objectivity of treatment, to perform retrievals of image data on the basis of such "object recognition-by-example"? Although still on the horizon, it is clear that the technical requirements for this have been tried and tested; and have succeeded in practical studies.

ACKNOWLEDGMENTS

I am grateful to Hans-Martin Adorf, ST-ECF, ESO, and to Jon Holtzman, Lowell Observatory, for inputs and discussion on topics dealt with here.

REFERENCES

Meurs, E. J. A. & Adorf, H.-M. 1991, in Proc. IIe Rencontres de Blois: Physical Cosmology, in press

FIGURE IV Cosmic ray hits (CRs) defined by the classifier. Flagged points represent CR object and do not indicate their extent.

Meurs, E. J. A., Bonifácio, V. H. & Lima, N. M. 1991, in 3rd ESO/ST-ECF Data Analysis Workshop, P. J. Grosbøl & R. H. Warmels, Munich: European Southern Observatory, 45

Murtagh, F. 1991a, Analyzing and Modeling Data and Knowledge, M. Schader, Berlin: Springer-Verlag, in press

Murtagh, F. 1991b, Neurocomputing, 2, 183

Murtagh, F. 1991c, Journal of Classification, 8, 115

Murtagh, F. 1991d, in Combinatorial Data Analysis, P. Arabie & L. J. Hubert, New York: World Scientific, in press

Murtagh, F. 1991e, Pattern Recognition Letters, in press

Murtagh, F. & Adorf, H.-M. 1991a, in Data Analysis in Astronomy IV, V. Di Gesù et al., New York: Plenum, in press

Murtagh, F. & Adorf, H.-M. 1991b, in 3rd ESO/ST-ECF Data Analysis Workshop, P. J. Grosbøl & R. H. Warmels, Munich: European Southern Observatory, 51

Murtagh, F. & Heck, A. 1987, Multivariate Data Analysis, Dordrecht: Kluwer

Prusti, T., Adorf, H.-M. & Meurs, E. J. A. 1991, Astron. Astrophys., in press

Richter, G. M. 1991, in 3rd ESO/ST-ECF Data Analysis Workshop, P. J. Grosbøl & R. H. Warmels, Munich: European Southern Observatory, 37

Richter, G. M., Böhm, P., Lorenz, H., Priebe, A., & Capaccioli, M. 1991, Astron. Nach., 312, 345

Van Moorsel, G. 1991, ST-ECF Tech. Rep.

Weir, N. 1991, in 3rd ESO/ST-ECF Data Analysis Workshop, P. J. Grosbøl & R. H. Warmels, Munich: European Southern Observatory, 115

THE DETECTION OF GLITCHES AND POINT SOURCES IN COBE/DIRBE DATA

H. T. FREUDENREICH, K. P. TEWARI
Hughes STX Corporation, 7601 Ora-Glen Drive, Suite 100, Greenbelt, MD 20770

T. KELSALL
NASA/Goddard Space Flight Center, Greenbelt, MD 20771

N. P. ODEGARD
General Sciences Corporation, 6100 Chevy Chase Dr., Laurel, MD 20707

INTRODUCTION

The Diffuse Infrared Background Experiment on the Cosmic Background Explorer is a 10-band, 16-detector liquid-helium-cooled photometer designed to survey the sky from 1-300 μm in search of a cosmological infrared background. (Hauser et al. 1990)

On a daily basis the data are processed in time-ordered format, then projected onto to a 393,216-pixel map. On a weekly basis the daily maps are merged. (Toller et al. 1992; Skard et al. 1992)

As is the case for most space-based instruments, the DIRBE software must be capable of recognizing discrete sources and rejecting glitches, most of which are caused by charged particle hits. The methods employed may be applicable to other experiments, particularly those in which a signal is sought among many unequal input channels.

The Difficulty of Detecting Errors

Glitches and true sources are often hard to distinguish in the DIRBE data. The DIRBE beam is large ($\approx$40 arcmin to a side), sharp-edged, and crosses nearly half a beam-width of sky per data sample. As a result, those bands in which discrete sources are common present a jagged appearance in the time domain. A discrete source appears as a two- or three-sample peak or plateau; charged-particle-induced glitches appear as peaks one–sometimes two–samples wide; and when the beam encompasses multiple sources, the "sky noise" is severe.

Differences between detector types, and between the nature of the sky as seen at different wavelengths, further complicate matters.

GLITCHES

Origin and Glitch Rates

Since the DIRBE's radiative signals are chopped, the sign of a glitch due to a charged-particle hit depends on the chopper phase at the particle's arrival. Such

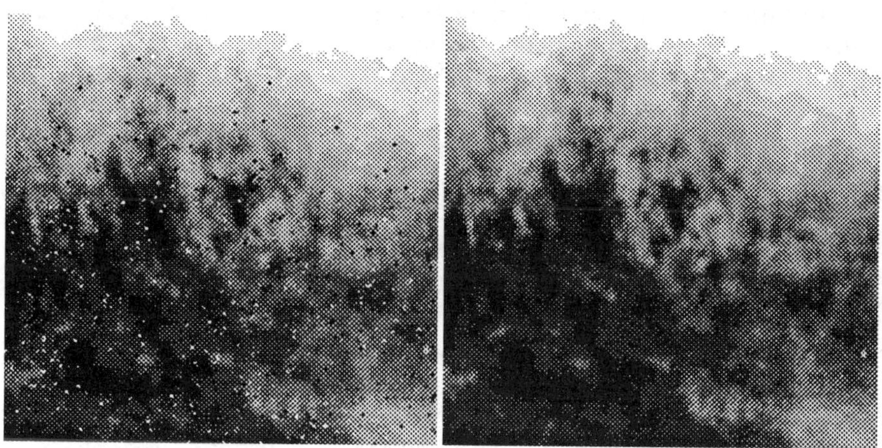

FIGURE I Sample of 100μm map, before and after deglitching. Range: l = 256° – 316°, sin b = .225 – .682. Histogram-Equalized image.

glitches may have magnitudes approaching detector saturation.

Other glitches are caused by an electronic "undershoot" that follows $\approx \frac{1}{4}$ of the bright source passages. The signal quickly recovers, so these glitches are rarely more than one sample wide.

We have four types of detectors. When the spacecraft is within the South Atlantic Anomaly, all data are irretrievably contaminated by proton hits, and are rejected. Outside the SAA, the charged-particle glitch rate is ~4/1000 data samples in the InSb, silicon photoconductor, and bolometer detectors (nominal wavelengths 1.25, 2.2, 3.5, 4.9, 12, 22, 140, 240 μm). In the Ge:Ga detectors (60 and 100 μm), which have larger active volumes, this rate is doubled. Below 4.9 μm, where bright star passages are frequent, the rates of charged-particle and undershoot glitches are similar.

Glitch Removal

Glitches are removed in two steps: first, through an algorithm applied to the time-ordered data; and second, by outlier-resistant averaging during the formation of the weekly maps.

The outlier-resistant merging of the daily maps alone is capable of reducing the number of glitches to negligible levels in most bands, but to form useful daily maps we must first deglitch the time-ordered data. Linear filters are unsuitable due to the jagged nature of the data, and more sophisticated approaches, such as non-linear filters and inter-band comparisons, are more computation-intensive. Instead, we created a data-driven glitch-detection algorithm. The algorithm applies a series of increasingly stringent tests to each data sample, all of which must be passed before the sample is labeled a glitch. Because the tests are individually simple, and few samples make it far through the gauntlet, this customized approach is no slower than the use of a linear filter.

The minimum total error rate of this method is <1/10000 samples. We chose our thresholds to the right of this minimum, trading a slightly higher rate of type II errors for a lower (negligible) rate of type I errors (deleting good data).

Robust averaging excludes essentially all remaining glitches from the weekly maps. The scheme we use rejects outliers on the basis of the median absolute deviation from the median (m.a.d.) before calculating the mean. This method offers a good compromise between efficiency and speed. The rejection threshold is currently set at three m.a.d.

Pixels with fewer than three samples (.7% of those seen per week) are blanked.

POINT SOURCE DETECTION

Once deglitched, the time-ordered data are searched for point source passages using a method analogous to the hardware configuration of

$$AMPLIFIER \Longrightarrow DISCRIMINATOR \Longrightarrow COINCIDENCE\ BOX$$

The *amplifier* convolves the data with the zero-sum filter:

$$(-1,-1,-1,2,2,2,-1,-1,-1),$$

whose major effect is to subtract the local baseline. In the filtered data a point source appears as a positive central lobe sandwiched by smaller negative lobes.

When the amplitude of the filtered data exceeds a threshold, the *discriminator* calculates the centroid of the positive lobe and adds a rectangular pulse to a matching byte array. The width of the pulse is a function of detector and source characteristics. The height is the comparative weight assigned the detector. The weight is lower in bands viewing many common sources and higher at longer wavelengths, where there is less overlap and sources are scarce.

Where the magnitude within the byte array exceeds a threshold, the *coincidence box* notes the detection of a point source. The source is identified, if possible, and its passage characterized and copied to a file for future use.

This method is fast, flexible, and easy to code, and could just as well be applied to 16,000 detectors as to 16.

ACKNOWLEDGMENTS

The COBE Project is supported by the Astrophysics Division of the Office of Space Science and Applications at NASA Headquarters.

REFERENCES

Hauser, M. G., et al. 1990, AIP Conference Proceedings, 222, 161

Skard, J. A., et al. 1992, this volume

Toller, G. N., et al. 1992, this volume

CO-ADDING IMAGES WITH DIFFERENT PSF'S

L.B. LUCY AND R.N. HOOK
Space Telescope - European Coordinating Facility,
European Southern Observatory,
Karl Schwarzschild Str-2,
Garching bei München, D-W 8046, Germany

INTRODUCTION

Co-adding or stacking of images and spectra has for many years been a standard procedure for obtaining high signal-to-noise ratios. In the case of ground-based images, however, there is the difficulty that each of the series of exposures to be co-added is likely to have a different PSF because of the different seeing. In this circumstance, a conventional co-addition results in an image whose resolution can be markedly inferior to the best in the series. Because of this, it is a not uncommon practice to retain only the exposure obtained in the best seeing conditions, discarding the remainder despite the major contribution that they could in principle make to the overall S/N.

Another circumstance where the ability to co-add images with different PSFs and not sacrifice resolution would be useful arises when images at the same wavelength are available from space-borne and ground-based telescopes. In space, we (ideally) obtain high resolution images but the photons are expensive. From the ground, on the other hand, the resolution is poor but the photons are cheap.

TECHNIQUE

Let $\tilde{\phi}_k$ with $k = 1, 2, ..., k$ denote the intensity distributions of the k images to be co-added, and let $P_k(x, y|\xi, \eta)$ denote the PSF of the kth image.

The co-added image corresponding to the PSF P_k is then (ideally) the limit as $r \rightarrow \infty$ of $\phi_k^r(x, y)$ given by

$$\phi_k^r = \iint \psi_k^r P_k d\xi d\eta \tag{1}$$

Here $\psi^r(\xi, \eta)$ is the rth estimate of the true intensity distribution given by the iterative procedure

$$\psi^{r+1} = C^r \psi^r \tag{2}$$

where the correction factor $C^r(\xi, \eta)$ is the means by which the observational data are imposed on the solution.

In the simplest case all the exposures are of the identical field; the correction factor is then

$$C^{r} = \sum w_k \iint \frac{\tilde{\phi}_k}{\phi_k^{r}} P_k dx dy \tag{3}$$

where the normalised weights w_k are proportional to N_k, the number of photons detected in the *kth* exposure.

COMMENTS

1. The above analysis provides a co-added image for each k. However, in almost all circumstances, the co-added image of interest is the one corresponding to the PSF with the smallest FWHM - ie, to the exposure with the highest resolution. This particular co-added image then achieves the optimal combination of resolution and integrated S/N.

2. Hidden in the machinery of this procedure is an iterative deconvolution (Richardson 1972; Lucy 1974). Nevertheless, artifacts associated with deconvolution - eg. speckling of low S/N backgrounds, excavation of the background near stars - are absent or negligible in the co-added image.

3. When $r \to \infty$, the co-added images reproduce noise fluctuations in the original images to the extent that they are compatible with the non-negativity constraint on $\psi(\xi, \eta)$. The data is therefore subjected to "object-feasibility filtering". This filtering has the very desirable property of removing pixel-to-pixel noise and thus greatly lessens the evidence of pixellation.

4. Because noise is retained except when incompatible with the fundamental constraint of object positivity, the co-added image can still be accorded the status of basic data. The retention of this status while removing pixel-to-pixel noise argues strongly for the procedure's wide use in the presentation of astronomical images in the archival literature.

AN EXAMPLE

Some early results of applying this method to one-dimensional spectral data have already been described (Lucy 1991). It has now been implemented under the IRAF system to work on image data and some trials have been performed on simulated and appropriately processed real images. In each case a low S/N image of high resolution is combined with a higher S/N having a broader PSF. Unfortunately there is only space to present one simple, but revealing, example in these proceedings. It is shown as Figure I.

IMPLEMENTATION AND PLANNED ENHANCEMENTS

Initial experiments with this algorithm were implemented as simple stand-alone FORTRAN programs. To increase ease of use and compatibility with other tools the code was subsequently interfaced to the IRAF system using the F77/VOS

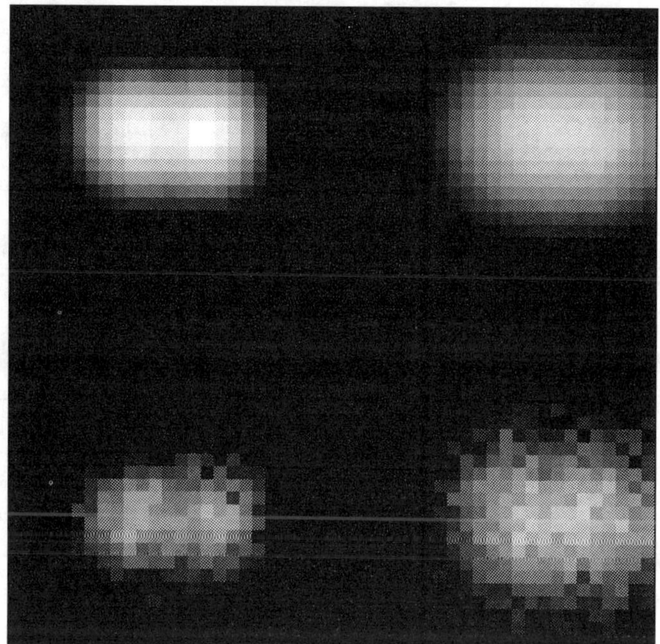

FIGURE I A simulated double star with equally bright components separated by 6 pixels. Each star in the lower left hand image contains 1000 photons and the PSF has a FWHM of 5 pixels. This is close to the Rayleigh limit for resolution. In the lower right hand image the PSF has been broadened to a FWHM of 7 pixels and the total number of photons in each star has been doubled. The upper images are the result of 5000 iterations of the co-addition method. The image at the upper left retains a FWHM of 5 pixels and the duplicity of the star is clear and the pixel-to-pixel noise is much reduced. The upper right image is the co-added result with the broader PSF.

library. The current version works with images which may have different scales and orientations. However it does not use Fast Fourier Transforms for the convolution part of the algorithm and also currently is limited to gaussian PSFs and a maximum of two images. However these limitations will be overcome in later versions and the intention is to have a general and easy to use tool for general problems of this type. A further possible enhancement is the incorporation of the 'speed-up' technique (Lucy 1991, unpublished) which greatly reduces the number of calculations required.

REFERENCES

Richardson, W. H. 1972, Journal of the Optical Society of America, 62, 55

Lucy, L. B. 1974, AJ, 79, 745

Lucy, L. B. 1991, ST-ECF Newsletter, 16, 6

AN ORTHOGONAL TRANSFORM FOR GEOMETRICALLY CORRECTING ASTRONOMICAL IMAGES, INCLUDING A NON-INTERPOLATIVE TOOL FOR PRODUCING A ROTATED IMAGE

MURIEL A. TAYLOR
Applied Research Corporation, GSFC

DALE J. FIXSEN
Applied Research Corporation, GSFC

MICHAEL E. VAN STEENBERG
NASA/National Space Science Data Center, GSFC

ABSTRACT The purpose of this paper is to show a technique for producing a rotation of an image matrix through the use of a pixel remapping procedure. This technique produces a file with virtually no smoothing. The rotation is usable in n-dimensions. One of our test cases is an IUE High Dispersion Image displayed in a horizontal reference for ease of extraction. The transform implementation displays a method for deconvolving any optically distorted surface to a correct geometry. The rotation is embedded in the transform.

INTRODUCTION

Our objective is to develop a rotation for astronomical images which preserves the internal data structure, including any relationship between the signal in an image and its underlying noise. A review of sampling theory uncovered few algorithms geared towards maintaining the fine structure in data. The ultimate goal of this research will be to construct an optimal transform similar in form to the optimal orthogonal Karhunen-Loeve Transform with a rotation of the matrix, and any other geometric corrections, embedded in the transform. The work presented here is a first step in that direction.

DISCUSSION

Construction of an optimal transform means adhering to certain criteria. An optimal orthogonal transform must (Elliot & Rao 1982):

- completely decorrelate any sequence in the transform domain.
- maximize the energy (or information) contained in a minimum of transform coefficients.

- minimize the mean square error between the original data and the reconstructed data for any specific bandwidth reduction.
- for any sequence, minimize the total degradation due to entropy.

In other words, the transform should reduce to near zero the merging of information on a per pixel basis.

One mathematical approach for a solution to this problem was to apply the transform to the following: Determining one irreducible one-dimensional representation of a real two-dimensional plane coincident with the unitary transformations of a single complex variable.

For example, for a complex coordinate $\mathbf{x}$, the rotation is given by:

$$\mathbf{x} \longrightarrow \mathbf{x}' = \epsilon \cdot \mathbf{x} \tag{1}$$

where $\epsilon = e^{i\phi}$ is a constant with unit modulus. The angle of rotation, ϕ, behaves additively on composition, (mod 2π). For our application, we obtain an approximate solution by the expansion of the Taylor Polynomial for f at x_0, y_0 to the $7th$ moment:

$$\begin{aligned} P_7(x,y) = f(x_0,y_0) + \partial f/\partial x(x - x_0) + \\ \partial f/\partial y(x_0,y_0)(y - y_0) + \cdots + \\ \partial^7\partial^7/\partial^7 x\partial^7 y f(x_0,y_0)(x - x_0)^7(y - y_0)^7 \\ = \sum_{i,j=0}^{7} \mathbf{A}_{ij} dx^i dy^j \end{aligned} \tag{2}$$

where $\mathbf{A}_{ij}$ is the image matrix, and

$$\begin{aligned} dx &= k - x cos\theta + y sin\theta, \\ dy &= l - x sin\theta - y cos\theta, \end{aligned} \tag{3}$$

for $P_7(k,l)$.

We fit to a model which follows from geometric theory of Gaussian optics - the Seidel optical aberrations - (Born & Wolf 1975) that have been shown to be present in this type of instrument. Distortion is the dominant aberration in SEC Vidicons. (Pilkington & Hartley 1972).

The resulting solutions will have the form

$$\begin{aligned} dx' &= dx + [optical\ corrections] \\ dy' &= dy + [optical\ corrections]. \end{aligned} \tag{4}$$

For our model we used laboratory measurements of IUE's reseau pattern (Turnrose & Thompson 1984), along with other applicable optical and spectrographic terms to correct the geometry. The dominant correction term comes from fifth order optical aberration theory.

When completed, this solution will be a functional solution based on a theoretical model. While the rotation is currently cast in integral form, it is not yet a transform. Furthermore, the re-pixelization procedure works very well for preserving the original signal to noise structure, but we are losing a small percent of the flux because the fit around the reseaux needs to be improved. An additional problem that remains to be solved is the aliasing that will be present in the resultant transform.

REFERENCES

Born, M., & Wolf, E. 1975, Principles of Optics (Oxford, Pergamon Press)

Bracewell, R. 1965, The Fourier Transform and Its Applications, McGraw-Hill, Inc.

Brigham, E. 1974, The Fast Fourier Transform, Prentice-Hall, Inc.

Busch, H. 1926, Ann. Physik, 81, 974

Elliott, D., & Rao, K. 1982, Fast Transforms Algorithms, Analyses, Applications, Academic Press, Inc.

Gonzalez, R., & Wintz, P. 1987, Digital Image Processing, Addison, Wesley, Inc.

Pilkington, J. D. H., & Hartley, K. F. 1972, Advances in Electronics and Electron Physics, 33A, 545 - 5th Symposium on Photo-Electronic Image Devices

Press, W., Flannery, B., et al. 1987, Numerical Recipes, Cambridge University Press

Reza, F. M. 1961, An Introduction to Information Theory, McGraw-Hill, N.Y.

Taylor, M. A., & Shaw, R. A. 1990, IUE/NASA Newsletter No. 42, "Electro-Optical Distortions in IUE Cameras", 75-84

Turnrose, B. E., & Thompson, R. W. 1984, IUE Image Processing Information Manual, Version 2.0, Computer Sciences Corp., CSC/TM–84/6058

Weyl, H. 1950, The Theory of Groups and Quantum Mechanics (Dover Publications, Inc.)

Zworykin, V. K., Morton, G. A., Ramberg, E. G., Hillier, J., & Vance, A. W. 1945, Electron Optics and the Electron Microscope (New York, Wiley & Sons)

MORPHOLOGICAL FILTERING OF INFRARED CIRRUS EMISSION FROM EXTRAGALACTIC FIELDS

JOHN P. BASART AND PAUL R. SIQUEIRA[1]
Department of Electrical and Computer Engineering, Iowa State University, Ames, IA 50011

PHILIP N. APPLETON
Department of Physics and Astronomy, Iowa State University

ABSTRACT Mathematical morphology has been applied to extragalactic fields of IRAS sources for the purpose of removing the Galactic infrared cirrus emission from the fields. Sieving was used to create a spatial structure spectrum for distinguishing the Galactic cirrus structure from the extragalactic structure. From the structure spectrum a filter function was formed and then used to greatly reduce the Galactic cirrus from the image. The results on the M81 field show a pronounced reduction in the local cirrus.

INTRODUCTION

Galactic infrared cirrus emission, especially at 60 and 100 μm wavelengths, can be so severe in extragalactic fields that is difficult to distinguish foreground structure from the extragalactic structure. Faint structure such as bridges between galaxies and plumes around galaxies are essentially impossible to detect if Galactic IR cirrus is distributed throughout the angular region of interest. Consequently, it is necessary to remove much of the Galactic cirrus if the extragalactic structure is to be detected. Spatial spectral techniques are not suitable for removing the Galactic cirrus because there is an overlap in the spatial frequencies of the two types of structure.

We are developing a technique for structure separation based on mathematical morphology (also called morphological image processing). Mathematical morphology is able to distinguish between structures based on their morphology. This is a reasonable approach to use for separating Galactic cirrus structure from the extragalactic source because of the different linear scales being sampled by the IRAS beam. For example, it might be expected that in local diffuse clouds, sources would show structure on many scales and may be quite fractal-like (Scalo 1990). Over the same range of angular scales an extragalactic source may be dominated by very different physical processes (e.g. spiral waves, bar

[1]Now at Department of Electrical Engineering and Computer Science, University of Michigan, Ann Arbor, MI 48109-2122.

driven star formation, tidal filaments) leading to structure with potentially different morphological characteristics from that of local cirrus. In this paper we briefly discuss a filtering method based on mathematical morphology and show the application of the technique applied to the M81 field.

MATHEMATICAL MORPHOLOGY

Mathematical morphology is expressed in the language of set theory. It starts out with a translation operator for binary images, and proceeds with the definition of dilation and erosion. Dilation expands an image while erosion contracts an image. (For the mathematical descriptions of erosion and dilation see the references at the end of this paragraph). From these two operations, opening and closing are defined. Opening and closing are frequently used operations in mathematical morphology. Gray-scale morphology is then built upon binary morphology. Introductory material on these topics can be found in Serra (1982), Giardina & Dougherty (1988), and Haralick et al. (1987). An application of mathematical morphology in astronomy is given by Lea & Kellar (1989).

All mathematical morphology operations are based upon a so-called structuring element which is usually a miniversion of an image. The structuring element contains structure of the type that is to be extracted from an image. The opening operation extracts all structure brighter than the background whose size is smaller than the structuring element and conforms to the shape of the structuring element. The closing operation extracts all structure darker than the background whose size is smaller than the structuring element and conforms to the shape of the structuring element. Opening is written as

$$\mathbf{A} \circ \mathbf{B} = (\mathbf{A} \ominus \mathbf{B}) \oplus \mathbf{B} \tag{1}$$

where $\mathbf{A}$ is the image and $\mathbf{B}$ is the structuring element. $\mathbf{A} \circ \mathbf{B}$ represents opening, $\ominus$ represents erosion, and $\oplus$ represents dilation. Similarly, closing is represented by

$$\mathbf{A} \bullet \mathbf{B} = (\mathbf{A} \oplus \mathbf{B}) \ominus \mathbf{B} \tag{2}$$

where $\mathbf{A} \bullet \mathbf{B}$ represents closing an image $\mathbf{A}$ by structuring element $\mathbf{B}$.

An operation called sieving is based upon opening and closing. Consider the opening of an image $\mathbf{A}$ by a structuring element $\mathbf{B_1}$ with width w_1. Since this operation eliminates structure smaller than w_1 we subtract the opened result from the original image to get the information that opening eliminated. We call this image $\mathbf{M_1}$. Mathematically, this is

$$\mathbf{M_1} = \mathbf{A} - \mathbf{A} \circ \mathbf{B_1} \tag{3}$$

Next we open the original image $\mathbf{A}$ with a structuring element $\mathbf{B_2}$ with width w_2 where $w_2 > w_1$. This gives $\mathbf{M_2}$ where

$$\mathbf{M_2} = \mathbf{A} - \mathbf{A} \circ \mathbf{B_2} \tag{4}$$

Then we subtract the two results which is a sieving operation. For this we use the symbol $\mathbf{G}$ for growth since a sequence of sieving operations will show the growth of the structure. Then

$$\mathbf{G_{21}} = \mathbf{M_2} - \mathbf{M_1} \tag{5}$$

This difference gives us the spatial structure of $\mathbf{A}$ in the range $w_1 \leq w \leq w_2$.

CIRRUS FILTER

The IR cirrus filter is created from a sequence of sieving operations. The preceding operations are repeated for a sequence of increasing sizes of the structuring element with a uniform step size. Since each growth step $\mathbf{G}_{ij}$ sifts out a narrow range of structure size, we have a sequence of spatial bandpass filters.

The next step in the creation of the filter is to place all the growth images into a cube ordered in size from images with small structure at the top to images with large structure at the bottom. Choosing a particular pixel in the top plane (smallest structure) we pass from top to bottom selecting the intensity at that pixel location from each plane. We then make a plot of intensity vs. growth step. Before plotting we normalize each intensity by the area under the curve. Choosing various pixels in regions containing the cirrus, we add many curves to the plot. At each growth location in the plot we have a spread of points in intensity since all slices down through the cube do not have identically the same growth curves. In our experiments thus far we have found that the spread of points about the mean at each growth step has a gaussian distribution. Our filter function is created from the sequence of means of the growth curves.

To apply the filter we start with a pixel at the origin of the top plane in the cube. From the area under the curve of this first vertical slice of growth we unnormalize the filter curve by multiplying the filter curve by the area. This curve is then subtracted from the vertical slice down the cube which completes the correction for the first pixel. The process is repeated for every pixel in the cube. When the filtering process is completed the final filtered image is obtained by adding all the corrected planes in the cube. The concept of adding the planes to create the final image can be tested by adding all the cube planes before filtering to see if the original image is recovered. If not, then the sieving process must be tuned to retain the flux of the image at each pixel. An example of the filtering process is given for a 4° x 4° extragalactic field centered on M81.

FILTER EXAMPLE FOR THE M81 FIELD

Figure I shows a contour plot of the M81 field at $\lambda 100\ \mu m$. This is a 360 by 360 pixel map of an IRAS image with a pixel scale of one arcmin/pixel. The FWHM of the IRAS point spread function varies somewhat with position in the map, but is approximately 4.6 x 4.2 arcminutes. The map is centered on $\alpha_0 = 146.0°$ and $\delta_0 = 70.0°$. The structuring element used for sieving was a circular gaussian whose initial size was 5 pixels FWHM (full width half maximum) with a step size of two pixels. The first growth map was made with 5 and 7 pixel widths. Pixel widths of the structuring element were increased up to a maximum sized growth map with structuring element widths of 35 and 37 pixels FWHM.

The cube was constructed with 16 growth maps. Forty-four slices were taken down through the cube through Galactic cirrus regions to form the filter curve of average cirrus growth. This curve is shown in Figure II. The average growth was applied to each pixel in the growth cube. After filtering, all the planes in the cube were summed to give the result in Figure III. This final image shows a considerable reduction in the intensity of the Galactic cirrus. The intensity of the Galactic cirrus in Figure III is reduced by a factor of about 15 from the original

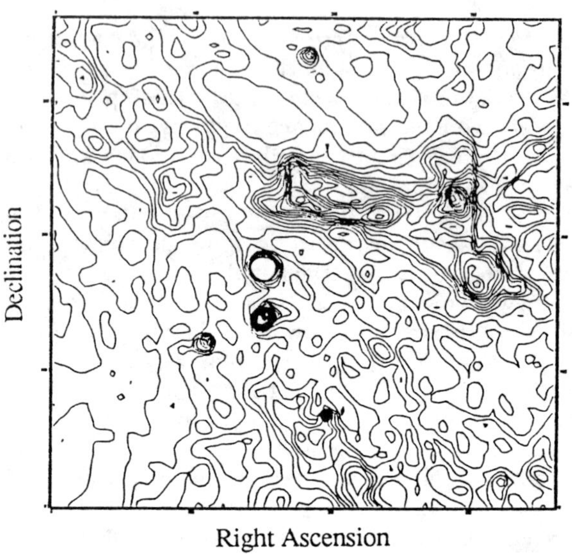

FIGURE I A 360 by 360 pixel map (one pixel = 1′) of the M81 field at λ100 μm. The contours range from 0.1 to 19.6 MJy in steps of 0.5 MJy.

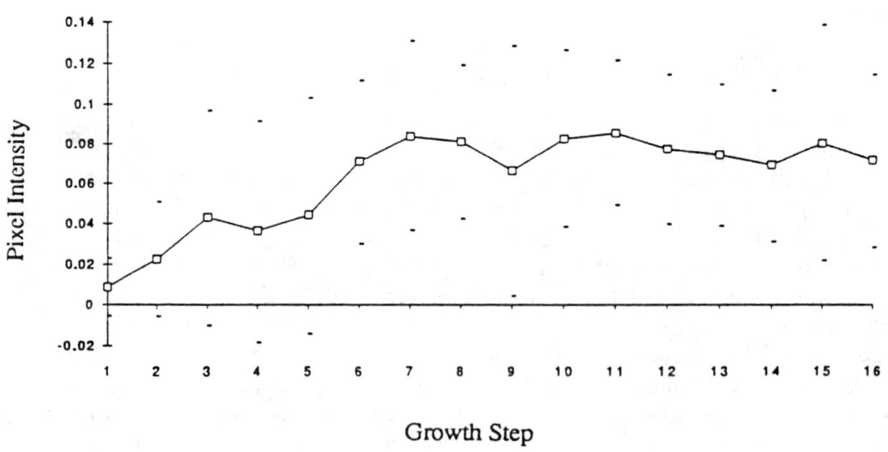

FIGURE II Average cirrus growth curve for the M81 field.

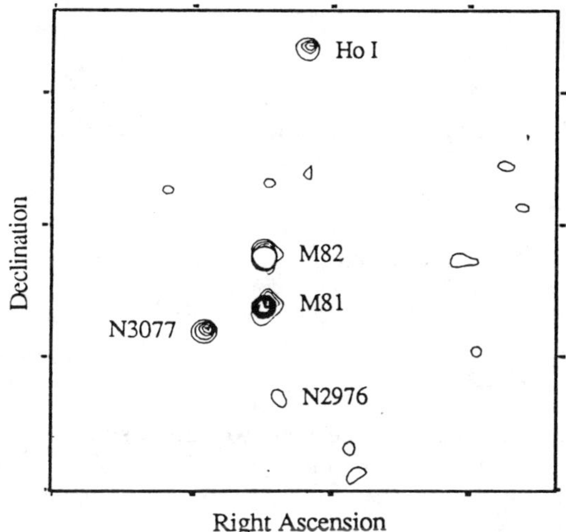

FIGURE III Final filtered map of the M81 field. Image size and contour levels are the same as in Figure I.

intensity. While the majority of flux removal is believable we must undertake more testing and adjusting of the filter method to have confidence that the small remaining structure surrounding M81 belongs to M81. Another feature of the filter is that small sources obscured by Galactic cirrus in the original image can become readily visible in the filtered image. Thus, in addition to aiding the recovery of structure from extragalactic fields the morphological filter has the potential of isolating small sources in the original image that may have interesting characteristics.

ACKNOWLEDGMENTS

We thank Mathew Chackalackal for helpful discussions. This work was partially supported by NASA-JPL Grant NAG 5-1186.

REFERENCES

Giardina, C. R., & Dougherty, E. R. 1988, Morphological Methods in Signal and Image Processing, Englewood Cliffs, New Jersey: Prentice Hall

Haralick, R. M., Sternberg, S. R., & Xinhua, X. 1987, IEEE Transactions on PAMI, 9(4), 532

Lea, S. M., & Kellar, L. A. 1989, AJ, 97, 1238

Scalo, J. 1990, in Physical Processes in Fragmentation and Star Formation, R. Capuzzo-Dolcetta, C. Chiosi & A. DiFazio, Dordrecht: Kluwer, 151

Serra, J. 1982, Image Analysis and Mathematical Morphology London: Academic

ANALYSIS OF AXIALLY SYMMETRIC EMISSIVITY DISTRIBUTIONS: CASE 1- SPHERICAL SYMMETRY

D.A. LEAHY
Physics and Astronomy Department, University of Calgary, Calgary, Alberta Canada T2N 1N4

INTRODUCTION

The problem of deconvolving an observed two-dimensional brightness distribution to derive a three-dimensional emissivity distribution is considered. The theory and methods are discussed in Leahy (1991) for the case that the emission is optically thin and is axially symmetric.

Here, the deconvolution process is discussed for the case that the emissivity is spherically symmetric. Two algorithms can be employed for deprojection. The first uses matrix techniques and is amenable to fast numerical evaluation. The second method used the Lucy deconvolution technique. This has the advantage that the derived emissivity is guaranteed to be non-negative. Both algorithms have been encoded and tested on simulated and real images. Below we discuss the matrix method.

MATRIX METHOD FOR DEPROJECTION

We consider a spherically symmetric emissivity distribution $\epsilon(r), 0 \leq r \leq R$. The emissivity is discretized over intervals of r: $r_{j-1} < r < r_j$, with $j = 1, 2...n, r_0 = 0, r_n = R$. The number of radius intervals and the width of each interval is arbitrary.

The observed brightness, i.e. the integration of emissivity along the line of sight, can be written as:

$$I_j = \sum_{i=1}^{n} V_{ji}\epsilon_i \qquad (1)$$

Here, I_j is the flux observed from the ring on the sky plane extending from projected radius r_{j-1} to r_j. V_{ji} is the volume of spherical shell i which is seen projected onto ring j. The emissivities, ϵ_j, can be recovered by using the inverse matrix to V_{ij}:

$$\epsilon_i = \sum_{j=1}^{n} V_{ij}^{-1} I_j \qquad (2)$$

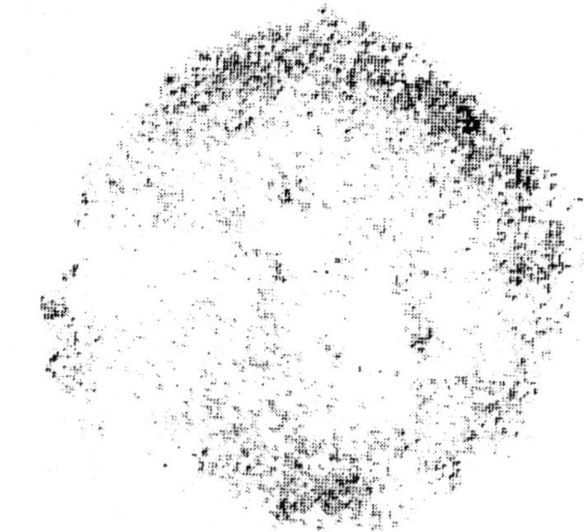

FIGURE I Einstein image of Tycho's SNR.

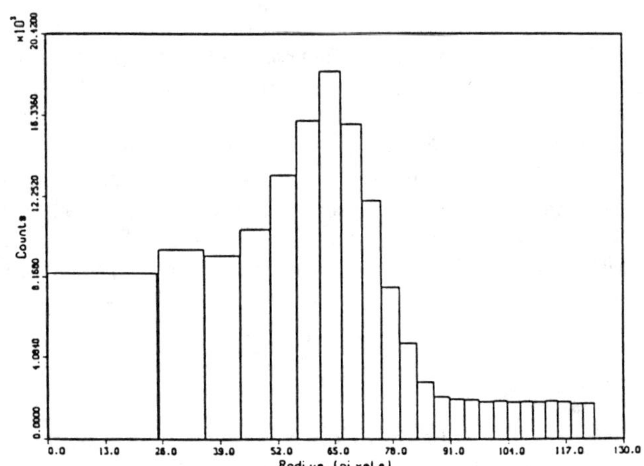

FIGURE II X-ray counts in 24 rings around the chosen centre

APPLICATION TO X-RAY IMAGES OF SUPERNOVA REMNANTS

The matrix method of deprojection for spherically symmetric emissivity distributions has been tested using simulated images constructed from given emissivity distributions and works well. It has also been applied to the images in the catalog of Einstein observations of supernova remnants (Seward 1990).

As an example, the results for Tycho's supernova remnant are given here. The image (Figure I) is approximately circularly symmetric, so a spherical deprojection for it is a reasonable approximation. Figure II shows the flux from the rings and Figure III shows the derived emissivity distribution. This is confined to a shell 60-80 pixels in radius, with zero emissivity inside.

The results of the deprojections will be compared with theoretical emissivity distributions, e.g. such as given by White & Long (1991). This will test

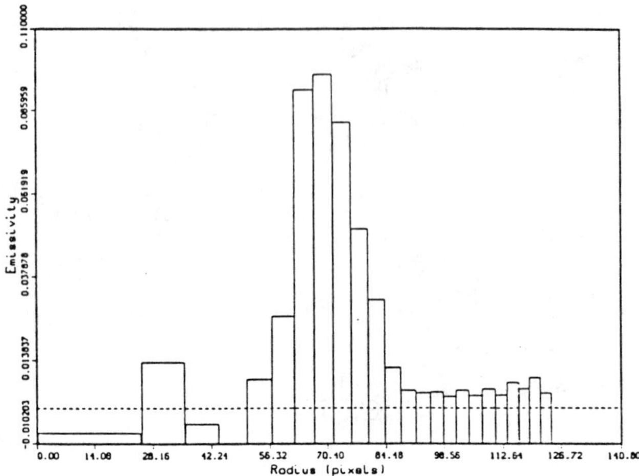

FIGURE III Derived radial emissivity distribution

whether the theoretical calculations can explain the range of observed brightness distributions, or whether other effects will have to be taken into account in theoretical work.

The main limitations of the current work are: the assumption of spherical symmetry; the matrix method does not guarantee the derived emissivity will be non-negative. The first assumption implies that the method should be applied only to objects which are nearly spherically symmetric. The second assumption can be corrected by using the Lucy deconvolution algorithm instead of the matrix method.

ACKNOWLEDGMENTS

This work supported by the Natural Sciences and Engineering Research Council of Canada.

REFERENCES

Leahy, D. A. 1991, A&A, 247, 584

Seward, F. D. 1990, ApJS, 37, 27

White, R. L., & Long, K. S. 1991, ApJ, 373, 543

USING A MATCHED FILTER TO IMPROVE SNR OF RADIO MAPS

NAILONG WU
Peking Astronomical Observatory, Academia Sinica, Peking 100080, China

ABSTRACT In this presentation the problem to be dealt with is described. Then, the concept of matched filter is briefly introduced and its performance in improving signal–to–noise ratio of radio maps is theoretically analyzed. Finally, results of computer simulation and real map processing are reported.

THE PROBLEM

The Miyun Synthesis Radio Telescope (MSRT) at Peking Observatory consists of 28 9m dishes in the East–West direction with a maximum baseline of 1164m, working at frequency of 232 MHz. Accordingly, radio maps are of quite low resolution (about 4 arcminutes), and most of radio sources appear approximately as point sources having roughly the same shape as the beam.

In searching these radio maps for point sources, the criterion for the occurrence of a source is $K/\sigma \geq m$, where K is the peak magnitude and σ is the noise standard deviation. Usually, $m = 5$.

To increase the number of detectable sources and improve the reliability of detection, it is crucial to improve the signal–to–noise ratio (SNR). This can be achieved in various ways at different expense. Using a matched filter is a cheap and optimal way in a sense.

MATCHED FILTER

Matched filters are widely used in signal processing to improve SNR. A detailed description in the context of general image processing can be found in Pratt (1978). The description is tailored for our purpose in the context of radio map processing.

A *matched filter* is a linear filter which (1) keeps the peak locations of point sources unchanged, and (2) maximizes the output SNR. The second requirement is general while the first one is particular to the application in radio map processing.

Suppose that a map has a point source at (x_0, y_0) in a noisy background. The image of the source is represented by

$$s(x,y) = Kb(x - x_0, y - y_0) \tag{1}$$

where $b(x,y)$ is the beam (its maximum is one), and K is the source peak magnitude. The Fourier transform (FT) of $s(x, y)$ is

$$S(u, v) = KB(u, v)\exp[-j2\pi(ux_0 + vy_0)] \tag{2}$$

where $B(u, v)$ is the FT of $b(x, y)$ (u,v are the spatial frequencies). For simplicity the noise is assumed to be white Gaussian with zero-mean and variance σ^2. Then, the input SNR (before filtering) is

$$SNR_i = K/\sigma. \tag{3}$$

Denote the impulse response or point spread function of the filter by $h(x, y)$. Its FT, *i.e.*, the transfer function, is denoted by $H(u, v)$. At the output of the filter, the image of the source and the variance of the noise are, respectively,

$$\begin{aligned} s_0(x, y) &= \iint_{-\infty}^{\infty} S(u, v)H(u, v)\exp[j2\pi(ux + vy)]\, dudv \\ \sigma_0^2 &= \iint_{-\infty}^{\infty} \sigma^2 |H(u, v)|^2\, dudv. \end{aligned}$$

The output SNR (after filtering) is

$$SNR_0 = |s(x_0, y_0)|/\sigma_0.$$

Using Schwartz's inequality it can be shown that

$$(SNR_0)_{\max} = K\sqrt{A}/\sigma \tag{4}$$

where

$$A = \iint_{-\infty}^{\infty} |B(u, v)|^2\, dudv = \iint_{-\infty}^{\infty} |b(x, y)|^2\, dxdy. \tag{5}$$

The condition for obtaining the maximum SNR_0 is

$$H(u, v) = CS^*(u, v)\exp[-j2\pi(ux_0 + vy_0)] = B^*(u, v) \text{ (by (2). Const } C = 1/K.) \tag{6}$$

Equation (6) indicates that in the spatial frequency domain the transfer function of the matched filter is the complex conjugate of the visibility corresponding to the beam. In the spatial domain

$$h(x, y) = b(-x, -y), \quad s_0(x, y) = s(x, y) * h(x, y) = s(x, y) * b(-x, -y)$$

which is the cross-correlation of the input map and beam.

PERFORMANCE ANALYSIS

Define the *SNR gain* by

$$G = (SNR_0)_{\max}/SNR_i = \sqrt{A} \qquad \text{(by (3) and (4)).} \tag{7}$$

In order to carry out theoretical analysis, the beam is assumed to be Gaussian-shaped, *i.e.*,

$$b(x, y) = \exp\left[-\frac{1}{2}\left(\frac{x^2}{a_x^2} + \frac{y^2}{a_y^2}\right)\right], \tag{8}$$

then

$$A = \pi a_x a_y, \qquad G = (\pi a_x a_y)^{1/2}. \tag{9}$$

G will be greater than one provided $a_x a_y > 1/\pi$.

In the discrete case the integration in (5) becomes summation, then

$$G = \left[\sum_{n_x=-\infty}^{\infty} \sum_{n_y=-\infty}^{\infty} \exp[-(\frac{n_x^2}{a_x^2} + \frac{n_y^2}{a_y^2})] \right]^{1/2} \tag{10}$$

which is always greater than one.

COMPUTER SIMULATION AND REAL MAP PROCESSING

In an MSRT map, the full half-power width, w (w_x, w_y), of the image of a point source is approximately 9 (1024×1024 point map) or 4.5 (512×512 point map) pixels. w (w_x, w_y) is related to the parameter a (a_x, a_y) by $a^2 = w^2/(8 \ln 2)$. (x – R.A., y – DEC.)

Computer simulation and real map processing are carried out on VAX-11/780. The white Gaussian noise for synthetic maps is generated from pseudo-random numbers. Filtering is accomplished in the spatial frequency domain. Specifically, a map and beam both Fourier transformed, then multiplied, and finally inverse Fourier transformed to obtain a filtered map. The results are as follows.

`Synthetic map:`

(1) 1024×1024 point map. $w_x = w_y = 9.0$.
$G_t = 6.77$. (Theoretical calculation by (9) or (10).)
$G_c = 6.75$. (Computer simulation.)

(2) 512×512 point map. $w_x = w_y = 4.5$.
$G_t = 3.39$. $G_c = 3.33$.

`Real map:`

(3) 512×512 point map. $w_x = 4.0$, $w_y = 7.0$.
$G_t = 3.98$. $G_c = 1.5$.

From the above we see that the results of computer simulation are close to those of theoretical calculation because the noise is nearly white. In contrast, the result of real map processing is not as good as expected. This is mainly because the assumption of white noise is far from the truth. Nevertheless, it is still worth doing since $G_c = 1.5$ is obtained at a little expense, about 4 minutes CPU time. Further improvement of SNR is possible by (1) taking the non-whiteness of noise into account, and (2) using a derivative matched filter. In this respect, more work is to be done.

REFERENCES

Pratt, William K. 1978, Digital Image Processing, John Wiley & Sons, Inc., New York

SECTION 7

IMAGE ANALYSIS SOFTWARE

MORE EXPERIMENTS WITH DAOPHOT II AND WF/PC IMAGES

PETER B. STETSON
Dominion Astrophysical Observatory, Herzberg Institute of Astrophysics, National Research Council of Canada, 5071 West Saanich Road, Victoria, BC V8X 4M6

ABSTRACT I give a progress report on experimental modifications to the stellar photometry program DAOPHOT II, and on the degree to which the current version of the program succeeds in deriving stellar photometry from the aberrated WFC (and PC) images.

INTRODUCTION

At this conference we are seeing many magnificent, grand, comprehensive software systems for dealing with enormous volumes of heterogeneous scientific data and complex data-analysis problems. Being myself unable to discourse on this lofty plane — on a par with, say, The Advanced Theory of Automotive Design — I am just going to crawl under the hood of my own personal car, yank out the carburetor, and spend my time contemplating the design details of one particular butterfly valve, in hopes of improving its efficiency under a specific set of driving conditions. Casting all metaphor aside, I will be discussing ways of encoding, storing, and evaluating a simple numerical model of the point-spread function for a digital image. I am trying to learn how to derive scientifically interesting stellar photometry from images obtained with the Hubble Space Telescope in its current optically aberrated state, but my experiments might also be of some interest for researchers using deconvolution techniques to obtain morphological information from HST or other digital images.

BACKGROUND

For those of you not familiar with the idea of profile-fitting photometry, let me assure you, it is quite straightforward: we wish simply to determine the brightnesses of stars recorded in some digital image. For isolated stars this is easy — we need merely add up the data-numbers in a subregion of the frame which includes some star, subtract off the contribution from the diffuse sky brightness (which may be determined from a nearby star-free region), and there we have it. However, when the profiles of several stars overlap, it becomes a little more complicated: we simply do not know *a priori* how many of the photons detected in pixel (i, j) should be attributed to star A and how many belong to star B. To resolve this dilemma, we exploit the fact that the actual profile-shape of the stars can be reliably estimated. Given a numerical model of the stellar profile, we use least-squares or other fitting techniques to match copies of this model, suitably shifted, scaled and summed, to the data for the blended stars.

This concept has been around for many years, extending back even to the era of analog image acquisition and analysis, and many algorithms have been developed to perform this function. Some examples from the days of analog imagery are found in Butcher (1977), Buonanno et al. (1979), and Stetson (1979) (software packages usually did not have clever, widely recognized names in those ancient days). Bridging the transition from analog to digital image-acquisition technology we find the program RICHFLD (Tody 1981), and these latter days have seen the development of quite a variety of profile-fitting packages: among them are ROMAFOT (Buonanno et al. 1983), STARMAN (Penny & Dickens 1986), WOLF (Lupton & Gunn 1986), DAOPHOT (Stetson 1987), HAOPHOT (Gilliland & Brown 1988), DoPHOT (Mateo & Schechter 1989), the LUND package (Linde 1989), and CAPELLA (Llebaria et al. 1989). I think it is probably fair to say that, while these packages differ in some minor details of numerical computation and in their approach to interfacing with the user, nevertheless they all share the same overall concepts and achieve results that are limited primarily by the quality of the data, rather than by any inadequacy of the algorithms.

There is one detail of algorithm design in which these packages differ which is probably not so minor. This is the butterfly valve I alluded to above, namely the way in which they encode and store the model stellar profile for later evaluation. As far as I know, there are only three ways to store the information required to compute the flux expected in a specific pixel within the profile of a particular star. The first of these may be called

The analytic point-spread function

One assumes that some mathematical formula is capable of accurately describing the intensity profile of a star image in a data array. Of course, this formula contains some parameters which may be adjusted to optimize the match between the formula and actual stellar profiles as observed in a particular image. Specific examples of programs which employ the analytic point-spread function are

ROMAFOT (Buonanno et al. 1986): PSF $\sim 1/(1+z^2)^\beta$
STARMAN (Penny and Dickens 1986): PSF $\sim$ Gaussian + Lorentzian
DoPHOT (Mateo and Schechter 1989): PSF $\sim 1/(1+z^2+z^4/2+z^6/6)$

(Note: $z^2 = r^2/\alpha^2$; or $z^2 = x^2/\alpha_x^2 + y^2/\alpha_y^2 + \alpha_{xy}xy$; or $z^2 = \ldots$)

The analytic point-spread function may be made as complex as desired simply by adding enough parameters; however, when many parameters are included to make a highly flexible analytic model, one can run into convergence problems in attempting to determine those parameters through comparison to actual stellar images. The simpler approach to highly complex model profiles is

The empirical point-spread function

One simply extracts from the original image one or more small subarrays, each of which contains a single bright, isolated star. If there are several subarrays they are averaged together, but in either case the results are stored in the computer as a two-dimensional digital data array, and subsequently the brightness at a particular location within a star's profile is estimated by numerical interpolation. Programs which adopt this approach include

WOLF (Lupton and Gunn 1986)
LUND (Linde 1989)
CAPELLA (Llebaria et al. 1989)

The empirical point-spread function is best suited to the oversampled case, where optical aberrations and tracking vagaries can lead to complex structures within the well-resolved image profile. These could be very hard to model with a limited number of parameters in some analytic function, but the empirical point-spread function has essentially one free parameter for every pixel in the digital table. However, the empirical PSF becomes less satisfactory when the data approach critical- or under-sampling, since under these circumstances practical interpolation techniques have a hard time producing the ~1% precision required in predicting a given star's profile. And remember, these interpolation techniques must be employed twice: once to align the various PSF stars' centers (to avoid artificial smearing of the profile by the random placement of the stars' centroids with respect to the pixel grid) and again when the model profile is applied to each program star. Conversely, the analytic model enjoys its greatest success with critically- or under-sampled data. Here, details of profile structure are simply not resolved — each star occupies only a small number of pixels, and a small number of parameters is adequate to describe everything that can be seen. The undersampling itself produces no interpolation errors in this case, because the analytic model can be numerically integrated over the area of each pixel before being compared to the raw image data.

The third method of encoding and storing a model point-spread function is, of course,

The flyspanker

This powerful method of profile-matching was developed in the days of analog image acquisition and analysis, but has not yet been adapted for use with digital images. Perhaps it should be.

In 1983 I began to develop DAOPHOT, which was (I believe) the first software package that attempted to retain the strengths of both the analytic and the empirical PSF by combining them. First, an analytic function is fitted to the observed data for several bright, isolated stars. Then these fitted profiles are subtracted from the original image data, and subarrays containing the subtraction residuals are extracted and averaged together, to yield a two-dimensional lookup table of the mean differences between the "true" point-spread function and the analytic first approximation. To compute the PSF at an arbitrary point in a stellar profile, first the analytic function is evaluated (note that both during the construction of the PSF and during its subsequent evaluation, the analytic function is carefully integrated over the area of each pixel), and then a correction to the "true" value is determined by interpolation within the table. This method works because for most reasonable astronomical images the analytic first approximation actually represents 90%, or 95%, or 98% of the intensity variations within a star's profile, and the analytic function can be evaluated with arbitrary precision. The lookup table of corrections has then only 10%, or 5%, or 2% of the total amplitude that it would have had were it required to represent the entire stellar profile, and errors due to inadequate interpolation are reduced in proportion. HAOPHOT (Gilliland & Brown 1988) is a second profile-fitting package which utilizes a hybrid analytic/empirical point-spread function.

Some years later, around 1985, I began to encounter CCD images where the point-spread function varied perceptibly across the frame. I tried to deal with this situation by allowing each element of the empirical lookup table of corrections to be a linearly-varying function of position within the frame. That is, instead of the PSF being

$$PSF(\delta x, \delta y) = \int\int_{\mathrm{pixel}(\delta x, \delta y)} \mathrm{Gaussian}\, dx\, dy + T_0(\delta x, \delta y)$$

it became

$$PSF(\delta x, \delta y) = \int\int_{\mathrm{pixel}(\delta x, \delta y)} \mathrm{Gaussian}\, dx\, dy + T_0(\delta x, \delta y) + X \cdot T_1(\delta x, \delta y) + Y \cdot T_2(\delta x, \delta y)$$

where (X, Y) represents the position of the star in the data frame. Thus, in addition to the analytic first approximation DAOPHOT's PSF now could have three lookup tables instead of one, the tables T_1 and T_2 representing a first-order Taylor expansion of the *change* in the stellar profile with location in the image.

THE HUBBLE SPACE TELESCOPE

With this history in mind, let us now recall what we know about the imaging properties of the Hubble Space Telescope (especially the Wide Field/Planetary Camera [*especially* the Wide Field Camera]):

- The core of the stellar profile is seriously undersampled.
- The point-spread function nevertheless has significant, complex extended structure without simple symmetries.
- The point-spread function varies significantly across the field.

At this point it begins to look like the hybrid approach to encoding the point-spread function is the obvious thing to try, and DAOPHOT should do about as well with HST images as any available package. However, we must also remember that

- HST's point-spread function is *BIG!* In itself, this would not be a serious problem; it would merely increase the amount of CPU time required to reduce a given frame. However, this particular property of HST does engender the following collateral damage:
    1. Every reasonably bright star may be expected to have at least one **cosmic ray event** somewhere in its profile.

2. Most bright stars will have some degree of **overlap** with other, often comparably bright, stars in the frame — truly isolated stars for the generation of even a provisional model PSF will be very scarce.
3. Unlike the case with most ground-based imagery, there may be **absolutely no "clear-sky" patches** within the image, where the underlying diffuse-sky brightness could be estimated — virtually every pixel in the frame may be involved in the profile of many individual star images.

Finally, the CCDs in WF/PC have a fairly narrow dynamic range, and astronomers, unreasonably enough, always seem to want to study *faint* things with HST. This has the obvious consequence that

- The best, brightest, and most isolated stars in the frame are all *saturated.*

Therefore, I have undertaken to try to modify DAOPHOT to deal with these unpleasant properties of the current, aberrated HST. A preview of some of these changes was presented by Stetson et al. 1989, where I discussed the following modifications; these are extensive enough that perhaps now it is worthwhile calling the program DAOPHOT II:

- The analytic part of the point-spread function is no longer assumed to be Gaussian. Instead the user has the choice of a Gaussian function, a Lorentzian, several Moffat functions (modified Lorentzians), or several sums of Gaussian + Lorentzian functions (which I call, generically, "Penny functions" after Alan J.). The Gaussian seems most efficient for well-sampled ground-based images; the Lorentz function seems to be the best first approximation for WFC images; the Moffat functions work well for undersampled ground-based data; and I have not yet encountered a situation where a Penny function seems required, but one may presume that they will do better than the others for ground-based images where the stellar profiles are particularly complex, due to erratic guiding or severe aberrations. The definition of the analytic function is now modular and well-separated from the rest of the code, so different analytic models may easily be incorporated.
- There are now four degrees of PSF complexity allowed, rather than the former two: there may be *no* lookup table of corrections (i.e., a purely analytic PSF, mimicking the operation of ROMAFOT, STARMAN, or DoPHOT), or one table, or three, or six (i.e. a PSF which varies quadratically with position in the frame). This part of the package, too, is now modular, so possible future expansion to higher-order variations (e.g. cubic variation with position, or variation with the *color* of the star) may easily be incorporated.

A first progress report on the application of the revised software to real HST images was presented earlier this year in Garching (Stetson 1991). There I commented on the first round of modifications intended to deal with the collateral damage produced by the large size of HST's PSF:

- During the construction of the lookup table(s) of corrections to the analytic PSF, apparently discordant pixels (as recognized by comparison among those pixels located at the same place within the profiles of the different PSF stars) are automatically down-weighted. This so-called "artificial skepticism" has always been included in the *fitting* of the PSF to program stars, but until now it has not been a part of the construction of the PSF. This simple step almost completely eliminates the damage caused by cosmic rays and overlap among the PSF stars.

- The local sky brightness may now be redetermined automatically during the profile fits, by a scheme resembling that of Parker (1990). This allows reasonably precise corrections for the contribution of diffuse sky light to a star's profile even in a frame with no clear-sky pixels.

- Given its new ability to disregard defective pixels within the PSF stars, the software has now been empowered to make its own selection of stars suitable for producing the point-spread function. As a result, the entire reduction process from finding stars in the original frame to generating the ultimate point-spread function and obtaining the final astrometric and photometric results may now be automated, with the astronomer having to do little more than type the image name into the computer. It is still advantageous to have the astronomer visually review the computer's choice of PSF stars at some point, perhaps rather far into the process; however, this need not take more than five or ten minutes of astronomer time per frame.

In the Garching presentation, I showed results from my first attempts to perform profile-fitting photometry in actual HST images — in that instance, Early-Release frames of the rich Magellanic Cloud cluster NGC 1850. Judging by the residuals left when the derived PSFs were used to subtract actual star images from the frames, I concluded that the individual stellar magnitudes should be internally precise to of order 3%, over a range of perhaps four magnitudes extending down from the magnitude at which the stars' centers become saturated. At that time I bemoaned the fact that photometry would be exceedingly difficult in cases where the frame contained many centrally saturated stars: first, it would be difficult to define an initial PSF in a frame where much of the image area and most of the brightest unsaturated stars were contaminated by the wings of still brighter stars; and second, the comparatively poor signal-to-noise ratio in the wings of the best PSF which *could* be generated, when scaled up in an attempt to match and subtract the wings of the saturated stars, would leave behind very messy patches of residuals which would obliterate the photometric information in a large fraction of the frame.

Since last Spring I have further modified the software. Now,

- The program is able to include saturated stars in the derivation of the average point-spread function for a frame. The saturated pixels in the cores of such stars are ignored. However, the unsaturated wings of those stars are scaled to a common magnitude by comparison with a provisional PSF formed solely from unsaturated stars, and then those wings are included in the outer parts of the lookup tables of corrections from the analytic first approximation to the "true" point-spread function. The signal-to-noise ratio in the outer parts of the model PSF is significantly improved, which now allows the wings of even saturated stars to be subtracted relatively cleanly (the cores are a total loss, of course), and at least some photometric information can be recovered for fainter stars veiled by the wings of saturated ones.

INSURING A CONSTANT PSF VOLUME

There is one subtlety involved in the variable hybrid point-spread function which has not yet, to my knowledge, been discussed in print: in order to maintain a uniform photometric zero-point across a frame, it is necessary to insure that the volume contained under the model point-spread function not vary with position. Given the way in which DAOPHOT II's variable PSF has been encoded, this boils down to requiring that each of the higher-order tables of corrections (viz. T_1 and T_2 for the linearly varying PSF, $T_1 \ldots T_5$ for the quadratically varying PSF) contain zero net volume. (The table of zero order corrections from the analytic to the best *average* PSF for the frame, T_0, may contain some non-zero volume.)

There are two ways in which the first estimates of the higher-order tables can acquire some net volume, and therefore two distinct steps are required to counteract their effects. First, we must recognize that with the severe overlapping of stellar images which is to be expected in HST frames, it is impossible to predict the local diffuse sky brightness underlying a given star with any sort of precision until *after* a good PSF has been constructed for the frame; in the initial stages of PSF construction, we must expect some errors in the assumed sky brightnesses. If there exists any reason why the *error* in the sky estimate may depend upon position in the frame — if there is a cluster in one corner of the image, for instance — then for most parts of the image the derived model stellar profile will not approach zero intensity at large distance from the centroid of the PSF. Rather, in some parts of the frame the stellar profile will appear to be sitting upon a stubby pillar of luminosity, while in other parts of the frame it will be sticking up out of a shallow hole. Since HST's PSF is so *BIG!* an error in the sky level of even a data-number or two can contribute a lot of volume to the integrated profile. I deal with this situation by examining the median brightness value around the rim of the higher-order correction tables, $T_2 \ldots T_N$: that constant value which is required to make the outermost edge of the corrections have a median value of zero is added to or subtracted from each of these tables. The same sort of correction is *not* made to the constant table, T_0, since it is possible (likely?) that the true stellar profile has not actually reached zero intensity at the limited radius (~ 25 pixels) to which we try to represent it. However, these corrections to the higher-order tables guarantee at least that the neglected flux does not vary with position in the frame.

The second source of a PSF-volume gradient lies in the fitting of the preliminary analytic profile to PSF stars which have some range of apparent magnitude. Recall that we wish to use the subarrays of residuals left after the subtraction of the analytic first approximation to define the tables of empirical corrections. Therefore, we must first scale that analytic profile to the actual brightness of each star and, after subtraction of this first approximation, the residuals must then be scaled to a common stellar brightness before they can be averaged together to make the correction tables. So what happens if, due to the fact that the PSF varies with position, the analytic model profile tends to be scaled too high on one side of the frame and too low on the other side? On the one side of the frame we extract tables of residuals which have little holes in the middle (due to the oversubtraction of the analytic profile) and we scale the residuals up too little (since we have overestimated the PSF star's brightness), while on the other side of the frame the residuals have a little peak left in the middle and they get scaled up too much. Again, the result is that the higher-order tables of corrections are left with some net volume. The solution in this case is to determine the actual volume contained within each of the higher-order tables by simple summation, scale the *constant* part of the PSF to the same volume, and then subtract it to leave the higher-order table with zero net volume.

In both cases — in correcting for both the sky error and the scaling error — whatever is subtracted from the higher-order tables is interpolated to the value the correction had at the centroid of the distribution of PSF stars in the frame, and added back in to the table of constant corrections, T_0. This leaves the point-spread function at the centroid of the PSF-star distribution unchanged, while forcing the total volume of the PSF to be independent of position.

NEW PHOTOMETRIC TESTS

Bill Baum and Bob Light have very kindly let me use four of their Wide-Field Camera visual (F555W) images of a field in Baade's Window for a new round of tests. These frames have exposure times of 800 sec (two of the frames) and 1000 sec (the other two), and each star reappears at the same place in the pixel grid in all frames to within a tolerance of better than 0.05 pixel. Baum and Light have also lent me four visual exposures of different fields within the open cluster NGC 188.

One of the questions that I have not yet seen definitively answered is how well one can do photometry in one frame using a PSF which has been generated from a different frame. The NGC 188 exposures gave me a good opportunity to experiment. These are comparatively shallow exposures, containing no saturated stars and, in fact, containing very few stars even within one magnitude of saturation. Each individual exposure contained only maybe a dozen stars from which one could hope to get any information at all useful, and this is simply too few for constructing a reliable quadratically-varying PSF. Therefore, I added the four exposures together to obtain a single frame with four times the number of useful stars. Of course, this also quadrupled the number of cosmic rays and doubled the read noise. Furthermore, each of the individual exposures may have had a significantly different history of tracking errors, so no one PSF model, however sophisticated, may be representative of all the stars in the summed frame. On the other hand, the average PSF for the frame contains some sort of "average" tracking history, and this may not be a Bad Thing.

Having used the PSF derived from the summed NGC 188 image to perform profile-fitting photometry for each of the individual Baade's Window exposures, I determined the precision of the results by intercomparing the magnitudes obtained for the same stars in the four exposures (after allowing for zero-point offsets between the frames). That is, for every star which was discovered in at least three of the four images, the standard error of a single magnitude measurement was estimated from

$$\sigma(m) = \sqrt{\frac{\pi}{2}\frac{n}{n-1}}\sum_{i=1}^{n}|m-\overline{m}|\Big/ n\,,$$

where $\overline{m}$ is a compromise between the mean and the median of the n individual magnitude measurements for the star. For a Gaussian distribution of photometric errors, this statistic gives essentially the Gaussian σ. For non-Gaussian errors, it is less dominated by extreme outliers than is the standard root-mean-square estimator, and gives a more realistic estimate of the error of a "typical" measurement — it will be less inflated, for instance, by cosmic-ray hits on individual stellar images, which would be detected and discarded in any reasonable photometric research program. It is important to note that whereas the precision estimates presented in the Garching report were inferred from the model's ability to represent the stellar profiles within a given frame, the present ones are based upon actual frame-to-frame photometric repeatability among independent images. Therefore the current tests provide a more direct estimate of the limiting capabilities of the combination (the present HST + existing software) for measuring, say, variable stars.

When the NGC 188 point-spread function is used for the four Baade's Window frames, the resulting photometric errors for 5318 three- or four-time repeat detections average about 0.07 mag per frame from about a magnitude brighter than the level at which stars first become saturated, to a level about three magnitudes below saturation. It is useful to remember that photometric errors of $\sim 7\%$ per frame are also achievable with the much simpler and more robust fly-spanker technology. At fainter magnitudes the errors increase, passing through 0.10 mag about four magnitudes below saturation and reaching a quarter magnitude at the detection limit maybe five and a half or six magnitudes below saturation.

When I derive a separate point-spread function for each of the Baade's window frames from some of the brightest stars within that frame itself, including some saturated ones, the results improve significantly. From a magnitude above saturation to three magnitudes below, the standard error per frame is constant at slightly better than 0.03 mag; four magnitudes below saturation it has increased to about 0.06 mag, and at five and a half to six magnitudes below saturation it is still only about 0.13 mag. It is interesting to note that in these reductions only 4100 stars were found in at least three of the four frames. I interpret this to mean that nearly 25% of the "stars" which were "successfully" reduced with the NGC 188 point-spread function were in fact garbage left behind when the inadequate, low signal-to-noise-ratio NGC 188 PSF was used to subtract out the brighter stars in the frames. Since the same PSF was used for the four frames, this garbage would have been duplicated in all of them. The more precise and accurate PSF derived from the program frames themselves permitted cleaner subtraction of the bright stars and more reliable detection — as well as better photometry — for the faint ones.

ACKNOWLEDGMENTS

I am grateful to Bill Baum and Bob Light for the loan of their WFC images.

REFERENCES

Buonanno, R., Coluzzi, R., Corsi, C. E., De Biase, G. A., & Ferraro, I. 1979, Mem. Soc. Astron. Ital., 50, 451

Buonanno, R., Buscema, G., Corsi, C. E., Ferraro, I., & Iannicola, G. 1983, A&A, 126, 278

Butcher, H. 1977, ApJ, 216, 372

Gilliland, R. L., & Brown, T. M. 1988, PASP, 100, 754

Linde, P. 1989, in Proceedings of the 1st ESO/STECF Data Analysis Workshop, P. J. Grosbøl, F. Murtagh, & R. H. Warmels, ESO: Garching, 201

Llebaria, A., Perichaud, L., Leporati, L., & Debray, B. 1989, in Proceedings of the 1st ESO/STECF Data Analysis Workshop, P. J. Grosbøl, F. Murtagh, & R. H. Warmels, ESO: Garching, 85

Lupton, R. H., & Gunn, J. E. 1986, AJ, 91, 317

Mateo, M., & Schechter, P. L. 1989, in Proceedings of the 1st ESO/STECF Data Analysis Workshop, P. J. Grosbøl, F. Murtagh, & R. H. Warmels, ESO: Garching, 69

Parker, J. Wm. 1991, PASP, 103, 243

Penny, A. J., & Dickens, R. J. 1986, MNRAS, 220, 845

Stetson, P. B. 1979, AJ, 84, 1056

Stetson, P. B. 1987, PASP, 99, 191

Stetson, P. B. 1991, in Proceedings of the 3rd ESO/STECF Data Analysis Workshop, in press

Stetson, P. B., Davis, L. E., & Crabtree, D. B. 1989, in CCDs in Astronomy, G. H. Jacoby, San Francisco: Astronomical Society of the Pacific, 289

Tody, D. 1981, RICHFLD Photometry Program Users Guide, Tucson: Kitt Peak National Observatory

AUTOMATED APERTURE PHOTOMETRY OF CCD IMAGES

MICHAEL V. NEWBERRY
Steward Observatory, University of Arizona, Tucson, AZ 85721

INTRODUCTION

A key project of the Steward Observatory CCD Transit Instrument ("CTI") is a search for supernovae in a sample of 10,000 galaxies. This is carried out by searching for "bumps" in the light curves of the galaxies, these possibly signifying the occurrence of supernovae. To minimize the frequency of false detections, the nightly magnitude measurements must be consistent within observational uncertainties and the internal and external errors must be estimated correctly. An aperture photometry method (see *e.g.*, Howell 1989; Stetson 1987, 1990) was determined to be best for this purpose, but the nightly volume of data is so large that the data reduction and analysis software, including the background estimation, must run without human interaction. I have developed a suite of algorithms, collectively named "autophot", which automatically determines the background and computes fluxes inside elliptical apertures. The sequence of operations is as follows:

1. Read the x and y coordinates of the target image.

2. Compute the mode of the pixel values on a line of data passing through the center of the image. Moments of the intensity distribution about the image are computed above this zeroth-order background value.

3. Fit an ellipse to the isophotes of the image at a semiaxis twice the radius of gyration. The fit parameters will be applied to a family of apertures used in determining the background.

4. The following equation defines the surface brightness, S, of a plane in cylindrical (r, θ) coordinates:

$$S(r, \theta) = a_0 + b_0 r \sin(\theta - \theta_0) , \quad (1)$$

where a_0 is the surface brightness of the plane at $r = 0$. Pixels sampled around different size apertures having the same center and located on the same plane should have the same values of a_0, b_0, and θ_0. If the local background can be adequately described by an inclined plane, then the run of central surface brightness $a_0(r)$ over a series of apertures of different size should become constant when the image profile merges into the background plane. To find the radius at which this occurs, equation [1] is fit to pixels in apertures spaced $\sqrt{2}$ pixels in radius and extending to the edge of the frame. The values $a_0(r)$ are compared with the mode of the distribution of $a_0(r)$ to determine at which radius the profile becomes statistically indistinguishable from a constant. In this sense, the algorithm "flattens" the planar background. To avoid sampling the tail of the image profile, a subset of 6 to 10 apertures outside this radius and spaced every $3rd$ aperture ($=3\sqrt{2}$ pixel increments) define the final background sample. The "good" pixels in these apertures are then fit with a plane in the x, y domain of the image raster. A powerful bad-data rejection algorithm effectively removes other images and wild pixel values from these fits.

5. (a) Fit a series of apertures to the image isophotes or (b) setup a series of similar ellipses at different radii and calculate the total flux inside each aperture. An exact analytic method is used to compute the areas of partial pixels. The computed area for each aperture is checked against the exact area for an ellipse, πab. Even for very large and very small ellipses, there have been no cases in which the computed area differs from the exact value by more than 10^{-3} pixels.

6. Finally, the differential aperture magnitudes are compared with previous nights' data to search for supernovae. A total magnitude is also computed from the asymptotically constant value of the growth curve.

The critical elements of *autophot* are the algorithms used for background determination and the exact integration of the aperture areas. Space here does not permit a detailed discussion, but this can be found in Newberry (1992).

TESTING THE AUTOPHOT ALGORITHMS

Synthetic CCD Frames

Simulated frames were generated using the *simstars* program written by M. Cawson. This program randomly distributes a specified number of stars in position and magnitude using the luminosity function of the Solar neighborhood, realistically simulating image crowding. The point spread function is specified by the user and all pixels are given Poisson noise and readout noise. The background can be inclined as desired. *Simstars* reports the coordinates and magnitudes, before the addition of noise, for all stars. Figure I compares *autophot* background estimates for stars in two simulated frames, one having a 5.55% background gradient in both x and y directions (left), and the other having a constant background (right). The y axis shows the magnitude residuals $V(autophot) - V(simstars)$ normalized by their uncertainty $\sigma(V)$. Neither plot suggests a systematic error with magnitude, and the proportion of residuals within $\pm 1\sigma$ is near the 68% of a Normal distribution. The ability of a background algorithm to find the "background" proper, rather than the shallow tail of the image profile, is crucial to the calculation of accurate magnitudes. Extensive testing was done to assure that *autophot* was locating its inner background radius appropriately.

Comparison with IRAF *apphot* on High Galactic Latitude Fields

Autophot was tested on real CCD data by comparing its results with those from the IRAF *apphot* software when both were run on the same pixel data. For this test, 565 stellar images were selected in a number of high Galactic latitude fields

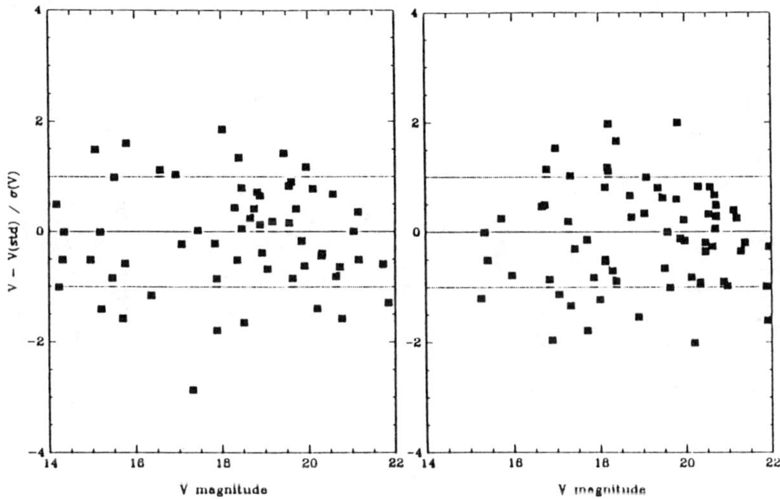

FIGURE I Normalized V residuals *vs* V magnitude for images on synthetic frames. Left: 5.55% background gradient; right: constant background. 1σ residuals are indicated by dotted lines.

observed by CTI. *Apphot* does not really "find" the background but, instead, uses the approximation $mode \approx 3 \times median - 2 \times mean$ to estimate the "most likely" pixel value inside a user-specified annulus around each object. *Autophot* was directed to use the same inner background radius as *apphot*, but was given complete freedom in the fitting and rejection procedures. The differences in the magnitudes computed by the two programs are shown as a function of magnitude in Figure II. The small systematic trend is consistent with *autophot* measuring the background $\sim$ 0.4 counts fainter, on average, than does *apphot.* This is substantiated by the histogram of the background differences in Figure III. One can infer that the backgrounds estimated by *autophot* are likely to be nearer the correct values: First, the mode approximation used by *apphot* is known to be biased toward higher values (Newberry 1992). Furthermore, histograms of the pixels used by *apphot* for background estimation were inspected and judged inconsistent with the value it computed. *Autophot* appears to give very good photometry using only the center coordinates of an image in a non-background subtracted frame.

REFERENCES

Howell, S. B. 1989, PASP, 101, 616

Newberry, M. V. 1992, PASP, submitted

Stetson, P. B. 1987, PASP, 91, 191

—. 1990 PASP, 102, 932

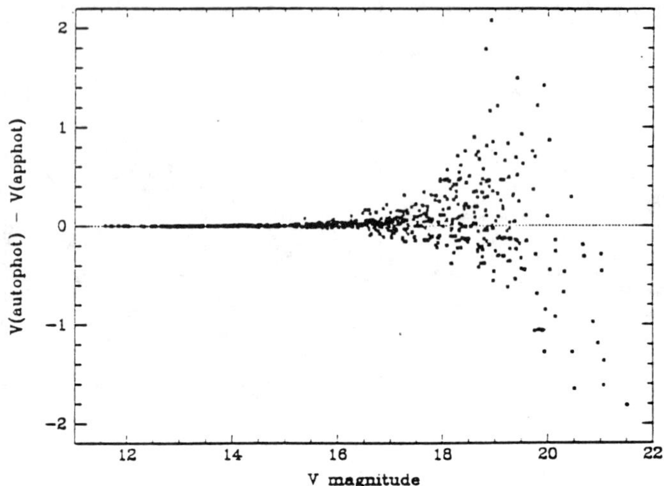

FIGURE II V(*autophot*) - V(*apphot*) *vs.* magnitude for 565 stars.

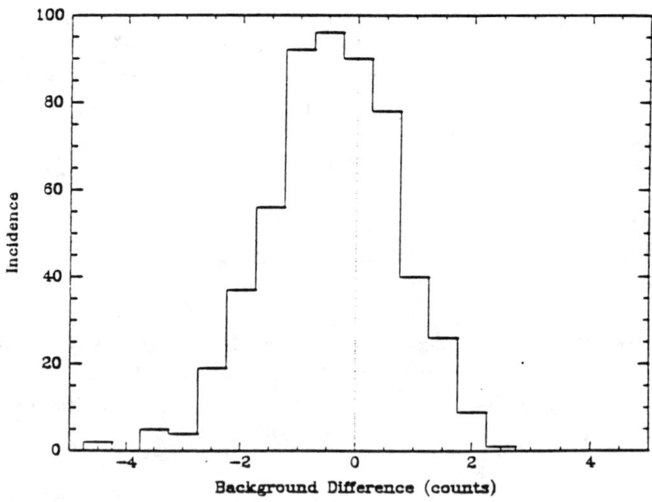

FIGURE III Difference in background, a_0(*autophot*) - mode(*apphot*). The rms noise of the background is ~ 7 counts pixel^{-1}.

NOISE MODEL-BASED COSMIC RAY REJECTION FOR WF/PC IMAGES

R.A. SHAW AND K. HORNE
Space Telescope Science Institute,[1] Baltimore, MD 21218

ABSTRACT We present two new STSDAS tasks that are useful in detecting and rejecting cosmic rays (CRs) from paired WF/PC images. The first determines the noise properties of the detectors, and the second employs those properties to distinguish CRs from noise in science images. We show this technique to be very successful at excluding even low-level CRs, while preserving the sharp inner cores of stellar PSFs on crowded WF/PC images.

INTRODUCTION

Cosmic ray (CR) events affect CCD images from the WF/PC at the rate of several per second, so that data will be lost from a substantial number of pixels for exposures of even relatively modest duration. This problem, which is common for many space-based scientific instruments, is particularly acute for WF/PC because the PSF is both large (due to the spherical aberration of the HST primary mirror), and undersampled. Thus many simple techniques, such as passing a spatial median filter over the image, do not exclude CRs that appear on stellar profiles, or they may remove the sharp inner cores of the PSFs. Yet excluding CRs is vitally important both to accurate stellar photometry and image deconvolution. Fortunately, the problem can be ameliorated by dividing the observation into two (or more) exposures and rejecting CRs by comparing the separate images. We have developed and implemented two new STSDAS tasks to reject CRs from WF/PC frames, and have applied these tasks to combine two WF/PC images of a crowded stellar field, NGC 1850.

FEATURES OF THE STSDAS TASKS

The first of these tasks, **wfpc.noisemodel**, provides the user with an interactive tool for matching the noise characteristics and gain of the detector with a three-parameter noise model. The detector noise is measured from the short-scale fluctuations in featureless images, such as a set of graded flat-field exposures. The three-term noise model used in this task is: $\sigma^2(X) = \sigma_o^2 + X/G + (SX)^2$, where X is the signal level in data numbers (DN), σ_o is the readout noise in

[1]STScI is operated by AURA, Inc. for NASA.

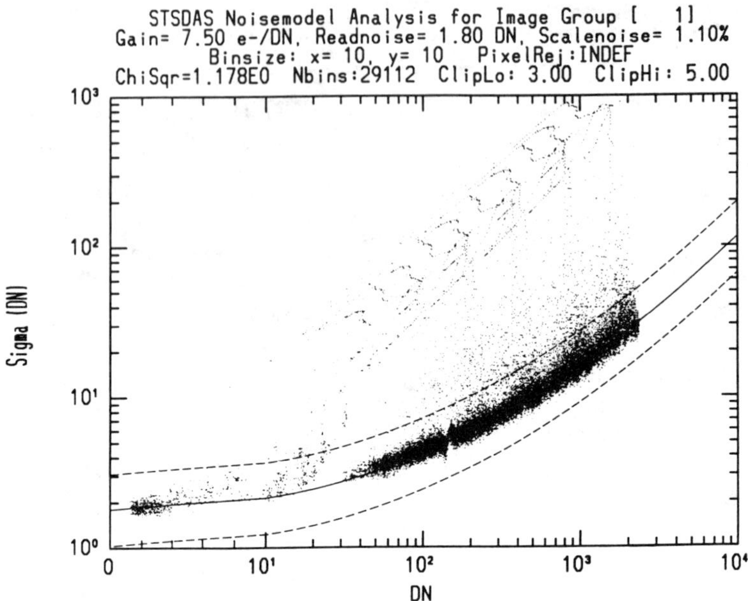

FIGURE I A plot of mean DN *vs.* sigma within 10 x 10 pixel bins on five WF/PC images of NGC 1976 (see text). The noisemodel prediction (*solid curve*) and the $+5/-3\sigma$ rejection limits (*dashed curves*) are summarized in the plot title of this interactive STSDAS task. The WF/PC CCDs are not fully illuminated by the image, causing the artifacts near the top of the plot.

DN, G is the gain in DN/photon, and S is the scale noise. The constant term accounts for additive noise, the term proportional to $\sqrt{DN}$ accounts for Poisson noise, and the term linear in DN accounts for noise that is multiplicative in nature, like the small changes in sensitivity from pixel to pixel which should be removed by applying the flat-field.

This interactive task first divides the input image(s) into bins of adjustable size within which it calculates the mean and standard deviation (sigma). The results are then plotted, along with a curve showing the predicted relation based upon the **noisemodel** parameters (see, e.g., Figure I). Purely Poisson noise corresponds to a straight line of slope +1/2 in the log-log portion of this plot. Readnoise is evident when the Y-intercept is non-zero, and scalenoise causes the slope to increase with increasing DN. Note that is very important to include images that span the full intensity range of the detector (0—4095 for WF/PC) in order to model the noise properties adequately. This task is not restricted to modeling CCD detectors, and a variant of this approach has been used successfully for optimally extracting IUE spectra (Kinney, Bohlin & Neill 1991).

The second STSDAS task, **wfpc.combine**, is an enhancement of the IRAF V2.9 **imcombine** task for averaging multiple frames with pixel rejection. This new task incorporates all of the pixel rejection features of **imcombine**, but adds the capability of masking pixels that are flagged in the WF/PC Data Quality

Files as well. A new option specifically tuned to reject CRs has also been added. The approach is to exclude all values that exceed the median pixel value by more than some factor, κ, times σ, or the uncertainty, at that pixel. Here "sigma" is determined from the noise model (described above) for that median pixel value. This approach differs from object classification techniques (Murtagh & Adorf 1991) in that CR rejection is performed on each pixel independently from its neighbors, which permits reliable CR rejections on both stellar profiles and the background.

RESULTS

We used a set of five graded exposures of the central portion of the Orion Nebula with the F555W filter (close to Johnson V) to measure the noise characteristics of the WFC. These images are relatively featureless, apart from a few field stars, and consist primarily of a slowly varying background of nebular [O III] + Hβ emission. The plot in Figure I shows that the noise and gain properties of the WF/PC CCDs are very close to pre-launch predictions as expressed in the WF/PC Instrument Manual: we derived a gain of 7.5 e^-/DN, a readnoise of 1.8 DN, and a scalenoise of 1.1%. We attribute the non-zero scalenoise to noise and other errors in the flat field used to calibrate these early-epoch images.

We employed these noise/gain characteristics to combine two 1100 s WFC images of NGC 1850 taken through the F555W filter, using the CRREJECT option described above and a 3σ rejection threshold. A blink comparison of the resulting image to the simple AVERAGE showed that even weak, "cometary" CRs were eliminated, while the sharp cores of the stellar PSFs were preserved. Figure II shows a side-by-side comparison of the simple AVERAGE of the images (left) with that using the CRREJECT option (right) in the **wfpc.combine** task.

We also analyzed the CR events themselves through the difference of the AVERAGE and CRREJECT images. A blink comparison of that difference with the CRREJECT revealed no particular correlation, or anti-correlation, between the position of CRs and stars on the image. That is, *CRs can be detected very reliably both in the background and on stellar profiles* with this technique. A histogram of CR energy per pixel (see Figure III) shows that affected pixels are detected consistently down to low DN levels. The resulting energy distribution (i.e., per CR event) has a power-law form: $N(E) \propto E^{-\alpha}$, with a slope $\alpha \approx 1.8$ which is not too different from the expected $\alpha = 1.6$ (Longair 1981). The integral of the energy distribution yields 17,000 detected CRs on these two 4-group images, which suggests an event rate about a factor of 3.3 higher than that stated in the WF/PC Instrument Manual. This difference is probably not significant, however, in that the CR event rate is highly variable with, e.g., position in the orbit of HST, solar activity, etc. One must also realize that, of the $\approx 30,000$ pixels identified as having been affected by CRs, approximately 6,200 would be expected solely on the basis of the 3σ rejection threshold. Since CRs can only be distinguished from noise on a statistical basis, the better the model of the noise, the better CRs can be excluded from science data.

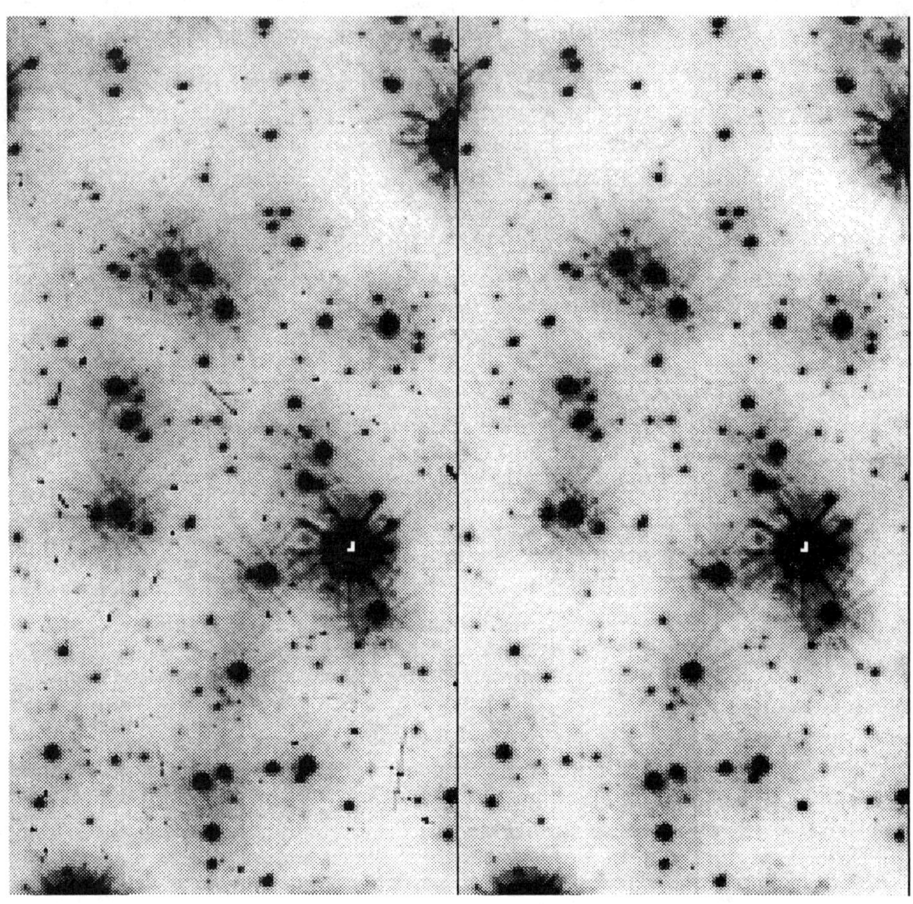

FIGURE II Expanded portion of the average of two WF/PC images of NGC 1850 without CRs removed (*left*), and with CRs removed (*right*) using 3σ rejection. Note that even weak, "cometary" CRs are eliminated, while the cores of the stellar PSFs are well preserved.

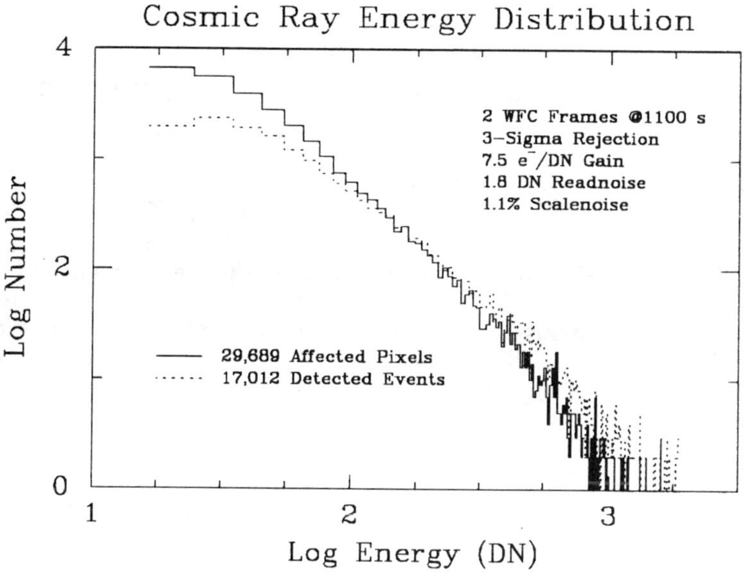

FIGURE III Cosmic ray energy distribution per pixel (*solid*) and per event (*dotted*) detected in two paired WF/PC images of NGC 1850. Note that many of the CR events affect more than one pixel, hence the difference in slope between the energy distributions.

CONCLUSION

We presented two new STSDAS tasks which can, respectively, model the noise in WF/PC (or any other) images, and use those noise/gain characteristics to exclude cosmic rays when combining images. This technique can only be applied when two or more images of the exactly the same field are available, but this is easily arranged for WF/PC observations if the observer specifys a "CR-Split," which is recommended for most observations exceeding $\sim$ 15 min duration. Our technique works extremely well at excluding even weak, "cometary" CRs and CRs that fall on the profiles of stars, while preserving the sharp inner cores of stars in WF/PC frames.

REFERENCES

Kinney, A., Bohlin, R., & Neill, J. 1991, PASP, 103, 694

Longair, M. 1981, High Energy Astrophysics, Cambridge: Cambridge Univ. Press

Murtagh, F., & Adorf, H.-M. 1991, in ST-ECF Newsletter No. 16, 4

SOFTWARE SIMULATION OF POINT-SOURCES IMAGED ON CCDS

W. J. MERLINE AND S. B. HOWELL
Planetary Science Institute, 2421 E. 6th St., Tucson, AZ 85719

ABSTRACT We have constructed a computer model for realistic simulation of point-sources imaged on two-dimensional detectors. This model can be used in planning for, and optimization of, photometric observations. We have successfully used the program to model and study various reduction methods for ground-based and spacecraft observations.

INTRODUCTION

The use of two-dimensional digital detectors, such as CCDs, for observational research is becoming the standard throughout the astronomical community. Widely available image processing packages used in astronomy (*e.g.* IRAF, MIDAS, AIPS) contain provisions to manipulate data collected using such detectors. There are some situations, however, for which these packages are not suited. What seemed to be missing was a simple, yet accurate, software package that would allow calculation of realistic models of point-sources for a wide range of input conditions. We therefore saw the need within the community, as well as for our own special research needs, to develop such a model. In the process, we have made an effort to make the model flexible enough so that it can be of general use to astronomers seeking to optimize observations of point-sources and associated reduction methods.

The purpose of this software is to produce synthetic CCD images of single stars. It *does not* take, as input, raw or reduced CCD frames, since it is not intended to be a data reduction facility. An attempt has been made to ensure that the model produces "data" that mimic, as closely as possible, real data taken with 2-D detectors. To be realistic, the simulations must include randomly generated noise of the appropriate type from all noise sources.

The software provides a tool for studying problems associated with imaging in normal situations, but also under conditions of low signal-to-noise ratios or undersampled data (big pixels), for which most existing software was not intended. It can also be used as a predictive tool for planning observations and as an interpretive tool for understanding, for example, noise components of observations already obtained.

FEATURES

Our software model is a stand-alone program intended to be used on a PC/AT-class machine and therefore need not be integrated into other sophisticated image analysis packages. We wanted the program to have easy-to-understand input and output data, to be fast, and to be portable. A typical model takes less than 30 seconds on a 20 MHz Dell System 220 PC/AT. The code was developed using Microsoft FORTRAN Version 5.0. It is also adaptable to other machines.

The point spread function (PSF) that we have adopted is discussed by Diego (1985, PASP, **97**, 1209). It was constructed by fits to actual 2-D images of stars, including the typical smearing effects of seeing, guiding, and telescope optics; it is approximately Lorentzian in shape. Integration of the PSF is performed over each pixel to ensure reliable simulation in cases of undersampled data (PSFs comparable in width to a pixel). We also treat the noise associated with the number of background pixels used and digitization noise, which can be substantial, but are often neglected.

HOW IT WORKS

The input to the program is a single, short ASCII file. The input parameters are easy to understand and modify. An exact copy of the input file is shown in Figure 1. Although few in number, the input parameters characterize fairly completely the conditions under which an image is obtained. Thus, the program is very flexible and almost any situation can be simulated. It is not designed to model a specific instrument or instrument/telescope combination.

The "detector" is "exposed" to the light of a "star", the distribution of light being determined by the PSF. Each pixel independently collects "light" from the usual generators of photoelectrons (source, sky, dark) and contains a bias "signal" as well, just as in a real instrument. In this way noise is added to the stellar signal on a pixel-by-pixel basis rather than being calculated wholesale for the integrated image. Each noise source can be turned on or off so that they can be studied separately as well as in combination. We have also derived, separately, an analytic noise equation that yields an expected photometric standard deviation for a particular image. Empirical errors, derived by many runs of our image-generation code with identical input conditions, are in good agreement with our analytic model.

The synthetic data are output as simple integrations (one-dimensional), as radial slices (two-dimensional), and as intensity-contour plots (three-dimensional). The program also provides a 2-D pixel-intensity map (the usual raw CCD output, in ADU), as well as similar grids for each of the noise components.

USES

The main use for the program is to "collect" observations without using telescope time. It is useful for planning observations when one would like, for example, the time required to reach a given signal-to-noise ratio or to simulate the effects of changing image position, image scale, telescope size, signal strength, detector characteristics, or data reduction procedures. Figure 2 shows an example of the model in a case where it was used to examine the effects of source brightness and detector noise. Specifically, the program has been applied to simulate the detected flux and study the image widths, photometric accuracy, and noise components of the Galileo Solid State Imager.

```
SOURCE  CHARACTERISTICS
A0     = SPTYPE         ! spectral type (CHARACTER)
18.0   = VMAG           ! V magnitude of object (mag)

ENVIRONMENTAL CHARACTERISTICS
21.6   = SKYMAG         ! sky brightness (mag/sq arcsec)
1.5    = FWHM           ! FWHM of seeing disk (arcsec)
0.0    = ZENITH         ! zenith angle of source (degrees) [.LT.0 for spacecraft]

TELESCOPE/INSTRUMENT/FILTER  CHARACTERISTICS
0.91   = DIAMTEL        ! clear aperture of telescope primary (m)
0.43   = DIAMSEC        ! diameter of secondary obstruction (m)
49.0   = EFFIC          ! efficiency of entire optical train (%)
28.7   = IMGSCALE       ! image scale (arcsec/mm) [= (arcsec/pix)*1000/PIXSIZE]
4400   = LAMBDA0        ! central wavelength of bandpass (A)  ; set for B filter
980.   = BANDPASS       ! effective width of bandpass (A)     ; set for B filter

DETECTOR  CHARACTERISTICS
55.0   = QE             ! mean quantum efficiency over BANDPASS (%); set to TI6 here
15.    = PIXSIZE        ! pixel size (micron)
8.0    = RN             ! CCD  read noise (e/pixel/read)
4.15   = GAIN           ! CCD  gain  (e/ADU)
550    = BIAS           ! DC  bias level (ADU)

INTEGRATION-DEPENDENT CHARACTERISTICS
50     = X0             ! centroid of star image, x coordinate (pixels)
50     = Y0             ! centroid of star image, y coordinate (pixels)
50     = XA0            ! x location of aperture center for integration (pixels)
50     = YA0            ! location of aperture center for integration (pixels)
3.5    = RANGE          ! radius of integration circle to use (units of FWHM)
50     = LIMRX1         ! starting location in x for RADIAL SLICE  (INTEGER pixel)
80     = LIMRX2         ! ending location in x for RADIAL SLICE  (INTEGER pixel)
120    = TIME           ! exposure time (s)

NOISE-SOURCE  OPTIONS
400    = NBKGD          ! # of pixels to be used in background (sky) determination
1      = ISEED          ! noise randomization ; standard = 0, randomize = 1
1      = NOISEB         ! sky noise ;     ON = 1, OFF = 0
1      = NOISES         ! source noise ;  ON = 1, OFF = 0
1      = NOISER         ! read noise ;    ON = 1, OFF = 0
1      = NOISED         ! digitize data ; (yes = 1, no = 0)
```

Figure 1. Copy of input file.

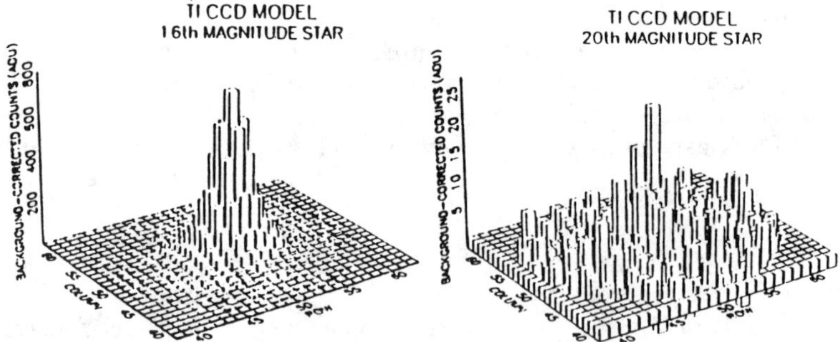

Figure 2. Effect of brightness and detector noise. The plots show the background-corrected intensity distribution of the star as imaged on the CCD. As the source gets fainter, the distinction between source counts and background counts becomes less clear. Note that the intensity distribution of the 16th magnitude star probably could be represented by a Gaussian, while that of the 20th magnitude star could not. The models are calculated only out to a given radius; therefore, in the 3-D figures, all pixels beyond this range will have zero counts.

PATCHING THE ODD GALAXY, NGC 4622

D. A. CROCKER[1] AND R. BUTA[1]
Dept. of Physics and Astronomy, University of Alabama Box 870324, Tuscaloosa, AL 35487

ABSTRACT Morphological studies of galaxies often require that an image be cleaned of all field star. We have used the star cleaning program, PATCH, which we describe here, to clean a variety of images. This patch program works best on smooth and slowly varying backgrounds.

We illustrate our success on NGC 4622 which is of particular interest as it is a candidate leading-arm galaxy. In the case of NGC 4622 we had to remove a star located on the steep intensity gradient that is the nucleus. Here we describe our efforts with the DAOPHOT PSF fitting routines now available in IRAF[2] to remove this problem star.

INTRODUCTION

We are doing a variety of studies where it is necessary to remove field stars from a galaxy image. We now do this semi-automatically with some help from IRAF. Stars are removed by fitting a cubic function to intensities in an annulus around a star (Figure I), using a program called PATCH (W. D. Pence, private communication). The "patch" is an interpolation of this cubic function over a circular affected area using a two-dimensional numerical mapping routine (Jones et al. 1967) with gaussian noise added. We add the noise to help the human judgment (by eye) of whether the patch has succeeded (i.e. whether the radius was large enough).

OPERATION

Instead of choosing one radius of patch large enough for all stars, it is quicker and less destructive to use a smaller patch for fainter stars. To do this, we run FIND (from the DAOPHOT suite of routines, Stetson 1987) on the whole CCD field, and we let PATCH use the output "magnitudes" and determine a half-width at half-maximum to set the patch radius ("r" in Figure I) empirically. There are problems, however: this does not work well with bright stars superimposed on

---

[1]Visiting Astronomer, Cerro Tololo Inter-American Observatory. CTIO is operated by AURA, Inc. under contract to the National Science Foundation.

[2]IRAF is distributed by National Optical Astronomy Observatories, which is operated by AURA, Inc., under cooperative agreement with the National Science Foundation.

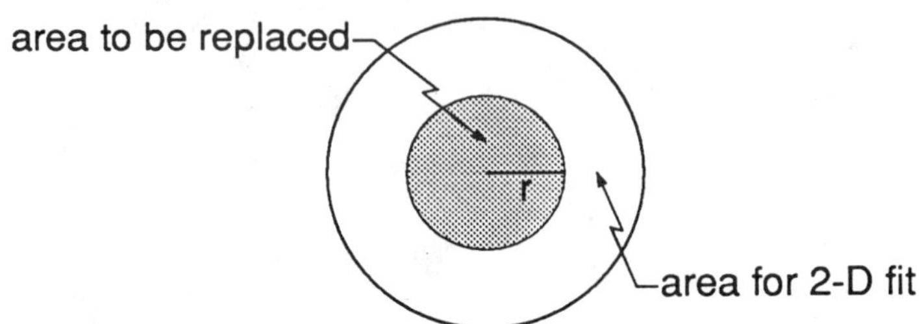

FIGURE I The fitting annulus and patch area

a steep, non-linear background. One such star is found within 10 arcseconds of the center of the southern spiral galaxy NGC 4622 (see arrow in Figure IIc).

We found that PATCH was capable of mostly removing this star, but not well enough to eliminate its effect from detailed Fourier images and amplitudes, especially in the B-band where the star is brightest. The following steps were used to remove this star: **Step 1.** We used an approximate bulge/disk decomposition model to remove most of the background light of the galaxy (Figure IIb). **2.** We made a point spread function from uncrowded off-galaxy field stars using the IRAF routine PSF. **3.** All field stars found by DAOFIND were cleaned off using routines PEAK and then SUBSTAR. **4.** The cleaned image was smoothed using IRAF routine MEDIAN (choosing a smoothing box large enough to smooth the area of the PSF but not so large as to obliterate the galaxy). **5.** The smoothed image was subtracted from the Step 1 result. This left only the faint residual of the galaxy (the high spatial frequency part) and field stars (Figure IIc). **6.** IRAF routine PSF was used on the offending star only. **7.** PEAK and SUBSTAR were used to remove only this star from the original image of the galaxy. The remaining stars were removed with PATCH the regular way. The galaxy field cleaned of all stars, both by the PSF method and the regular patch, is shown in Figure IId.

Why not do PSF on the offending star to begin with? Because we would have the same problem of getting a good background (i.e. sky level) just as with PATCH.

CONCLUSIONS

Why would we want to remove all the field stars? There are two procedures which we often use on galaxy images, and a third which is of interest to many morphologists. 1. We do Fourier analysis in the inner regions. In the case of NGC 4622, this is where a rare leading arm is found (see arrows in Figure IId). 2. We obtain star-corrected integrated magnitudes and colors. 3. For automatic feature identification and galaxy typing.

We have a wish list for the future that we hope we might implement soon. 1. An interactive version of a *circular* PATCH routine fully integrated into IRAF. 2. The ability to use a mouse to change the radius of the patch as seen on an image display. 3. The ability to undo an unsatisfactory patch.

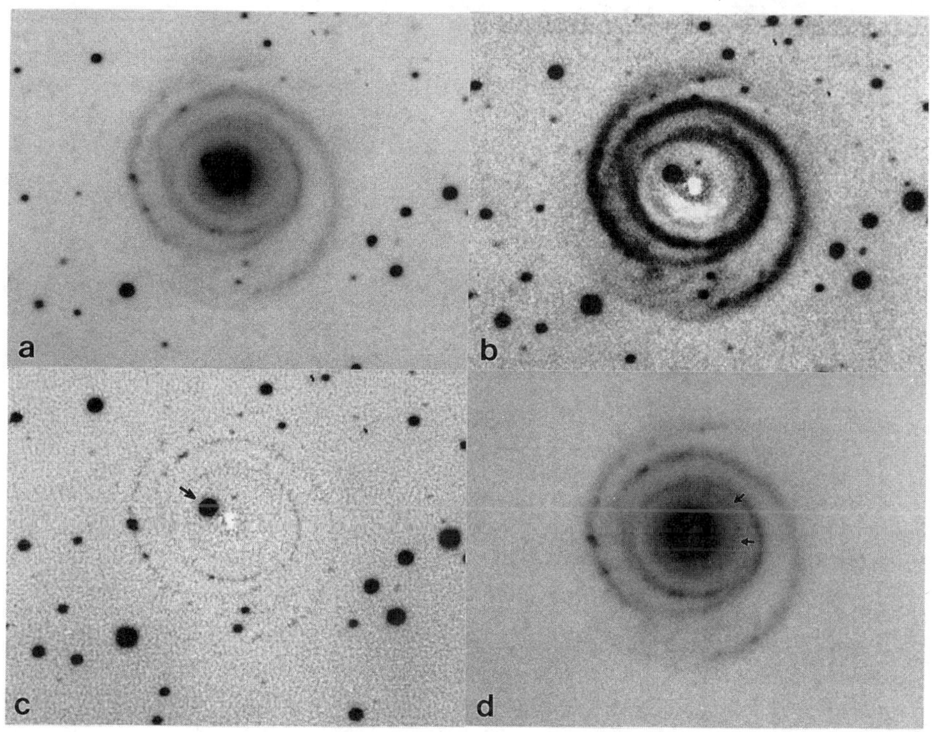

FIGURE II NGC 4622 before (a) and after (d) all stars have been removed from the field. Intermediate steps of bulge-disk subtracted model (b) and fields stars only (c) are also shown.

A full analysis of this galaxy will be presented in an upcoming journal article in preparation entitled "The Leading Arm, Ringed Spiral Galaxy NGC 4622: A Diagnostic Case of a Retrograde Tidal Encounter" by R. Buta, D. Crocker and G. Byrd.

ACKNOWLEDGMENTS

We thank Dr. W. D. Pence for generously supplying his PATCH program.

REFERENCES

Byrd, et al. 1989, Cel. Mech, 45, 31

Jones, et al. 1967, Pub. Univ. Texas Austin, Vol. I, No. 8

Stetson, P. B. 1987, PASP, 99, 191

SPIRAL STATUS : PRESENT AND FUTURE

S. ICHIKAWA
National Astronomical Observatory, Osawa, Mitaka, Tokyo 181 Japan

M. HAMABE
Kiso Observatory, University of Tokyo, Mitake-mura, Kiso-gun, Nagano 397-01 Japan

ABSTRACT SPIRAL is one of the most useful software systems for surface photometry in Japan. SPIRAL now runs on Sun workstations under the IRAF environment as well as on a super minicomputer and a mainframe computer. We report the present status of SPIRAL and discuss the future prospects.

INTRODUCTION

SPIRAL (Surface Photometry Interactive Reduction and Analysis Library) was originally developed for photographic surface photometry of galaxies. Now the objectives of SPIRAL have been extended to CCD surface photometry, spectral data reduction, and various analyses. SPIRAL is used by many astronomers in Japan because SPIRAL has many excellent routines, especially those for surface photometry.

HISTORY

Development of SPIRAL was started in early 1985, when the super minicomputer FUJITSU S-3500 started to run at Kiso Observatory. In the middle of 1986 SPIRAL version 1 was released, and in the middle of 1987 version 2 was released. SPIRAL version 2, the current version, has tasks for CCD surface photometry and spectral data reduction, while the target of version 1 was only photographic surface photometry. Porting to the mainframe computer FUJITSU M780/10S at the National Astronomical Observatory (NAO), Mitaka, was done from late 1987 to the middle of 1988 by FUJITSU. A major result of SPIRAL was the *Photometric Atlas of Northern Bright Galaxies* (1990). This atlas includes photometric parameters for about 800 galaxies. Porting to Sun workstations under the IRAF environment was started in early 1991.

PRESENT STATUS AND OUTLINE OF SPIRAL

SPIRAL was developed on the super minicomputer FUJITSU S-3500 at Kiso Observatory. SPIRAL now runs on the mainframe computer FUJITSU M780/10S at NAO, Mitaka, and on Sun workstations under the IRAF environment at several institutes in Japan. For details of SUN/SPIRAL, see the following paper in these Proceedings. Before porting to the Sun workstations, SPIRAL ran only on two computer systems (S-3500 and M780/10S), so users would visit Kiso Observatory or NAO, Mitaka. Only a few astronomers or engineers could take part in the development of new tasks in SPIRAL. Now many people can develop new tasks under the IRAF environment at their institutes, using SPIRAL as a template or as subroutine libraries.

SPIRAL has the following features;
(1)Basic parameters defining the image and the history of processing are recorded automatically as a House Keeping Block (HKB; Figure I) with the image.
(2)SPIRAL has excellent interactive performance with an image display and a graphics display.
(3)A menu scheme is adopted, so that beginners can use it easily. On the other hand, each routine runs alone, so experts can use it simply and developers can test it quickly.
(4)SPIRAL has a highly modular structure, so that one can develop new tasks easily using subroutines and template programs.
(5)SPIRAL is written in FORTRAN77, one of the most popular computer languages, so that older people can develop new programs easily as well as young people.

FUTURE PLANS

Workstations (mostly Suns) were introduced to a lot of institutes in Japan, and IRAF was installed on them. Now IRAF has become the standard astronomical software system in Japan. Many astronomers (optical, IR, radio, X-ray, etc.) will develop their software under the IRAF environment. An IRAF managers' group was organized in Japan. The group wrote an "IRAF CookBook" written in Japanese including a "Beginners' Guide", a "CookBook for Japanese Instruments", a "Developers' Guide", and a "Managers' Guide". Because of this situation, we decided to port SPIRAL to the Sun workstations under the IRAF environment. For details of SUN/SPIRAL, see the next paper in these Proceedings. There are many convenient and interesting routines and subroutines in SPIRAL. Source code for SPIRAL is available, and documents for development are written in Japanese also. Combining them, we can develop new tasks easily. SPIRAL on the Sun workstations will be a base or template for development.

SPIRAL on the S-3500 and the M780/10S also helps us to develop new software. AIMS (Automated Image Measuring System; Doi et al. 1991) uses many subroutines in SPIRAL. AIMS runs on super computers (for example, FUJITSU VP-200E) as well as mainframe computers. SPIRAL on the mainframe computer will be a base of development of the next-step software systems as well as SUN/SPIRAL.

```
OBJECT    :N3226/3227          D:N3227 R 8FRAMES
SOURCE    :KCB 8FRAMES          :89/04/28
BAND      :R                    :
EXPOSURE  : 600.0           0.0 :EXPOSURE TIME(SEC.), SEEING(")
IMGSIZE   :  0.0     0.0    0.0 :NOT YET DETERMINED
POLARI    :                     :NONE
RECTYPE   :VB       R*4         :FUJITSU M780/10S (NAO MITAKA)
SCALING   : 1.00E+00  0.00E+00  :BSCALE, BZERO (VAL.=CCD COUNT)
MATSIZE   :      512        512 :MATRIX SIZE (IEXT,JEXT)
PIXSIZE   :     1.50       1.50 :PXSI("), PXSJ(")
IMGCENTER:    256.50     256.50 :CENTER OF THE FRAME
ORIENT    :      0.0          N :ANG. BETW. UP & N (NOT YET DETERMINED)
SKYMAG    :    -1.000           :(MAG ARCSEC**-2) UNDETERMINED
TELESCOPE:KISO 105CM SCHMIDT    :ISHIDA AOKI ICHIKAWA
DETECTOR :TI-CCD (1024X1024)    :BINNING=2*2    GAIN=?    TEMP=-83.5C
CCDLOADT              19890709  :CREATE HOUSE KEEPING BLOCK FOR CCD IMAGE
IFF                   19890712  :BIAS SUBT. & FLAT-FIELDING
ISKY                  19890712  :ORDER=0: SKY SUBTRACTION
CCDSHIFT              19890717  :    2   68
IMGADD                19890717  :  8 FRAMES ADDED
                                :
                   <Binary Image Data>
```

FIGURE I House Keeping Block recorded as a header of an image.

ACKNOWLEDGMENTS

One of the authors (S.I.) thanks the Japan Society for the Promotion of Science for a travel fund.

REFERENCES

Ichikawa, S., Okamura, S., Watanabe, M., Hamabe, M., Aoki, T., & Kodaira, K. 1987, Annals of Tokyo Astron. Obs., 21, 285

Kodaira, K., Okamura, S., & Ichikawa, S. 1990, Photometric Atlas of Northern Bright Galaxies, Univ. of Tokyo Press

Doi, M., Kashikawa, N., Okamura, S., Tarusawa, K., Fukugita, M., Sekiguchi, M., & Iwashita, H. 1991, in proc. of 2nd Meeting on Digitised Optical Sky Surveys

Hamabe, M., & Ichikawa, S. 1992, this volume

DEVELOPMENT OF THE SURFACE PHOTOMETRY PACKAGE SPIRAL UNDER THE IRAF ENVIRONMENT

MASARU HAMABE
Kiso Observatory, Institute of Astronomy, University of Tokyo, Mitake, Kiso, Nagano 397-01, Japan

SHIN-ICHI ICHIKAWA
National Astronomical Observatory, Mitaka, Tokyo 181, Japan

INTRODUCTION

SPIRAL (**S**urface **P**hotometry **I**nteractive **R**eduction and **A**nalysis **L**ibrary; see preceding paper in this volume) is one of the most efficient interactive astronomical image processing systems for surface photometry of extended objects such as galaxies.

However, SPIRAL had a disadvantage: the system depended strongly on the operating system (FACOM's OS) and expensive hardware such as an image display device. Therefore, users of SPIRAL had to come to Kiso or Mitaka where the system was available. This also prevented us from distributing SPIRAL to other countries and from exchanging software with other systems.

At the request of many Japanese users that they wanted to do surface photometry in their home institutes, we have decided to port SPIRAL to the IRAF environment. Here, we briefly report the present status, problems and future plans of the development of SPIRAL under the IRAF environment.

MEANING OF SPIRAL UNDER THE IRAF ENVIRONMENT

The development of SPIRAL has the following important meanings: (1) We can do *surface photometry* without very expensive hardware (*e.g.*, main-frame computers or superminicomputers and image displays) by using SPIRAL together with IRAF. The tasks currently available are listed in Table I. (2) We can easily *exchange useful routines* with other systems running under UNIX. (3) The subroutine library of SPIRAL (*KISOLIB*, including *KISOimage* developed by T. Ichikawa, see Table II) enables us a *quick and easy way to develop new application programs* under the SPIRAL/IRAF environment.

HARDWARE AND SOFTWARE ENVIRONMENT

The target machines are limited to Sun Microsystems' workstations (SPARC stations) with 16 megabytes of memory or more. Most programs can be executed with 8 megabytes of memory but some limitations may exist. SPIRAL/IRAF assumes the use of the X-Window System (X11R4 or X11R5) and MONGO

TABLE I Available tasks and packages.

Task	Function
Package spiral	
apinteg	Make simple aperture integration [under development]
badcol	Correct bad columns in an image
badrow	Correct bad rows in an image
efits	Visual editor of FITS file header
imgdisp	Simple image browser
imregist	Transform an image from one coordinate system to another
isophotes	Make an isophotal map of an image
jplocal	Local package for Japanese and users of SPIRAL/FACOM
oshirase	Read revision history and current version number
photog	Photographic data reduction package (for PDS data)
profs	Luminosity profile package
setsumei	Japanese explanation (Manual in Japanese)
skysub	Estimate and subtract sky background from an image
skysub2	Same as the above (semi-automatic) [not completed yet]
smooth	Perform local smoothing of images
unitmag	Estimate the intensity unit of an image
vhedit	Visual editor of IRAF image header
Package jplocal	
rspiral	Read SPIRAL/FACOM-formatted image
ruh88	Read UH88 NSF1 CCD raw data
trimrca	Trim over-scan pixels of Okayama RCA CCD raw data
mtread	Read file(s) on a magnetic tape
Package photog	
pdsload	Read PDS scan data and convert them to IRAF format
wdkiso	Read PDS scan data (Kiso wedge) and convert them to IRAF format
wdread	Derive mean density in each step of an wedge image
ccfit	Make a photographic characteristic curve
pds2i	Convert PDS density images to intensity images
Package profs	
eqprof	Make an equivalent luminosity profile and derive parameters
mmprof	Calculate luminosity profiles along the major- and minor-axes
vprof	Calculate luminosity profiles by ellipse-fitting method*
vplot	Plot the output of vprof

*This is a modified version of the PROFILE program developed by Dr. Lauer (1985).

TABLE II Subroutine library KISOLIB.

File	Contents
subastro.f	astronomical calculations (coordinate conversion etc.)
subfit.f	fitting routines & calculation of polynomials
subguty.f	general utility (code conversion, sorting)
subimagef.f	display image
subimagec.c	FORTRAN callable subroutines for image access (KISOimage)
subio.f	read/write IRAF-formatted image
submongo.f	routines to use MONGO library easily
submtxop.f	matrix operations
subphoto.f	routines to do luminosity integration within an area
subsmth.f	routines for smoothing
substat.f	routines for statistics
subetc.f	routines not included in the other files

(Lick Observatory version). Tasks that do not display images can be used under SunView or the OpenWindows environment. We have no plans to develop the routines for the SunView environment.

Most of the programs are written in (Sun) Fortran and some are in C. All programs have `cl` scripts and can be executed as tasks under IRAF. The IMFORT library and Lick MONGO are used for the file I/O and the graphics, respectively. For display and interactive access of the images, a newly developed simple subroutine library *KISOimage*, which is a Fortran callable C library, is used. The subroutine library *KISOLIB* (Table II), which was developed along with the original version of SPIRAL, was also rewritten and incorporated.

PROBLEMS AND FUTURE PLANS

The most important part of the original SPIRAL (SPIRAL/FACOM) has been ported to the IRAF environment. However, some small problems remain. (1) The current version of SPIRAL/IRAF depends strongly on the MONGO graphic library and we cannot redistribute it. We hope to develop our own Fortran callable subroutine library for graphics in order to make SPIRAL easy to install. (2) The original version (SPIRAL/FACOM) has more useful programs for surface photometry than SPIRAL/IRAF. For instance, the tasks for extracting luminosity profiles and photometric parameters must be extended. (3) One of the most important features of the original SPIRAL/FACOM has not yet been reproduced. The automatic recording of the process history and the relevant parameters will be supported in future versions. (4) Most tasks and subroutines have manuals written in English but they are not complete nor consistent. After completing these documents, SPIRAL/IRAF will have the possibility to be a platform for easy development of new programs.

ACKNOWLEDGMENTS

We are grateful to Drs. T. R. Lauer, S. L. Allen, J. Kormendy, J. Souviron, T. Ichikawa, and T. Sasaki for providing us useful routines. We also thank Drs. T. Horaguchi and S. Yoshida for their help. We acknowledge the users and developers of the original SPIRAL for their helpful comments.

REFERENCES

Ichikawa, S., Okamura, S., Watanabe, M., Hamabe, M., Aoki, T., & Kodaira, K. 1986, Ann. Tokyo Astron. Obs. 2nd Ser., 21, 285

Lauer, T. R. 1985, ApJS, 57, 473

A COOKBOOK FOR REDUCING KPNO INFRARED ARRAY DATA

F. A. RINGWALD
Dept. of Physics & Astronomy, Dartmouth College, Hanover, NH 03755

Infrared astronomy has been revolutionized by solid-state array detectors. These enable imaging in the IR, in contrast to the previous generation of single-channel detectors, which had to be slowly raster-scanned to map the sky to build up images of questionable photometric accuracy. Even for photometry of point sources, array detectors have advantages, as centering an object in the aperture becomes no problem, plus measuring sky brightness can be done in the same frame as the target and therefore simultaneously with it.

One well-known array detector is Kitt Peak National Observatory's Infrared Array Imager, or IRIM. A manual available from NOAO (Probst 1988) has a detailed description of this device, an observing cookbook, and a description of the camera-control software (Gillies 1988), which writes the 58×62 pixel images in FITS format, so they can be reduced with IRAF. The present paper shows one way to use IRAF (*v.* 2.9.3) to do this reduction. While no cookbook can cover all possible observing programs, it is hoped this will suffice for most common ones, by outlining general methods useful for IRIM and other arrays.

To do accurate photometry with any solid-state digital detector, bias, or the DC level on which signals ride, must be subtracted; unevenness in the pixel-to-pixel response of the chip must be evened out by dividing by a flat-field exposure; and the dark count must be found and subtracted. Furthermore, the sky signal must be measured and subtracted. In the IR, the sky is usually the strongest signal present, so one may subtract it from the data first, before flat-fielding. This also takes care of bias subtraction, since sky and bias manifest themselves as DC levels on which small and spatially-varying signals ride.

Even with imaging in the IR, raster-scanning is still not obsolete. IRIM has two modes of operation, grid mode and stare mode. In grid mode, the telescope is repeatedly offset by preset amounts to make a grid of observations of an object. This is useful for mapping out extended objects, as IRIM has a narrow field of view, and it is indispensable for accurate sky subtraction, since an odd number of offset exposures of the same object, all taken roughly simultaneously, can be median-filtered together to make a sky frame to subtract from the data. This can also suppress negative star images made by simply subtracting one sky frame from the target frame. Stare mode can involve chopping, which is of limited use for an integrating detector such as IRIM; it can also be used with wobbling, but this essentially is grid mode.

As described in the IRIM manual, often the only calibrations for IRIM data one needs to take are dark frames. Flat fields and sky (and bias) frames, for each waveband, can be made by median-filtering the data themselves together. This has the advantage that they were taken in exactly the same way as the data to be corrected. If observing crowded fields, or extended objects larger than IRIM's field of view, one should, however, take some blank sky frames as well, from areas of the sky close to one's targets (see Probst 1988).

For photometry, one should of course also take IR photometric standards, e.g., those of Elias et al. (1982), although be aware they are still problematical, being bright and involving large extrapolations to faint levels. Try to take standards for each waveband at least hourly, to evaluate sky transparency and background, as well as to ensure accurate flux calibration. It may be better not to sky-subtract or dark-subtract the standard star exposures, as they are so bright this would just be making unnecessary noise. It is also helpful to take data so that the brightest pixels in the targets and in the standards are near the same level, to minimize the effects of detector nonlinearities when extrapolating from short to long exposures.

Starting the reduction: filling display window, grouping images together

IRIM images are small. To avoid displaying tiny images surrounded by acres of white space, log into IRAF, keep the default **set stdimage=imt512**, and type

`cl>` **epar display**

and change the **fill** parameter to **yes**, to replace the default **no**. Exit EPARAM with a CTRL/Z; only a small amount of white space will peek through at the top of the screen, as IRIM images are not quite square.

Read in the FITS images from tape, using an **rfits** command with parameters depending on the hardware for reading them and how the frames are recorded. For grid mode observations, group the images together for median filtering. Much here depends on the details of one's observing program, but @ files can be helpful, e.g., to save some typing, type

`cl>` **files** *imagename****.imh** > *listname*

and edit this file, now filled with the names of the individual images, to produce many files, each containing the names of a group of images. The **ccdgroups** command may alternatively be useful, if the frames have unique names.

Sky subtraction

To make sky frames from the data, median filter each set together with

`cl>` **newimcombine** @*listname* **m//**@*listname* **combine=median**

prepending an 'm' to the new sky file names, to avoid confusion. Also type

`cl>` **imarith** @*listname* – **m***listname* **s//**@*listname*

to do the subtractions for each set. It may be wise to check this by experimenting with different ways of median filtering different images together (e.g., in differing groups of 3 or 9), to avoid rapid sky brightness variations.

Dark subtraction

Dark frames should be taken with exposure times corresponding to each different exposure time of the data. To suppress noise from the individual darks, average groups of darks with the same exposure time together with

`cl>` **newimcombine** @*darklist* **av//**@*dark* **combine=average**

and for each different exposure time of the data, subtract the dark count with

`cl>` **darksub** @s*listname* **av** *dark* **ds//**@*listname*

and if one is unfortunate not to have taken darks for each exposure time used for the targets, one must scale them with this command's **exp = exposure** option, albeit with less accuracy in the subtraction.

Flat fielding

With flat fields, is important to eliminate read noise, by averaging together many frames. Normalize the original frames to their sky level with

cl> **normalize** *flatfiles flat* **sample=[3:48, 3:38]**

where the sample selected here is chosen to avoid bad pixels on IRIM. Median filter these normalized frames in each waveband together, as before, to make the flats, and apply them to the data (overwriting them) with

cl> **flatten @ds***listname flat* .

It is also important to evaluate the flats. One way to do this is to display histograms down rows or columns of the flats, and another way is to type

cl> **imstat** *flatname* **[*, 1]**

which will evaluate pixels statistics, for example for an image section selecting row 1. One might also use the region **[3:48, 3:38]** for this. Various statistics can be evaluated by selecting options with this command: the mean should $\doteq$ one pixel, with standard deviation $\ll$ one pixel. Also, divide flats from different parts of the run by each other, and see if they really do flatten each other.

Mosaicing and summing fields together

Since IRIM and other IR arrays have narrow fields of view, and also since grid-mode observations take many images, it may be useful to mosaic them together to increase the field-of-view and S/N. Set up with

cl> **irmosaic @ds***listname mosaicfile database nxsub nysub*

(use 2 and 2 or 3 and 3 for **nxsub** and **nysub**, as appropriate), and use

cl> **center** (and mark features with the cursor)

to interactively identify stars common to the fields to be joined together. Typing

cl> **iralign** *mosaicfile images* database coords*

will align the images, to be co-added together with

cl> **newimcombine** *images* imsum* **option=sum** .

At this stage, the images should be ready for photometry. IRAF's aperture photometry routines (Massey & Davis 1990) will want, as inputs, IRIM's system gain (54 electrons/ADU) and readout noise (400 electrons rms; for sky-subtracted frames, multiply this readout noise input by $\sqrt{2}$).

REFERENCES

Elias, J. H., Frogel, J. A., Matthews, K., & Neugebauer, G. 1982, AJ, 87, 1029

Gillies, K. 1988, Infrared Array User Software Reference, Version 1.6, NOAO

Probst, R. G. 1988, User's Manual for the KPNO Infrared Array Imager, NOAO

Massey, P., & Davis, L. E. 1990, A User's Guide to Stellar CCD Photometry with IRAF, NOAO

COMBINING GRAPHICS AND IMAGE DISPLAY IN IRAF

J. D. EISENHAMER
Space Telescope Science Institute,[1] Baltimore, MD 21218

INTRODUCTION

IRAF gained the ability to combine image and graphics with the introduction of the **imdkern** graphics kernel in version 2.9. A number of tasks have been developed at STSCI to take advantage of this functionality and give the observer the tools necessary for planning and understanding HST data. The **imdkern** task will be described. Tasks developed for STSDAS will be presented and a discussion of how already-existing tasks can be used effectively with **imdkern**. Basic techniques for combining graphics and images and future development will be presented.

IMDKERN FUNCTIONALITY

The **imdkern** graphics kernel is a task that overlays IRAF graphics output into any IRAF image display, including both IMTOOL and SAOImage. The task can be run explicitly as a task or implicitly by specifying the graphics device to be the **imdkern** device. Table I lists the graphics device names for **imdkern**. Technically, **imdkern** renders the graphics by overwriting pixels values in the display frame buffer. This process can be repeated as many times as possible. However, to erase the graphics, the user must redisplay an image. Since **imdkern** operates on the frame buffer directly, the quality of the graphics depends on the configured size of the display. For example, graphics rendered in IMTOOL which is configured for a 800 by 800 frame buffer, will look better than the same graphics rendered in IMTOOL configured for a 64 by 64 frame buffer.

The **imdkern** is capable of producing graphics in sixteen different colors. The graphics devices listed in Table I allow direct access to the primary colors. In order to use the other available colors, the user must either add their own devices to the standard list, or redirect the graphics output of a task to a file and use **imdkern** explicitly.

The most important feature of **imdkern** is its generality. **imdkern** is part of IRAF's Graphics Input/Output (GIO) interface, which all tasks within IRAF system use. Thus graphical output from any task can be directed to **imdkern** and viewed in an image display.

---

[1]STScI is operated by AURA, Inc. for NASA.

TABLE I Standard **imdkern** Graphics Device Names

Device	Color
imd—imdkern	Same as imdg
imdw	White
imdr	Red
imdg	Green
imdb	Blue
imdy	Yellow

TASKS

A number of tasks have been introduced in STSDAS that are explicitly intended to be used to overlay their graphics onto the image display. The task, **stplot.wcslab**, through the use of the Mini World Coordinate System (MWCS) interface, takes the world coordinate information from the header of an image (present in all HST images and easily added to any other image), and plots labels around the image, specifically a right ascension/declination grid indicating what region of the sky is covered.

To assist observers in planning and understanding their HST observations, the tasks **stplot.siaper** and **gasp.targets** have been introduced. Like wcslab, siaper uses an image's WCS information to correctly scale and orient a plot of the HST's science apertures over an image. The targets task takes information from the Guide Star Catalog and overlays the positions of guide stars into the image display.

Other pre-existing tasks can also take advantage of graphics overlays. Tasks such as **plot.contour** or **stplot.newcont** are well suited for overlaying contours onto images. For example, one can overlay contours of radio data onto an optical image of the same region. Two dimensional spectral data can also benefit by overlaying a plot of the one-dimensional extracted spectrum onto the original two-dimensional form of the spectrum. The Interactive Graphics Interpreter task, **stplot.igi**, can be used to added titles and special annotations and graphics that may be necessary to emphasize features of a displayed image.

TECHNIQUES

There is one shortcoming in the current implementation of the **imdkern** interface. The task **images.tv.display** allows the user to arbitrarily position an image within the frame buffer, but, as yet, there is no way to communicate this information to an arbitrary graphics task. Except for wcslab, the user has to specify this placement explicitly by calculating and setting the graphics viewport parameters of the graphics task. To make this simple but cumbersome calculation easier, STSDAS provides the task **playpen.imdisp_pos** to perform this calculation.

The example below demonstrates the use of imdisp_pos. An image, dev$pix, is displayed in the upper right corner of the image display and a contour plot of the same image is then overlayed onto the image.

```
cl> display dev$pix 1 xcenter=.75 ycenter=.75 \
     xsize=.4 ysize=.4
cl> imdisp_pos dev$pix
cl> newcont dev$pix device=imd left=imdisp_pos.left \
     right=imdisp_pos.right bottom=imdisp_pos.bottom \
     top=imdisp_pos.top
```

As tasks are modified or written explicitly for overlaying graphics, the need for imdisp_pos should gradually subside.

FUTURE DEVELOPMENT

STSDAS plans to further combine imaging with graphics through the development of **psikern**, a dedicated PostScript graphics kernel interface. The psikern kernel will provide IRAF's GIO interface with direct access to PostScript functionality. This will enable IRAF to take advantage of windowing systems either based on PostScript, such as NEXT, or windowing systems that have PostScript extensions, such as the X-based OpenWindows or DECWindows. Also, psikern will provide both imaging and graphics through the GIO interface, alleviating the need for tasks such as imdisp_pos described above or special code within tasks to handle both graphics and images.

CONCLUSION

A brief discussion of how images and graphics can be combined using the **imdkern** graphics kernel and how current graphics tasks within the IRAF system can take advantage of this functionality has been presented. A number of tasks developed specifically by STSDAS to be used with image overlays have been introduced. Though some care needs to be taken to properly position the graphics viewport within the display frame frame buffer, tools are provided to make such bookkeeping manageable. For the short term, STSDAS is developing the psikern to help unify imaging and graphics.

ACKNOWLEDGMENTS

Zoltan Levay for his expertise and support.

FEATURES OF THE PROS/IRAF IMCONTOUR TASK

JANET DEPONTE, DIANA M. WORRALL
Smithsonian Astrophysical Observatory, Cambridge, MA 02138

INTRODUCTION

IMCONTOUR is a task which contours an image and optionally overlays a sky or pixel coordinate grid (Figure I). A companion task, TVIMCONTOUR, displays an image onto the tv device and overlays the results of IMCONTOUR (Figure II). The tasks are fully integrated into the Image Reduction and Analysis Facility (IRAF; Tody 1986) and are available as part of the X-ray analysis Post-Reduction Off-line Software (PROS; Worrall et al. 1991) in the 'xray.xplot' package.

Graphs can be made easily and quickly with IMCONTOUR. The task includes features which allow user customization, and it is the flexibility and integration of these features which makes this software effective.

CONTOUR LEVEL SPECIFICATION

A flexible interface is implemented for user specification of contour levels. The user may choose between logarithmic and linear scales, and may specify contour levels in units which are absolute ('pixel'), a percentage of the peak flux ('peak'), or relative to the uncertainties, if known ('sigma'). The choice of 'sigma' will result in the contouring of an array formed by dividing the Input Image by the square-root of an Error Image. (Errors are stored as the squares of the uncertainties in the PROS.) If an error array element is zero, the result is forced to be zero. The contour levels are optionally listed as a legend to the plot.

SCALING

The plot scale-size may be specified for each of the X and Y axes; the task will default to compute a scale to fill the output window, assuming square pixels. A user may request a specific Y scale in arcseconds/mm when displaying a Sky Map, or pixels/mm when displaying a Pixel Map. The X/Y ratio may be changed from the default value of unity.

A user may plot to the graphics window using a scale appropriate to either the graphics window itself or the hardcopy plotter device. The latter option is included so that an IRAF 'snap' of the graphics window will produce hardcopy output which is scaled to specification, and it is selected by setting the IMCONTOUR parameters 'graph device = stdgraph' and 'scale device = stdplot'.

SOURCE POSITION OVERLAY

Source positions may be overlayed onto a plot with the user's choice of Marker Symbols. Marker symbols can be a number, a character, or a combination of both.

The Source list can be in either ASCII or ST Table format. The format of the ASCII list is right ascension and declination (defined in rh:rm:rss.s, dh:dm:dss.s). The format of the ST Table (right ascension and declination defined in degrees) is specified to interface with the output from other PROS tasks. The coordinates of the source data and image are permitted to be of different epochs.

CAMERA READY PLOTS

Minor customizations are available to aid in readying a plot for publication. These include the choice of grid-type, the positioning and form of axes labels, and whether or not the legend or title should appear.

TVIMCONTOUR

TVIMCONTOUR is a script which runs the IRAF DISPLAY task followed by IMCONTOUR. The output from the DISPLAY task is loaded into the image display and the graphics from IMCONTOUR are overlayed onto the image.

The advantage of TVIMCONTOUR is that two tasks are run as a single task, and the parameter choices are made automatically for display on the image window device. Common parameters shared by DISPLAY and IMCONTOUR are specified once. Parameters that place labels and borders in IMCONTOUR are set to defaults appropriate for the image window; and parameters in DISPLAY that determine how the data fills the window are set so that the image matches spatially with the contour lines.

ACKNOWLEDGMENTS

We thank Zolt Levay and Jonathan Einsenhamer of the Space Telescope Science Institute for providing sky-coordinate labeling code that was incorporated into IMCONTOUR.

REFERENCES

Tody, D. 1986, in Instrumentation in Astronomy VI, SPIE, 627, part 2

Worrall, D. M., et al. 1991, in Data Analysis in Astronomy IV, eds. V. Di Gesu, et al., (Plenum Press), in press

Briel, U. G., et al. 1991, A&A, 246, L10

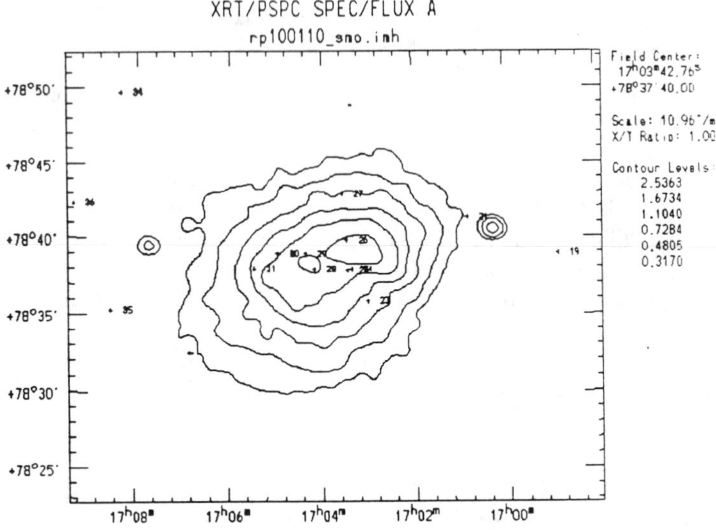

Figure I. Example output of IMCONTOUR. Data are the central region of the ROSAT PSPC observation of Abell 2256 (Briel et al. 1991). Numbers and crosses mark positions of objects in the field found in the SIMBAD database and input to IMCONTOUR as a table file.

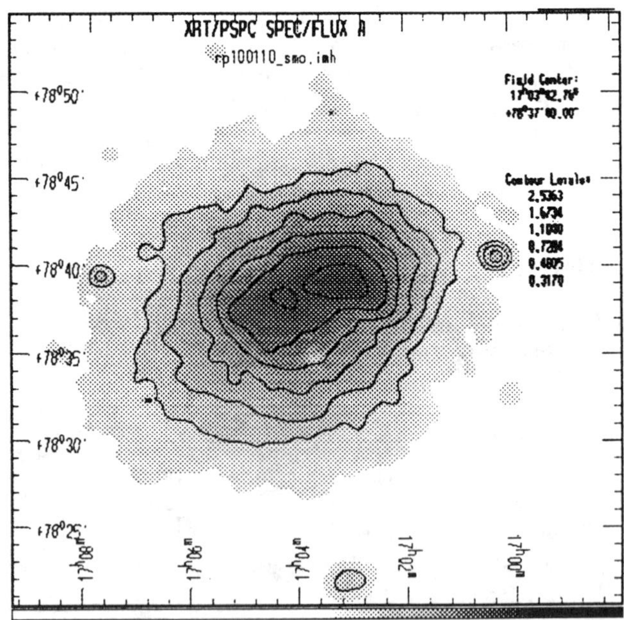

Figure II. Example TVIMCONTOUR output for data of Figure I.

INTERACTIVE GRAPHICS INTERPRETER: *IGI*

Z. G. LEVAY
Space Telescope Science Institute, 3700 San Martin Drive, Baltimore, MD 21218

INTRODUCTION

igi is a program that allows a user to enter commands to read data, draw plots, and manipulate those plots interactively. It is a task in the STSDAS package. As an IRAF application, *igi* is portable to any host architecture and graphics device supported by IRAF. It uses built-in graphics cursor interaction and produces standard IRAF graphics metacode. Therefore, *igi* is fully compatible and complementary with other IRAF graphics tasks. *igi* includes language features such as named macros, command files, and expression evaluation in addition to common graphics capabilities from simple move and draw commands to more powerful features that permit drawing a full plot with a few straightforward commands. Input data may be any IRAF data format. A brief example illustrates some of the capabilities.

igi is implemented in the IRAF SPP language. It incorporates a custom command line interpreter to parse commands and arguments. The user may operate on internal data vectors with arbitrary expressions, implemented using the **evexpr** SPP/VOS procedure. Fonts are implemented using a custom machine-independent subset of the Hershey fonts using the machine-independent integer facility of SPP. All graphics instructions in *igi* are implemented using the IRAF graphics I/O library procedures. Additional features of IRAF's general file I/O facilities used in *igi* include multiple I/O streams on a stack for command input, internal memory "files" for stored commands, and pushback to the input stream to implement user-definable macros.

While *igi* is useful in its current implementation, several enhancements are envisioned for a more complete facility. A useful and natural extension is to render multi-dimensional input data by drawing contour maps or wireframe surface plots. Changes to the command interpreter would improve the robustness and provide more programming language capabilities. Many users would find it easier to interact with *igi* if the user interface were more visual. Finally, most of the capabilities of *igi* should be available to any IRAF graphics application, not only from the cl. That is, we will provide a library of procedures that provide the same capabilities as most of the interactive plotting commands of *igi*.

RUNNING *IGI*

The full capabilities of *igi* are described in the reference document, formatted as IRAF help text, available in the cl by typing "**help igi**".

Commands input to *igi* may be typed interactively, redirected from a file, or piped from another application. Commands are case *in*dependent and may be minimum-matched. A line may contain more than one command. Any line beginning with a # will be ignored and may be used as a comment. Commands typed interactively may be stored in a file and this command file may then be edited and subsequently input as commands. The user may define macros using the `DEFINE` command. Arguments are defined in the body of the macro and replaced on invocation by the positional macro arguments. The user may "escape" to the enclosing environment, the IRAF cl, by preceding a cl command with the ! escape character. Since the cl also recognizes this character as an escape, commands may therefore by escaped to the host operating system by prefixing with a second !.

Plot data input to *igi* may be in any data format used in IRAF:

- Text (ASCII) tables — including comments (#) and blank lines
- Binary STSDAS tables — columns specified by name
- IRAF images (OIF, STF or QPOE) — including sections and groups

Subsets of input data files may be specified, such as a range of rows in a text file or table, or a section of an image file. If an input image file contains "world coordinate" information, it may be used to construct the independent variable axis. The input data buffers may be subsequently modified arbitrarily using expression evaluation syntax.

Several different coordinate systems are used to provide flexibility in page layout. When interacting with data, world or user coordinates are usually the most convenient. Virtual page or viewport coordinates are useful for locating one or more graphs on the page or screen. Normalized device or screen coordinates may be used to position features of a plot explicitly at a particular position on the screen or page regardless of the data scale.

Several commands permit plotting data axes and control the appearance of them. A high level command permits drawing a full set of axes with labeled ticks. A more primitive command permits more control over the attributes of a single axis such as its angle and the appearance of the tick labels.

Data may be plotted in several ways. Data points may be connected simply by straight lines or with "stepped" horizontal segments connected by vertical lines. In addition, a "histogram" may be drawn, with the vertical segments dropped to the bottom axis. Symmetrical or asymmetrical error bars, upper or lower limit symbols, or a centered symbol may be drawn at each data coordinate.

igi has flexible text drawing facilities, specifically for labeling axes and plot titles as well as arbitrary text. Many attributes of the text such as font or italics are specified by embedded "escape sequences," single characters preceded by the \ character. Other attributes are "global" parameters changed by a command and controlling other drawing primitives as well.

The full range of capability of the IRAF graphics cursor is available from *igi* for interacting with a plot. In addition, all of the *igi* commands are available in cursor mode as "colon commands." Any command taking coordinates as arguments use the cursor position as the data coordinate when used in cursor mode.

EXAMPLE

The *igi* commands shown in Figure II produce the relatively simple plot in Figure I. The input data file referred to in the `YSECTION` command is an IRAF image and is not shown here. In producing the command file, some cursor interaction was used to position the identifications.

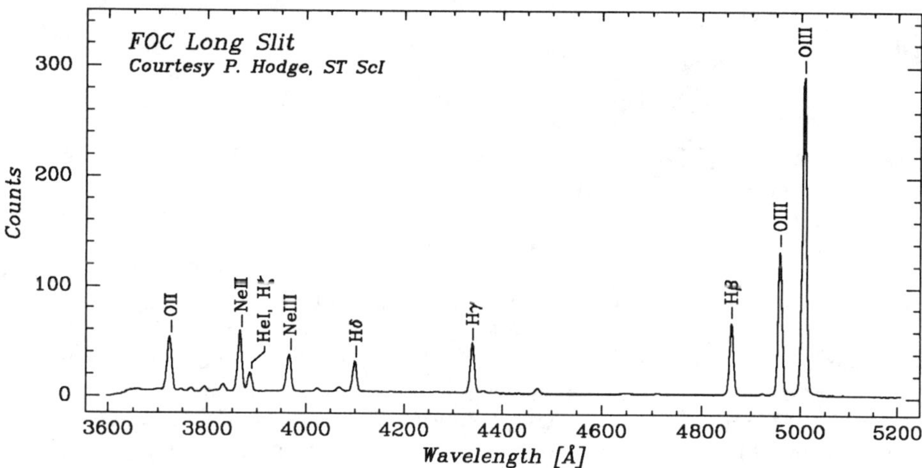

FIGURE I Sample *igi* plot

```
imgwcs;  YSECTION n6543.hhh[200:1105]
vpage .1 .9 .25 .75;  location .15 .95 .15 .95;  expand .9
LIMITS 3555 5245 -15 365;  ticksize 50 200 20 100;  BOX
CONNECT;  xlabel "\\iWavelength [\gV]";  YLABEL "\\i Counts"
justify 3;  expand 1;  VMOVE .18 .9;  LABEL "\\iFOC Long Slit"
EXPAND .75; VMOVE .18 .83; LABEL "\\iCourtesy P. Hodge, ST ScI"
justify 6;  angle 90;  expand .8
move 3727 58;  draw 3727 68;   move 3727 71;  label OII
move 3869 63;  draw 3869 73;   move 3869 76;  label NeII
move 3889 25;  draw 3889 30
draw 3915 40;  draw 3915 45;   move 3915 48;  label HeI, H\gz
move 3969 42;  draw 3969 52;   move 3969 55;  label NeIII
move 4102 36;  draw 4102 46;   move 4102 49;  label H\gd
move 4340 52;  draw 4340 62;   move 4340 65;  label H\gg
move 4861 73;  draw 4861 83;   move 4861 86;  label H\gb
move 4959 139; draw 4959 149;  move 4959 152; label OIII
move 5007 297; draw 5007 307;  move 5007 310; label OIII
```

FIGURE II Sample *igi* commands

MIRA: SOPHISTICATED IMAGE PROCESSING FOR THE MICRO-COMPUTER

MICHAEL V. NEWBERRY
Steward Observatory

OVERVIEW

MIRA ("Microcomputer Image Reduction and Analysis") software brings powerful astronomical image processing capabilities to microcomputers running MS-DOS on Intel 80286, '386 and '486 processors. MIRA contains a rich assortment of powerful image processing functions which permit astronomical image data to be reduced and analyzed. All functions support short integer (2-byte) and short real floating point (4-byte) pixel data and can write the result of an operation as either an integer or real type image. Functions can freely combine images having different data types to produce a result of a selected type and most functions can accept either 1- and 2- dimensional images. MIRA converts standard FITS or non-FITS image files into a MIRA specific format. If desired, images can be exported in FITS format for transport and use by other software.

MIRA operates in a high resolution 256 color video mode through an apparent 3-dimensional graphical user interface, or "GUI", having windows, buttons, and menus that are controlled by a mouse. The windows intercommunicate when necessary—for example, plotting a 1-d section of a displayed 2-d image only requires pressing a single graphics button. Every effort has been expended to make this interface aesthetic and efficient to use.

DESIGN OBJECTIVES

1. Provide extensive image processing capabilities for the reduction, analysis, and calibration of astronomical image data.
2. Provide an aesthetic graphical user interface that is easy to use, responsive, and efficient.
3. Execute mathematical operations at speeds limited by the 16 bit nature of DOS running on a 286 or better processor with modest hardware.
4. Use graphical input for image processing functions whenever possible.
5. Read and write standard FITS format image data.

HARDWARE REQUIREMENTS

MIRA requires a relatively modest base of computer hardware; however faster hardware gives a corresponding improvement to processing speed and makes the graphical interface more responsive. The following items are required:

1. 80286, 386, or 486-based computer running DOS version 3.3 or higher.
2. 80287, 387, or compatible floating point coprocessor.
3. A minimum of 2 MB of memory; 3 MB suggested.
4. A commonly available Paradise SuperVGA video card with 512KB of memory. MIRA is currently being modified for 1024x768x256 color display.
5. A 3-button Logitech mouse, model C-7, C-9, or MouseMan.
6. Some means of getting images into the computer such as a tape drive or a network connection to a computer that has one. Floppy disks may be used but are inefficient.

THE MIRA GRAPHICAL INTERFACE

The MIRA graphical user interface was designed to hold more than one "screen" worth of information within easy reach of the user. A number of dedicated windows are used for displaying and working with 1- and 2-dimensional images, managing text input and output, displaying on-line help, editing function parameters, and selecting image processing functions. The 1- and 2-d data windows have dedicated crosshair and rectangle cursors that continuously read data values in the coordinate system of the data. Text output uses the standard "–more–" buffering protocol and can be written in parallel to a log file. As with all useful GUI's, windows may be moved, closed, foregrounded, or resized as needed, and hidden windows are updated as new information arrives. Related image processing functions are organized using a menu structure. Functions are actuated by "clicking" on graphical buttons and, in a few cases, using the keyboard.

IMAGE PROCESSING FUNCTIONS

General file operations:

1. File directory, image directory, and disk space.
2. Rename, copy, or delete images, or convert the data types of images.
3. Make grayscale plots on an HP LaserJet; dump x, y plot data to a file.

FITS file utilities:

1. Read FITS or non-FITS images of byte, short integer, or floating point data type, and write FITS images in integer or floating point format.
2. List FITS headers or individual keyword lines from images.
3. Analyze keyword values from a series of images (modify values, fit, smooth, etc.) and copy new values back into the image headers.

4. Edit (change or add) or make an x, y plot of keyword values.

5. Rename header keywords (e.g., EXPT → EXPTIME).

Arithmetic:

1. Operations between an image and a constant: +, −, *, /, polynomial.

2. Operations between two images: +, −, *, /, interpolate.

3. Operations between a 2-d image and a 1-d image: +, −, *, /.

4. Unary operations on an image: $\log_{10}$, byteswap, change sign, reciprocal.

5. Mean, median, mode, std. dev., min, and max over a region of an image.

6. Normalize (offset or scale) images to a reference value.

7. Directional gradient and edge enhancement.

Data reduction:

1. Combine images using the mean, median, mode, or sum of pixels, with or without additive or multiplicative normalization to the mean, median, or mode of a target image.

2. Column bias with interactive bias calculation and output image trimming.

3. Flat field division and dark frame subtraction.

4. Smoothing (median, boxcar, Gaussian, etc.).

5. Repair bad pixels and columns (can be interactive).

6. Make subimage (2-dimensional or mean/median collapse to 1 dimension).

7. List pixel values inside an image region.

8. Mosaic m x n 1-d or 2-d images.

9. Resample with a sinc function filter.

Image display:

1. Display one or more 2-d images with selectable magnification, centering, color palette, and linear, log, or gamma transfer function over 151 levels. Contrast and brightness can be interactively adjusted using the mouse.

2. Plot a histogram, 1-d image, or a collapsed 1-d section of a 2-d image.

3. Use an alternate x, y coordinate system.

Photometry:

1. Sky fitting: locate, edit, delete, and obtain statistics on sampling regions, fit and subtract a 2-d polynomial, or create an image from the fitting function.

2. Elliptical aperture photometry: isophote fitting, growth curve photometry, object removal, reduction to standard values.

BENCHMARKS

The following timings have been made by the author using the following hardware: a 20Mhz Dell 386 computer, MS-DOS 5.0, a Cyrix FasMath 80D387 coprocessor, and a fast 212MB Conner 3204F IDE hard disk. MIRA does not make use of 32 bit instructions which would increase computation speed on the 386 processor.

TABLE I Benchmarks for some selected MIRA operations. Time required for disk read/write and header decoding is included.

Size	Input type[a]	Operation	Output type[a]	Time (sec)
384x576	int	/ flat field (real)	int	8.7
384x576	int	+ image (real)	int	7.8
384x576	int	− dark frame (real)	real	8.9 [b]
384x576	real	× real constant	int	6.4
384x576	int	log_{10}	real	6.9
384x576	real	3-term polynomial scaling	real	8.3
384x576	int	subtract 4-term x, y polynomial	real	8.6
384x576	real	directional gradient filter	real	12.7
384x576	real	smoothed using 5x5 boxcar filter	real	9.8
384x576	int	mean, std. dev., min, and max	to TEXT window	5.5
384x576	2(int)+2(real)	scaled and mean-combined	real	25.0
384x576	4 real	scaled and mean-combined	int	32.5 [c]
384x576	4 real	scaled and mode-combined	real	46.3
800x800	int	convert to real data type	real	15.0
800x800	int	display, log, zoom=−2	to IMAGE window	3.0
800x800	real	display, log, zoom=−2	to IMAGE window	7.2 [d]
800x800	real	mean, std. dev., min, max	to TEXT window	14.0
800x800	int	subtract column bias (real)	int	17.0

[a]data type of pixel: "int"=2 byte integer, "real"=4 byte float.

[b]Dark frame pixels were scaled by exposure time; computations were performed in floating point.

[c]4 MB of data were moved to and from disk during this operation.

[d]The image occupies 2.56 MB of disk space. The image was uniformly sampled for graylevel autoscaling and then displayed at a demagnification of 2 times using a logarithmic transfer function.

PC-IPS: INTERACTIVE SYSTEM FOR ASTRONOMICAL IMAGE PROCESSING

O.M. SMIRNOV
Institute of Astronomy, USSR Acad. Sci., 48 Pyatnitskaya st., Moscow 109017 USSR

N.E. PISKUNOV
Obs. and Astrophys. Lab. Univ. of Helsinki, Tähtitorninmäki, SF-00130 Helsinki, Finland

V.P. AFANASYEV, A.I. MOROZOV
Moscow Engineering Physics Institute, 31 Kashirskaya hwy., Moscow 115409 USSR

INTRODUCTION

PC-IPS was designed as a low-end facility for different sorts of interactive scientific image processing, with emphasis on user-friendliness and a flexible environment for software development. It integrates tools for examining and reducing data in a single menu-driven environment. A wide range of applications for astronomical data reduction was written to run under the system: basic arithmetics, geometric transformations, a CCD package, and a spectral package that includes Echelle frames processing.

GENERAL CAPABILITIES

PC-IPS will run on any AT or compatible with a 287/387 math coprocessor and an EGA card, under DOS 3.3 or higher. For better performance, at least 4 Megs of memory are recommended. The system can process one- and two-dimensional images of several numeric types: 8-, 16- and 32-bit signed and unsigned integer, 4- and 8-byte floating point and 16-byte algebraic and trigonometric complex. The maximum size of an image depends on its type, from 1024×16384 for complex, to 16384×16384 for 8-bit integer.

Because all applications are external, the system is not limited in the kind of image processing it can perform. As new applications are written, they are easily added to the application library without rebuilding the system.

PC-IPS uses expanded memory as much as possible for temporary storage of images. When all expanded memory is allocated, the remaining images are stored on disk and accessed through a disk buffer. This is completely transparent to the user and to the applications currently running, apart from a decrease in processing speed.

THE USER INTERFACE

The user's side of PC-IPS is the Integrated Environment (**IE**). It is responsible for maintaining a mouse- and keyboard-driven menu through which all operations are performed. The menu supports "hotkeys", so that experienced users don't have to waste time navigating it. For deeply nostalgic users, a simple command line interface is also available. All applications are started from this menu. On-line context-sensitive help can always be obtained at the press of a key.

The IE provides a set of tools for loading and examining images. When loaded, images go into the image stack, and the top (most recently loaded) image is displayed automatically. One-dimensional images may be zoomed by an arbitrary factor. Two dimensional images are displayed in 16 colors. Using the menu, the range of data values covered by the colors is easily changed. The IE can plot one-dimensional images over other one-dimensional or two-dimensional ones.

The IE also maintains a "magnifying glass" that constantly displays a small part of the screen under the mouse pointer zoomed by a factor of two or four.

For images that are bigger than the screen size, the user can either select a portion to be displayed, or specify a shrink ratio, in which case several points of the image are averaged and displayed as one pixel. This sacrifices some detail in exchange for displaying a larger portion of the image or even the whole image.

THE IMAGE DATABASE

PC-IPS keeps all its images in its internal image database (**IDB**) that provides fast and reliable access to images. The IDB supports user IDs and passwords, with users organized into groups. It is divided into *directories*, which have selectable user-group-world access attributes for reading, writing and deleting.

Within directories, images are identified by 16-character names and version numbers. All external images are imported into the IDB before they are processed; export/import applications for a variety of formats are included in the application library.

Besides image data, the IDB stores comments, coordinate scales, coloring ranges, etc. It also supports variables for storage of scalar values and/or compound data.

ARCHITECTURE

The Main Problem Encountered

PC-IPS was conceived as a general-purpose system, which led us to decide that no image processing capabilities would be built in, but would be provided by applications instead. Each application would be a separate program. This called for a powerful application program interface (**API**). When developing it, we kept in mind the fact that one day somebody was bound to try writing a custom application, armed with nothing but our instructions, so ease of use was of top priority. With the necessity of supporting a wide range of data types and living up to the "user-friendly" label, the API became quite a project.

This brought us to the main internal conflict of PC-IPS. By nature, the system must serve two clients—the user and the application, or, if we consider the application to be an extension of the programmer's thought, the programmer. They have irreconceivable interests. The user wants a convenient interface with menus and a mouse, all the time, while the programmer has image processing in mind, and does not want to clutter his head with menus and, least of all, with mice.

Another side of this conflict is best illustrated with an example. Consider an elementary application, one that adds two images. As long as they are of the same size and type, it's infinitely simple. Different size is a slight nuisance to the programmer, and different types are a curse. So he designs the application to handle images with matching type/size only. The user, on the other hand, will surely need to add different images sometimes. Neither of them can compromise here.

The Solution Found

An image processing session under PC-IPS has five components, the first four of which make up the system: the IE, the API, the IDB, the applications, and, most important, the user. The core program (PC-IPS itself) consists of the IE and the API. All the other components interact with the core only, *never with each other.* This is the key to resolving the user-programmer conflict.

The user and the programmer communicate completely through PC-IPS. The programmer tells PC-IPS what sort of input he needs, or what sort of output he's generating. PC-IPS then interacts with the user, in its usual friendly manner, complete with menus and a mouse, and reports the user's answers back to the programmer. In this position, PC-IPS can assume the role of the wise judge whenever conflicts arise, and resolve them before either side is aware of it, or in extreme cases it can at least offer to the user a compromise or a solution.

A typical (and most common) conflict is "type/size", outlined above. It is resolved easily. The programmer tells PC-IPS what images he will handle, in our example, it's matching ones. PC-IPS then takes the user's probably unmatching images, converts them to matching type and size, passes them to the programmer, and when the result is ready, it can again convert it to whatever the user needs. Type/size conversion is a very important part of PC-IPS, it insures that any application is automatically compatible with any image.

Another benefit of this approach is the potential, though not yet implemented, capability of using applications as building blocks for constructing complex *metaapplications*, or image processing procedures. Because each application interacts with PC-IPS only, it never knows if its input is coming from an actual user or from another application.

THE APPLICATION INTERFACE

Once we settled on the architecture described above, designing a powerful and programmer-friendly API became much easier.

On the lowest level, applications communicate with PC-IPS via a software interrupt. Since interrupts are rather programmer-unfriendly, an interface library was developed for C, FORTRAN and Pascal. When it is used, the API becomes a set of functions. Using the simple subset, the programmer may start writing applications almost immediately. Once he is familiar with the API, he can use the advanced subset to vastly improve the performance and capabilities of his applications.

The most important part of the API is parameter and image transfer. The programmer does this by requesting parameters or images from, or reporting results to, PC-IPS. It is no harder than printing them to the screen or getting them from the keyboard the way it was done in the pre-mouse days, because PC-IPS handles all interaction with the user itself, and that is usually the messiest part of the program. This way, all applications have a consistent and user-friendly interface, with no efforts on the programmer's part.

Besides the standard data types, PC-IPS supports such compound types as window coordinates, etc. From the programmer's point of view, getting window coordinates is done exactly like getting a parameter of any other type, for example, an integer number.

When requesting images, the programmer must determine the image type and size he is prepared to handle. He can specify a concrete type and size, or something as complex as "two 16-bit integer images of the same size, an image of any type and of the same size as the first two, and two images of the same type, not necessarily the one before, but with the same size as before". This is made possible by the concept of *image request groups*, inside which images must have the same type and/or size. PC-IPS performs type/size conversion for images that do not meet the programmer's specifications.

Requests may be organized into *lists* so that they are presented to the user together in a single window. A list of requests, once formed, may be activated over and over again by calling a single function.

The API also lets an application display images, and access images either point by point, or page by page (a *page* is large chunk of the image). The images may be either in memory or in the IDB, but this is invisible to the programmer.

APPLICATIONS

We have made six groups of applications which are supposed to simplify the process of getting acquainted with PC-IPS. These include arithmetics, geometric transformations, import/export procedures, a CCD package, spectroscopy, etc.

The CCD package includes a CCD frame "quick look" for fast inspection of the frame, statistical analysis of flat field frames, and cosmic ray spike removal from 1D and 2D data. For spectroscopic analysis we offer an automatic search for Echelle orders on a CCD frame and geometric correction of each order. Applications for computing line centers and dispersion curves as well as fitting the continuum level enable the user to make further reduction and obtain spectral line profiles.

FUTURE DEVELOPMENT

Future development of PC-IPS is expected to take several directions. The first, expansion of the application library, does not require any modifications to the system and is in progress even now.

The DOS version of PC-IPS will probably grow to acquire pseudo-3D graphics, automatic image compression, a PostScript(R) hard copy driver, improved resolution and color capabilities, etc.

We also plan to port PC-IPS to a multitasking platform, such as OS/2 or Unix. This version of PC-IPS would have facilities for multitasking applications and interapplication communication. It is here that we want to implement the "building blocks" approach, described earlier, for constructing metaapplications.

FEEDBACK

We welcome all comments and questions about PC-IPS. You can reach us at the addresses on the title page, or by E-mail (Internet):

N. Piskunov piscounov@cc.helsinki.fi
O. Smirnov iaas@node.ias.msk.su

ACKNOWLEDGMENTS

We thank the Institute of Astronomy of the USSR Academy of Sciences for providing the hardware, and Observatory and Astrophysics Laboratory of the University of Helsinki for financing this project.

We would also like to thank the Smithsonian Astrophysical Observatory for the financial support which made this presentation possible.

SECTION 8

ANALYSIS OF ALL-SKY DATA

THE ROSAT EXTREME ULTRA-VIOLET SKY SURVEY

C. G. PAGE & M. DENBY
X-ray Astronomy Group, Physics & Astronomy Department, Leicester University, Leicester LE1 7RH, UK

ABSTRACT The wide field camera on ROSAT has carried out an all-sky survey in the previously unexplored EUV band (60 – 200 Å) resulting in the detection of several hundred sources. Some novel problems were faced in the data reduction, both in sorting data from a scanning imaging instrument and in processing the data in parallel at five different sites.

THE ROSAT WIDE FIELD CAMERA

ROSAT was launched on a NASA Delta II rocket on 1990 June 1 into a circular orbit at an altitude of 575 km, giving an orbital period of just over 90 minutes. The inclination of 53° allows control from the ground station at Weilheim (near Munich) of the German Space Operations Centre (GSOC). After the initial check-out and calibration period, ROSAT started its sky survey phase on 1990 July 30. This was scheduled to last six months but was cut short by a few days when attitude control was lost for a time near the end of January. The small gap which remained has partly been filled in during the summer.

ROSAT carries both a German X-ray telescope and a British EUV telescope called the Wide Field Camera (WFC). The latter was designed and built by a consortium of five British research groups: Leicester University, Birmingham University, Mullard Space Science Laboratory (part of University College London), Imperial College London, and Rutherford Appleton Laboratory (RAL). The WFC has three concentric gold-plated aluminum mirrors in Wolter-Schwarzschild type I configuration, giving a total geometrical area of 475 sq.cm. The detector is a curved micro-channel plate with a CsI coating and resistive plate read-out. The field of view is 5° in diameter, with angular resolution of 1 arcminute at the centre of the field, falling off to about 3 arcminutes at the edge.

This paper describes the initial reduction of the EUV sky survey which was carried out jointly by the five institutes; at this stage it was more important to detect sources rapidly than to obtain the ultimate in sensitivity.

SKY SURVEY ANALYSIS REQUIREMENTS

The sky was scanned by rotating ROSAT on an axis once per orbit so its telescopes always looked away from the Earth. The resulting scan path was a series of great-circles passing over both ecliptic poles and crossing the ecliptic at a fixed angle to the Sun, nominally 90°. Thus the scan path advanced ≈1° per day to cover the entire sky in six months. Sources on the scan path could be seen by the WFC for up to 80 seconds per orbit; those near the ecliptic plane were scanned for five days, this period increasing towards the ecliptic poles which were scanned every orbit for six months.

Microchannel plates have no intrinsic spectral resolution so a filter wheel was provided; during the sky-survey two filters, known as S1 and S2, were used on alternate days. The approximate band-pass of these is shown below.

filter	material	energy	wavelength	frequency
S1	C/Lexan/B	90 – 206 eV	60 – 140 Å	21 – 50 PHz
S2	Be/Lexan	62 – 110 eV	110 – 200 Å	15 – 27 PHz

There were several unusual aspects to the sky-survey analysis. Firstly the EUV band was almost unexplored: only a dozen sources were known at the outset and the effects of interstellar absorption were highly uncertain. It was possible to predict only that the WFC might detect between a hundred and a few thousand sources.

Uncertainties in the background rate were almost as large because microchannel plates are also sensitive to charged particles and the effectiveness of our magnetic diverter was difficult to assess before launch. Fortunately it turned out that background rates in most parts of the orbit were well below the telemetry saturation limit of 200 events/second. Data collection was of course disabled while the satellite passed through the auroral zones and the South Atlantic Anomaly; overall data were collected for about 70% of the time.

In order to involve as many astronomers as possible in data analysis, and to make best use of our limited computing resources, the processing was carried out in parallel at all five consortium sites. Fortunately all groups were using VAX/VMS equipment and had good network connections, but during the design phase (and until shortly before launch) these were mostly VAX-11/750s already heavily used for general astronomical data analysis. The overall software design emphasised efficiency, since it was expected that it would have to be capable of sorting up to 10^9 events into 3×10^8 pixels (1 arcminute resolution in two bands).

DATA REDUCTION

Data from the WFC were recorded on-board on tape and transmitted to GSOC once per day. The position of each event in the detector was encoded as two 9-bit numbers, and its slot in the telemetry frame timed each event to 32 ms, sufficient to unfold the scanning motion.

At GSOC the telemetry data were sorted into time-order and combined with the appropriate house-keeping, orbital, and star-tracker data. The data volume, typically 45 Mbytes/day, precluded transmission to the UK over the network, so the data were written to 8mm (Exabyte) tapes each day and posted to the Rosat UK Data Centre established at RAL.

It was however feasible to send the house-keeping data and about 20% of the WFC scientific data over a network connection to our quick-look facility (QLF) established at MPE Garching. At the QLF a cut-down version of the full data processing software was run to ensure correct operation of the instrument and to allow a rapid response to any problems.

At the RAL Data Centre the incoming data tapes were subjected to comprehensive quality checking, and star-tracker data used to determine the aspect of the WFC as a function of time.

All sky surveys have to work out how to represent the spherical sky using images formed from square pixels. Because of the special significance of the ecliptic poles in the scanning scheme an ecliptic spherical-polar coordinate grid was used. The sky was divided along lines of ecliptic longitude and latitude giving cells up to $2° \times 2°$ across. There were fewer cells per longitude strip above $\pm 60°$ latitude to avoid over-sampling the polar regions, giving in total 13,560 cells in each spectral band.

The raw detector coordinates of each event were linearized and transformed to the ecliptic frame. The cell number of each event was computed, together with its offsets in the tangent plane from the centre of its cell; by this means only 2×10 bits were needed for a resolution of 7.5 arcseconds. It was decided to retain the raw detector coordinates of each event in case any hot-spots arose (though fortunately none did); linearised telescope-frame coordinates were also retained to allow selection of the appropriate point-spread function for each event. These sorted files then required the following storage, amounting to 11 bytes per event:

Raw detector X, Y	2×9 bits
Linearised X, Y	2×8 bits
Ecliptic X, Y offsets	2×10 bits
Time-tag (32 ms)	32 bits

Sorted events were accumulated in records of 1024 bytes. As records became full they were written to a direct-access file. The events from a single orbital scan covered $5° \times 360°$ but fell into fewer than 1000 cells, so the sorting could be done in less than 2 Mbytes of memory, avoiding paging overheads. The records in the output file were chained into a doubly-linked list, with an index area to hold pointers to the start and end of each of 13,560 chains (plus a free-list). These sorted event files were written to Exabyte tapes and sent each day to the five consortium sites for further reduction. Each site was sent the complete data-set but had the primary responsibility for covering a particular area of sky (with small overlaps to avoid boundary problems).

Each processing site maintained a reservoir file of events which had an identical linked-list structure. Each day the chains of records for each cell on the incoming data-tape were copied to disc and inserted in the corresponding chain in the reservoir file, and the links updated. On completion of this merging process it was possible to identify cells which had just left the trailing edge of the scan path and which were therefore fully exposed. For all these areas of sky an image was formed in each spectral band. The x and y coordinates of each event could be computed just by scaling the tangent-plane offsets already present in the event file. No further trigonometry was generally required, the exceptions being a small guard-band around the edge of each cell in which events were imported from adjacent cells, and the areas around the polar caps in which many cells had to be combined to form a square image. After analysis was complete, the records from each completed and analysed cell could be extracted to another file and archived to tape. The space in the reservoir file was then recovered by linking these chains of records to the free-list.

The main point-source detection method involved passing a circular sliding box over the image and using a Poisson significance test. The background level was estimated initially with a technique involving median-filtering; when a possible point source was found the local background was re-estimated from an annulus around the source. An averaged PSF was then fitted using a maximum-likelihood technique. A more sophisticated matched-filtering detection algorithm was also used which had slightly greater sensitivity, but it proved less reliable until rather late in the survey processing phase. Positions of point-sources could generally be determined to a few tens of arc-seconds; these positions were cross-correlated with standard catalogs of sources in other wavebands. The preliminary source-lists were then fed into the optical identification program.

Images were also scanned visually for defects and extended sources. Apart from the Moon, which confused the point-source detection algorithm at regular intervals, only two extended sources were found, the Vela and Cygnus Loop SNRs.

SOFTWARE ENGINEERING AND RESULTS

As far as possible the system made use of existing software, especially packages and libraries from the Starlink software collection. The Starlink Hierarchical Data System (HDS) was used extensively for calibration files, images, time-series, etc. Image processing was performed with the ASTERIX package, and details of detected sources were stored in the SCAR relational database. But project-specific software, together with the integration, and testing of the system, absorbed around 100 person-years of effort. Software packages developed at each of the five institute were used at each of the others; this required good communications and configuration control. Our systems were all linked by DECnet over JANET (the UK equivalent of Internet); we also used EARN/BITNET and SPAN for links with Germany and the U.S.

This system was generally successful in keeping up with data flow, and our initial processing was complete in 1991 March, less than 6 weeks after last survey photon was detected. The first catalog of $\approx$400 sources is now being prepared for publication; Figure I shows these sources in an Aitoff projection. Planning is now well advanced for the complete reprocessing of the survey data next year; this will have the benefit of improved software and better understanding of the instrument and the components of background flux. With greater detection sensitivity we hope to produce a catalog of $\approx$1000 EUV sources.

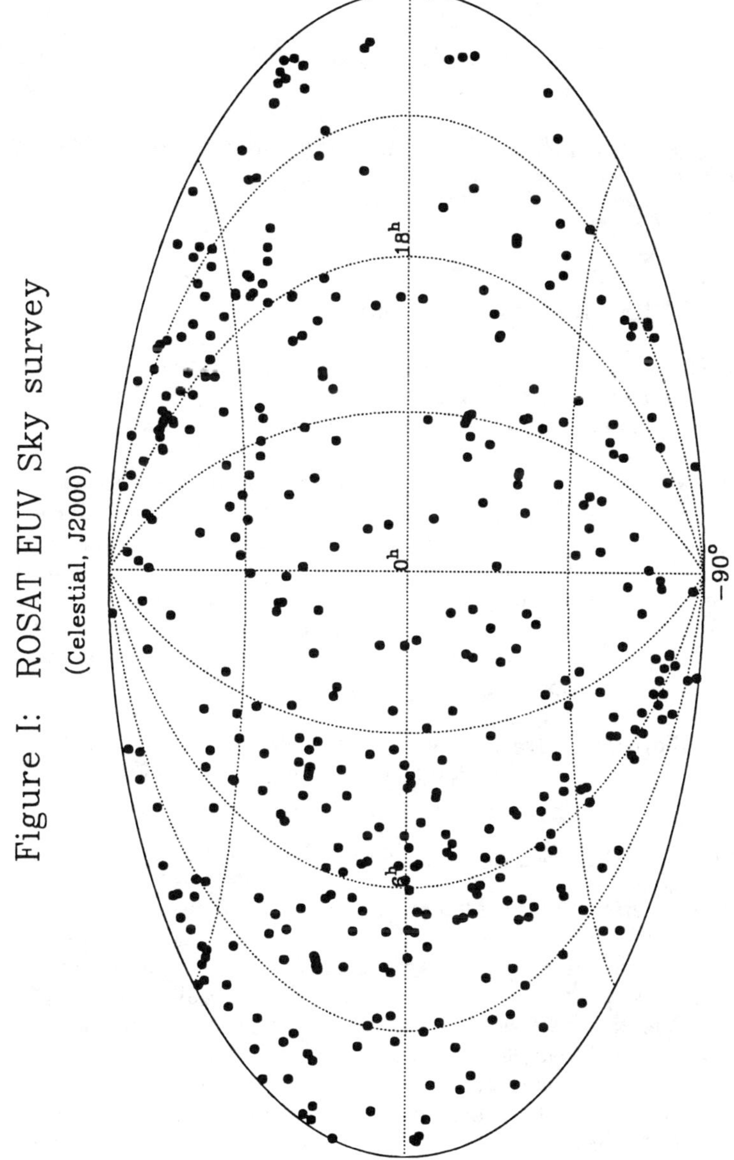

Figure I: ROSAT EUV Sky survey
(Celestial, J2000)

EINSTEIN SLEW SURVEY: DATA ANALYSIS INNOVATIONS

M. ELVIS, D. PLUMMER, J. SCHACHTER AND G.FABBIANO
Center for Astrophysics, 60 Garden St. Cambridge MA 02138 USA

ABSTRACT Several new methods were needed in order to make the *Einstein* Slew X-ray Sky Survey. These may be useful for other projects.

SCALE OF THE SLEW SURVEY PROJECT

The *Einstein* X-ray observatory was not intended to make a survey of the sky. However in moving between targets the instruments were left on, so that exposure was accumulated on most of the sky. Eventually this led to the *Einstein* Slew Survey containing 818 bright X-ray sources, 40% of which were unknown (Including almost doubling the number of known BL Lac objects).

The idea of a sky survey made from *Einstein* data was alluring. However the project faced a number of problems: the software was on obsolete 16-bit machines; the data had never been looked at, was poorly documented, and was dispersed on ~2500 magnetic tapes; the star tracker data was unusable so there was no known aspect system; the gyro aspect data needed extrapolation over ~ 120° slews; the data set was fairly large, ~ 400 Mbytes; we needed an arcminute, all-sky exposure map ; and the 'sliding box' detect has $\sim 4\times10^7$ cells over the sky. No manager could commit large resources to such an uncertain a project. Fortunately technology made the Slew Survey feasible. Below we summarize the innovations which enabled the Slew Survey to be done.

EXPERIMENTAL APPROACH TO LARGE PROJECTS

By greatly lowering the investment in time and manpower needed to process a satellite data set modern computer hardware allows an 'experimental' approach to large data analysis projects. Instead of deriving optimum analysis methods from first principles the data can be processed with 'quick and dirty' algorithms, and reprocessed many times until an successful result is obtained. This resembles *rapid prototyping.* A small team (one full time programmer and one part-time scientist) were able to carry out the whole Slew project in less than two years.

The two keys are: (1)**Large on-line storage** (a 1-Gbyte disk in our case; optical WORMs for larger archives) has removed the manual overhead and time involved in loading tapes. To process the original 2000 magnetic tapes of *Einstein* data even once was a major undertaking. Now the data are all on WORMs, on-line; (2)**Large amounts of CPU** available without being noticed by other users of the system mean that a complete reprocessing becomes a small decision. We had ~100 times the CPU power available to the original *Einstein* project.

The processing time was thus reduced from months to just days. The whole Slew Survey was re-processed at least a dozen times to reach an acceptable solution.

This approach is only possible for a data product, such as a survey. Projects with a software deliverable cannot operate this way. Our software is not intended for public distribution, is almost undocumented, and is certainly unsupported. The quality assurance lies in the scientific paper describing the survey results, which includes many checks on quality.

PARALLEL PROCESSING ON A LAN

A local area network of workstations was used in a quite simple and primitive way to run up to 10 parallel streams of processing. A complete re-analysis of the survey could in this way be accomplished in *3 days instead of 3 to 4 weeks.* This, as noted above, changes the whole approach to the project.

The main problem in implementing this parallel system was the control list since occasionally two machines would attempt access simultaneously. Unix does not handle file locks well, so we resorted to 10 separate lists. Several off the shelf packages now exist for SUNs that manage parallel processing more flexibly. These should enable the whole SAO High Energy LAN of ~60 4-Mflop SPARC 1 machines to be combined. Depending on the application such a 'network supercomputer' may reach of order 200Mflop (roughly 2/3 of a Cray Y/MP).

PERCOLATION SOURCE DETECTION

Standard source detection algorithms in X-ray astronomy have been of the sliding-box type. They are thus 'sky-centered' *i.e.* every part of the survey area is examined and there are 150×10^6 box positions on the sky. To keep compute time reasonable we were forced to a different approach. In the Slew Survey the sky is sparsely populated with photons. Since there are 3×10^6 photons in the Slew Survey a photon centered-approach is some 50 times more efficient than a sky-centered approach.

We developed a method to find photons which were too closely associated with their neighbors for mere chance. This approach turned out to be identical to that used to locate groups of galaxies in the CfA redshift survey (Huchra & Geller 1984), and is a simple version of the class of 'percolation algorithms'.

'MINIMUM ACTION' IDENTIFICATIONS

Archive material

Two thirds of the Survey sources have been identified using archival resources (NED, SIMBAD, z-cat). Of the remaining sources 75% have plausible counterparts in one or more of the existing digitally accessible sky surveys: the IRAS Point Source and Faint Source Catalogs; the 5GHz 300ft Green Bank (87GB) radio survey; the HST Guide Star Catalog (GSC); the ROE/NRL and Minnesota catalogs of the optical sky surveys.

These resources now make it possible to produce spectral energy distributions for many X-ray sources in the Slew Survey without any new observing.

We can isolate the most likely counterparts. Source classes can be assigned with reasonable probability based on these distributions. In this way ~96% of the sources have some likely counterpart. The follow-up observing time needed on optical telescopes is thus minimized. Only a couple of dozen high Galactic latitude fields have no counterparts and need be observed with large telescopes.

Statistical Sieveing

Many of our identifications will be with objects bright enough to be in the HST GSC, but within our 2 arcmin radius error circles there are typically several GSC objects. A maximum likelihood method (de Ruiter et al. 1977; Prestage & Peacock 1983) is powerful for picking out the correct identifications. This calculates the likelihood that an optical candidate in the error circle is correct given its position in the error circle and the local background optical object density at that magnitude. This yields 57 unambiguous identifications with GSC objects that had no previous counterpart (Schachter et al. 1992). This method readily allows the inclusion of other factors, such as X-ray to optical ratios, or a stellar/non-stellar flag to separate galaxies from stars.

RAPID DISSEMINATION OF THE WHOLE DATA BASE

Naturally astronomers working on on a large project tend to milk all the interesting information from it before making it public. Instead The Slew Survey source catalog and the complete photon and timing data used to construct it were released on CD-ROM (Plummer et al. 1991) as soon as the survey was complete (Jan '91) and well before the paper describing the survey was submitted to the Ap.J. (Sept '91) or published (April '92). As software is developed it is placed on-line to assist a user's own Survey analysis. This will encourage others to work on the data and the Slew Survey will become more widely known, used, and referenced. Self-interest and public interest can converge!

Our aim is to get the Slew Survey 100% identified as soon as possible, so we can begin population studies (*e.g.* AGN evolution). The best way to do this is to get others to help. An on-line, regularly updated, source identification list is maintained at CfA (on *einline*, Karakashian et al., this volume). We will put any proposed identifications sent to us into this data base with proper credit to the discoverers, providing them with an incentive to get there first.

REFERENCES

de Ruiter, H. R., Willis, A. G., & Arp, H. C. 1977, A&AS, 28, 211

Elvis, M., Plummer, D., Schachter, J., & Fabbiano, G. 1992, ApJS, in press

Huchra, J. P., & Geller, M. 1984, ApJ, 257, 423

Karakashian, T., et al. 1992, this volume

Plummer, D., et al. 1991, CD-ROM issued by SAO

Prestage, R. M., & Peacock, J. A. 1983, MNRAS, 204, 355

Schachter, J., et al. 1992, in preparation

THE INFRARED SKY ON-LINE

PETER R. ROELFSEMA AND DO J.M. KESTER
Space Research Groningen, Postbus 800, NL-9700 AV, Groningen, The Netherlands

ABSTRACT To allow easy access to all IRAS data, they are stored on an optical disc jukebox at Space Research Groningen. Currently these data can be accessed in a limited fashion through a mail server. The server provides data extraction, calibration and imaging for survey and LRS data. In the future also more interactive access to the database will become possible. The system can be distributed to a number of different sites with expertise in IRAS data processing.

IRAS

The mission

In 1983 the infrared astronomical satellite (IRAS) performed an almost complete all-sky survey. The infrared sky brightness was measured at 12, 25, 60 and 100 μm for $\sim$95% of the sky. The satellite was in a polar orbit, precessing by $\sim 1°$ per day thus bing able to observe the whole sky in a few months time (see Figure Ia). During standard operation the telescope scanned the sky along small circles.

For each wavelength band the focal plane contained two arrays of detectors (see Figure Ib). These arrays were laid out such that the second array scanned the same area of sky 5 to 10 seconds later than the first array, giving redundant coverage on a seconds time scale. Since subsequent scans on the sky were partly overlapping, also confirmation of source detections on an hours time scale was obtained (for more details on IRAS see IRAS-ES 1988). This physical design and observation strategy were chosen in order to optimally fulfill the prime goal of IRAS: obtaining a complete survey of celestial infrared point sources.

Apart from the standard scans, also a large number of Pointed Observations (PO, also known as Additional Observations, AO) were carried out. These were typically raster scans covering small areas ($2° \times 0.5°$) repeatedly.

Data products and wishes

After the IRAS mission it was found that apart from yielding accurate point source positions and fluxes, the data quality allowed imaging of the infrared emission. As a consequence projects were started e.g. at the IRAS Processing and Analysis Center (IPAC) and at Space Research Groningen (SRG) to generate calibrated infrared images of portions of the sky. At IPAC the emphasis was on generating a standard set of large plates (the Infrared Sky Survey

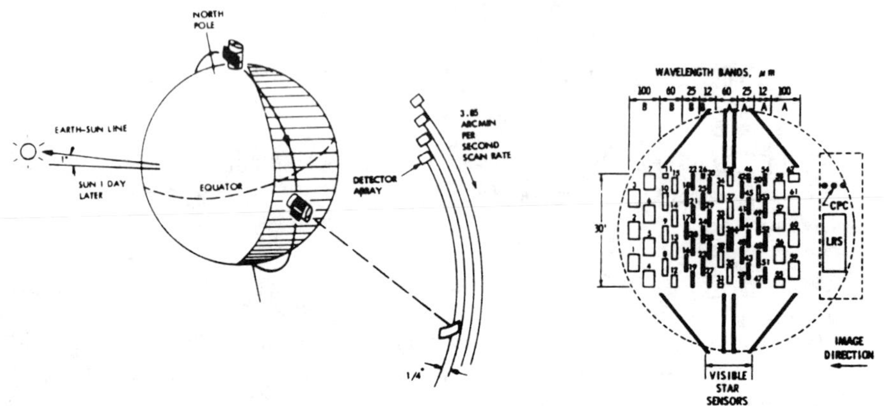

FIGURE I IRAS.
a) Geometry of the IRAS orbit. **b**) IRAS focal plane layout.

Atlas, ISSA), in Groningen most of the effort is directed towards special processing (high quality calibration, enhanced spatial resolution imaging, coadding of scans, LRS spectra).

FURTHER PROCESSING OF IRAS DATA

Until recently only standard IRAS products (e.g. the ISSA plates and the LRS catalogue) have been available. However, in many cases an astronomer is interested in only the small area of the sky containing his favourite object. For this area of sky he wants access to *all* the IRAS data products (i.e. survey, LRS and PO's) in an interactive fashion. In Groningen a system now has been set up to allow such access from an interactive system for analysis of astronomical data.

The data reduction process

The reduction and analysis of IRAS data roughly proceeds along the following lines: given an on-sky area of interest to the astronomer, the IRAS data relevant for that area are extracted from the IRAS data base. Subsequently these data need to be calibrated and imaged (that is images in celestial coordinates for the survey data and spectra for LRS data). These steps may take some iteration, since the highest quality images are desired. In some cases superresolution ($\sim 1'$) may be needed. Both calibration and imaging should be done using the raw IRAS data as input. Only when the raw data are used it is possible to use newer, more accurate techniques for calibration and imaging.

Once the user is satisfied with the generated images, image analysis tools will be required for e.g. position measurements, flux measurements, line profile fitting etc. In some cases the image analysis results will point towards residual calibration problems, requiring the user to go back to the calibration/imaging step.

Raw IRAS data within GIPSY

To be able to perform calibration, imaging and analysis of IRAS data, firstly the raw data have to become accessible within the data analysis system. The raw IRAS data with all the ancilliary information is not just a simple array of numbers. At SRG a 4-dimensional GIPSY data structure has been developed allowing a very convenient way to have access to raw IRAS data in a fully interactive processing system.

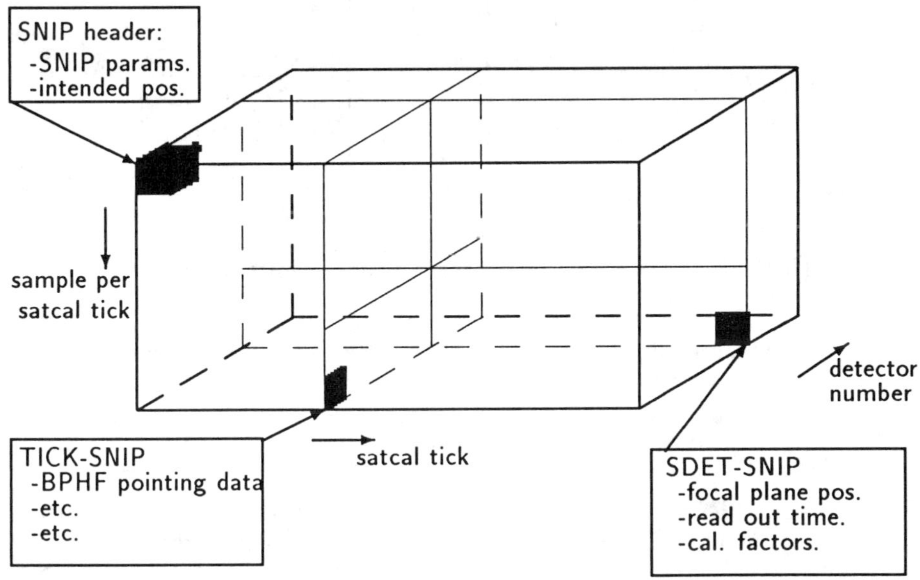

FIGURE II GIPSY implementation of a single snip

Due to the way in which IRAS has operated, the raw data for each detector consist of a series of data samples taken at constant time intervals at a rate of $N_{samples}$ per second or, more accurately, per satcal tick (the IRAS internal clock; one tick$\sim$1.00005 second). Once for every satcal tick housekeeping information such as the pointing of the satellite is obtained. Thus for a section of a single scan (such a section is called 'SNIP' in the IRAS-GIPSY system) of N_{ticks} satcal ticks for a given observing band (12, 25, 60, 100 μm or LRS) with N_{dets} detectors, the data is a cube of $N_{samples} \times N_{ticks} \times N_{dets}$ pixels. Such a 3-D data structure is shown in Figure II. Note that the detector axis SDET doesn't have anything like a gridspacing or coordinate transform associated with it, it is an axis used only to order the data for the different detectors.

The SNIPs for a single wavelength band covering a given on-sky area can be collected in a 4-D structure (Figure III). Such an IRAS Data Structure (IRDS) then contains all data (samples *and* position *and* calibration information) required to compute images, or, for LRS, spectra. The SNIP axis, like the SDET axis, is used only to order the snips.

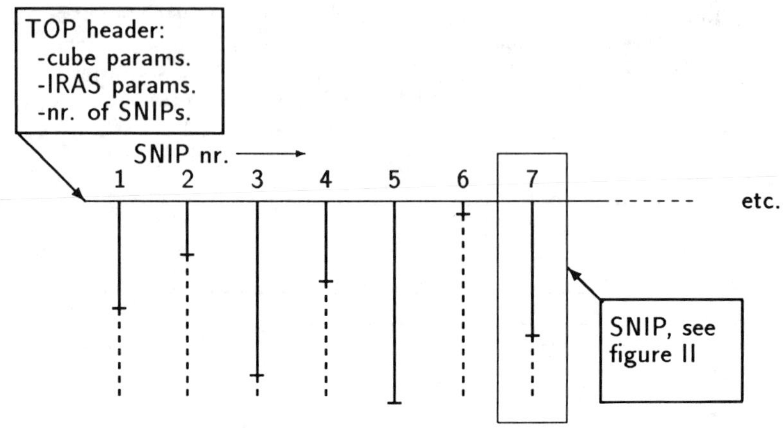

FIGURE III A GIPSY IRAS Data Structure (IRDS)

Note that since an IRDS is a fully valid GIPSY data structure, *all* programs available under GIPSY can in principle be used on raw IRAS data.

THE IR SERVER

The total raw IRAS data base (survey, LRS and PO) comprises some 15 Gigabytes (originally stored on ~450 magnetic tapes). Thus accessing the data for a given on-sky region requires a flexible access system. For this purpose a jukebox with magneto-optical discs is used for long term storage. The jukebox (total capacity ~20 Gbyte) functions as a standard UNIX file system, and thus in principle the data can be accessed using basic UNIX commands. The full raw IRAS scans of ~ 180° length, are stored in a compressed format as FITS files on the optical discs. For selection, access and decompression of these files a series of subroutines are available. These routines can be used to select and transfer the relevant IRAS data from the optical disc system to a local magnetic disc.

For astronomical research purposes a mail server system has been set up which can be used to generate standard products such as calibrated images or spectra.

Software

Access to the data base and calibration/reduction facilities occurs using a series of programs running under the GIPSY (van der Hulst et al. 1992) environment. These programs can be run interactively from the various computers available in Groningen.

Users not at SRG can send mail requests for IRAS data products to the server. Such a mail request consists of a sort of script specifying which GIPSY task(s) need to be run. Following a request from a user, the mail server actually starts up a batch version of GIPSY and subsequently runs the different programs specified in the request. When the request is completed the system sends a mail message back to the requestor, telling him/her where the constructed data set can be obtained or, in the case of a processing or other error, what went wrong and if possible why it went wrong. Following a successful request, the user can fetch his data through anonymous ftp.

RESULTS AND PLANS

Currently the mail server is being tested in-house, and limited access is provided for outside users. The experience to date shows that standard images ($\sim 1° \times 1°$ on-sky) are obtained through the mail server with a turn-around time of some 20–30 minutes.

Interactive use of the system allows even shorter turn-around times; the same image can be generated (with full calibration) interactively in less than 15 minutes. Similarly a fully calibrated LRS spectrum for any given source can be obtained within 10 minutes.

In the near future more elaborate experiments with outside users requesting larger amounts of data are anticipated. Also the reduction facilities, i.e. GIPSY and the IRAS processing programs which do not require the optical disc system will be made available to the community. Thus the mail server can be used to obtain the raw data, and calibration and imaging can be done at the users home institution.

Possibly at a later stage (mid-1992) the entire system, software *and* data will be exported to a few sites which have inhouse experience with IRAS data processing.

More information about the system described above and about the availability of GIPSY can be requested through E-mail from "pjotr@sron.rug.nl" or "do@sron.rug.nl".

REFERENCES

IRAS-ES; Joint IRAS Science Working Group 1988, NASA Reference publication 1190; Infrared Astronomical Satellite (IRAS), Catalogs and Atlases, vol. 1, Explanatory Supplement, NASA Scientific and Technical Information Division, Washington D.C., eds.: Beichman, C. A., Neugebauer, G., Habing, H. J., Clegg, P. E., & Chester, T. J.

van der Hulst, J. M., Terlouw, J. P., Begeman, K. G., Zwitser, W., & Roelfsema, P. R. 1992, this volume.

Wesselius, P. R., de Jonge, A. R. W., Kester, D. J. M., & Roelfsema, P. R. 1991, Les Houches 1991

IMAGE ANALYSIS AT IPAC

T. N. GAUTIER
Infrared Processing and Analysis Center, California Institute of Technology, Pasadena, CA 91125

ABSTRACT The Infrared Processing and Analysis Center (IPAC) supports maintenance and analysis of archival data from the IRAS survey conducted in 1983. Ongoing development of IPAC capabilities will provide fast access to the IRAS image products and make available powerful image analysis and display software. Remote use of the IPAC data bases and archives is planned via national computer network connections.

INTRODUCTION

The Infrared Processing and Analysis Center (IPAC), established in 1984 as a data reduction and analysis support center for the data obtained by the Infrared Astronomical Satellite (IRAS), has nearly completed all planned IRAS data products. IPAC is working to support on-line access and analysis of the entire IRAS data archive. Significant improvements to IPAC hardware and software are being made to enable quick and easy access to archival IRAS data and to provide powerful analysis tools for IRAS images and catalogs. Connections to national computer networks allow remote access to the IPAC archives and access to data at other institutions. Much of the image analysis software developed at IPAC will be available for export to external users.

IMAGE DATA SETS SUPPORTED AT IPAC

The IRAS Sky Survey Atlas (ISSA, previously called Super Sky Flux) and the IRAS Zodiacal History File, Version 3.1, (ZOHF V3.1) will be the principal image data sets maintained at IPAC. The ISSA images are the result of reprocessing of the IRAS survey data into sky maps with substantially higher signal to noise and greater usability than the original IRAS Sky Flux maps (Wheelock et al. 1992a). ISSA consists of 430 12.5°x12.5° fields with 1.5′ pixels. Sixteen images are associated with each field; four sky coverages (three survey coverages plus a coaddition of the three) in each of the four IRAS bands. About 40% of the fields have large enough dynamic range to require 4 byte data numbers while the other 60% need only 2 byte numbers, yielding a total data set size of about 5 gigabytes.

The ZOHF V3.1 is the latest version of the time ordered history of the IRAS survey obtained by averaging the IRAS survey array data into a 0.5°x0.5° beam (Wheelock et al. 1992b). ZOHF V3.1 contains 1.2×10^7 seconds of IRAS survey data with one sample in each of the four IRAS bands every 8 seconds, yielding a data set size of about 24 megabytes.

The original Sky Flux images (212 500x500 images in 4 bands with 3 coverages yielding 2 gigabytes) and the Faint Source Survey images (110 gigabytes) will also be available at IPAC as miscellaneous image data sets.

The current and future plans for storing and accessing these image data sets are shown in Table I. IPAC currently has enough magnetic disk storage to contain the coadded ISSA and ZOHF data sets as well as a large amount of astronomical catalog data. A hierarchical archive server is being purchased to provide about 200 gigabytes of near-line data storage with essentially unlimited capacity for off-line storage. The hierarchical archive server will manage three tiers of successively slower but higher capacity data storage. Less frequently accessed data will automatically be moved to slower tiers with the ultimate location being off-line storage on 8mm magnetic tape or optical disks.

A large fraction of the archive server capacity will be used for the IRAS Level 1 Archive containing calibrated detector data and pointing data for the entire IRAS mission. The archive server is being specified with growth options of 1 terabyte and more to accommodate future projects.

TABLE I IPAC Image Data Set Storage and Access

Data Set	Now	June 1992
ISSA	mag. disk	Hierarchical archive server using cache, winchester
ZOHF	mag. disk	disk and an optical disk jukebox.
Misc	tape	200 gigabyte near-line storage

DATA ANALYSIS AND DISPLAY TOOLS

IPAC will support analysis of image data by maintaining familiar integrated analysis packages such as IRAF, IDL and MIDAS and by providing a "tool kit" of easy-to-use image processing, analysis and display programs and subroutines for custom applications. The integrated packages are useful for general purpose applications and instrument or site specific applications and offer the advantage that there is no learning time for knowledgeable users.

The "tool kit" approach will provide users with OS level (e.g. unix shell) utilities for science application development. These utilities include image tools for easy image data access, image arithmetic and reprojection, an image display utility (s.X for instance, see below) and simple image processing utilities. Processing is accomplished by assembling the utilities with shell scripts. Canned scripts for common tasks will be provided. User supplied custom utilities, written using subroutine versions of the image tools, could be included when appropriate.

Image display software is planned to include SAOimage and s.X, an X-window version of IPAC's display utility "s". Coming releases of s.X are greatly enhanced over "s" and deserve some description.

S.X is a "smart" image viewer which accepts common astronomical image file formats, handles several geometric projections and coordinate systems, recognizes blank pixels and supports a variety of image measuring commands. By policy, s.X itself does not do any image data manipulation. External processes can be called by s.X to manipulate image data. Currently s.X supports a command line interface and a mouse directed image pointer interface. Enhancements to the s.X interfaces are in development to allow attachment of other command input processors and to improve the ability to attach other processors for image manipulation. The s.X software is intended for export to other sites as well as for internal IPAC use. The standard, low-level X-windows library, X-lib, is used to maintain source level compatibility among various X-window implementations. Eight bit and twenty four bit color frame buffers are supported, if s.X is run with a 24 bit X-server. A new version of s.X incorporating many of the enhancements mentioned above should be released in late January 1992 with the subsequent version planned for mid 1992. Further details on the operation of s.X can be found in Ebert et al. (1991).

The IRAS data archive and the IPAC analysis and display software are supported by a network of file servers and processors which will include the hierarchical archive server mentioned above. Two processors are dedicated to support remote user access to IPAC facilities. User workstations at IPAC are SPARC station 1s, 2s, IPCs and IPXs. One SPARC 1 is equipped with a SUN GS twenty four bit frame buffer. IPAC maintains two twenty eight bit Jupiter color frame buffers attached to SUN 4 processors.

FUTURE PLANS

In the future IPAC intends to support analysis of images and catalogs from the proposed 2 Micron All Sky Survey (2MASS) (Kleinmann 1992), proposal preparation for U.S. observers on the Infrared Space Observatory (ISO) and, eventually, analysis of data from the Space Infrared Telescope Facility (SIRTF). 2MASS support will include data extraction and analysis from J, H and K images, access to about 3 terabytes of image data and catalogs of up to 100 million sources. The ISO proposal preparation software will support instrument specific preparation software and will permit easy access to on-line catalog and image data bases. SIRTF data analysis is planned to include image data from small and large area mapping observations, spectroscopic analysis and production of deep survey catalogs. 2MASS production software is in the early design and testing stages at IPAC. Prototype versions of IPAC's proposal preparation software for the ISO mission will be available for evaluation in the spring of 1992.

REFERENCES

Ebert, R., Hartley, B. & Laughlin, G. 1991, IPAC User's Guide, 6th ed., Section IV, JPL Publication D-2416, Jet Propulsion Laboratory, Pasadena, California

Kleinmann, S. G., et al. 1992, Bull. AAS, in press

Wheelock, S. L., et al. 1992, IRAS Sky Survey Atlas Explanatory Supplement, Jet Propulsion Laboratory Document, in preparation

Wheelock, S. L., et al. 1992, IRAS Sky Survey Atlas Explanatory Supplement, Appendix G, Jet Propulsion Laboratory Document, in preparation

COBE - THE SOFTWARE

E. CHENG
NASA/Goddard Space Flight Center
Laboratory for Astronomy and Solar Physics
Infrared Astrophysics Branch, Code 685
Greenbelt, MD 20771

ABSTRACT The essential scientific products from the *COBE* mission will be a set of all-sky maps spanning four decades in wavelength with a variety of spatial and spectral resolutions. The instruments which generated these data are briefly reviewed and the concepts governing the design of the software system are described. We discuss several notable challenges encountered during development of the software system, and the chosen solutions.

INTRODUCTION

The Cosmic Background Explorer *(COBE)*, NASA's first satellite dedicated to cosmology, will produce a set of all-sky maps spanning the wavelength range from 1 cm to 1 μm. These maps are the legacy of the *COBE* mission, and will provide crucial information on the physical conditions in the early universe, from the time the cosmic microwave background radiation (CMBR) last interacted with matter in the the universe through the time of galaxy formation. In order to extract this information, a variety of local astrophysical sources must be modeled and removed, including emission from interstellar dust, interstellar electrons, and emission and scattering from interplanetary matter.

To accomplish these goals, the final data sets must be as complete and accurate as possible. Detailed attention during instrument design assures that a variety of systematic effects can in fact be measured. The long integration times available from a space platform allow for taking the appropriate systematic test data at many relevant conditions. Ultimately, the ground data system must remove all instrument signatures and systematic effects from the data, perform the necessary calibrations, and produce a set of sky maps which best reflect the physical radiation impinging on the instrument apertures from the sky. The ability to control and measure systematic effects cannot be overstressed, and is the primary reason for much of the complexity in the software. These measurements are notoriously difficult, and historical attempts from other, less optimal, platforms have yielded unexpected and often suspect results (Boughn et al. 1981; Matsumoto et al. 1988).

In this overview, the mission characteristics and instruments are briefly discussed and the design concepts of the ground software system are covered.

Detailed discussion of some of the major software components can be found in related contributions to this conference.

MISSION OVERVIEW

The goal of the *COBE* mission is to obtain quantitative data which provide insight into the physical conditions in the early universe (Gulkis et al. 1990; Mather et al. 1991). The observational quantities which have been chosen for study are the CMBR spectrum, the CMBR spatial distribution, and the Cosmic Infrared Background (CIB).

The keystone of "Big Bang" cosmology is the CMBR spectrum, predicted to be of Planckian form and at a characteristic temperature of roughly 3 Kelvins. The observed origin of this radiation is postulated to be at a surface of last scattering (at a redshift of approximately 10^3), which is when the CMBR ceased to interact strongly with the rest of the universe. Although the Planckian spectrum is well motivated for a universe in thermal equilibrium at early times, it is expected that large scale energy release occurring at later periods (after a redshift of approximately $10^{6.4}$, when the photon number becomes fixed) would modify this distribution, causing deviations from the pure Planckian spectrum. Such deviations have yet to be detected, and would provide important clues to the nature of energy distribution and evolution in the early universe.

Similarly, the spatial distribution of the CMBR provides a probe for any clumpiness in the universe at the earliest times. Such enhanced density structures are believed to be the seeds which evolved into the rich structures seen in the universe today. It is already thought from previous measurements that the CMBR is in fact "too" smooth, and that there has not been sufficient time for gravity to create the observed structures from primordial fluctuations. This observation has motivated the dark matter theories, which propose that various forms of unobserved matter, all possessing the tendency to clump, assist the formation of galaxies by providing centers of gravitational attraction. These theories predict levels of primeval anisotropy of the CMBR which are testable and, if confirmed, would add an interesting component to the forces in the early universe.

Working from the observed universe backwards in time, it is known that there is an overabundance of heavier elements in even the oldest stellar systems. Since the early universe contains mostly hydrogen and helium, it is not likely that the currently observed structures are direct descendants of the clumping of primeval matter. To get around this dilemma, an earlier generation of objects has been proposed. In order to produce the heavier elements, these objects must have produced a significant amount of energy, which is predicted to be observable in the near or far infrared, and is the major component of the CIB.

The approach taken to address these questions is to generate a set of background limited sensitivity, all-sky maps. The spacecraft orbit is configured so that full-sky coverage is complete in six months. Since the wavelength span required is large, three instruments are used, each tailored for one of the topics listed above, with their associated local foreground, calibration, and systematic error issues.

DIRBE

The Diffuse Infrared Background Experiment (DIRBE) covers the wavelength range from 1 to 300 μm in ten bands with a beam size of 0.7°. The optical system is a baffled, off-axis Gregorian telescope. Calibrations are provided by an internal reference source and celestial objects, yielding photometry at the 1% level. The line of sight is 30° off the spin axis of the spacecraft in order to observe the same position at various solar elongation angles.

The major sources of foreground radiation for DIRBE are starlight and scattering from interplanetary dust in the short wavelength bands, and emission from the interplanetary and interstellar dust in the long wavelength bands. Since the radiation from interplanetary dust varies with solar elongation (and thus the season), it provides a time variable foreground which must be removed.

Early scientific results based upon quicklook data processing have been reported by Hauser et al. (1991).

DMR

The Differential Microwave Radiometer (DMR) is a set of three redundant Dicke-switched microwave receivers at 3.3, 5.7 and 9.6 mm wavelength. The field of view is 7°. The instrument measures the flux (temperature) difference in the sky from two beams separated by 60° with the spacecraft spin axis bisecting this angle. Calibration is provided by internal sources, the Moon, and the modulation of the dipole CMBR anisotropy as the Earth moves around the Sun. Sensitivity to fluctuations in sky temperature on large angular scales are at the $\Delta T/T \approx 1 \times 10^{-5}$ level for a one year data set.

Local foregrounds in this wavelength range are dominated by galactic synchrotron and bremsstrahlung radiation. These are modeled using the data from all three bands combined with other measurements.

The extreme precision of these measurements demands that careful attention be paid to contamination by Earthshine, uneven sky coverage, magnetic sensitivities, and a host of other instrumental and spacecraft effects which must be removed by the software. Computational challenges in processing this data include solving for the pixelized sky temperatures (to within a constant offset) given only the difference temperature measurements.

Early scientific results have been reported by Smoot et al. (1991).

FIRAS

The Far Infrared Absolute Spectrophotometer (FIRAS) is a Michelson interferometer which covers the wavelength region from 0.01 to 1 cm with 1% resolution and observes the sky with a 7° beam along the spin axis. Absolute calibration is provided by an external calibrator, which can be commanded into the instrument aperture and substituted for the sky emission. This calibrator presents a Planckian spectrum to the instrument which matches the theoretical spectrum to better than 0.1% (Mather et al. 1990). The expected accuracy of FIRAS spectrum measurements is 0.1% of the peak intensity of the CMBR spectrum.

The dominant foreground for the FIRAS is interstellar dust emission and lines from interstellar gas (Wright et al. 1991). These must be adequately modeled and removed since they can be misinterpreted as deviations from a Planckian spectrum, and are the primary reason that the FIRAS spectral coverage extends to 0.01 cm even though the CMBR flux becomes negligible around 0.05 cm.

Calibration is the key issue in FIRAS data processing, and has been the focus of an intensive effort to understand both the on-orbit calibrations as well as the prelaunch test data. This modeling is now almost complete, and the results will be incorporated into the initial data sets.

Initial scientific results based upon a small data sample were reported by Mather et al. (1990).

Mission Lifetime

The *COBE* was launched on November 18, 1989, beginning a cryogenic operations period during which all instruments were returning science data. The cryogen was depleted on September 21, 1990. After that time, the FIRAS instrument was turned off. The DIRBE continues to operate in the 1 to 5 μm spectral range. The DMR continues full operation since it is not dependent on the cryogen.

Data Volume

The overall data volume for the mission is summarized in Table I. The processing space indicated is that which is dedicated to a particular subsystem for processing, validation and analysis. Comparison should be made carefully since continued operation imposes a certain open-endedness to the resources required for the affected subsystems. The individual subsystems are discussed in more detail below.

TABLE I Data volume.

Subsystem	Raw Bit Rate bits/sec	Raw Volume Gbytes/year	Processing Space Gbytes	Expansion Factor
Spacecraft [a]	837	3.3	15.0	4.5
DIRBE [b]	1713	6.8	30.0	4.4
DMR [c]	216	0.9	15.0	16.7
FIRAS [d]	1330	5.2	12.0	2.3
ADB [e]	–	–	5.0	–

[a]Includes attitude processing for two years.

[b]Partial data (10 out of 16 detectors) from post-cryogen period.

[c]Full operation in the post-cryogen period.

[d]Operation terminated in the post-cryogen period.

[e]Astronomical Data Bases, see Software System Overview.

SOFTWARE SYSTEM OVERVIEW

The major components of the ground software system are illustrated in Figure I. This system is developed and operated in a VAX/VMS environment, the hardware for which is described in a later section.

The raw telemetry flows into *ingest* processes which decommutate and reformat the data for the individual subsystems. Attitude data are also extracted and processed in this first step.

This information is then fed into four *pipelines* which process data for the three instruments plus the spacecraft engineering quantities. The three instrument *pipelines* are covered in separate articles. These programs perform all the necessary computations to convert the incoming data into *Project Data Sets* (PDS's) which are the official products from the mission.

The PDS's contain data in one of two formats. *Sky maps* are stored in an indexed version of the quadrilateralized spherical cube pixelization which is discussed further in the paper by White. Each pixel may contain one or more records with the same field structure. Time series data (*ancillary data*) are stored as sequences of records with the same field structure and are indexed by time tags.

The *sky map* data from *COBE* are merged into a larger collection of data known as the *Astronomical Data Base* (ADB). This is done in order to create a homogeneous set of data, accessed by common tools, which allow for comparison and modeling amongst instruments and using *external data sets.* As a consequence, the *external data sets* are reformatted into the *COBE* pixelization and coordinate system before insertion into the ADB (see the paper by Freedman).

Access to both *ancillary data* (time series) and the ADB is made through a network-based interface called the *data server.* It is through this mechanism that data are retrieved for validation and analysis.

Configuration Control and Testing

Throughout the software development, production and maintenance phases, structured software configuration control procedures are required. The method used for *COBE* has worked well, and is discussed in the paper by Raugh. Regardless of the actual configuration mechanism chosen, the most valuable aspect has been the automated error reporting and tracking system which has been in use for the past six years. This level of automated support and change requirement documentation is essential in large development efforts.

Closely related to configuration control is the testing of released software *(verification).* This is done in a controlled manner by running a new release on benchmark data sets in a predefined series of tests. For each release, the developers can choose to augment the standard suite of tests in order to more fully test modified code.

While this concept of regression testing has merit after a software system is mature, it is not clear that this is the most effective method during the development phase. Typically, during development, requirements and functionality can change dramatically between releases, and the time required to execute regression testing can occupy a significant part of the release cycle. In addition, a significant amount of disk and CPU resources are required for this effort. It may be more beneficial under these circumstances to release partially tested software earlier, allowing for more timely user testing, provided that the coupling between the users and the developers is sufficiently tight. This also assumes, of course, that the development environment is sufficiently partitioned so that individual

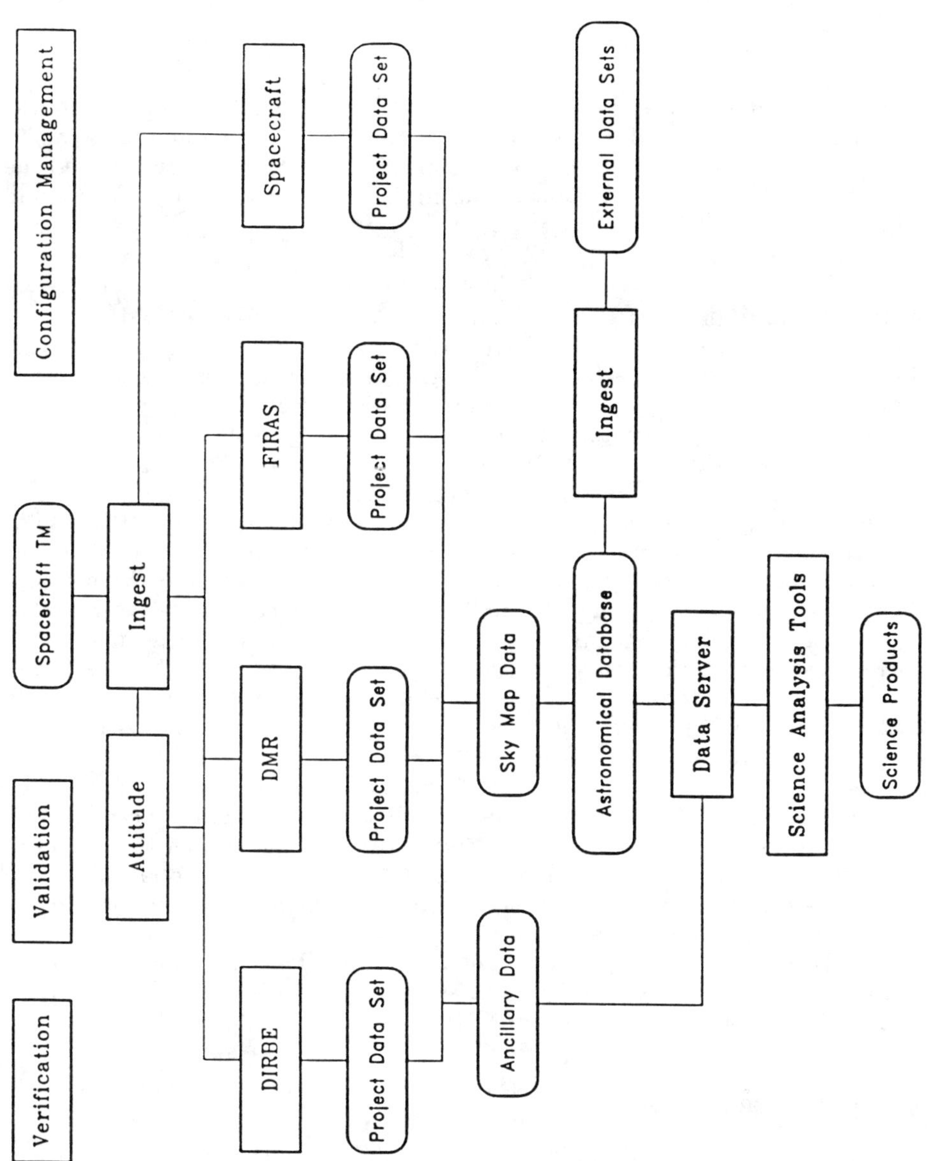

FIGURE I Main data flow for the ground software system.

developers can complete sufficient testing to ensure that there are no egregious errors.

Validation

The process of validation attempts to assure that the software produces the intended results. This will not be discussed further except to note that a combination of end-to-end simulations, benchmark data set testing, brute force code checking and science analysis and modeling is used to check the software and data products.

Core Software

Not shown in Figure I is the *core* software subsystem which provides the environment in which the software system is built. Components include software libraries with common conversion and formatting routines, graphics routines, numerical routines, logical name definitions which create useful development and processing environments, and the database management system.

At the lowest level, all *COBE* data sets are managed by COBETRIEVE, a database management system developed specifically for this purpose. This system is mostly custom built, with occasional usage of features in DEC DATATRIEVE for query functions. The system currently performs quite well, but its development was plagued with difficulties which rippled into the development of the major subsystems. An additional difficulty is that the system relies heavily on the VAX/VMS environment, and is thus not easily portable. With current technology, the best approach may be to adapt or directly use one of the multiple platform database management systems which are commercially available. However, this was not sufficiently clear at the time the *COBE* system was designed over ten years ago.

In order to support multiple platforms in a timely way, we have opted for layering a network interface on top of COBETRIEVE, the *data server*. This software responds to SQL-like requests from the network for either time-series or sky map data, and will invoke the necessary COBETRIEVE functions to return the requested data. Based on a client-server model, the server resides on the main *COBE* computer system with the data sets. Currently supported clients include DEC VAX/VMS platforms and DEC RISC/ULTRIX. These functions can be accessed from high-level languages or interactively using IDL, the most commonly used analysis package for *COBE* (see the paper by Eplee).

PIPELINE STRUCTURE

Integral to the design of the *COBE* software is the concept of pipeline processing. In this context, pipeline processing implies that the final version of the data sets are produced by running a set of programs in a documented and repeatable way. To achieve this end, all programs are invoked with command line qualifiers and script files (setup parameters read by the programs on initialization) which control the details of the processing. However, as software development progressed, it became clear that this interface had become so complex that the programs were inaccessible to the *COBE* scientists who are essential to the validation process. Indeed, with few personnel other than the developers able to operate the system, realtime quicklook processing during the mission also became a potential problem.

A number of approaches were taken to solve this problem. First, experienced scientists on the team developed the detailed knowledge necessary to operate the software. In addition, a menu system was developed which could execute routine processing by simply entering a time range and pressing a button. This menu system was crucial to the success of the realtime quicklook processing which was routine, continuous, and had to be handled by a large number of less experienced personnel. However, since the menu system was rather inflexible for non-routine processing, many scientists have found it to be of limited usefulness. This has resulted in a split in processing philosophies amongst the instrument teams. One has chosen to develop the menu so that it can be used for the final production runs, one has eschewed the menu in favor of knowledgeable people creating and running command files, and one uses elements of both these philosophies. In general, using higher level people to run the software results in faster turnaround, and setting up a system to use less experienced personnel may improve the productivity of the higher level staff. However, when the software system is under development and may not function as intended, it seems that the involvement of more experienced personnel is critical and produces better results overall.

DATA MANAGEMENT

Since the pipeline software produces a great many intermediate data sets, it is not possible to supply sufficient disk space to store all the data at once (estimated to be greater than 350 Gbytes). Consequently, it is necessary to move data between online disk and some offline storage medium. This is the responsibility of the Database Administrators who, using the COBETRIEVE catalogs and a set of data migration tools, respond to requests from the production group and users to online/offline specified data sets. With production well under way, it is clear that this data migration is labor intensive. Various methods are being investigated to automate this process. However, the explosive improvement in processing capability at negligible cost has made it possible to revisit the tradeoff between keeping these intermediate data sets or regenerating them as the need arises.

Currently available offline media include 4 mm tapes, 8 mm tapes (including two jukeboxes) and Sony optical drives. The reliability of our 8 mm tapes has been a problem in the past, with data retention times being no more than two to three years. However, the manufacturer claims that recent developments in tape technology have overcome this problem.

HARDWARE SYSTEM

Figure II illustrates the structure of the *COBE* VAXcluster. It is physically located at the Cosmology Data Analysis Center (CDAC), close to NASA/GSFC in Greenbelt, Maryland. It supports approximately 100 users in the daytime, production work at all hours, and system management and monitoring activities.

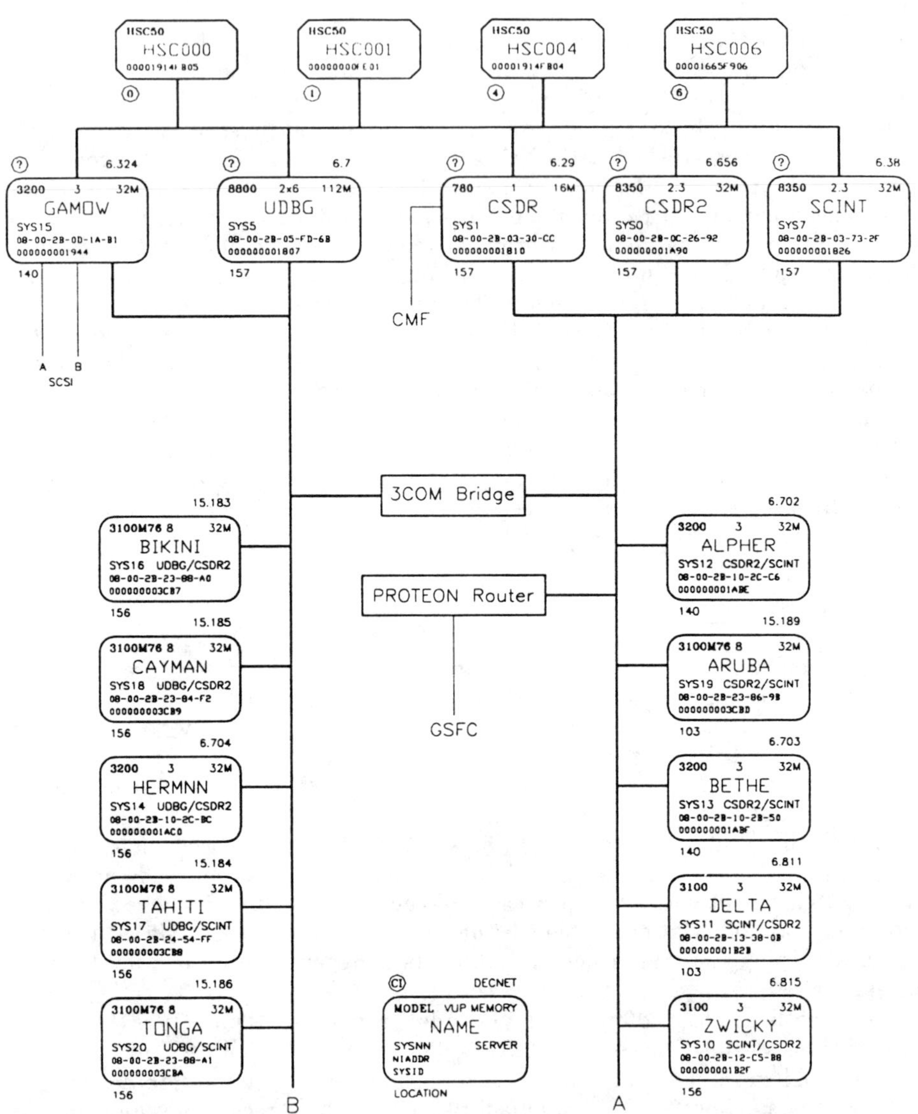

FIGURE II Hardware complement.

The HSC's serve 100 Gbytes of magnetic disk to the cluster, which consists of four mainframes and eleven workstations. Interactive development and analysis work is done on the workstations, which provide approximately 70% of the CPU power in the cluster. The mainframe machines are reserved for disk serving and batch processing. With this configuration, network loading needs to be carefully monitored. In fact, the VAXcluster now spans three Ethernet ribs and provides acceptable performance. The addition of a high speed multiport bridge in the near future will enable a split into four ribs and provide sufficient margin for growth.

With the advent of truly high performance workstations, the I/O demands are also increased and disk serving has become a critical load to all but the most powerful mainframes. In order to address this limitation, we have tested a commercially available Qbus to CI adapter which enables VAXstation 3200's to act as HSC servers to the local area VAXcluster. Initial indications are that this device performs exactly as specified, and we will soon be serving the entire cluster from four VAXstation 3200's (one for each Ethernet rib) which are equipped with this product. This will allow for retiring the 780 and 8350 class machines which have become difficult to maintain.

The availability of RISC workstations has presented both a major opportunity and a problem. The experience base of most of the Project staff is with VAX/VMS systems, yet the benefits of this new resource are extremely alluring. Since the *COBE* science community is already somewhat familiar with UNIX systems, and recognizing that the existing software is too costly to port, we have started a pilot program to use these cost effective machines to do science validation of the data products. The data sets generated by the *pipelines* are available through the *data server*, and can be manipulated using IDL which is already in widespread use on the VAX/VMS platforms. Two DECstation 5000 workstations are currently available in the CDAC and are linked to the VAXcluster using NFS.

SUMMARY

The development phase of the *COBE* ground data system is nearing completion, and the initial production phases have started. The *COBE* Initial Products (IP's) are planned for release to the NSSDC in June 1993, at which time the first Guest Investigators will be arriving at the CDAC. The IP's will contain the galactic plane portions of the DIRBE and FIRAS sky maps, in addition to the first year DMR all-sky maps.

ACKNOWLEDGMENTS

It is a pleasure and a privilege to acknowledge the *COBE* team of computing professionals, many of whom were at this conference, and the dedication which they have shown to turning vague scientific notions and conflicting requirements into a fully functional system which produces real results.

The National Aeronautics and Space Administration/Goddard Space Flight Center (NASA/GSFC) is responsible for the design, development, and operation of the *Cosmic Background Explorer (COBE)* under the guidance of the *COBE*

Science Working Group. GSFC is also responsible for the development of the analysis software and for the production of the mission data sets. The *COBE* program is supported by the Astrophysics Division of NASA's Office of Space Science and Applications.

REFERENCES

Boughn, S., et al. 1981, ApJ, 243, L113

Gulkis, S., et al. January 1990, Scientific American, 132

Hauser, M. G., et al. 1991, After the First Three Minutes, AIP Conf. Proc. 222, College Park (Holt, S. S., et al. eds.), Amer. Inst. of Phys., New York

Mather, J. C., et al. 1990, ApJ, 354, L37

Mather, J. C., et al. 1991, Adv. Space Res., 11, (2)181

Matsumoto, T., et al. 1988, ApJ, 329, 567

Smoot, G. F., et al. 1991, ApJ, 371, L1

Wright, E. L., et al. 1991, ApJ, 381, 200

THE QUADRILATERALIZED SPHERICAL CUBE AND QUAD-TREE FOR ALL SKY DATA

RICHARD A. WHITE(1) AND SALLY W. STEMWEDEL(2)
(1) NASA/Goddard Space Flight Center, Code 936,Greenbelt, MD 20771
(2) Applied Research Corp., 8201 Corporate Dr., Landover, MD 20785

ABSTRACT The quadrilateralized spherical cube (Chan & O'Neill 1975 and O'Neill & Laubscher 1976) is a geometrical projection well suited for mapping all sky data, as well as to the use of the quad-tree nearest neighbor (rather than rasterized) storage scheme for the the archiving and retrieval of data. The celestial sphere is projected onto the six faces of a cube in a tangent plane projection. The lines of latitude and longitude on each face are then curved such that, when the cube face is divided into equally spaced rows and columns of pixels, each pixel is equal in area on the sky to every other pixel. The maps thus preserve photometric integrity. The changes in lines of latitude and longitude result in only minimal distortion of the shape of objects on a single face. The quad-tree pixel numbering scheme, being hierarchical, permits arbitrarily high resolution and allows comparison of maps at differing resolutions in a particularly elegant and straight forward manner.

We present the basic geometry of the quadrilateralized spherical cube, the form of the quad-tree structure, and suggest FITS structures and keywords for use in displaying, storing, and transporting data in this form.

THE QUADRILATERIZED SPHERICAL CUBE

The sphere of the sky (or of the earth) is projected onto the 6 faces of a cube. (Figure I) Initially the sphere is projected onto the cube in a tangent plane projection. However, the tangent plane projection results in pixels at the edges and corners representing smaller areas on the sky than those in the centers of the cube faces. The quad-sphere approach is to distort the tangent plane projection such that equally spaced pixels of equal area on the cube face also represent equal areas on the sky. The cube faces can now be rasterized and written easily in standard FITS image format. While all pixels now are of equal area, they have a variety of shapes when projected back on the sky. Thus beam shape or point response functions will differ, depending on location in the quad-cube face, but of course such problems exist in any flat representation of a curved surface.

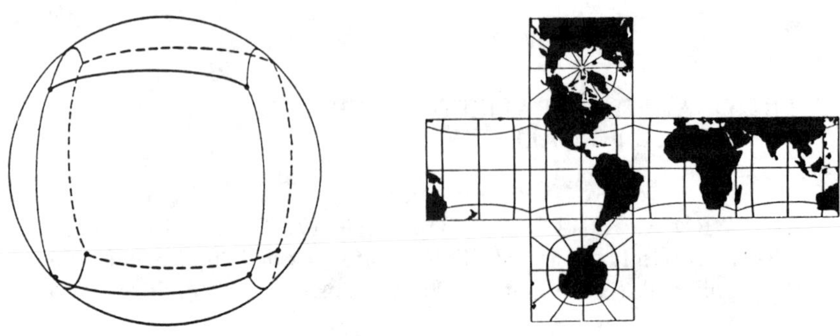

FIGURE I The sphere is subdivided into six equal sections. The earth is projected onto the 6 cube faces.

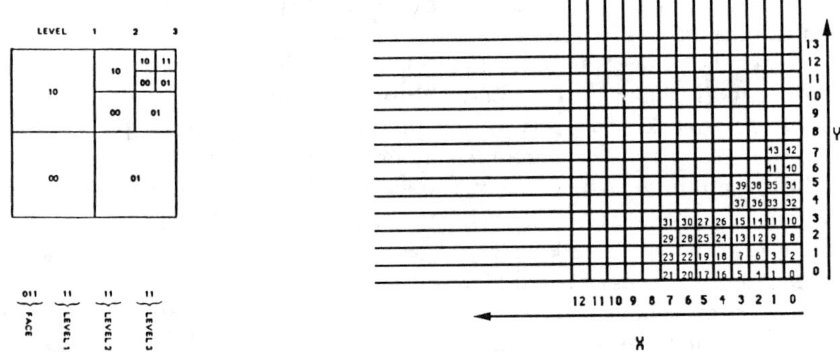

FIGURE II The quad tree numbering scheme and its relation to x, y positions.

THE QUAD-TREE PIXEL NUMBERING AND STORAGE STRUCTURE

The quad-tree (basically a binary tree) is an alternative approach to the typical row by row rasterization; it numbers pixels and stores data in a nearest neighbor sequence, based on binary numbers. It, therefore, has a binary hierarchical capability; each pixel can be subdivided into four smaller pixels, and those four each into four more to any arbitrary level desired. Each time the pixel number is changed by adding two binary bits. In the left of Figure II the binary number shown is the pixel number for top right pixel. In using the quadrilateralized cube, each face is numbered (from 0 to 5). To illustrate the ease of conversion from quad-tree pixel number to x, y rasterized coordinates, pixel 24 (in the right of Figure II) as written in binary is 011000. Assigning alternate binary digits to y and x gives y=010 and x=100; converting to decimal shows the rasterized position of the pixel to be (4,2), where counting starts with 0.

COBE has adopted both the quadrilateralized spherical cube geometric projection and the quad-tree pixelization scheme as its fundamental system. The DIRBE data use cube faces 256 by 256, giving pixels of about .35 degrees, appropriate to the intrinsic resolution of the DIRBE instrument. FIRAS and DMR each use cube faces 32 by 32 (3 levels coarser than DIRBE). (White & Mather 1991)

PROPOSED FITS STRUCTURES FOR QUAD-SPHERE AND QUAD-TREE

For images, the six faces of the cube would be stacked into a data cube. The first dimension of the cube would be X, the second Y, and the third would be labeled FACE. Relevant header entries for COBE/DIRBE include:

```
NAXIS1  =                    256 / 256 pixels on x axis of each face
NAXIS2  =                    256 / 256 pixels on y axis of each face
NAXIS3  =                      6 / 6 faces from the cube are present
CTYPE1  = 'ELON--QS'             / x axis is ecliptic longitude, quad-sphere
CTYPE2  = 'ELAT--QS'             / y axis is ecliptic latitude, quad-sphere
CTYPE3  = 'FACE--QS'             / z axis is face number of the quad-sphere
```

The standard formulae given in the referenced texts could be integrated into any of the standard image processing systems to allow conversion from x, y and face to sky coordinates. This version of FITS format makes no use of the quad-tree pixel numbers, although such numbers could of course be constructed from the face, x, and y coordinates. Alternatively, a companion image could also be added with the BUNIT being the quad-tree pixel number.

For data where the quad-tree ordering is important, and where the data associated with each pixel may be many and with different physical units, the BINTABLE extension would be ideal. The header would include, TFORMn, TTYPEn, and TUNITn, etc. for each field contained in the table. The first column would give the quad-tree pixel number, with successive columns providing all the data associated with that pixel.

```
TFORM1  = '1J       '  /  Field 1 is a 4-byte integer
TTYPE1  = 'QSPIXEL  '  /  Field 1 is a quadsphere pixel number
TUNIT1  = '         '  /  Pixel numbers are unitless, and begin with 0
```

REFERENCES

Chan, F. K., & O'Neill 1975, Feasibility Study of a Quadrilateralized Spherical Cube Earth Data Base, EPRF Tech. Rep., 2-75 (CSC)

O'Neill, I. M., & Laubscher, R. E. 1976, Extended Studies of a Quadrilateralized Spherical Cube Earth Data Base, NEPRF Tech. Rep., 3-76 (CSC)

White, R. A., & Mather, J. C. 1991, "Databases from Cosmic Background Explorer (COBE)", in Databases and On-Line Data in Astronomy, ed. M. Albrecht & D.Egret, p.29-34, Kluwer, Dordrecht, the Netherlands

COBE DIFFERENTIAL MICROWAVE RADIOMETER (DMR) DATA PROCESSING TECHNIQUES

P.D. JACKSON[1], G.F. SMOOT[2], C.L. BENNETT[3], J. AYMON[2], C. BACKUS[1], G. DE AMICI[2], G. HINSHAW[4], P. B. KEEGSTRA[1], A. KOGUT[4], C. LINEWEAVER[2], L.A. ROKKE[1], AND L. TENORIO[2]

ABSTRACT The Differential Microwave Radiometers (DMR) experiment measures temperature differences in the microwave sky at frequencies of 31.5, 53, and 90 GHz. We discuss the software system used to calibrate and invert these difference measurements to produce full-sky maps of the large-scale distribution of the cosmic microwave background. Tests indicate that any spurious features introduced by the data processing are negligible compared to the instrument noise.

INTRODUCTION

The DMR instrument aboard NASA's Cosmic Background Explorer (*COBE*) measures the large-scale temperature distribution of the microwave sky. We model the sky as 6144 independent pixels with fixed amplitude and employ a software system to convert the measurements of temperature differences to maps of the distribution of temperature over the full sky. Torres et al. (1990) describe the software system in place near the Nov. 1989 launch. Janssen & Gulkis (1992) provide further detail on the techniques used to invert the difference measurements to produce a sky map. In this paper, we update the current status of the software and discuss on the effects on the sky maps of signals that are not well modelled by fixed sources of constant amplitude.

DMR DATA PROCESSING

The DMR data consist of a series of measurements of the temperature difference between two 7° portions of sky separated by 60° (Smoot et al. 1990). The spacecraft spin and orbit combine to sample half the sky every orbit (103 minutes), with full sky coverage every six months. We test the data quality and

[1]Hughes STX Corp. 4400 Forbes Blvd., Lanham, MD 20706

[2]Lawrence Berkeley Lab. and Space Sciences Lab. MS 50-232, UC Berkeley, CA 94720

[3]Lab. for Astronomy and Solar Physics, NASA/GSFC, Code 685, Greenbelt, MD 20771

[4]Universities Space Research Assoc. Code 610.3, NASA/GSFC, Greenbelt, MD 20771

reject data known or suspected to be contaminated (see the discussion by P. B. Keegstra et al. in this volume).

We remove the instrumental offset by fitting a slowly-varying baseline to the uncalibrated data. We need to evaluate any baseline fitting procedure with great care since we hope to recover sky information that is less than one thousandth the rms of the raw data, . As a cross-check, baselines are routinely fitted three different ways: a cubic spline, 12th order Legendre polynomial, and a running mean. Calibration is obtained using internal noise sources and external astronomical phenomena (Bennett et al. 1992).

We correct the calibrated data for known instrumental and systematic effects. Current corrections include the Doppler effect from the *COBE* velocity about the Earth and the Earth's velocity about the solar system barycenter; the magnetic susceptibility of the instrument; emission from the Moon picked up in the antenna sidelobes and emission from the planet Jupiter.

The calibrated, corrected data are pixelized according to the directions of the two antennas using the quad-cube representation (see discussion by White and Stemwedel in this volume) in which 6144 pixels of 6.7 square degrees cover the entire sky. In one year, each DMR channel provides more than 60 million differences between pairs of pixels, providing a highly redundant set of equations in pairs of the 6144 pixel temperatures. The redundancy is removed by forming the chi-squared sum of the measurements involving each pixel, and forcing the derivative of the chi-square with respect to the pixel temperature to be zero. This yields 6144 equations in the pixel temperatures, which we call normal equations.

The coefficients of the normal equations form a 6144 x 6144 matrix of the pixel temperatures. The constraint of 60° separation between the observed pair of pixels ensures that this matrix will be mostly empty, hence the term 'sparse matrix' is used. In principle, the pixel temperatures can be obtained by inverting the sparse matrix, but the nature and size of the matrix favors an iterative type solution. Starting from an initial (zero) model, solutions are iterated using a Gauss-Seidel procedure whereby the new solutions computed during an iteration are used in later computations within the same iteration. The convergence is good: by comparing the results of the Gauss-Seidel technique with an exact solution (both solutions obtained at lowered angular resolution, 384 pixels), we find that the iterative solution agrees with the exact solution to $\Delta T/T < 10^{-8}$.

SIMULATION TESTS

The DMR software models the sky as a collection of fixed sources of constant amplitude. We use simulations to test the software response to input signals from sources that vary in time or position. These simulations use the actual DMR antenna pointing directions to create a time-ordered series of differences from an input model sky map, to which we add random noise and various non-random systematic effects. We then run the DMR mapping software over the simulated data set and compare the output to the input sky map. For ideal data (5 mK noise, no systematic effects), the DMR software correctly converges to the input sky pattern to precision $\Delta T/T = 7 \times 10^{-7}$.

Signals inconsistent with fixed celestial sources create a complicated pattern in the maps. Figure I shows the simplest case, a spike added to a single datum,

FIGURE I A full sky map showing the effect of a 100 K spike added to a single datum in a 16-day simulation. The largest effect is 0.1 K in the pixel observed when the spike occurred (center pixel); all other structure is < 0.01 K.

which can be generalized to signals with more complicated time dependence. The output map shows a convoluted pattern characterized by a series of intersecting positive and negative temperature rings of 60° radius. The amplitude and shape of the pattern are determined by the exact sky coverage; as expected, the amplitude of the pattern is diluted by successive unperturbed measurements of the same pixels. The amplitude in a one-year map scales to $<< 10^{-6}$ of the spike amplitude, implying a limit $\Delta T/T << 10^{-7}$ in the DMR maps for the effects of a single spike just under the 5σ rejection threshold. In summary, we have developed a software system to invert DMR difference measurements to produce full-sky maps of the microwave temperature distribution. Tests indicate that uncertainties introduced by the data processing contribute negligibly to the large-scale features in the maps.

ACKNOWLEDGMENTS

COBE is managed by the Goddard Space Flight Center and sponsored by NASA's Astrophysics Division, Office of Space Science and Applications, Washington, DC.

REFERENCES

Bennett, C. L., et al. 1992, ApJ, in press

Janssen, M. A., & Gulkis, S. 1992, in Proc. Les Houches, Infrared and Submillimetre Sky After COBE, in press

Smoot, G. F., et al. 1990, ApJ, 360, 685

Torres, S., et al. 1990, in Data Analysis in Astronomy III, ed. V. Di Gesu et al., (New York:Plenum), p. 319

DISPLAY AND MANIPULATION OF COBE/DIRBE ALL-SKY DATA

G.N. TOLLER (1), M.G. HAUSER, T. KELSALL, R.F. SILVERBERG (2), H.T. FREUDENREICH (3), P.M. MITRA (1), A.R. PANITZ (3), F.S. PATT (1), J.A.J. SKARD (3), W.J. SPIESMAN (1), S.W. STEMWEDEL (4), J.L. WEILAND (1), R.J. HOLLENHORST (2)
(1) General Sciences Corporation, Laurel, Md., 20707
(2) NASA/Goddard Space Flight Center, Greenbelt, Md., 20771
(3) Hughes STX Corporation, Lanham, Md., 20706
(4) Applied Research Corporation, Landover, Md., 20785

ABSTRACT The organization, display, and manipulation of all-sky data from the Diffuse Infrared Background Experiment (DIRBE), a 10 band photometer on the Cosmic Background Explorer (COBE) mission is depicted. The information carried along with each pixel-based map record is described. The sky data are projected onto the 6 faces of a cube and stored in an array. We discuss the tools used to retrieve, manipulate, and display the map data. Finally, the problems encountered and lessons learned in building this large astronomical software system are presented.

The DIRBE instrument on the COBE mission is a 16 channel, 10 spectral band absolute photometer, spanning 1 - 300 μm. Launched Nov. 1989, DIRBE is still operating in the four shortest wavelength bands (1 - 5 μm). The DIRBE measures sky brightness and seeks to discriminate and characterize each brightness component. Approximately half the sky is viewed each day. The DIRBE instrument and program have been described by Hauser et al. (1991).

The time-ordered data flow through a VAX/VMS Fortran "pipeline" (Skard et al. 1992). Quality-checking of the instrument state, spacecraft attitude, detector noise, and glitch rate (Freudenreich et al. 1992) occurs. Suspect data are flagged or rejected. An instrumental offset is subtracted from each detector sample and data are rectified to a fixed electronic gain. Data are rectified to a fiducial detector temperature, or rejected if the temperature exceeds specified limits. Short-term (.25 to 12 hours) variations in detector responsivity are removed using an internal calibrator.

Using aspect data from the attitude solution files in the ancillary archives, the data are converted into sky maps. Daily maps contain a separate record for each observation of a pixel. Weekly maps contain data averaged per pixel over a week. Standard map fields are: pixel number, time of observation, photometry values, scan direction, number of observations per pixel, variances, various sky angles, scan rate, satellite position, and data quality indicators. Map trending using AIPS (Freedman et al. 1992) assists in quality control. Supplementing the standard "pipeline" map production programs, other software is used for specific analyses and as a check or forerunner of the standard products. This software can create maps from any of the time-ordered datasets, re-pixelize existing maps, or compare map data over time or as a function of transit angle.

In the S/W pipeline, map data are stored in quadtree format (White & Stemwedel 1992; Chan & O'Neill 1975; O'Neill & Laubscher 1976). Data are projected onto the quadrilateralized spherical cube . The projection is curvilinear and preserves area. DIRBE pixel resolution is 256x256 pixels per cube face, each pixel approximately 0.32 degree on a side. The coordinate system for map data storage is Geocentric Ecliptic J2000.

The Fortran capabilities available for access and manipulation of sky maps include conversion from coordinate to pixel number or vice versa for user-specified map resolution, interpolation or extrapolation of intensities to a user-specified solar elongation, and mathematical operations (sum, difference, product, quotient, linear combination, exponential, logarithm, color indices). Customization is possible with user-specified selections and operations on fields in the input map, or user-specified record structure for input or output files. Robust averaging can be applied to multiple records, records under a pixel can be treated separately by scan direction or merged, and output can be optionally produced in FITS format. Available projections are Aitoff, Global Sinusoidal, or QuadCube. Coordinate systems are Ecliptic, Equatorial, Galactic, Galactic (l=180 centered), or user-specified center and pole. Data quality flags exist for each observation; values in the map can be replaced with user-defined sentinels. Among other capabilities are the selection of specified information and/or specified regions, sky coverage and scan pattern displays, and human-readable dumps of specified map records. Figure 1 displays one week of DIRBE data.

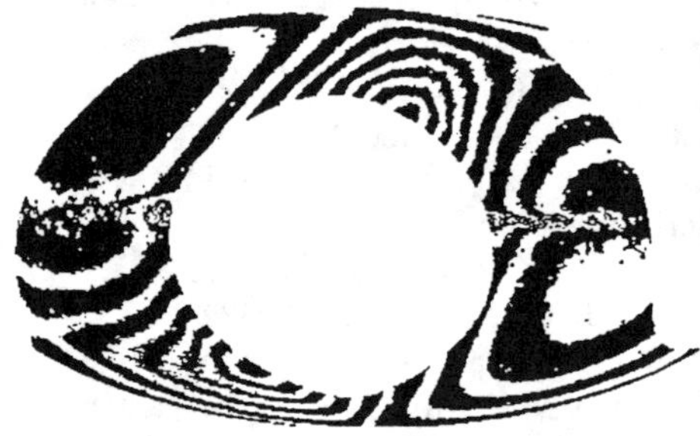

Figure 1. 12 micron image of one week's data in a Galactic Aitoff projection. Galactic and ecliptic plane emission dominate the sky brightness. Produced using a Mac connected to a VAX through public domain NCSA Software.

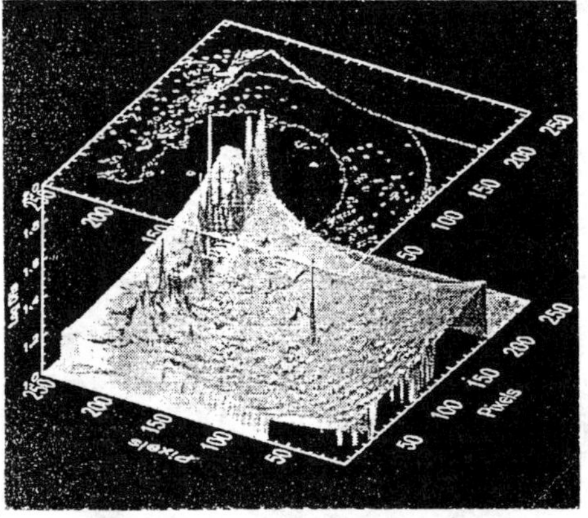

Figure 2. 12 micron surface plot of one week's data. The ridge on the left is the galactic plane.

In addition to Fortran tools, a library of IDL version 2 tools is available to enable access and manipulation of all-sky data. In the time-ordered domain, IDL tools read the data archive, extract science data, and retrieve time strings of user-specified map quantities. To image map data, IDL tools read and display user-specified fields, reproject and subset data, repixelize data, select color or monochrome, adjust the image scale and scaling method, set subimage by cursor, zoom, rotate the image, apply a filter for source removal, perform image or edge enhancement, display 1-D profile of array row, column, or cursor-specified slice, display and overlay a contour plot, display a 3-D surface plot, compute statistics for cursor-specified regions or portions of an array, overlay coordinate grids, blink images, and read cursor position and intensity from the screen. IDL tools are used for conversion among equatorial, ecliptic, galactic, pixel number, and rasterized image coordinates; conversion among time formats; precession or ephemeris calculations; producing FITS format I/O routines; 1-D and 2-D function fitting; residual calculation and display; linear combinations and map ratios; and comparison with selected Astronomical Data Bases. Figure 2 displays an IDL-based representation of DIRBE data.

Several problems were encountered which influence the data analysis. Source intensity values vary from observation to observation due to pixelization, scan coverage, and beam profile effects. Therefore, photometry on discrete objects will require mission integrated maps. Occasional spin rate: orbital period phase coherence causes variations in the mean time of observation across a map. This, coupled with time variations in the brightness of the zodiacal emission foreground, produces "brushstrokes" in the weekly maps. Pixelization noise and the statistical nature of DIRBE photometry also impact analyses (Mitchell et al. 1992). Finer pixel resolution (superpixelization) is needed for beam profile determination, corrections to the pointing of individual detector channels, and analyses of regions dominated by point sources or large gradients. We also encountered difficulties with new releases of imaging S/W not always being upward compatible.

Many lessons were learned in the development of this large S/W system. First, use canned, supported S/W whenever possible. Pick a well developed, flexible image processing system. Organize analysis tools in a common area and supply help files. Real data always stress the S/W in unexpected ways. S/W requirements should be developed in cooperation with the end-user. Lastly, scratch space is needed in large quantities.

This work is supported by the Astrophysics Division of the Office of Space Science and Applications at NASA Headquarters.

REFERENCES

Hauser, M. G., et al. 1991, AIP Conf. Proceedings, 222, eds. Holt, Bennett, & Trimble, AIP, NY, 161

Skard, J. A. J., et al. 1992, this volume

Freudenreich, H. T., Tewari, K. P., Kelsall, T., & Odegard, N.P. 1992, this volume

Freedman, I., et al. 1992, this volume

White, R. A., & Stemwedel, S. W. 1992, this volume

Chan, F. K., & O'Neill, E. M. 1975, Feasibility Study of a Quadrilateralized Spherical Cube Earth Data Base, CSC, EPRF Tech. Report, 2-75

O'Neill, E. M., & Laubscher, R. E. 1976, Extended Studies of a Quadrilateralized Spherical Cube Earth Data Base, CSC, NEPRF Tech. Report, 3-76

Mitchell, K. J., et al. 1992, this volume

COBE/DIRBE SKY MAP TRENDING

I. FREEDMAN, J.A.J. SKARD
Hughes STX Corporation, Suite 100, 7601 Ora-Glen Drive, Greenbelt MD 20770

T. KELSALL
Code 685, NASA Goddard Space Flight Center, Greenbelt MD 20771

CHALLENGE

The Diffuse InfraRed Background Experiment (DIRBE) is a liquid-Helium-cooled 10-band photometer covering the wavelength range 1 to 300 microns. It was flown on the Cosmic Background Explorer (COBE) satellite, and the main mission covered the time from 21 November 1989 to 21 September 1990. The goal of the DIRBE program is to do a deep all-sky survey looking for the remnant radiations from the epoch of galaxy-formation.

The "pipeline" data processing software produces a goodly number of plots, tables and text reports based on the processing of the time-ordered data. While these products reveal a great deal about the performance of the instrument, the results are not easily interpreted in terms of photometric results projected onto the sky. How to do sky-map projections efficiently while requiring a modest amount of CPU time and on-line storage to produce useful human-readable review results had to be solved.

REPLY

Trend the sky maps by visual comparison of successive grey-scale and contour plots observed on time-scales from 1 day to full mission.

High-resolution quality assurance images are produced for DIRBE sky maps at several stages of processing in several coordinates and projections. A wide dynamic range of large-scale features (e.g. time-variable Zodi light, static Galactic cirrus and Galactic plane) is displayed. Bright stars are de-emphasized together with noise and glitches.The display is autoscaled to accommodate a wide change of detector responsivity under cold operation (~1.5K) to warm operation (~50K after the helium ran out).

The processing is rapid enough so as not to disturb the data processing from all three COBE instruments and small demands are made on disk storage and CPU while retaining flexibility.

FIGURE I Grey-scale plot for annual average intensity of 100 μm band

SOLUTION

Constrained to use the AIPS image processing system, we determined the trade-offs between plot quality and speed. Figure I shows a typical plot.

Results indicated each plot should contain 512 by 256 pixels (1/4 full resolution). Grey-scaling is provided by a bit-mapped character font generated by the STARLINK algorithm using a sine-wave mask with 13 dots-per-pixel on both axes. Contours are provided at 1/16 full resolution using bilinear interpolation for smoothing and compact plot files. Outliers are trimmed with plot-specific static flux windows to exclude bright stars, noise, glitches and "blanking"values. Relative flux limits are found by scaling the dynamic flux windows by the average intensity inside the static windows. Flux limits are determined for large-scale feature enhancement from the map data distribution function. Plot-specific image-to-printer transfer functions are chosen to maximize the information content of the prints. Standards of comparison are set via fixed contrast stretch parameters. All projections of a plot share the same contrast stretch.

IMPLEMENTATION

The map-generation software has a sound and flexible structure, based on menu-driven command procedures, supporting batch or interactive processing. FITS files created via the sky map manipulation tool (Toller et al. 1992) are used to create plots from reliable, familiar AIPS procedures. Temporary files transfer parameters from the VMS-DCL procedures to AIPS. Contrast-stretch parameter files are run-time-selectable AIPS procedures.

A variety of free parameters define selectable detector bands, linear or logarithmic grey-scale wedges, arithmetic or geometric contour spacing and the choice of autoscaling method.

The plot labelling is sufficient to describe the product - sky map name, band names from time-dependent multiplexed detector sampling sequence, plot type (intensity, approach-vector difference, observational coverage), coordinate system (Ecliptic, Galactic), projection (Aitoff, Unfolded Cube, Stacked Cube) and time coverage via Julian start and stop times. Labelling difficulties are resolved by regarding each image as an overlay of contour or grey-scale plots over a blank image and adjusting the headers of both images.

LESSONS LEARNED

Though the map trending software is operational, it was not completed via a trouble-free route and we learned that AIPS is difficult to use for this application. AIPS startup command files need modification for access from private and operational accounts. AIPS "memory" files need **write** permission and we chose to copy them at run-time from a protected area. Source-size limitations initially caused some data-driven failures for large plotting requests which we resolved by tightly packing the code. Multiple concurrent AIPS jobs with the same user number are permitted by adjusting process names at run-time. Cryptic error messages from AIPS inhibit testing and the learning curve was very steep even with 1 man-year of development. 500 lines of AIPS script yield more than 35 000 lines of log file which set limits to the speed of debugging as did the turnaround time of 2 hours for a complete standard set of 16 plots.

PERFORMANCE

The production of the plots has been very satisfactory and does not require undue resources so as to disturb the processing of the data from all three COBE instruments. Execution is fast - 50 minutes wallclock time for 16 bands on a shared VAX 8820. Plot files are small - 160 000 block sky maps are reduced to a 325 block printer driver. Printing is rapid - 4 minutes per plot on a Talaris 800 QMS printer.

ACKNOWLEDGMENTS

This research is supported by the Astrophysics Division of the Office of Space Science and Applications at NASA Headquarters.

REFERENCES

Toller, G., et al. 1992, this volume

SECTION 9

SPECTRAL ANALYSIS METHODS

THE DEVELOPMENT OF AN IMAGING FABRY-PEROT REDUCTION SYSTEM WITHIN IRAF

JONATHAN BLAND-HAWTHORN AND PATRICK SHOPBELL
Rice University, Houston, TX 77251-1892

GERALD CECIL
University of North Carolina, Chapel Hill, NC 27514

INTRODUCTION

Powerful imaging Fabry-Perot interferometers are now available for use at observing sites in Australia, the Canary Islands, Chile and the United States. In a single night, these instruments permit accurate spectrophotometry in a million pixel positions at a resolution limited only by the seeing. Most observational astronomers have found the enormity of the data reduction problem too daunting, not least because of the lack of a suitable software environment. In certain respects, there is some overlap with techniques used by radio astronomers who have developed the `AIPS` environment for handling their vast data problem. In practice, the Fabry-Perot problems are specific enough that little use can be made of `AIPS` until very late in the process. It is commonly believed that Fabry-Perots would find more widespread use within astronomy if generalized data reduction techniques and associated documentation were widely available.

For the past six years, we have developed a wide range of software methods for analyzing Fabry-Perot data within the `Zodiac` environment at the University of Hawaii. A collaboration is now under way between Rice University and Kitt Peak National Observatory to transport these techniques to `IRAF`. The data analysis involves a suite of 3D transformations operating on large 2D sets of periodic spectra where the objective is to reconstruct spectral line profiles with photometric quality at each pixel position. Here we address basic issues that hamper our attempts to generalize Fabry-Perot analysis. We outline the primary stages of data calibration and reduction, and summarize some of the analytic techniques unique to imaging Fabry-Perot spectrophotometry. (See also Shopbell et al. 1992; Cecil 1990.)

THE GENERAL PROBLEM

It is particularly difficult to generalize Fabry-Perot data analysis since even small departures from the most primitive optical system can lead to radically different software requirements. Our reduction path assumes the simplest possible optical arrangement that is needed for spectrophotometry with an imaging Fabry-Perot system. The heart of the Fabry-Perot is the etalon which comprises two parallel plates of glass with highly reflective coatings applied to the inner surfaces. The etalon performs as a tuneable filter within a restricted wavelength range which

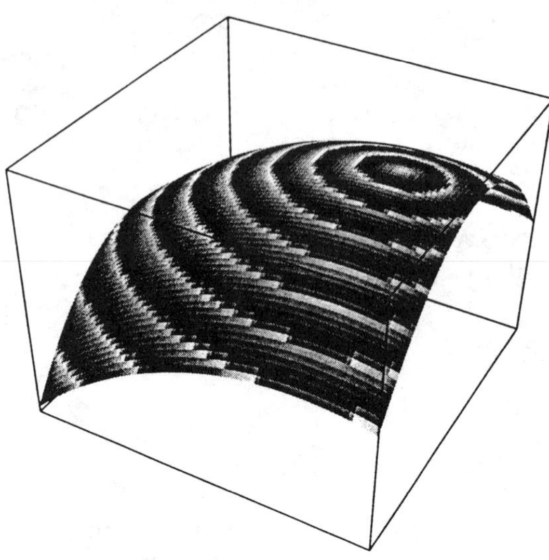

FIGURE I The monochromatic Airy surface: the horizontal plane defines the spatial axes and the concentric rings arise from discrete, consecutive etalon gap spacings.

is normally selected by a fixed narrowband filter in the optical path. Nowadays, the etalon is tuned by varying the plate separation. We obtain a series of images at different wavelengths by making an observation at each unique spacing: in this way, a spectrum is built up at each pixel of the detector.

In the basic system, a Fabry-Perot etalon is placed within the collimated beam of a standard focal reducer arrangement (Bland & Tully 1989, hereafter BT). Experience shows that the amount of analysis in software is strongly correlated with how well the system has been optimized to achieve a particular observation. Crudely speaking, each departure from the optimal system doubles the workload in software. By way of example, the narrowband filter normally placed at the entrance aperture is usually chosen to be roughly half the free spectral range of the etalon. In this way, we restrict interorder confusion between different emission lines from the neighboring orders. However, at many institutions, well optimized filters do not exist and the various spectral features need to be sorted in software. Another example is the choice of imaging detector. Instrumental distortion commonly observed, say, in photomultipliers greatly complicates the data analysis. While CCDs have good linear properties, some are particularly susceptible to cosmic ray events or suffer from cosmetic defects. Suffice it to say, an optimal system ideally requires that the imaging detector is both photometrically and spatially linear.

In even the simplest Fabry-Perot arrangement, there exist at least ten reflecting surfaces (e.g., CCD window, entrance filter). Internal reflections lead to spurious "ghost images" at the detector which are easily confused with real spectral features! It is important to realize that ghost images haunt all Fabry-Perot devices, even with proper application of anti-reflective coatings (BT). The first-order reflections can be deflected out of the field of view by tilting the etalon with respect to the optical axis. This has an important consequence since now the center of the Airy ring pattern is shifted such that a large part of the Airy surface is no longer imaged at the detector, as illustrated in Figure I. This simple adjustment complicates several stages of the basic reduction path. Because tilting the etalon is always recommended in photometric work, we consider this part of the "general problem".

The position of the (shifted) optical axis with respect to the imaging detector *and the sky* should be preserved as carefully as possible during the complete observation of a particular celestial object. If the optical system is not sufficiently rigid or the telescope is allowed to drift between exposures, the different wavelength frames have to be corrected to the same effective optical axis. Quite apart from requiring an additional reduction step, the induced non-linear shift often requires that the *entire* data cube is resampled to a linear grid.

A related problem arises from temperature drifts in Fabry-Perot systems that are not thermally controlled. All of the etalon parts are essentially cylindrical optical flats. For each component, the heat transfer as a function of position is given by the diffusion equation. This equation can be solved to determine the different time constants for the four basic parts of the etalon arrangement. Typically, these vary from roughly five minutes for the gap capacitors to about three hours for the outer plates. Inter alia, the thermal drifts induce changes in the optical gap. While these can be monitored during the course of the observations, they induce non-linear effects into the spectral sampling that, once again, require the data be resampled to a linear grid.

THE REDUCTION PATH

In the last section, we attempted to show the intimate relationship between the optical arrangement and the software requirement. There are several other important considerations that have been discussed elsewhere (BT). In this section, we outline the most rudimentary reduction path that assumes a primitive Fabry-Perot arrangement. In order to remove the various artefacts from the data, it is essential to observe various calibration sources. The following data structures are required by the calibration process:

(i) monochromatic light **(m)** - we observe two or more emission lines through at least two orders which allows for phase and absolute wavelength calibration;
(ii) white continuum light **(w)** - this maps the filter *and* the detector response as a function of position and wavelength;
(iii) standard star **(s)** - this provides the photometric standard and maps most of the ghost reflections.
(iv) monochromatic rings **(r)** - these are taken sporadically and monitor the instrumental stability during the night;

A complete description of the calibration data is given in BT. The main stages of the reduction path are enumerated below (the relevant calibration data are indicated in brackets):

1. calibration of optical system constants **(m)**;
2. cosmetic (glitch, bad column, ghost) corrections;
3. spatial and spectral drift correction **(r)**;

4. monochromatic (phase) correction (**m**);
5. whitelight (flatfield) correction (**w**);
6. photometric correction (**s**);
7. line profile parametrization and fitting;
8. data visualization.

The IRAF Fabry-Perot package currently under development will tackle each of these stages as stand-alone tasks. Stages 1, 3 and 4 require that the properties of the incomplete Airy surface be understood to high accuracy. In the next section, we broach some of the analytic techniques that are used for this purpose. The overlap with interferometric analysis in other wavebands occurs in the last two stages although even then, preexisting AIPS algorithms, for example, would need to undergo substantial modification. These steps are closely related in that meaningful parametrization requires substantial familiarity with the data. They constitute the major bottleneck in the reduction process although recent advances in imaging workstation accelerator boards may help to alleviate the problem.

NUMERICAL TECHNIQUES

The general form. For an input beam with intensity I_0, the instrumental profile is given by the Airy function,

$$I(\theta, l) = \frac{I_0}{1 + \frac{4\mathcal{R}}{(1-\mathcal{R})^2} \sin^2\left(\frac{2\pi\mu l \cos\theta}{\lambda}\right)} \tag{1}$$

where θ is the off-axis angle of the incoming ray, $\mathcal{R}$ is the reflectivity of the coating between the glass plates, and μl is the optical gap. The 3D nature of the Airy function becomes clear after writing θ in terms of the detector pixel position (x, y) and the gap spacing l in terms of the observed frame number z. It can be shown that the Airy function takes the general form

$$A(x, y, z) = \frac{1}{1 + \alpha^2 \sin^2(\beta z \cos(\gamma\sqrt{(x - x_0)^2 + \epsilon^2(y - y_0)^2}) + \delta)}. \tag{2}$$

which involves seven constants that measure properties of the system. In practice, it is important that one restricts the number of system constants to be determined in each fit. It usually helps to fit each constant separately and use this as an initial condition to a multi-parameter fit.

Gradient search optimization. Robust algorithms require an evaluation of the first few terms of the Taylor series, with or without user interaction. These computations are very expensive (M $\sim 10^{4.5}$). If we wish to determine $\underline{u} = \underline{u}[\alpha, \beta, \gamma, \delta, \epsilon, x_0, y_0]$, then we can write

$$\chi^2 = \sum_i^{\mathrm{M}} \frac{1}{\sigma_i^2} (A_i - A(\underline{u}))^2 \tag{3}$$

$$A(\underline{u}) = A(\underline{0}) + \underbrace{\sum_j \frac{\partial A}{\partial u_j} u_j}_{\nabla A(\underline{0}).\underline{u}} + \frac{1}{2!} \sum_{j,k} \frac{\partial^2 A}{\partial u_j \partial u_k} u_j u_k \quad \cdots \tag{4}$$

Forced cylindrical symmetry ($\epsilon = 1$). This form attempts to circumvent the problem of incomplete surface sections. However, it is essential to determine the position of the optical axis (x_0, y_0) very accurately before a full optimization is attempted. Thus we write

$$A(r, z) = \frac{1}{1 + \alpha^2 \sin^2(\beta z \cos(\gamma(r - r_0)) + \delta)} \tag{5}$$

We have assumed that $I(r, \theta)$ is constant which is rarely true due to vignetting, back reflection, and so on. For the simpler case where $\underline{u} = \underline{u}[\alpha, \beta, \gamma, \delta]$, we see that even the second Taylor term requires many expensive computations, viz.

$$\nabla A = \begin{pmatrix} -2\alpha A^2(r, z) \sin^2(\beta z \cos(\gamma r + \delta)) \\ -2\alpha^2 z A^2(r, z) \cos(\gamma r + \delta) \cos(\beta z \cos(\gamma r + \delta)) \sin(\beta z \cos(\gamma r + \delta)) \\ 2\alpha^2 \beta r z A^2(r, z) \cos(\beta z \cos(\gamma r + \delta)) \sin(\gamma r + \delta) \sin(\beta z \cos(\gamma r + \delta)) \\ 2\alpha^2 \beta z A^2(r, z) \cos(\beta z \cos(\gamma r + \delta)) \sin(\gamma r + \delta) \sin(\beta z \cos(\gamma r + \delta)) \end{pmatrix} \tag{6}$$

Small θ approximation. While this form is the simplest to implement, it discards a lot of useful z information and yields only the first-order phase effect, i.e.

$$P(r) = K_z(r - r_0)^2 + \delta \tag{7}$$

where K_z is the curvature of the Airy surface (BT). This method is usually adequate when the physical angle subtended by the detector is less than a few degrees because then the Airy surface is roughly paraboloidal in shape. However, this approximation will be inadequate for Fabry-Perot systems used with the new generation of large format CCDs (2048×2048, 4096×4096, etc.) in a typical f/8 Cassegrain arrangement.

J.B.H. and P.L.S. are partly funded by NSF grant AST 88-18900 that supports the HIFI 'Nearby Active Galaxies Program'. We wish to thank Frank Valdes and the IRAF team for their strong encouragement in this project.

REFERENCES

Bisseling, R., & Kosloff, R. 1985, J. Comp. Phys., 59, 136

Bland, J., & Tully, R. B. 1989, AJ, 98, 723

Cecil, G. 1990, in Workshop on CCDs in Astronomy II, D. Phillip (ed.), in press

Shopbell, P., Bland-Hawthorn, J., & Cecil, G. 1992, this volume

A QUANTITATIVE STUDY OF OPTIMAL EXTRACTION

F. VALDES
IRAF Group, NOAO[1], PO Box 26732, Tucson, AZ 85726

INTRODUCTION

Optimal extraction of spectra from two dimensional formats consists of summing pixels across the spatial profile of the object using weights based on the noise statistics. This can lead to an improvement in the signal-to-noise ratio (S/N) of the extracted one dimensional spectra as compared to not using weights and simply summing the pixels. The statistical theory and some algorithmic implementations of optimal extraction have been described by Robertson (1986), Horne (1986), Marsh (1989), and Kinney et al. (1990). In these papers there are general statements on the amount of improvement to be expected in certain cases based on either mathematical approximations or examples from real data. The mathematical analysis only gives limiting statements for restricted idealized cases while the examples based on real data only cover a small range of situation. Furthermore, in real data it is difficult to decouple other effects, such as flat field noise or the effects of spectral features, in evaluating the S/N improvements.

In this paper we explore a range of S/N levels and sampling conditions using artificial data, where the true signal and noise properties are well known, to yield quantitative comparisons of the extracted S/N between optimal extraction and simple extraction.

THE EXPERIMENT

A series of artificial, two dimensional long slit spectra of varying signal-to-noise ratio, gaussian profile width, and slopes relative to the image columns are created using the following steps. The true one dimensional spectrum is created with constant flux and no spectral features. The constant, featureless aspect of the model spectra allows the signal-to-noise to be easily and unambiguously determined by a simple mean and standard deviation. The flux levels are increase by factors of 2 to cover the range from low to high S/N in approximately uniform logarithmic intervals. The one dimensional spectrum is convolved with a gaussian profile of specified width and the profile center relative to the image columns varied smoothly with wavelength using a linear trajectory of specified slope. The gaussian profile is then sampled into pixels and a constant sky added. The last step is to add a combination of constant gaussian noise, corresponding

[1]National Optical Astronomy Observatories, operated by the Association of Universities for Research in Astronomy, Inc. (AURA) under cooperative agreement with the National Science Foundation

to detector read out noise, and poissonian photon noise at some detector gain. Since the object of this study is ratios at various signal-to-noise levels the actual signal levels, readout noise, and gain parameters are all scalable and so not important.

The two dimensional spectra are extracted using various aperture widths, sky widths, and algorithms. In each case the spectra are extracted with both optimal weighting and no weighting.

The signal-to-noise ratio for both the weighed and unweighted extractions is computed as the average extracted flux divided by the standard deviation. This simple measure is precise since the true spectrum has a constant value and there are no spectral features or cosmic rays. The artificial spectra have 1000 dispersion points so the precision of the individual measurements is 3%. In addition to the S/N other parameters, such as the actual flux values and the accuracy of the derived uncertainty estimates, are measured but are not reported here.

THE SOFTWARE

The study was done entirely with V2.10 IRAF[2]. The artificial data was generated with scripts calling the tasks **mk1dspec**, **mk2dspec**, and **mknoise** in the **artdata** package. The extractions were done with the **apextract** package. The S/N measurements were made with a simple IRAF program written for the project. The IRAF software is freely available and the specific software used in this study may be obtained from the author.

A complete description of the IRAF spectral extraction software is outside the scope and page limits of this paper. Below is a very brief outline of the software. The spectral extraction consists of defining a fixed width apertures along the image axis closest to the spatial axis. The center of the aperture is defined by a smooth function of the dispersion coordinate. Fixed width sky apertures, relative to the center of the aperture, are defined and used for determining the sky in various ways. Sky subtracted aperture spectra are extracted at each point along the dispersion image dimension; extraction is along the image axes even if lines of constant wavelength are not exactly aligned with the axes.

There are three extraction algorithms: simple unweighted summing of the pixels in the aperture with partial pixel weights at the aperture edges, weighted extraction using profile smoothing along image lines or columns parallel to the dispersion axes and a read out and gain model of the noise (called *algorithm 1*), and weighted extraction using a profile smoothing function described by Marsh (called *algorithm 2*). In addition it is possible to use a separate image for defining the profile rather than the data being extracted. While there are differences in some details one may think of algorithm 1 as that described by Horne (1986) and algorithm 2 as that described by Marsh (1989).

---

[2]Image Reduction and Analysis Facility, distributed by the National Optical Astronomy Observatories

THE RESULTS

A large number of spectra were extracted with the parameters varied. The number of parameters sets would be extremely large if all parameter space was sampled and, more importantly, it would be very difficult to present and interpret the results. Instead we define one set of parameters to be the standard and then vary individual parameters one at a time. The standard set of parameters have spatial profiles with a full width at half maximum (FWHM) of 3 pixels, a profile center slope relative to the image columns of 1 pixel in 1000 pixels, extraction apertures extending to ± 3 times the gaussian sigma, two sky apertures (one on each side) each of width equal to the object (so that the sampled sky has twice the number of pixels as the object), extraction using algorithm 1 with no pixel rejection, and the profile determined from the data being extracted.

There are a number of quantities which can be compared in this study. In this paper we consider only the S/N of the weighted and unweighted extraction. It is tempting to only look at the S/N improvement, as measured by a ratio or difference. However, then we could not tell whether a S/N improvement is due to a change in the weighted or unweighted extraction. In particular, the ratio could increase either because the weighted S/N increased or the unweighted S/N decreased. For that reason we will look at both the ratio of the weighted to unweighted S/N (or the noise ratio because the extracted fluxes are always nearly identical) and the S/N of the weighted extraction alone. Since we are interested in intercomparisons, the later quantity is normalized by $S/\sqrt{S+1000}$ where S is the true input flux. This factor is an empirical function which approximately normalizes the weighted S/N at all S/N levels.

Figure I graphically presents all the results. Each graph is plotted as a function of the extracted weighted S/N on a logarithmic axis. The panels on the left plot the ratio of weighted to unweighted S/N; i.e. the optimal extraction improvement. The panels on the right plot the normalized weighted S/N. The pairs of panels from top to bottom each vary one parameter relative to the standard parameters. The different parameter values are represented by different symbol types. The reference line is the standard model extracted using a high S/N reference image for determining the weighting profile as opposed to determining the profile from the data being extracted.

CONCLUSIONS

The reader may draw their own conclusions from Figure I. The conclusions I reach from the data are as follows with supporting panels referenced in brackets. Optimal weighted extraction always improves the S/N, though at high S/N it is insignificant [a-f]. The improvement is maximal around a (weighted) S/N of 2 and actually decreases at lower S/N [a-f]. The peak improvement of about 30% (70% improvement in effective exposure time) for 3σ aperture limits [a-f] agrees well with the expected value derived analytically by Horne (1986).

The improvement in S/N is essentially the same at each S/N level for different spatial profile widths when the extraction aperture is set to the same point in the profile, the 3σ point in panel [a]. However, in this case the narrower profiles achieve a higher actual S/N [g]. This is to be expected since the nar-

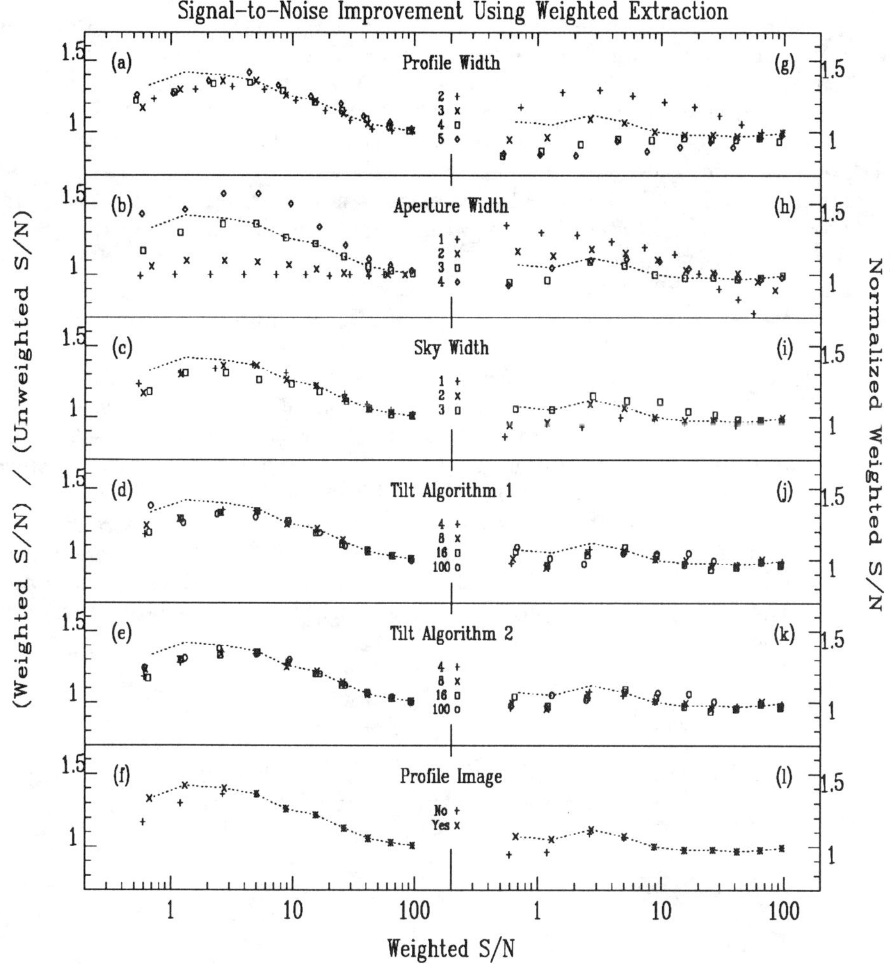

FIGURE I S/N improvements using weighted extraction
A single parameter is varied in each figure relative to the "standard" model and parameters described in the text. The dotted line in all figures is the standard model and paramters with a high S/N image used in determining the weighting profile in place of the profile from the data being extracted. [a,g] Variation of the spectrum profile FWHM in pixels. [b,h] Variation of the aperture width in sigma relative to the profile. [c,i] Variation of the sky width in apertures. [d,j] Variation of the total tilt in pixels from end-to-end using the 1D fitting algorithm. [e,k] Variation of the total tilt in pixels from end-to-end using the Marsh fitting algorithm. [f,l] Comparison of using a separate profile image.

rower profiles put more signal into the central pixels and have a higher contrast relative to the sky.

For a fixed profile width the improvement in S/N appears to increase strongly with larger apertures [b]. However, this is due to the increase in noise in the unweighted extraction as more background noise is added. The actual S/N is higher for narrow apertures when the S/N is low but becomes poorer than wider apertures (for 1 sigma apertures) at S/N greater than about 20 [h]. Thus, for very low S/N, less than 1, very narrow apertures with either weighted or unweighted extractions are best, while at high S/N wider apertures with weighted extraction are indicated. The point that using narrow apertures yields comparable S/N between weighted and unweighted extractions was made by Robertson (1986).

There is little difference in the S/N improvement with the size of the sky apertures [c] but the actual S/N (for low S/N) is improved with wider sky apertures as is expected [i].

There is little difference between using algorithm 1 (based on one dimensional column smoothing) and using algorithm 2 (based on coupled polynomials) and little dependence on slope up to a slope of 100 pixels per 1000 pixels [d,e,j,k]. The good performance of the simpler column fitting is somewhat surprising since the Marsh algorithm was designed for the highly sloped case.

When a high S/N profile can be used for the profile model in low S/N cases, the extraction is improved by a modest amount (14% S/N improvement over the weighted extraction without a profile image at a S/N of 0.6) and becomes negligible at a S/N of 3 [f,l]. Use of a separate profile image is largely, though not exclusively, limited to fiber spectra.

ACKNOWLEDGEMENTS

The mathematical core of the tilted spectrum profile fitting algorithm implemented in IRAF was ported from code kindly provided by Tom Marsh.

REFERENCES

Horne, K. D. 1986, PASP, 98, 609

Kinney, A. L., Bohlin, R. C., & Neill J. D. 1990, in Proceedings of the 2nd ESO/ST-ECF Data Analysis Workshop: ESO Conference and Workshop Proceedings No. 34, ed D. Baade & P. J. Grosbol, European Southern Observatory, 73

Marsh, T. 1989, PASP, 101, 1032

Robertson, J. G. 1986, PASP, 98, 1220

PATTERN MATCHING IN *COBE* SPECTRAL ANALYSIS

RICHARD ISAACMAN
General Sciences Corporation, Code 685.3, NASA/Goddard Space Flight Center, Greenbelt, MD 20771

SHIRLEY M. READ
Hughes STX Corporation, Code 685.9, NASA/Goddard Space Flight Center, Greenbelt, MD 20771

WILLIAM BARNES
Massachusetts Institute of Technology, Rm. 20F-001, 77 Massachusetts Ave., Cambridge, MA 02139

ABSTRACT We demonstrate how two simple pattern matching techniques have allowed the Far Infrared Absolute Spectrophotometer (*FIRAS*) aboard the Cosmic Background Explorer (*COBE*) satellite to help us achieve unprecedented photometric accuracy of our spectra while contending with enormous volumes of often heterogeneous data. The first is a robust averaging technique that reduces the volume of data by identifying outlying spectra that in a statistical sense do not "belong" to the ensemble being coadded. The second technique is a variant of the venerable CLEAN algorithm, and is used to remove the effect of cosmic ray hits on our detector. The method is an improvement over the conventional CLEAN in that the algorithm continuously updates the choice of CLEAN profile during the CLEANing process.

INTRODUCTION

The *FIRAS* instrument aboard *COBE* collected over 2 million spectra during its 10 month lifetime and has improved by orders of magnitude the accuracy with which the spectrum of the Cosmic Background Radiation (CBR) is known. Nonetheless, in order to constrain many models of the Big Bang and of galaxy formation in the early Universe, we must work at the sensitivity limits of the instrument. This of course requires extensive averaging of the data, which, though improving the signal-to-noise ratio (S/N), carries some risk. There are at least two dangers: first, there is the danger that anomalous data, taken outside the range of desired operating conditions, may escape inspection and make it into the final average. Some "intelligent" software is required in order to ensure that all of the data to be coadded are consistent, *i.e.*, belong with some statistical certainty to the same parent distribution.

A second danger of coaddition is that cosmic ray hits whose effects may be

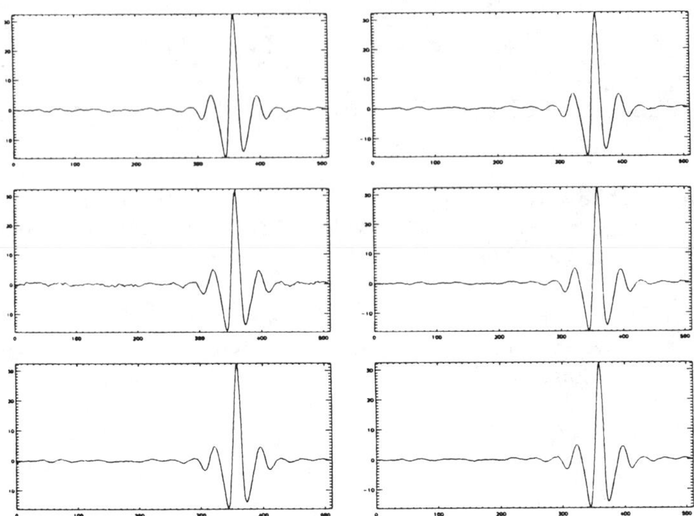

FIGURE I Six supposedly homogeneous IFGs selected for coaddition. In fact the middle left IFG is twice as noisy as the others and was rejected by the consistency checker.

removable from individual spectra may be washed out in the averaging process and so appear as an elevated random noise level. The power-law distribution of cosmic ray energies requires that some kind of optimal filtering be applied to remove a substantial fraction of all hits ("glitches") before they are lost to view.

CONSISTENCY CHECKING

Systematic errors may dominate detector noise if either variations in instrument operating conditions or spatial gradients in the sky emission (*e.g.*, near the galactic plane) cause heterogeneous data to be averaged together.

Our selection process operates on data which are collected in the time domain in the form of interferograms ("IFGs"). These are the modulated infrared signals obtained by scanning a moveable mirror platform in the *FIRAS*, which is a Michelson interferometer. Each IFG is the Fourier transform of the average sky spectrum seen by the instrument during the $\sim$ 45 sec collection period.

The IFGs are first sorted by pixel and nominal instrument state (*e.g.*, mirror scan speed). Each sorted group contains an ensemble of anywhere from 5 to 100 IFGs which we hope form a statistically homogeneous set that can be safely coadded. One such ensemble is shown in Figure I. Each IFG is 512 points long.

We check consistency by first forming a "template", *i.e.*, a robust average of the ensemble in the form of a point-by-point midaverage. That is, if there are k IFGs $I_j{}^i, i = 1...k, j = 1...512$, then the jth element of the template is the average of the middle two quartiles (containing $k/2$ points) of the set $\{I_j\}_k$. The difference between each IFG and the template is then a set of k "residual

IFGs" ${R_j}^i, i = 1...k, j = 1...512$ containing noise, residual signal due to spatial or temporal gradients, and any anomalous signal due to other causes.

Because of our coarse pixel size ($\approx 2.6°$) there is a strong gradient in the Galaxy's contribution to the signal within IFG ensembles that span pixels at the edge of the plane. This contribution is expected to vary in amplitude but not in shape (to first order) and so can be recognized and removed. We extract the eigenfunction of the galactic signal by the following procedure.

1. The sign of the peak signal in each residual IFG R^i is checked; if negative, the ith residual R^i is inverted and a flag set.

2. A modified midaverage is performed, using the third to seventh "octiles" (rather than the middle two quartiles); the result is taken to be the eigenfunction $D_j, j = 1...512$ of the galactic dust spectrum.

3. The best-fit scale factor α^i of D_j (the "secondary template") to the ith residual is determined from the dot product $R^i \bullet D$. This dot product is taken over the range of index j where the modulated signal is greatest.

4. The set of final residuals is defined to be $F^i = R^i - \alpha^i D$ for all indices j.

It is the set of noise-like residuals F^i upon which we perform the statistical consistency checking. For the ith residual we calculate an RMS value σ^i, then in turn form a characteristic overall σ equal to the median of the σ^is. Any IFG whose σ^i differs by more than some threshold (currently 50%) from σ is rejected. Figure I shows the outcome of this test to one small set of IFGs.

A second criterion by which IFGs can be rejected is based on the non-Gaussian (but still well understood) amplitude distribution of the j points within each F^i. We take to be outliers in F^i all the points j where $|{F_j}^i| \geq 5\sigma^i$. More than 5 such points implies either that the noise in this IFG comes from the "wrong" parent distribution or that there is some residual infrared signal.

DEGLITCHING

The aggregate effect of untreated charged particle hits on the detectors can reduce the signal-to-noise of the data by up to an order of magnitude at frequencies near the peak of the CBR spectrum. In Figure II, the lower ("Prelaunch") and upper ("Raw Orbit") curves respectively show the noise levels of glitch-free lab data taken prior to launch and untreated data taken in orbit.

Glitches are due to hits on the detector by charged particles (mostly high energy protons) and so their pulse heights obey the ($energy^{-3}$) distribution of cosmic rays. Consequently, a large fraction of the individual glitches are at or below the detector noise limit. Complicating the problem is the fact that the observed shape of the glitches is determined by the transfer function of the onboard electronics and so each glitch rings through an entire IFG. Fortunately, the shape of the profile is known. The upper curve in Figure III shows a large glitch as well as some smaller ones that are not easily resolvable by eye.

We have based our removal technique on the CLEAN algorithm beloved of radio astronomers (Högbom, 1974 A&A *Suppl.*, **15**, 417). Though developed to

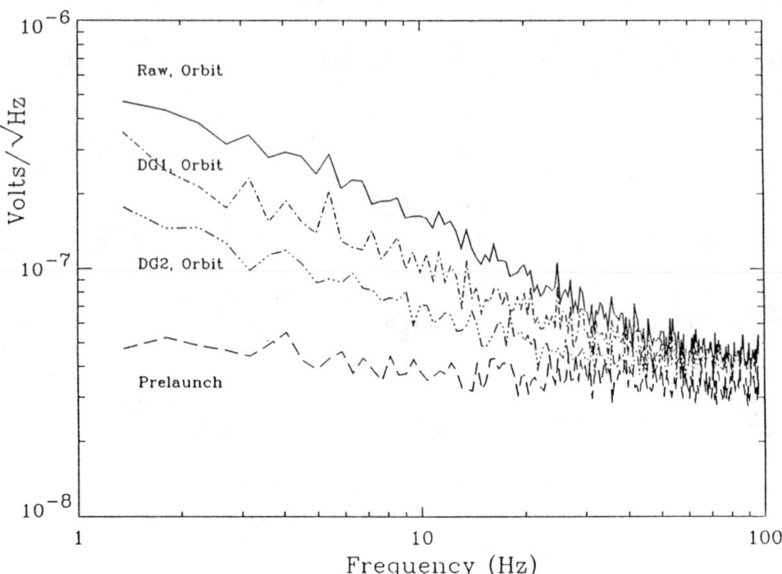

FIGURE II Noise power density spectra. Prelaunch data are essentially glitch-free due to shielding in the lab. Untreated data from orbit (top curve) are dominated by glitches. Curve DG1 is orbit data deglitched with the traditional CLEAN; DG2 shows improvement after using the "smart" CLEAN.

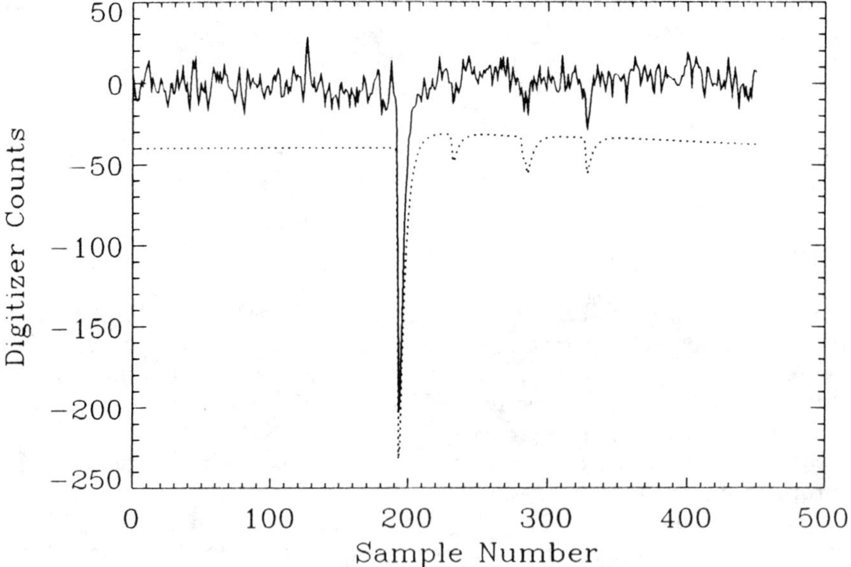

FIGURE III Upper curve: an unusually large glitch, as well as some smaller ones, in an IFG residual after the template has been subtracted. Lower curve: The summed CLEAN components show a near-perfect match after dynamically updating the glitch profile (curve offset for clarity).

deconvolve two-dimensional telescope response functions from sky maps, we can apply it to our IFG time series by treating the glitch profile as a one dimensional "beam profile". A major difference with the conventional CLEAN, of course, is that we are interested only in the removal of the components; we have no desire to reconvolve the glitches back into the data afterwards! Thus, we must be conservative in CLEANing so as to remove as much of the glitch power as possible while ensuring that misidentifications do not leave spurious residuals.

To reconcile these conflicting demands we first modified CLEAN to use a variable "loop gain" Γ, *i.e.*, the fraction of the profile, centered on and scaled to the current maximum, that is subtracted from the time series. CLEAN uses a constant Γ for every iteration, but we have modified it so that Γ decreases with pulse height. For example, Γ for a large peak may start at the conservative value of 0.10, then increase to *e.g.*, 0.8 as the feature gets "eaten" down into the noise. Curve DG1 in Figure II illustrates that this technique gives noticeable but hardly impressive improvement over un-deglitched data.

The problem here is that CLEAN uses a fixed profile that will in general never match exactly the true glitch. Glitches strike at random times and are of course not synchronized with our onboard sampling clock. Even worse, we compress the data by decimation before telemetering it: that is, the IFG as received on the ground is actually an average of consecutive groups of M points, where M can range from 2 to 12 depending on the operating mode of the instrument. A fixed CLEAN profile will always embody a (generally incorrect) assumption about the arrival time of the glitch with respect to the beginning of the averaging window. Superposing this profile on the true glitch will thus result in a misregistration whose residual after CLEANing will resemble the derivative of the profile. (The problem of CLEANing "sources" which do not fall on sampling gridpoints is discussed by D. Briggs and T. Cornwell elsewhere in these Proceedings.)

We have attacked this problem by making CLEAN smarter. Instead of looking only at the amplitude of a given peak, it looks at the feature's shape as well, fitting a parabola to the peak and its surrounding points and estimating the "true" (*i.e.* noninteger) position of the glitch. The algorithm then uses this "true" position to choose from a library of M profiles, the library having been generated using various arrival phases for the nominal profile with respect to the decimation window. The choice of profile used in the subtraction is updated for every CLEAN iteration.

The results show dramatic improvement. Curve DG2 in Figure II shows that roughly two-thirds of the noise due to glitches has been eliminated. It is doubtful that much additional improvement can be obtained, since the power law energy distribution ensures that most glitches have $S/N \leq 1$. The lower curve in Figure III shows that the CLEANed components sum to an excellent approximation of the true glitches, uncovering some low-level events and one double glitch that are otherwise nearly unobservable.

ACKNOWLEDGMENTS

This research is supported by the Astrophysics Division of the Office of Space Science and Applications at NASA Headquarters.

DATA ANALYSIS ISSUES FOR THE HIGH ENERGY TRANSMISSION GRATING (HETG) ON AXAF

D. DEWEY AND T. ISOBE
Center for Space Research, M.I.T., Cambridge, MA 02139

THE HETG INSTRUMENT AND ITS SIMULATION

Unique data analysis issues are raised by dispersive imaging instruments in which the spatial location of a photon encodes both spatial and spectral information. One such instrument is the High-Energy Transmission Grating (HETG, Schattenburg et al. 1991) which will fly on the Advanced X-ray Astrophysics Facility (AXAF, Winkler et al. 1991).

In operation, the HETG is inserted following the AXAF mirror, producing an energy-dispersed image in the focal plane (Figure I). This image is detected by the AXAF CCD Imaging Spectrometer (ACIS), giving the position and the energy of each detected photon. The ACIS energy resolution ($E/\Delta E_{\rm FWHM}$) is at best of order 60; however, the position information of diffracted photons can be converted into a wavelength through the grating equation: $\sin\theta = n\lambda/p$, where θ is the (measured) dispersion angle of an $n^{\rm th}$-order diffracted photon and p is the period of the grating; p is 2000Å and 6000Å for the HETG instrument. This dispersive technique allows a spectral resolution as high as 1000 to be achieved in the 0.4 keV to 10 keV energy range.

We created an instrument simulation to test and evaluate analysis schemes;

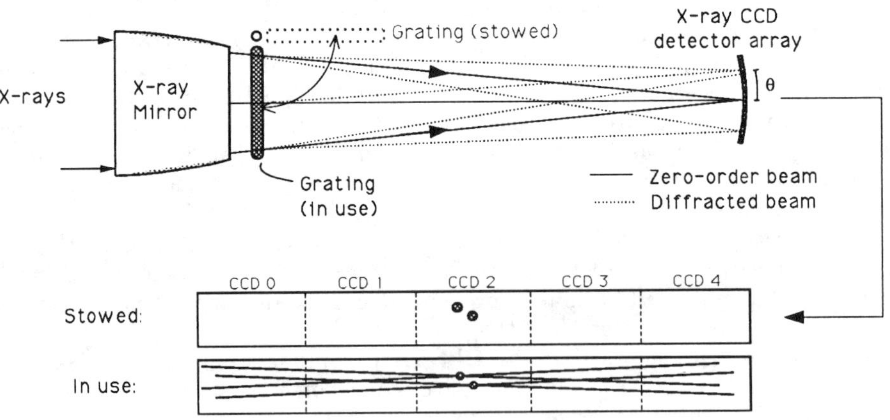

FIGURE I Layout of the HETG grating spectrometer on AXAF. When in-use, two types of gratings with dispersion directions oriented 5° apart produce an "X" pattern of dispersed X-rays on the ACIS imaging detector.

our end-to-end simulation flow can be summarized diagramatically as:
XSPEC → Ray-trace Simulation → Analysis (FORTRAN) → XSPEC . A source spectrum is generated with XSPEC (An X-ray Spectral Fitting Package V.7.0 from Goddard Space Flight Center, 1990) and is input to our ray-trace code (built on IRT routines, Interactive RayTrace© from Parsec Technology Inc., P.O. Box 4332, Boulder, CO 80306) running in the IDL programming environment. The output list of ACIS detected events is analyzed by a FORTRAN program which applies the grating equation to assign each photon a high-resolution energy. These energy data are then analyzed with XSPEC using appropriate resolution functions; for the 2000Å period grating the resolution function is 0.9 eV FWHM at 1 keV. The simulation results, Figure II, demonstrate that the grating dispersion data, (c), does achieve the desired high-resolution.

DATA ANALYSIS ISSUES

Dispersive X-ray instruments have been flown on the *Einstein* (Mewe et al. 1982) and *EXOSAT* (Paerels et al. 1988) missions; using experience and results from these missions as a starting point we identify a variety of analysis steps, under development, needed for optimal grating spectral extraction:

Determine zero-order location(s). Use CCD energy and spatial information to classify events, e.g., nth order or background. Remove spurious contaminating sources. Apply optimal aspect correction: the dispersion direction is fixed w.r.t. the CCD array and thus standard post-facto aspect roll corrections will "roll" the spectra about the object. Remove CCD-pixel quantization effects which can produce artificial non-uniformity in binned dispersion data. Perform consistency checks between plus and minus orders. Deconvolve spatial-spectral information in cases of multiple-point or extended sources (e.g., using a direct image as a spatial template). Optimally include higher-order spectral information. Optimally combine spectral data obtained with different instruments.

In addition to the processing steps above, a grating analysis package (e.g., PROS) will have to provide tools for high-spectral resolution model generation and parameter analysis in order to extract parameters of astrophysical interest.

ACKNOWLEDGMENTS

We thank our many colleagues at the Center for Space Research. This work was supported in part by NASA contracts NAS8-38249 and NASW-4372.

REFERENCES

Mewe, R., et al. 1982, ApJ, 260, 233

Paerels, F. B. S., et al. 1988, ApJ, 329, 849

Schattenburg, M. L., et al. 1991, Optical Eng., 30, 1590

Winkler, C. E., et al. 1991, SPIE, Space Astronomical Telescopes and Instruments, 1494, 301

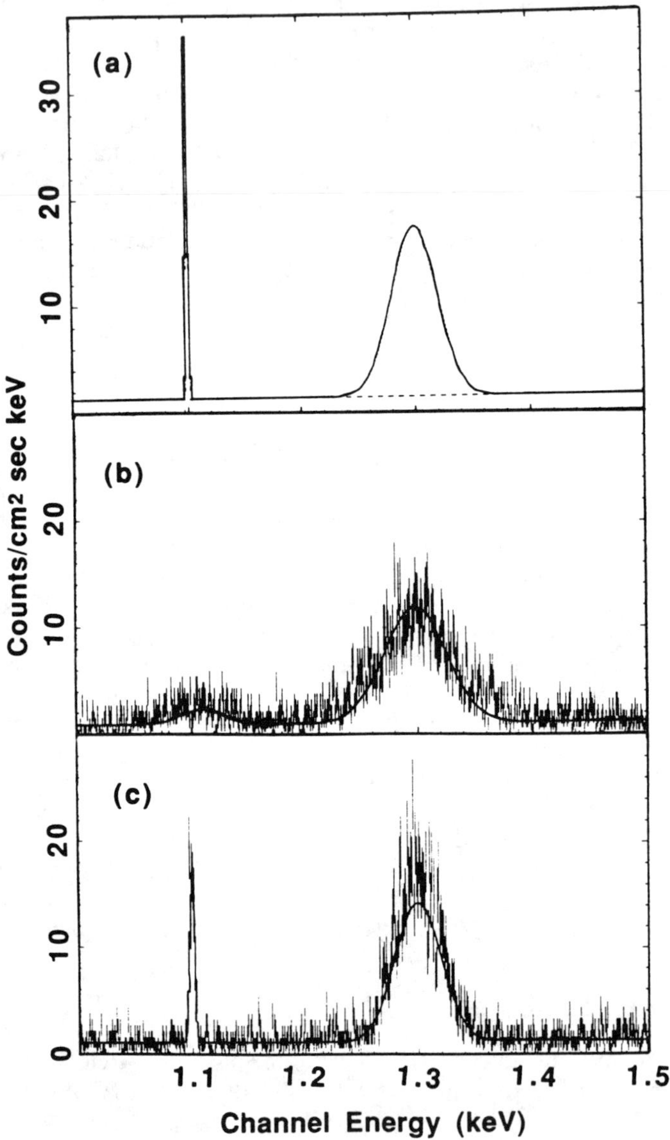

FIGURE II HETG Simulation Results. (a) XSPEC-generated input spectrum. (b) Result of simulated observation using only the CCD-measured photon energies. (c) The same observation analyzed using the position of each diffracted photon to assign a high-resolution photon energy.

A MODEL INDEPENDENT METHOD FOR X-RAY SPECTRA DECONVOLUTION

V. F. POLCARO
Istituto di Astrofisica Spaziale, CNR, Frascati, Italy

ABSTRACT It is proposed a method to obtain an easily manageable detector matrix and to deconvolve the spectra obtained by means of any kind of X-ray device without "a priori" hypothesis on the input spectrum. The effect of the noise on the deconvolution process is also examined.

INTRODUCTION

The new generation of X-ray telescopes is characterized by increasing sensitivity and spectroscopic resolution, both in soft and hard X-ray range. On the other hand, the deconvolution of the measured spectra, is still obtained by means of best-fit procedures of theoretical spectral laws through the detector matrix (e.g. Lampton et al. 1976). This procedure is mainly due to the difficulty in the detector matrix inversion, but is extremely dangerous giving, in practice, only the solutions that have been "a priori" supposed to be the right ones. This fact makes it difficult to recognize the presence of emission processes that have not been previously hypothesized. This problem can become serious when no counterpart of the detected source is known and thus no emission mechanism can be suggested by the previous experience, as it is the case of new telescopes, reaching, in a given energy range, a sensitivity level much higher than the previous ones. Furthermore, the best-fit deconvolution procedures are computer time expensive and thus difficult to apply when many new sources are detected in a short time by a new telescope.

A MODEL-INDEPENDED SPECTRAL DECONVOLUTION METHOD

We have proposed some years ago (Polcaro et al. 1984) a method to obtain an easily manageable detector matrix and to deconvolve the spectra of any kind of X-ray spectroscopic device just by inversion of this matrix (and thus without "a priori" hypothesis on the input spectrum). The method starts from the basic equation of the spectroscope:

$$\mathbf{D} = \mathbf{A} \times \mathbf{S} \tag{1}$$

where $\mathbf{D}$ is the detected spectrum, $\mathbf{S}$ is the input spectrum, both considered as a n-component vector and $\mathbf{A}$ is the (n,n) detector matrix. It is clear from equation (1) that, once the detector mathematical model has been generated, the detector matrix can be easily obtained just processing through this model a sequence of ideal input spectra having a single component equal to 1 and all the

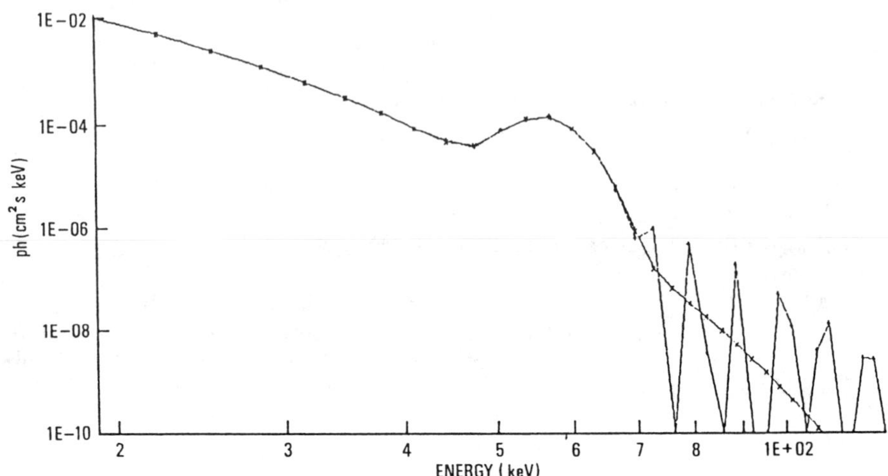

FIGURE I The effect of 50 times repeated process of convolution and deconvolution on an input spectrum through the matrix of a Xenon Multiwire Proportional Counter. Only the lower counting rate channels are affected by numerical noise.

other equal to 0: the outputs of the model will give the corresponding detector matrix columns. In fact, equation (1) can be written in explicit form as:

$D_1 = a_{11}\, S_1 + a_{12}\, S_2 + \dots + a_{1i}\, S_i + \dots + a_{1n}\, S_n$

...

$D_i\; = a_{i1}\, S_1 + a_{i2}\, S_2 + \dots + a_{ii}\, S_i + \dots + a_{in}\, S_n$

...

$D_n\; = a_{n1}\, S_1 + a_{n2}\, S_2 + \dots + a_{ni}\, S_i + \dots + a_{nn}\, S_n$

It is thus clear that putting as input a vector **S**, whose components are

$S_{ij} = 0$ if $i \neq j$

$S_{ij} = 1$ if $i = j$

we will obtain a solution corresponding to the jth column of the matrix **A**.

We have proved (Polcaro et al. 1984) that the matrix obtained in this way is surely non-singular and the system is extremely stable, if the mathematical model of the detector carefully reproduces the physical process whose the incoming radiation is transformed through in the outcoming signal. As an example, Figure I shows the effect of 50 times repeated processes of convolution and deconvolution of a kT=5.3 KeV thermal spectrum plus a gaussian line with E_o=56 KeV and σ=4 KeV FWHM through the matrix obtained in the previously described way from the mathematical model of a Xenon Multiwire Proportional Counter described in Bazzano et al. (1983).

On the other hand, if the mathematical model or the detected spectrum are not reliable, it is not surprising that the deconvolved spectrum is meaningless, no matter of the method used to obtain it.

A common problem comes from the attempt to deconvolve spectra with a number of channels higher than allowed by the actual spectral resolution of the detector, all over its energy range: it is clear that, if we try to distinguish between photon energies that are not physically resolved, we will obtain a random result, that will be translated in a singular matrix, i.e. in a matrix giving random numerical output to any input.

A more tricky problem arise from the uncertainties on the detected spectrum: it is well known that any detected spectrum is obtained by background subtraction and is affected by noise of different origin. The worst effect is given by an underevaluation of the background, that will generate an increase of the counting rate respect to the actual one. Given that the detector matrix is usually far from diagonal, this distorsion of the detected spectrum, usually visible only in the low counting rate channels, will propagate all over the deconvolved one, even in case of a relatively small bias. This problem could be partially overpassed by using a "Random Walk" matrix inversion procedure, that is driven by the input statistics, but this solution is definitely expensive from a computer time point of view. When a simplier (e.g. Gauss-Seidel) inversion method is used, the precise background evaluation becomes crucial.

A last problem arise from the presence of the noise: it is obvious from equation (1) that, if we will substitute the actual input vector $\mathbf{D}$ by the vector

$$\mathbf{D_n} = \mathbf{D} + \mathbf{N} \tag{2}$$

where $\mathbf{N}$ is the noise figure, again considered as an n-component vector, we will obtain as output

$$\mathbf{S_n} = \mathbf{A}^{-1} \times \mathbf{D} + \mathbf{A}^{-1} \times \mathbf{N} = \mathbf{S} + \mathbf{A}^{-1} \times \mathbf{N} \tag{3}$$

that can be significantly different from $\mathbf{S}$. This problem is relatively easy to solve: in fact most of the noise must be of statistic origin. Thus a realistic choise of the channel width in the detected spectrum will make $\mathbf{N} \simeq 0$ and $\mathbf{S_n} \simeq \mathbf{S}$.

CONCLUSIONS

The model independent spectral deconvolution methods have been believed up to now unreliable. We have shown that the detector matrix inversion procedures are just the mathematical translation of the physical capability of the detector to give a definite output signal to any different source spectrum. We have also shown how the deconvolution problems arise because of human errors, as the overevaluation of the detector resolution, the bad background estimation and the presence of noise on the detected counting rate. It is clear that this kind of problem does not depend on the procedure employed for the deconvolution. Thus we believe that it is now time to switch from the model-dependent procedures to the model-independent ones, as that described in this work, having much greater heuristic value.

REFERENCES

Bazzano, A., et al. 1983, Nucl. Instr. Meth., 214, 481

Lampton, M., Margon, B., & Bowyer, S. 1976, ApJ, 208, 177

Polcaro, V. F., et al. 1984, Nucl. Instr. Meth., 221, 41

SECTION 10

SPECTRAL ANALYSIS SOFTWARE

THE IRAF SPECTROSCOPY REDUCTION PACKAGES AND TASKS

F. VALDES
IRAF Group, NOAO[1], PO Box 26732, Tucson, AZ 85726

ABSTRACT A variety of generic and instrument and data specific spectroscopy reduction packages are available in the NOAO suite of packages in IRAF. The instrument specific packages contain specialized routines written as IRAF command language scripts using complex and generic image processing and spectroscopy programs. The reduction routines connect all the reduction operations needed to fully reduce spectral data from particular instruments and customize their parameters. The package organization and the capabilities of the specialized reduction packages is presented.

INTRODUCTION

There are many types of spectroscopic instruments and data in astronomical use. The spectroscopic data, now usually in digital form, requires many reduction steps. In addition, there are many reduction recipes, scientific goals, and computer aptitudes among astronomers. This leads to the need for a diverse set of spectroscopic data reduction tools. This paper describes the approach taken in the NOAO[1] suite of IRAF[2] packages for handling this diversity.

The approach is to provide both a large set of generic tools which address a particular reduction operation in a very general way, a set of packages devoted to particular instruments or types of data, and specialized reduction tasks (programs) which integrate the reduction steps for particular instruments or types of data into a single command. This approach is applicable to other software systems although the software described in this paper is specific to IRAF.

Generic tools perform specific operations on spectra; for example determining dispersion functions from arc calibration spectra. High priority is given to generality in the type of spectra and algorithms. This is done by providing many parameters and, often times, multiple algorithms to allow tailoring the task to the requirements of the data. Emphasis on generality and modularity has a number of benefits. The most obvious is that a variety of spectral data is addressed once rather than writing new or modified versions for each new

---

[1]National Optical Astronomy Observatories, operated by the Association of Universities for Research in Astronomy, Inc. (AURA) under cooperative agreement with the National Science Foundation

[2]Image Reduction and Analysis Facility, distributed by the National Optical Astronomy Observatories

type of data encountered. Addressing individual reduction operations in separate programs allows new algorithms to be added without modifying other code and makes maintenance easier. General and flexible tools can also be exported and used at other institutions. I also believe that emphasis on generality and feedback from application to a variety of types of data also leads to better software. Finally, flexible, generic tools can be easily and quickly tied together in a command language scripting environment to produce specialized functions.

The most interesting, and recent, components of the IRAF spectroscopy software are the specialized data reduction packages and tasks. These are IRAF command language (CL) scripts selecting and combining the generic tools for the reduction of a particular type of spectroscopic data. The instrument specific packages collect only the generic tools needed for a particular instrument or type of data. The specialized reduction tasks provide a capability missing from the generic tool approach; namely, a complete and focused task for reducing spectroscopic data. The following sections of this paper describe the selection of spectroscopy packages and the capabilities of the specialized reduction tasks.

Specialized data reduction scripts offer several advantages over compiled tasks. The first is that scripts provide a very quick prototyping capability. They are also easier to maintain and modify for specific situations. To emphasize this last point, there have been instances where users were able to modify the scripts for their own purposes and, in one case, an instrument developer was able to build his own package and reduction task for his instrument. For compiled tasks users would need to learn SPP (the IRAF programming language) and how to compile, link, and install IRAF tasks. This leads to the point that scripts are architecture independent (although the commands in them will be architecture dependent), so there is no need to worry about multiple object libraries, binary directories, etc. Finally, command language scripts are easier to update as improvements are made to the generic tasks. For instance the recent addition of a nonlinear dispersion function capability required only the addition of another parameter and slight modification of the command invoking the modified generic task.

SPECTROSCOPY PACKAGES

IRAF commands are organized into packages. A package, when *loaded*, defines a set of commands, called *tasks*. Multiple packages may be loaded at the same time and commands from all packages are available. Typically, general image processing packages are always loaded while specific data reduction or analysis packages, such as the spectroscopic packages, are loaded as needed.

In this paper we discuss only the spectroscopy packages developed at NOAO and contained in the **noao** package. Other IRAF packages are are being developed at a number of institutions and these packages may also contain generally useful spectroscopy software.

The NOAO spectroscopy packages divide into two types, generic and instrument specific. The generic packages contain tasks for a variety of general spectral reduction operations. They typically have many parameters and algorithms to allow application to the widest possible types of data. Table I lists the current generic spectroscopy packages in the IRAF **noao** package.

TABLE I Generic Spectroscopy Packages

apextract	- Aperture spectra extraction package
longslit	- Long slit reduction package
onedspec	- One dimensional spectral reduction and analysis package
rv	- Radial velocity analysis package

The **apextract** package extracts one or more, one dimensional spectra from two dimensional images by summing data values within an aperture across image lines or columns. This assumes that the spatial and dispersion axes are close to the image axes; how close depends on the spatial extent of the spectral profile and the needs of the user. If significant distortions are present or very extended profiles are to be analyzed the **longslit** package provides tools for making geometric distortion corrections and wavelength and flux calibrations in two dimensions. The **onedspec** package contains all the tools for calibrating and analyzing one dimensional spectra. The radial velocity package provides tasks for measuring radial velocities from spectra.

The large number of tasks and task parameters in a generic package, plus the need to use tasks from several generic packages, can be confusing to novice users. The IRAF command environment allows creating new packages containing a set of the generic tasks from multiple packages, defining different default parameter values, and possibly adding additional tasks which are relevant to a particular instrument or type of data. The specialized spectroscopy reduction packages, listed in Table II, have been created in this way.

TABLE II Specialized Spectroscopy Packages

Fiber	argus	- CTIO ARGUS reduction package
	foe	- KPNO Fiber Optic Echelle reduction package
	hydra	- KPNO HYDRA (and NESSIE) reduction package
	fiber	- KPNO coude three fiber reduction package
	msred	- GENERIC multiple spectra reduction package
Slit	echelle	- GENERIC echelle reduction package
	goldcam	- KPNO GOLDCAM reduction package
	slit	- KPNO coude slit reduction package
	specred	- CTIO spectrophotometric reduction package
Scanner	iids	- KPNO IIDS reduction package
	irs	- KPNO IRS reduction package

The packages may be classified into three categories; optical fiber spectra, slit spectra of stellar objects, and one dimensional, beam-switched scanner spectra. The reduction operations required are similar in each category. Echelle spectra, both with fibers and with slits, are not categorized separately because, apart from the form of the dispersion function, the reductions can be treated as multiple fiber or slit spectra. These are not the only possible classes of instruments and data types. Others will be added in the future.

Because the data for each category tend to be quite similar one should appreciate that data from similar, non-NOAO, instruments may also be reduced with one of the reduction packages and the specialized reduction tasks after possibly minor parameter changes. Alternatively, one of the these packages may be taken as a starting point for developing IRAF-based, reduction packages for other instruments.

The fiber and slit categories of spectral data generally begin with two dimensional CCD images from which one dimensional spectra are extracted. This means that the generic tools collected in the package come from both the **apextract** and **onedspec** packages. In principle one could include the basic CCD reduction commands (from the IRAF **ccdred** package) and also include CCD reductions as part of the integrated reduction tasks; currently this is not done. Generally, observers separate the basic CCD operations, which may be done as the data is collected or as a separate data reduction session, from the spectral data reductions.

In the first category are spectrographs using optical fibers. These types of instruments have become increasingly common because they have many important features, such as obtaining many spectra per exposure, including a mix of object, sky, and arc spectra, and using very stable spectrographs located off the telescope. There are a number of such instruments within NOAO as well as at other observatories.

Argus is a multifiber instrument at Cerro Tololo Interamerican Observatory (CTIO) which positions fibers with a "fishing pole" style system in which each pole has an object and sky fiber with a fixed separation. *Hydra* is a multifiber instrument at Kitt Peak National Observatory (KPNO) which positions fibers with a robot positioner. *Nessie* is an older KPNO multifiber system using manually wired plugboards. The *Fiber Optic Echelle* is a two fiber (simultaneous object and arc spectra), echelle spectrograph on loan to KPNO. The image format consists of pairs of echelle orders. The **fiber** package refers to a three fiber configuration at the KPNO coude feed where two fibers can be used for arc spectra while one fiber is used for an object. The **msred** package provides a generic package for dealing with multiple spectra in a 2D format. While this is not limited to fibers it is most often used for that type of spectra.

The second category of slit reduction packages is specifically for stellar slit spectra which have been separately distortion corrected or which do not require distortion corrections. This class of spectra are very similar in their reduction requirements and the packages and tasks differ mostly in default parameters. The **echelle** package falls in this category because the reduction operations are basically the same as for single slit spectra but the operations are performed over multiple orders and, of course, an echelle dispersion function is used.

The scanner category consists of instruments which record one dimensional aperture spectra directly. These differ from the other categories mostly in requiring special beam-switch operations. At NOAO the two instruments are the *IIDS* and *IRS* scanners.

THE SPECIALIZED SPECTROSCOPY REDUCTION TASKS

For some astronomers having the appropriate tools in a package is sufficient. They prefer to reduce their data one step at a time with control over all the options of the general tools. However, other astronomers prefer to have a single program which handles all the data reductions in a standard way with a minimum of user interaction. Some reasons for this are for preliminary (often called quick-look) reductions, for processing large amounts of data uniformly and quickly, for repetitive operations which require many steps using the general tools, and for beginning or infrequent users.

Most of the instrument and data specific packages now have an integrated reduction task combining individual reduction steps and providing some degree of automation. These tasks are IRAF command language (CL) scripts for the reasons described previously. The reduction tasks have the following goals:

1. Perform all the reduction operations in an integrated task
2. Guide the user through the reduction steps
3. Address instrument and data specific requirements
4. Collect and limit the number of parameters to those specific to the problem
5. Perform interactive reference calibrations first
6. Perform bulk of processing automatically and noninteractively
7. Perform record keeping and processing checks

Though the tasks provide a complete reduction path (excluding basic CCD calibrations), these integrated tasks are still quite flexible. Task parameters allow selecting just one operation, a set of operations, or all operations. Thus, one still has the option of doing the reductions a step at a time or all at once.

An important point about instrument and data specific reduction tasks is they can contain functions for dealing with specific features or peculiarities of a particular type of data or instrument. Examples of this are computing fiber throughputs or merging pairs of arc observations because a single arc may not have sufficient lines or fully illuminate all the fibers in a multifiber spectrograph. Though these specialized functions may not be addressed by any single generic task they can be performed using a combination of general image processing tasks and parts of the generic spectroscopy tasks. The reduction scripts naturally allow including these specialized, compound commands saving the user many separate steps.

A necessary aspect and virtue of the very powerful and general generic tasks is that the user is given a choice of many parameters to specifying data dependent properties and algorithm options. For some this makes the tasks overly complex; even with the IRAF parameter mechanism which gives most parameters recommended default values. Furthermore, when there are many such tasks involved in the data reduction process the number of parameters, distributed among different parameter sets, to be kept track of becomes quite large. The integrated, instrument and data specific reduction tasks can reduce and simplify the parameters by collecting all the parameters in one place, combining parameters which are common to several tasks into a single parameter, and hiding parameters which have fixed values or which are not applicable to the particular instrument or data.

One of the features of data reductions when there is a lot of data is that once calibration information is set up much of the work can be done noninteractively or even as a background or batch process. Also, for a consistent set of data some of the calibration data can be defined noninteractively by reference to other similar calibration data. The prime example of this is wavelength calibration using arc lamp spectra. Once the arc lines have been identified in one arc spectrum other arc spectra will be sufficiently similar that the computer can find the lines and determine new wavelength calibrations. In the data reduction scripts the user is guided to do the required interactive operations first and only once; if a step has been done previously then it is not repeated. When all the required interactive calibration steps have been accomplished, the bulk of the data can be processed noninteractively and the script may even submit a background or batch job.

When doing many reduction operations on many observations, keeping track of what has been done and what is left to be done can become a significant problem. This problem is compounded by instruments which multiplex many observations simultaneously. The best example of this is multifiber spectrographs which can produce a hundred or more spectra per exposure. The integrated reduction tasks can be designed to do this record keeping consistently and to include processing checks to prevent reprocessing or out of order operations.

SPECTRAL REDUCTION OUTLINE

This section discusses the types of reduction operations performed by the specialized reduction tasks excluding the one dimensional scanner instruments. The description roughly follows the actual flow of the programs. Space does not allow much usage and algorithmic detail. Table III shows the basic reduction operations provided by the tasks. These basic operations generally consist of a number of steps. In addition, there are setup and display steps.

TABLE III Basic Spectroscopy Reduction Operations

Fibers	Slits	Scanners
Extract	Extract and sky subtract	Flat field
Flat field	Wavelength calibrate	Wavelength calibrate
Wavelength Calibrate	Flux calibrate	Beam switch
Sky Subtract		Flux calibrate

Setup Steps

The setup steps consist of setting parameters and preparing input lists. The parameters are easily reviewed and set with the IRAF parameter editor. The parameters likely to be adjusted are those which vary due to the CCD used, such as pixel scales and orientations, and selecting the operations desired. The types of input lists which might need to be prepared are long lists of input images, aperture identifications for multifiber data, and arc assignment lists if the default assignment by observation time is not desired.

The aperture identification list for multifiber data is a very valuable feature. This list defines the aperture numbers and object titles for each fiber. It also defines which fibers are object, sky, arc, or other type of spectra. For data with many fibers, record keeping becomes a major concern and the initial preparation of this list allows the identifications of the spectra to be maintained through any sorting and analysis of the spectra. In the future there may be instruments which produce this file automatically for the user.

Aperture Definitions

The first step in all the reduction tasks is to define a reference set of apertures. Once done the reference apertures are automatically used to set the apertures in all other spectra. The reference apertures may be used unchanged or as a starting point for automatic recentering, resizing, or even retracing (determining the path of the spectrum along the dispersion through the 2D image). While the intent is that only the initial reference apertures have an interactive component it is also possible to review the apertures for each spectrum if desired. For multifiber data with hundreds of spectra it is clearly important not to have to do this for each observation.

An initial set of apertures is determined automatically based on task parameters such as the number of spectra expected. The user may then review the definitions graphically. Since this step is usually only performed once it would be wise to check such things as the aperture assignments, the background regions, and the apertures widths. For multifiber instruments, which may suffer from missing (broken) fibers, this is important to avoid misidentifications of spectra.

The positions of the spectra along the dispersion are fit by a smooth function. The user may review the the function fits to the measured positions and adjust the fitting parameters. While this may be done for all apertures it is usually only necessary to do this for one aperture and then apply the same parameters to all other apertures. Again, for large numbers of fibers, it is important to not have to look at each fiber except if truly desired.

Flat Fielding

A flat fielding step is part of the fiber data reduction tasks. This step is optional if one wants to flat field the CCD images separately. However, flat fielding of fiber data by two dimensional division is very sensitive to even a small fraction of a pixel shift in the fiber positions. When there are no shifts it is entirely equivalent to divide the extracted spectra by the extracted flat field spectra. While it is not equivalent when there are small shifts, the approximation is close and preferable to dealing with artifacts caused by dividing misaligned flat field and object fiber spectra.

A flat field is also used for determining fiber throughput corrections. Alternatively, a sky flat field may be used for the throughput corrections if the illumination of the fibers is different than the flat field illumination. Only the relative total flux through each fiber is used so a high S/N sky exposure is not required for this step.

When the flat field spectra are extracted the user has the option of fitting a smooth function to the average of the flat field fiber spectra and then dividing the individual spectra by this smooth mean. This removes the large scale shape of the flat field spectrum and, thereby, approximately preserves the pixel statistics of the object images.

The operations performed in this step are to extract the flat field fiber spectra, possibly remove the average spectrum shape, extract the sky flat field spectra if given, compute throughput corrections to be incorporated in the final flat field spectra, and finally normalize so the mean flat field through all fibers is unity. Only the fitting of the average flat field shape is possibly interactive. Once the flat field has been derived this step is not done again when the task is repeatedly executed.

The fiber flat fielding algorithm, which requires many operations using the generic image processing tasks, is an example of special operations that can be implemented for data and instrument specific packages and reduction tasks.

Wavelength Calibration

Virtually all spectroscopic data needs to be wavelength calibrated. In the specialized reduction tasks this step is divided into a initial setup step, which is partly interactive, and then new calibrations are derived and applied as needed when the object spectra are processed.

The initial calibration setup consists of extracting one arc calibration reference spectrum and having the user identify the lines (only a subset need be identified and further lines are added automatically) and define the dispersion function to be fit to the wavelengths as a function of pixel position. For multifiber spectra the other fiber arc spectra are extracted and automatically used to find lines and determine dispersion functions. This step has the option of interactive review if it appears arc lines are misidentified. At this point a common linear dispersion system is defined. All spectra will then have the same dispersion sampling. Note that in V2.10 it is not necessary to linearize the spectra though it might still be desirable.

In the instrument and data categories, echelle data was not identified as a separate class. This is because the only significant step which is different is computing and applying an echelle dispersion function (a function of pixel position and order number) instead of a one dimensional function. In the reduction scripts it was found that one could simply substitute the echelle dispersion task and otherwise the logical flow of the processing remained unchanged.

For fiber spectra there is the possibility of using a different type of calibration to monitor zero point shifts in the wavelengths. This is an another example of a special purpose algorithm. For the *Nessie* instrument, because it is time consuming to take many arc lamp exposures due to manually moving the fibers to a small region in a special plugboard for illumination, users can take quick exposures of scattered light from the dome lights (mercury lines) or possibly sky lines, and then select a set of such lines to be monitored and used for making wavelength zero point corrections.

Another special purpose algorithm for *Nessie* data is to combine two or more arc exposures in which only some fibers are illuminated in each exposure. This is because the arc illumination covers only a small region of the field. By creating a simple file giving which fibers in each exposure are to be used this instrumental peculiarity is dealt with.

Flux Calibration

The slit instrument reduction tasks include flux calibration as one of the processing options. This option is not included in the fiber tasks because there has been no demand for it; it could be added if desired.

Flux calibration consists of extracting *all* the standard star observations first and computing a sensitivity function from the ratio of observed verse reference calibration fluxes. The standard star extractions follow the same steps as the object processing except they are done before any objects and, of course, they are not initially flux calibrated. After each standard star is extracted a query is made for the name of the standard star. It would be nice to have this done automatically. Currently it is up to the observer to identify which objects were observed since image header information is subject to error. There is an optional interactive step to allow reviewing and modifying the bandpasses used for the calibrations. Typically, the same bandpasses as used in the reference calibration observations are used so that no interpolation is required. However, if the dispersion is a poor match to the calibration data it may be necessary to define additional bandpasses.

The ratios of the observed fluxes through a set of bandpasses to the fluxes found in the calibration data is recorded in a file. After all the standard stars have been processed the entire set of ratios is fit by a smooth function of wavelength. This is a very interactive step in order to evaluate the photometric conditions and define an acceptable sensitivity function. The sensitivity function is recorded as a spectrum which is then applied to object spectra as needed.

Object Processing

After the various calibration steps have been completed, the object spectra are processed. One goal of the reduction tasks is to keep track of the processing state of each observation. Thus, if a specified input spectrum has been previously processed it will be skipped or reprocessed depending on a user set processing option. The ability to skip processed spectra is particularly useful in allowing wildcard input lists so that the user doesn't have to keep track of which spectra have been processed.

Another goal of the reduction tasks is to allow the object processing to be done completely noninteractively; though with options to allow interactive review and intervention if desired. The tasks keep track of whether interactive steps have been selected and allow these steps to be turned off during processing. When all interactive steps have been turned off the task may automatically submit a background or batch process if the user has selected this option.

In the following the processing steps are described in the order in which they are performed. Most of the steps have parameters which allow selecting or skipping each step. The first step is assigning the apertures. This is done by inheriting the reference apertures and providing task parameters to select recentering of the apertures based on a part of the spectrum, resizing the aperture widths to accommodate seeing changes, interactively reviewing the apertures, and retracing based on the object being extracted. Whether these steps are done depends on the stability of the instrument and characteristics such as the strength of the spectra; for example, it would be worse to retrace very weak spectra than to assume it is the same as the reference spectrum.

The spectra are then extracted, with sky subtraction in the case of slit spectra. This step may also include options such as optimal weighting and cosmic ray detection. If a 1D fiber flat field has been computed it is divided into the extracted object spectra.

If wavelength calibration has been selected, calibration arc spectra are assigned. This is usually the two nearest in time but any selection may be made with an input table. The assigned arc spectra are extracted if needed. Note that extraction of arc spectra is only done as needed and later objects using the same arc spectra may not need to extract them again. However, for slit spectra one generally wants to use the same apertures as the object spectra. Therefore, if the object apertures are recentered or retraced the user may choose to have the arc spectra be reextracted with the new apertures. For fiber spectra the reference aperture is generally used.

The arc lines are automatically reidentified and new dispersion functions are fit. For the special algorithm using night sky or dome vapor lamp lines for tracking zero point shifts it is these type of spectra that are extracted and the new dispersion function is the original arc dispersion function with a zero point shift applied. Alternatively, some fiber instruments provide one or more fibers to take simultaneous arc spectra. In this case the spectra from these fibers are compared against the same fibers in the master arc calibration to determine zero point changes while keeping the same dispersion shape.

The one or more dispersion calibration spectra are then interpolated either in time or spatially in the case of simultaneous arc fibers. The final interpolated dispersion functions are then applied to the object spectra. This application consists of either resampling the spectra to a common linear dispersion sampling or, in IRAF V2.10, the actual function may be stored with the spectrum and no interpolation done.

Sky subtraction with fiber spectra requires selecting the sky spectra, combining them, and subtracting the final sky from the object fibers. The sky spectra selection is done based on aperture or beam numbers. It has an interactive option to plot all the sky spectra and allow deletion of those which might be contaminated by objects. The combining options include various choices of algorithms such as medians or sigma clipping.

Flux calibration is a straightforward, noninteractive, application of extinction corrections and the sensitivity functions.

An interactive option, which might be selected if the reduction tasks are used at the telescope for quick reductions is to automatically plot the final spectra using an interactive spectra plotting task.

REFERENCES

There are a variety of documents covering the IRAF NOAO spectroscopy packages. There are overviews of the packages, detailed descriptions for each task, and user's guides for each of the specialized instrument reduction tasks.

AUTOMATED ANALYSIS OF SPECTRA

T.N.WILKINS
Institute of Astronomy, University of Cambridge, Madingley Road, Cambridge, CB3 0HA, United Kingdom

D.J.AXON
Nuffield Radio Astronomy Laboratories, University of Manchester, Jodrell Bank, Macclesfield, Cheshire, SK11 9DL, United Kingdom

ABSTRACT We describe LONGSLIT, a sophisticated FIGARO program designed to automatically analyse long-slit spectra, fibre array spectra and velocity cube data obtained with instruments such as TAURUS or radio interferometers.

The power of the algorithm relies upon a powerful Hierarchical data base system, allowing detailed control over the optimisation algorithm and clear presentation of the results to the user. Using a series of modern robust statistical techniques allows the analysis of even complex multi-component line spectra in a fully automatic manner to a high degree of precision. We describe both the novel features of our algorithm and illustrate these with real world examples from a variety of astrophysics areas.

INTRODUCTION

In order to interpret spectroscopic data, it is often useful to parameterise it by fitting an analytical model to it which describes both the line and continuum properties of the data.

There are basically three stages to fitting a model:

1 Provide a first approximation to the fit parameters—*guesses.*

2 Converge on a solution, for example using a least squares algorithm.

3 Decide whether to accept the solution

Many programs have been developed to fit various parameterised line profile models in the form or multiple Gaussians for both astronomical and laboratory spectra. Most of these programs require the user to perform the first task, probably providing some means of manipulating the guesses by hitting various keys. Numerical libraries will perform the second task, leaving the user to perform the third. For large data sets this requires the astronomer to spend a considerable time performing what is, most of the time, a fairly mundane task.

Any means of automating the process will therefore allow much quicker analysis of data.

OUR SOLUTION

One of the basic principles of LONGSLIT in to maximise the return the user gets from their time in using the program. LONGSLIT is therefore designed to perform as much of the work as is possible offline—while retaining the ability to perform interactive fits for more "difficult" cases. To achieve this a considerable investment has been made in allowing the user to re-use information the program has already learned about the task in hand, applying this to other related data. In addition an assessment of the success or otherwise of the fits can be carried out automatically. The availability of the Hierarchical Data System allows LONGSLIT to accomplish these tasks. This permits the operation of the program to be controlled mainly by information stored with the data.

The Guesses

One of the most important time-savers in such optimisation problems is to obtain the best possible guesses to start with for as small a cost as possible. LONGSLIT can *INHERIT* the results from previous fits to use as guesses for subsequent fits. This is useful where the line profile evolves relatively slowly along a slit. In addition LONGSLIT can *TRANSFER* fits from one line to use as guesses for another, allowing models to weak lines to start from the fits to stronger lines. These two modes are illustrated schematically in Figure I.

We can even *CLONE* the results from one data set to use as a starting point for another, and re-fit data after addition of further data (or flux calibration) starting from previous fits scaled appropriately.

Optimisation

To converge on the solution we use the Nag library or modified Levenberg-Marquardt (Minpack), although other routines could easily be slotted in as required.

Do We Believe It?

One of the most useful facilities is the ability to create feasibility space bounds on fit parameters, either during the fitting process or post priori by applying TOLERANCES to the fit parameters. LONGSLIT provides a very flexible scheme for both these methods allowing many combinations of constraints at fit time and the ability to apply and modify TOLERANCE rejection criteria to flag invalid results without the need to re-fit any data if the criteria are subsequently relaxed or tightened.

This works quite adequately for simple cases such as galaxy rotation curves, but for more complicated cases, showing considerable variation in line prfiles throughout the data set, a more sophisticated method is required if we are to perform the task automatically We need some objective method of deciding the number of components which takes this into consideration.

A suitable test is Akaike's Information Criterion (here denoted C), which modifies the normal least squares criterion by adding a term increasing with the number of free parameters. This is given by:

$$C = M \times \ln(S) + 2 \times N$$

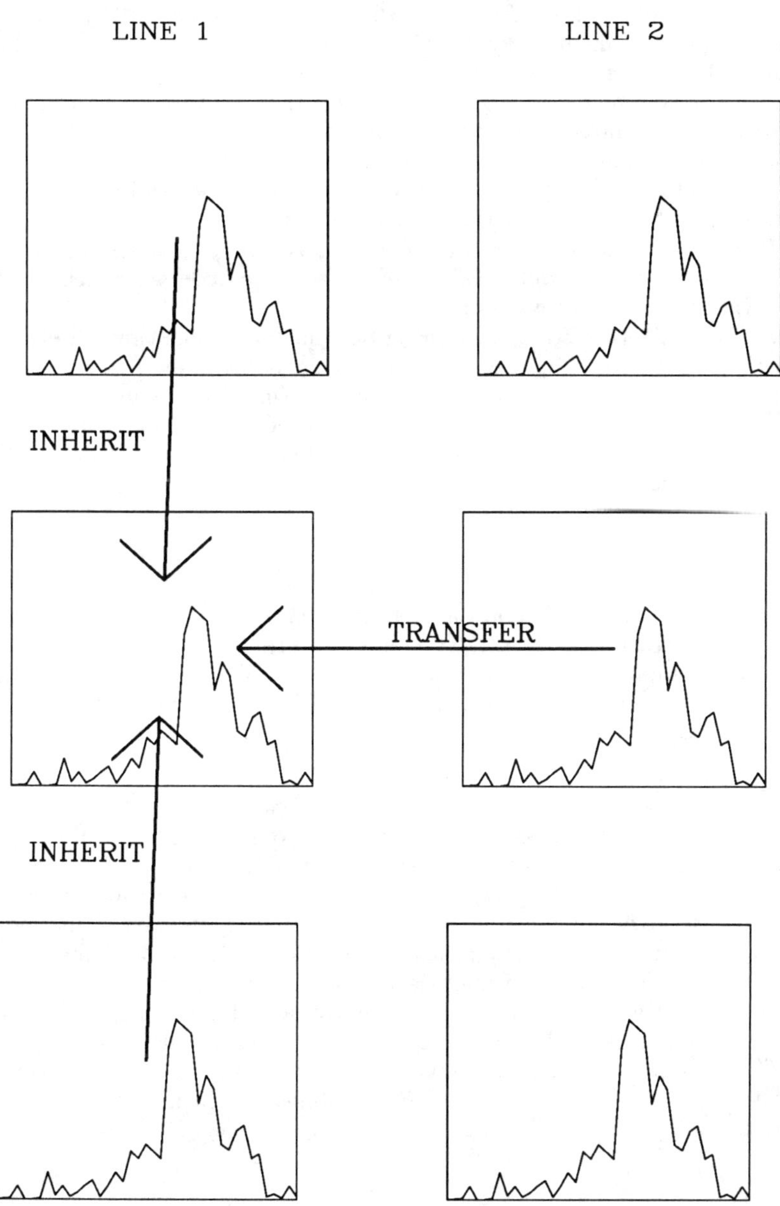

FIGURE I Schematic representation of INHERIT and TRANSFER modes.

where M is the number of data points, N the number of fit parameters and S the weighted sum of squares (Akaike 1973). This is a penalised likelihood method. The lowest value of C represents the best result. An example of the results from applying this is given in Figure II. The values of the criterion for the fits shown, in increasing order of the number of components, were 659, 512, 490 and 494, giving a minimum at three components, as would be chosen by visual inspection (fitting a flat base would give a value of 714). Experience has shown this test to work well with real data.

Even with these tests (or any other we may add), the astronomer is unlikely to trust the results without actually seeing how they compare with the data., so we provide a number of ways of doing this.

All these processes require storing information about the fits, etc., between runs of the software. By using an hierarchical data files (as provided by FIGARO), we have been able to easily store information in the file to control the various processes, for example to define the model we want to fit, and to store the results. This allows us to concentrate further work only on regions that actually require it.

THE FUTURE

At present LONGSLIT is good at fitting lines, but not so good at complicated continua and work has recently started on improving this. The fitting routines are very modular and we plan to add the capability for multiple Lorentzian functions, and to experiment with further optimisation algorithms.

CONCLUSIONS

The procedure of fitting analytic line profiles to emission and absorption lines, which normally would involve a large amount of user interaction, has been automated to a considerable degree. A separate structure within the data file is used to control the fitting process and for storing the results. This means that the fits can be performed in batch mode automatically, and inspected later. Unsatisfactory fits may then be rejected, and where necessary fits can be performed interactively. Several options are available for fitting. With one option a peak-searching algorithm gives good guesses for the optimisation procedures, and a statistical test is used to determine an appropriate number of components to fit without user intervention. The principles are easily extended to include additional models.

REFERENCES

Akaike, H. 1973, 2nd int. Symp. Information Theory, p.267, Eds. Petrov, B. N., & Csaki, K., Akademai Kiado, Budapest

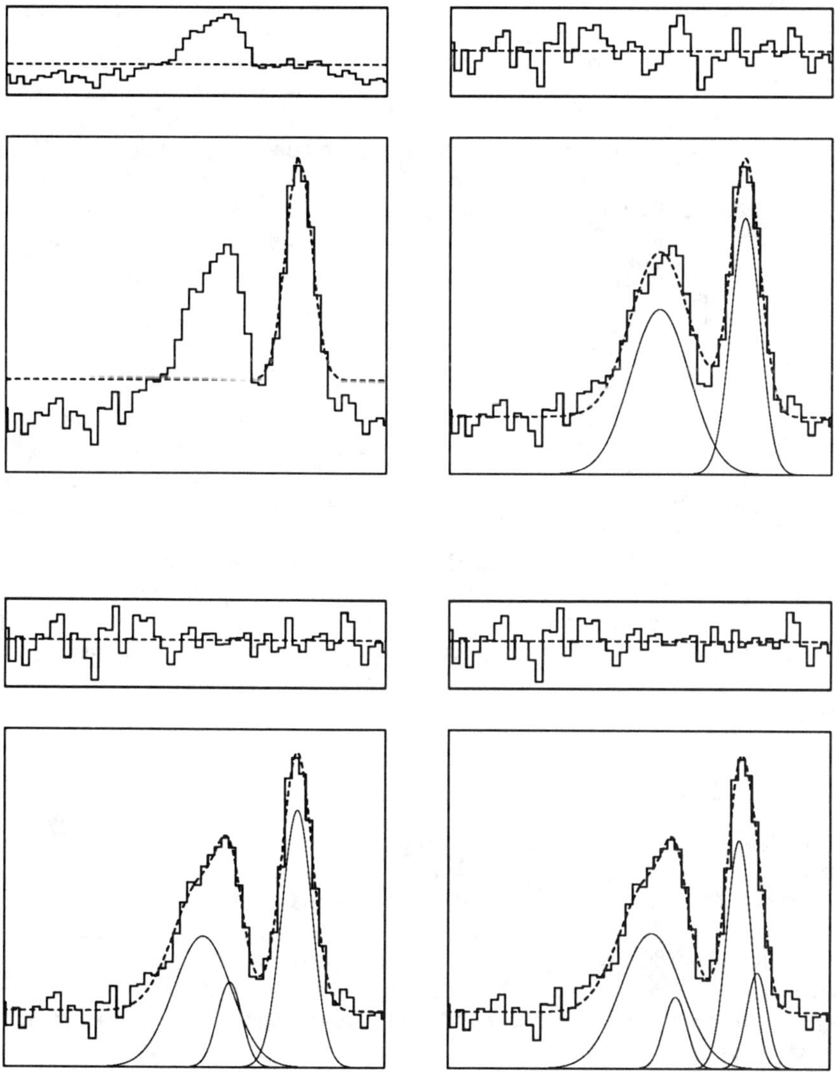

FIGURE II Use of Akaike's information criterion—example fits. The lower plots show line profiles with the fits superimposed, the upper plots show the residuals on the fits.

XCSAO: A RADIAL VELOCITY PACKAGE FOR THE IRAF ENVIRONMENT

MICHAEL J. KURTZ, DOUGLAS J. MINK, WILLIAM F. WYATT, DANIEL G. FABRICANT, AND GUILLERMO TORRES
Harvard-Smithsonian Center for Astrophysics, Cambridge, MA 02138

GERARD A. KRISS
Department of Physics and Astronomy, Johns Hopkins University, Baltimore, MD 21218

JOHN L. TONRY
Department of Physics, MIT, Cambridge, MA 02139

ABSTRACT XCSAO is a software package for obtaining radial velocities of stars and galaxies from optical spectra using the cross-correlation method. It has been written to work within the IRAF environment, and is compatible with other packages being developed at SAO for optical spectra. We discuss the individual components of the package in terms of the decisions required to set up the package for a new instrument/observing program. Using artificial spectra we show the procedures necessary to optimize the set up parameters, and demonstrate the robustness of the package when applied to spectra from both stars and galaxies.

INTRODUCTION

During the decade of the 1980s all radial velocities produced at SAO, both stellar and extra-galactic, were reduced using a suite of FORTH routines on NOVA computers written by John Tonry (Tonry & Davis 1979) and maintained and extended by Bill Wyatt (Tonry & Wyatt 1988). Moving our reductions onto modern computers in the IRAF (Tody 1986) environment required a total rewriting of the code; this has resulted in the RVSAO package, the main parts of which are XCSAO, for cross-correlation velocities, and EMSAO for emission line velocities.

Copies of the source and IRAF installation scripts can be obtained by anonymous FTP from sao-ftp.harvard.edu as pub/iraf/pkg/rvsao/rvsao.tar.Z. A README file is also in the directory. XCSAO has been extensively tested on Decstations and Sparcstations and is expected to experience no major changes. EMSAO is still undergoing substantial revision. Both tasks have been in production use at SAO for the past year, and are currently in production use by several large redshift and radial velocity programs worldwide.

XCSAO has its origins in the XCOR program written by Gerry Kriss, as well as the FORTH code, and a FORTRAN rendition of it by Tonry. Eventually we included the continuum estimation methods from Mike Fitzpatrick's FXCOR. Our code was written and is being maintained by Doug Mink.

This paper presents a very brief overview of the capabilities of XCSAO, and some of the testing procedures we have developed for it. An extensive description of the program is in preparation for PASP (Kurtz et al. 1992).

ELEMENTS OF XCSAO

The following subsections describe briefly the elements of XCSAO in the order in which they occur in the code. We indicate those elements which we believe should receive critical attention when setting up XCSAO for a new instrument.

Input Formats

XCSAO takes lists of 1-dimensional wavelength calibrated spectra in the standard format used by the ONEDSPEC task in the NOAO package. Spectra may be either log or linear lambda binned. Two lists are required, one for the unknowns and one for the templates; each unknown will be compared separately with each template. Figure I shows the parameter file for XCSAO.

Rebinning

Spectra are rebinned into the specified number of log wavelength bins (NCOLS) so that there is the maximum possible overlap between the unknown and the template in the rest frame. This is achieved by iterating on the radial velocity a total of NZPASS times; in addition the user can choose an initial guess CZGUESS by setting CZINIT=yes. The error in the estimate of radial velocity can often be halved on the second pass.

Note that this binning suggests that it is worthwhile to have the template cover the broadest wavelength range possible, at a dispersion equal or better than the unknown's, the rebinning will then choose the proper wavelength region for maximal overlap, with the resolution determined by the unknown. We find that it is in all cases better to rebin to the next larger power of two than the number of actual pixels, rather than the next smaller.

Apodization

The spectra are apodized by a cosine bell over a specified percentage of its extent. The size of the apodized region should be kept as small as possible while still preventing ringing in the fourier transform. We have found 5% is satisfactory for most spectra, but this should be set for each new instrument following some testing.

Continuum and Emission Line Suppression

The continuum is subtracted from each spectrum after being calculated by routines derived from the ICFIT task, and only slightly modified from their implementation in the FXCOR task in the RV0 package. There are a large number of possibilities here, different observing programs require different settings. This is

```
                          I R A F
            Image Reduction and Analysis Facility
PACKAGE = rvsao
   TASK = xcsao

spectra =  550045-300.template  List of input spectra
(templat=      550045.template) List of template spectra
(tempdir=                     ) Directory for template spectra
(obj_plo=                   no) Plots the raw object data
(xcor_pl=                   no) Plots the filtered correlation function
(emchop =                   no) Chop out emission lines
(bell_wi=                 0.05) Fraction of spectrum to taper
(low_bin=                    5) Bin number of low frequency cutoff
(top_low=                   20) Bin number where low frequency reaches one
(top_nru=                  125) Bin number where high frequency starts to slope
(nrun   =                  250) Bin number of high frequency cutoff
(tshift =                   0.) night to night zero point velocity shift
(czinit =                   no) Set initial velocity (yes or no
(czguess=                   0.) Initial velocity guess in km/sec
(nzpass =                    0) Number of iterations shifting template to match
(curmode=                   no) Wait for keyboard response after each object (ye
(st_lamb=                   0.) Starting wave length of spectrum
(end_lam=                   0.) Ending wave length of spectrum
(tempfil=                   no) Template transform already filtered (yes or no)
(contfit=                  yes) Continuum fit: IRAF (yes) or polynomial (no)
(contdim=                    5) Continuum fit dimension if polynomial
(renorma=                   no) Renormalize spectrum before transform (yes or no
(smooth =                    0) Smooth hardcopy spectrum (o=no else number of ti
(pkmode =                    1) Peak-fitting mode (1-parabola 2-quartic 3-cosx/1
(pkfrac =                  0.5) Fraction of peak or number of points for peak fi
(pksrch =                   25) Half-width of search region for graphically set
(maxshif=                 1000) Maximum allowable pixel shift
(minshif=                -1000) Maximum allowable negative pixel shift
(ncols  =                 8192) Number of cols into which to rebin
(cvel   =                INDEF) Center velocity of output graph
(dvel   =                INDEF) Velocity half-width of output graph
(svel_co=                 file) Spectrum velocity correction
(tvel_co=                 file) Template velocity correction
(interp_=                poly5) Rebin interpolation mode
(report_=                    2) Report mode (1=normal 2=one-line)
(save_ve=                   no) Save result in IRAF data file header (yes or no)
(archive=                   no) Save results in binary archive record (yes or no
(displot=                  yes) Display results on <device> (yes or no)
(device =            stdgraph) display device
(hardcop=                   no) Auto hardcopy of results
(plotter=              stdplot) Hardcopy output device
(debug  =                   no) Displays intermediate results
(contsub=                   no) Plots the continuum-subtracted data
(apodize=                   no) Plots the apodized data
(fft_plo=                   no) Plots the fft of the data
(uxcor_p=                   no) Plots the unfiltered cross-correlation
(phase_p=                   no) Plots the phase of the correlation function
(logfile=            xcsao.log) List of log files
(cursor =                     ) Graphics cursor input
(mode   =                   ql)
```

FIGURE I Parameter File for XCSAO.

the area which we believe deserves the most experimentation when setting up a reduction.

If EMCHOP=yes XCSAO will remove emission lines and replace them with the value of the continuum before performing the cross correlation. The emission lines are found using ICFIT, with parameters settable as for the continuum suppression; the method for finding emission lines here is similar to that used in the sister task EMSAO. Removing emission lines is critical to finding absorption line redshifts from galaxies with emission lines, if the spectrum is too noisy for the emission line removal, or if it has sharp spurious absorption features from poor sky subtraction we recommend that the offending regions be exised by hand, using SPLOT.

Fourier Filtering

Before correlating the unknown with the template each is put through a band-pass filter in the fourier domane (unless TEMPFIL=yes, then the template is considered to have already been filtered). We have done a number of experiments with the fourier filter, some of the main results are:

1. The type of ramp used in the bandpass filter is not important (assuming it is a reasonable choice, a step function is not reasonable). We have therefore removed all the options for filter type from XCSAO, and use just a cosine bell.

2. The exact choice of the high pass roll-off is unimportant. in so far as it does not impinge on the main data region.

3. The exact choice of the low pass roll-on is critical to the reduction, it should be chosen with great care. This also effects the relation of the r statistic with the actual error, see below.

4. Use of a frequency weighting scheme which can be shown in the limit of infinitely wide spectra to produce the maximum likelihood estimation for the lag (radial velocity) (e.g. Hassab & Boucher 1979) is not an effective technique for spectra similar to ours.

Cross Correlation

The normal fourier cross correlation, the transform of one spectrum times the conjugate of the transform of the other, is performed.

Peak Fitting

The user can set the function to use to fit the peak with PKMODE, normally this makes no significant difference in the velocity. The user also sets the fraction of the peak height to fit the function to with PKFRAC, .5 is the default choice.

Because the spectrum is quantized in pixels while the fractional power point to fit to is not we perform two fits, one two pixels wider than the other, and report the weighted mean as the result of the fit.

Error Calculation

Following Tonry & Davis (1979) we calculate an error as a function of their r statistic. It can be shown analytically that for sinusoidal noise with a half width of the sinusoid equal to the halfwidth of the correlation peak that the mean error

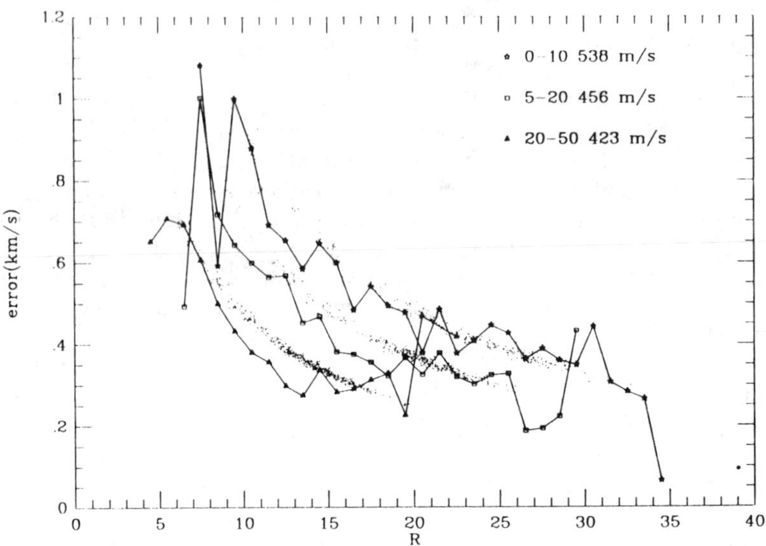

FIGURE II Individual error estimates (dots) versus actual mean errors (solid lines) for 1200 synthetic echelle spectra. The three sets are for three different fourier filter roll-on points; the filter parameters, and the mean errors for each filter are shown on the figure.

in the estimation of the peak of the correlation function is

$$error = \frac{3}{8}\frac{w}{(1+r)} \tag{1}$$

where w is the FWHM of the correlation peak, and r is the ratio of the correlation peak height to the amplitude of the antisymmetric noise.

For large observing programs a better error estimate may be obtained. The coefficient $\frac{3}{8}w$ is for many situations a constant, and may be estimated by taking the mean of a number of measures. The error estimate then becomes a constant divided by $(1+r)$. This is not in all cases possible, for example when the low pass fourier roll-on is changed the relation of r with error changes. $\frac{3}{8}w$ tracks this properly. Figure II shows the calculated errors (dots) versus the actual mean errors (solid lines) as a function of the r statistic for three different low-pass filters; note that the calculated errors do track the actual errors, but that the relation of r with error changes.

Output Formats

While XCSAO may be run totally as a batch procedure, it may also be run in an interactive mode. Intermediate stages in the reduction may be plotted, and the correlation peak to chose may be specified. At the conclusion of a reduction one may choose to redo the reduction interactively, to view a plot of the main absorption lines, to edit the emission lines, to remove spectral regions, ...

Figure III shows the standard final plot, and Figure IV shows an absorption

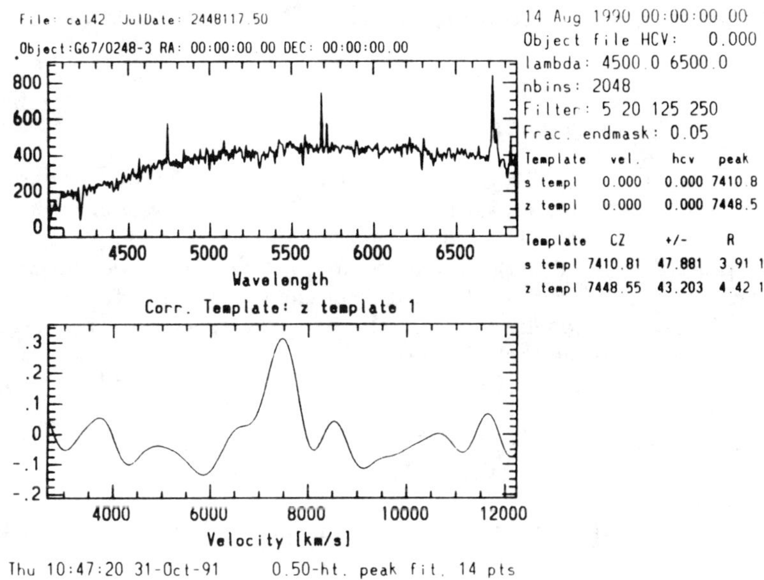

FIGURE III Primary output plot from XCSAO

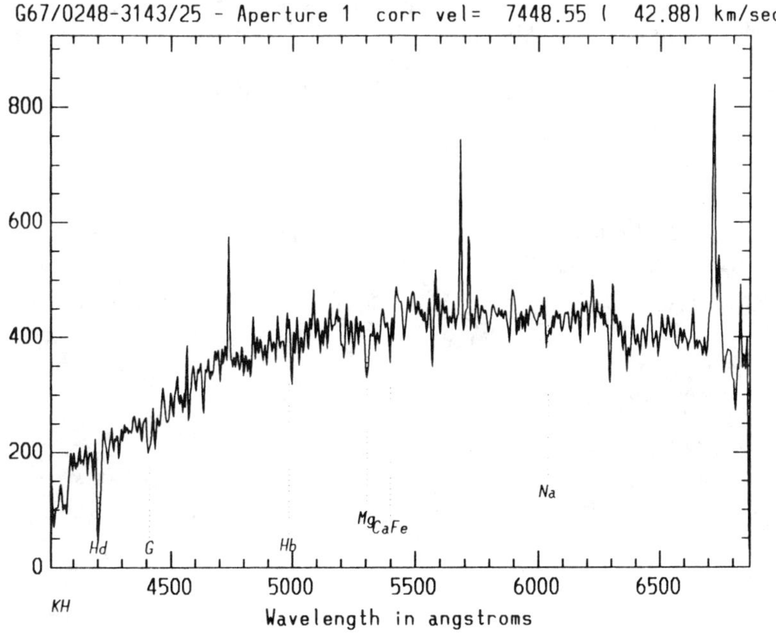

FIGURE IV Output plot showing absorption lines labeled

line plot, both for a galaxy spectrum observed by M. Ramella with the ESO 1.5m telescope.

TESTING

We have tested XCSAO with synthetic spectra created to have a specified number of counts sampled from a parent distribution, which is a template spectrum. For tests of echelle spectra we used Kurucz models for the templates, for tests of galaxy spectra we used high S/N observations. Note that we can know the true radial velocity difference between the unknowns and the templates either exactly in the case of models, or where the template spectrum was exactly the one used to create the synthetic spectra, or with very high accuracy, as when the radial velocity template is different from the template used to create the synthetic spectra.

The data used to create Figure II were 1200 synthetic spectra created from a Kurucz model of a 5500K dwarf, and having between 5000 and 30,000 total counts. The template used for cross correlation was exactly the same template as was used to create the synthetic spectra. A complete description of the testing procedures will be found in Kurtz et al. (1992).

ACKNOWLEDGMENTS

Steve Levine and John Morse made important contributions to the development of the code. Dave Latham, Alejandra Milone, Susan Tokaraz, and John Huchra have contributed mightily of their expertise. Much of the IRAF code we have used was written by Mike Fitzpatrick and Frank Valdes.

REFERENCES

Hassab, J. C., & Boucher, R. E. 1979, IEEE Trans. Acoust. Speech Signal Process., ASSP-27, 922

Kurtz, M. J., Mink, D. J., Wyatt, W. F., Fabricant, D. G., Torres, G., Kriss, G. A., & Tonry, J. L. 1992, PASP, to be submitted

Tody, D. 1986, in Instrumentation in Astronomy VI, ed. D.L. Crawford, Proc. SPIE 627, 733

Tonry, J. L., & Davis, M. 1979, AJ, 84, 1511

Tonry, J. L., & Wyatt, W. F. 1988, CFA Z-Machine Data Analysis Software, Cambridge: Smithsonian Astrophysical Observatory

A PRODUCTION SYSTEM FOR RADIAL VELOCITY MEASUREMENTS

DOUGLAS J. MINK AND WILLIAM F. WYATT
Harvard-Smithsonian Center for Astrophysics, Cambridge, MA 02138

ABSTRACT A system has been developed at the Center for Astrophysics to mass-produce and archive radial velocity measurements. One-dimensional spectra arriving as FITS files are reduced and stored in an extensible archive format which contains complete information about the reduction process. In order to keep over 150,000 reduced spectra on-line for analysis, the format combines information compactly in variable-length ASCII and binary records. Each spectrum in the archive occupies one file which, for each instrument, has a unique identifying number. A collection of stand-alone Fortran programs allow examination of the archived spectra, translation to a portable format, and co-addition of spectra, and several other functions. A library of Fortran-callable access subroutines allows scientists to use the archive with their own programs. IRAF tasks to import and analyze archived spectra have been integrated with the Fortran and C programs to take advantage of IRAF's data analysis and display capabilities.

INTRODUCTION

At the Center for Astrophysics, over 150,000 spectra have been collected at telescopes in Massachusetts and Arizona to be used to compute radial velocities of stars and galaxies. In the past, spectra were reduced using a software system written in Forth on Data General Nova computers (Tonry & Davis 1979; Tonry & Wyatt 1988). The results were then archived in a compact binary format on one of a roomful of removable disks. To examine one spectrum, its disk had to be retrieved and mounted on the computer. Generating reports directly from the archive could require much disk changing.

As the Novas began to wear out, a decision was made to replace them with a hardware-independent archive and a portable data reduction and analysis system. The hardware selection was simple–a Unix system with lots of disk space. Reduction and database software could then be written in Fortran or C, and the archive could be made accessible across the local network via the Network File System for processing or analysis on other computers. The radial velocity analysis software, however, was written in the IRAF environment (Tody 1986) so that it could be used on a range of data beyond that taken on Smithsonian Astrophysical Observatory telescopes.

DATA ARCHIVE

The new archive had to meet the several criteria. As the archive would have to hold all present and future spectra in a reasonable amount of disk space, the format had to be as compact as possible. This precluded the standard FITS format. The storage format would have to contain all of the information about the spectra in the Nova archive format. It should be easy to change any information about any spectrum or add new information about any spectrum. Lastly, the format should be machine-independent, so it may be easily accessed from several types of computer.

An archive format was designed with one file per spectrum, organized into a Unix directory tree by instrument, telescope, and sequence number. Each file is made up of multiple variable-length records which may be ASCII, integer, IEEE floating-point, or some combination of the three in network (big-endian) byte order. Each record within a file has a 48-byte ASCII header beginning with a record name and length in bytes. Programs which access the archive can skip over records they don't need or recognize, but must know the format of the records they are using.

An assortment of reporting software was developed to display the contents of the archive. ZPRINT prints the contents of a single archive file in a variety of formats. PSPEC plots the spectrum in one file in wavelength or pixel space. ARCHEDIT allows editing of existing archive files, performing consistency checks to avoid damage. SUMSPEC co-adds spectra, with appropriate wavelength shifts, to improve signal-to-noise.

The pre-existing Nova reduced data files were converted to this format to keep the entire available on line. The current archive has about 150,000 spectra totaling 1.5 Gbytes of disk storage.

DATA REDUCTION

Initial reduction is done under the Unix operating system using standalone C and Fortran routines which duplicate the actions taken by SAO's time-proven Nova software. The primary task of the reduction is to create a normalized (via flat-field incandescent) object spectrum and to compute the wavelength dispersion polynomial used in later analysis. In general the latter computation can be easily automated by using an earlier polynomial as the starting point and figuring shifts from it.

In addition, enough information is stored to permit the exact re-reduction of the data should that be necessary. All control parameters, including names of raw data files, are stored in a file known as the 'skeleton'. This file is later copied into the archive file as one of its records. Since most of the parameters rarely change, they are stored on a 'template parameter file' to which each skeleton points. These files are saved forever; when a change is needed, a new version becomes current, but the old one remains as the record of past reductions.

The variable part of the skeleton is unique to each spectrum, but a system script helps the data aide create the entire series of skeleton files for each night's and each lunation's data. Barring data difficulties, the initial reduction of several hundred objects from a week or so of observing takes only a few hours.

DATA ANALYSIS

ARCHIVE, an IRAF package, is centered on RARC, a task which converts archive files to IRAF files using the dispersion correction calculated outside of IRAF. IRAF Command Language scripts and supporting SPP tasks in the ARCHIVE package allow access to groups of reduced archive files from within IRAF by their instrument and sequence numbers, so users do not need to know the full file names. Radial velocities are computed using the XCSAO task in the RVSAO package (Kurtz et al. 1992), which cross-correlates the spectra against a set of templates, and ESPEC, which fits the emission lines in a spectrum. ESPEC, a Fortran program which works directly on the archived spectra, is declared to be a foreign IRAF task so it can be run in CL scripts with XCSAO. An IRAF SPP version, EMSAO, is currently in the testing stage.

The results from the radial velocity computations, with enough supporting information to rerun them under identical conditions, are appended to the archive files. Summaries of the results are obtained using the standalone program RSPEC which tabulates radial velocities and other identifying information for a list or file of archive files.

FUTURE

The archive is currently being extended to include more than the five original instruments. Modifications are being made to accept data reduced entirely or partially inside the IRAF environment, and new IRAF tasks are being written to automate the transfer of multi-fiber spectroscopic data to file-per-spectrum archive files.

ACKNOWLEDGMENTS

Robert Davis, John Huchra, David Latham, and Susan Tokarz helped define the data archive format and debug the software.

REFERENCES

Kurtz, M. J., Mink, D. J., Wyatt, W. F., Fabricant, D. G., Torres, G., Kriss, G. A., & Tonry, J. L. 1992, this volume

Tonry, J. L., & Davis, M. 1979, AJ, 84, 1511

Tonry, J. L., & Wyatt, W. L. 1988, CFA Z-Machine Data Analysis Software, Cambridge: Smithsonian Astrophysical Observatory

Tody, D. 1986, in Instrumentation in Astronomy VI, ed. D. L. Crawford, Proc. SPIE, 627, 733

THE THEORY AND APPLICATION OF FABRY-PEROT PHASE CORRECTION WITHIN IRAF

PATRICK SHOPBELL AND JONATHAN BLAND-HAWTHORN
Rice University, Houston, TX 77251-1892

GERALD CECIL
University of North Carolina, Chapel Hill, NC 27514

INTRODUCTION

Work is currently under way to implement an imaging Fabry-Perot data reduction package within IRAF[1]. Over the course of the past 6 years, a vast array of software tools for this purpose have been developed by us within the ZODIAC environment, a mathematical parser originating with George Miyashiro at the University of Hawaii. We are transporting this code to IRAF, re-working the algorithms and re-structuring the data paths to produce a coherent data reduction and analysis package. Spectrophotometric Fabry-Perot data is a merger of CCD imagery and spectra, and the package structure will therefore be made to closely resemble current approaches within IRAF to both of those aspects. For a discussion of the important stages of Fabry-Perot data analysis, see Bland-Hawthorn et al. (1992).

This poster treats a complex area of reductions unique to Fabry-Perot data: the removal of "phase surface" effects arising from the inherent instrumental profile of a Fabry-Perot etalon. In order to obtain a cube of individual CCD frames which are monochromatic, this three-dimensional structure present in all Fabry-Perot data must be parameterized. In a manner directly analogous to the dispersion correction of longslit spectra, these surfaces are measured, calibrated, and removed. This process is known as "phase correction."

THEORY OF PHASE CORRECTION

The instrumental response of a Fabry-Perot etalon to a monochromatic source is given by the Airy function. In terms of the actual observed data structure, or data "cube", it is written as:

$$I(x,y,z) = \frac{I_0}{1+\alpha^2 \sin^2(\beta z \cos(\gamma\sqrt{(x-x_0)^2+\epsilon^2(y-y_0)^2})+\delta)}, \qquad (1)$$

where I_0 is the intensity of the input beam, (x_0, y_0) defines the optical axis, and $\alpha, \beta, \gamma, \delta$, and ϵ are properties of the optical system (e.g. plate detector scale,

[1]IRAF is distributed by the National Optical Astronomy Observatories, which is operated by the Association of Universities for Research in Astronomy, Inc. (AURA) under cooperative agreement with the National Science Foundation.

wavelength dispersion). Note that this function is periodic in all dimensions. Proper selection of the instrumental parameters, however, can be used to limit the resulting monochromatic surfaces to a single period of the function. The resulting form technically obeys a $(\sec\theta - 1)$ relation, although it is approximately paraboloidal, with the coefficient of curvature independent of wavelength and dependent solely on instrument characteristics (Bland & Tully 1989). It should be noted also however, that in order to reduce "ghosting" from internal reflections, the optical axis is usually placed at the edge of the field, resulting in only partial Airy surfaces, which greatly complicates the correction scheme.

The desired result is a cube of images, each monochromatic and sequential in wavelength, which can thereby also be interpreted as a two-dimensional array of spectra. This package seeks to parameterize and remove the Airy surface distortion in a rigorous manner, transforming the raw data into just such a form. The problem is dealt with in two stages. First, the surface itself is fitted and removed in three dimensions. Second, temporal changes in the surface are monitored by fitting the two dimensional rings which arise from the intersection of the surface and the z frames of the cubes.

APPLICATIONS OF PHASE CORRECTION

Three Dimensional Phase Surface Fitting

The fitting of the incomplete monochromatic Airy surface, described by Equation 1 above, has been designed as a set of four IRAF tasks. These tasks exactly parallel, and are in some cases identical to the four tasks involved in the dispersion correction of two-dimensional spectra within IRAF (Anderson 1987). The names given to the tasks correspond accordingly.

The overall algorithm is to determine the surface from a fit of the Airy formula to a given line in a calibration lamp cube. This fit is then used to remove the surface, both in the calibration cube as a check, and in the actual data cubes.

IDENTIFY: The initial phase correction task allows the user to identify the lines within a calibration lamp cube. The user denotes the wavelengths of a few lines appearing in a single spectrum from the cube. The task then attempts to identify the other major lines appearing in the spectrum using its internal tables. The user is allowed to verify the identifications.

REIDENTIFY: Assuming the spectral offsets are small from spectrum to spectrum across the cube (i.e. the Airy surface curvature is small), the next task automatically attempts to locate the identified lines in other spectra of the cube. Again, user verification is allowed. In this manner, the spectral lines are followed through the cube, tracing out the form of the Airy surface.

AIRYFIT: Having traced the Airy surface, it is then fit using the IRAF interactive non-linear fitting routines, allowing optional user interaction for discarding outliers, etc. As verification, the computed fit parameters may at this point be compared directly to physical constants of the optical system, as noted earlier.

PHCORRECT: Using the generated fit, the Airy surface distortion may be removed from the cubes. Usually the surface is first subtracted from the calibration lamp cubes, resulting in perfectly flat frames at the wavelengths corresponding to lamp emission. Prior steps may be iterated until the calibration cube frames are satisfactorily flat, at which point the task is performed on the actual data cubes.

Two Dimensional Ring Fitting

Parameterization of changes in the Airy surface makes use of the two dimensional rings of emission resulting from the intersection of the Airy surface with a single z frame. Periodically throughout an observing run, a single calibration ring corresponding to a particular etalon setting (i.e. wavelength) is taken. Each ring can be fitted with an ellipse and changes in the parameters and corresponding "constants" of the optical system can be examined over time. The fitting of the rings has been implemented as a single IRAF task.

RINGFIT: Again using the interactive non-linear least squares tools provided within IRAF, an ellipse can be fit to the shape of each ring. The resulting ellipse parameters give valuable information on the position of the optical axis, the wavelength response of the etalon, and the geometric stability of the system. The most crucial measurement is that of drifts in the ellipse dimensions, which usually reflect undesired changes in the optical gap between the etalon plates, often the result of changing humidity, pressure, etc. In addition, the approximately Lorentzian cross section of the ring is also fit, yielding the "ring sharpness", a measure of the instrumental resolution.

CONCLUSIONS

This poster has outlined the Fabry-Perot phase calibration software currently under development by the authors, under collaboration with the IRAF group, most specifically with Mr. Francisco Valdes. This software will be part of a general IRAF Fabry-Perot data reduction package, constructed in parallel with the `onedspec` and `twodspec` spectral reduction packages already within IRAF. When completed, this package should provide the astronomical community with a coherent set of routines for the rigorous reduction and analysis of photometric Fabry-Perot data.

The authors would like to express their thanks to the National Optical Astronomy Observatories, and the IRAF group in particular, for their continued support in this project. P.L.S. and J.B.H. are partially funded through NSF grant AST 88-18900.

REFERENCES

Anderson, E. 1987, Reduction of Longslit Spectroscopic Data Using IRAF, IRAF documentation

Bland, J., & Tully, R. B. 1989, AJ, 98, 723

Bland-Hawthorn, J., Shopbell, P., & Cecil, G. 1992, this volume

'TAUCAL' — A SOFTWARE PACKAGE FOR CALIBRATION OF FABRY-PEROT SPECTROSCOPIC DATA

JAMES R. LEWIS & S.W. UNGER
Royal Greenwich Observatory, Madingley Road, Cambridge, CB3 0EZ, United Kingdom

ABSTRACT The advantages of Fabry-Perot spectroscopy over slit spectroscopy are often ignored owing to the difficulty of calibrating the resulting data. The calibration process involves a complex 3–dimensional rebinning of a spectral line data cube. In this paper we describe a new approach to this problem, which breaks down the original 3-dimensional problem into two more manageable 1-D and 2-D problems. The new algorithms have been implemented in the 'TAUCAL' software package. The software was originally written with the TAURUS instrument in mind, but in principle can apply to any astronomical Fabry-Perot system.

PHASE CALIBRATION OF FABRY-PEROT DATA

Fabry-Perot spectroscopy is a powerful technique for determining the morphology and velocity field of extended emission-line objects; HII regions, planetary nebulae, supernova remnants and galaxies. A Fabry-Perot interferometer can be regarded conceptually as a tunable filter; by varying the spacing of the Fabry-Perot etalon we can vary the wavelength transmitted. By taking a series of images at slightly different wavelengths and stacking them together we can generate a spectral line data cube.

The problem is that the wavelength transmitted is not just a function of etalon spacing, but also varies across the field of view. For small angles of incidence on the etalon, the condition for constructive interference is given by:

$$n\lambda = 2\mu\ell\left(1 - \frac{R^2}{2F^2}\right)$$

where:

n is an integer (order number)
λ is the wavelength
μ is the refractive index of air
ℓ is the etalon gap
R is the radial distance to the optical axis on the detector
F is the focal length of the camera used

If we set λ equal to a constant this becomes the equation of a paraboloid of revolution. This equation defines the surfaces of constant wavelength in a

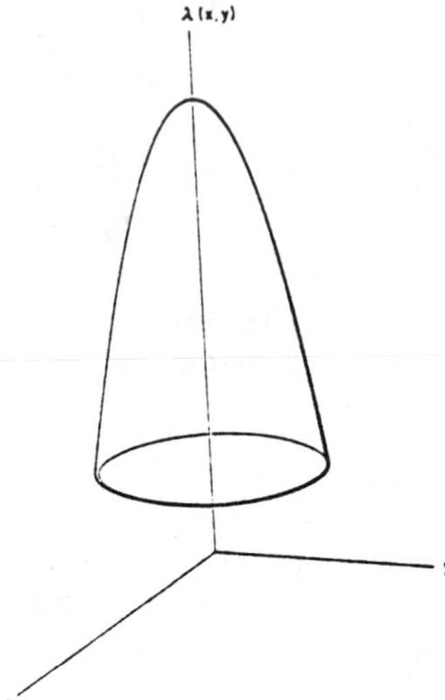

FIGURE I A sketch of a Fabry-Perot phase surface (e.g. surface of constant wavelength) centred on its optical axis. The two horizontal axes are spacial. The vertical axis corresponds to either the gap width or wavelength, although the latter is clearly a function of the (x, y) position.

raw spectral-line data cube produced by a Fabry-Perot interferometer (see also Figure I). It can be seen that the spectral line data cube has spatial and spectral axes which are not separable. Before any useful information can be obtained from Fabry-Perot data it is necessary to rebin the data cube, applying a shift to each spectrum so that the surfaces of constant wavelength become planes of the data cube. The amount of the shift as a function of position in the image is referred to as the phase-map, and the process of determining the shifts is referred to as 'phase-calibration'.

In order to do this phase calibration the various parameters of the phase map need to be defined. This involves fitting the surface defined above to the data in a 3-dimensional raw data cube. Fortunately it is possible to simplify the problem, by first considering each plane of the data cube individually. For a constant etalon gap ℓ and constant wavelength, equation (1) reduces to the equation of a circle (R = constant). The TAUCAL package therefore first fits a ring of radius R, centered on the optical axis, to the data in each plane of the data cube. The next step is to plot R^2 vs. the frame number (i.e. etalon gap). The result should be a family of straight lines, one for each interference order. The gradient of these lines and the intercepts of these lines give us all the parameters of the phase surface. This procedure thus breaks a difficult problem in three dimensions into two much simpler problems of two and one dimensions, and solves for all of the necessary parameters of the phase surface without loss of any of the calibration information. Once these parameters are determined, it is a straight forward matter to correct the data such that the wavelength scale is independent of the (x, y) position within the cube.

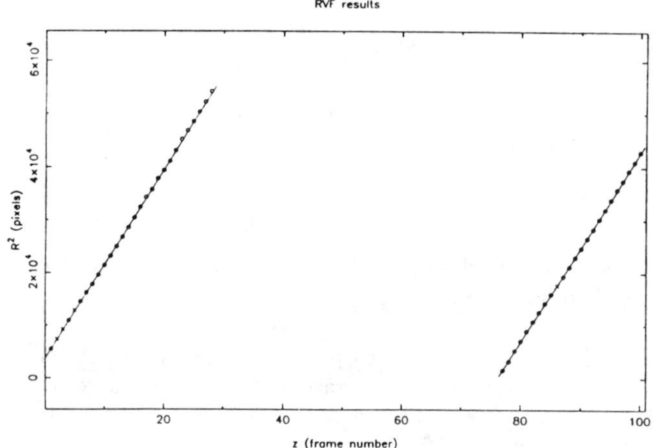

FIGURE II A plot of a typical Radius vs. Frame relation. Notice that data from more than one order are represented on the plot.

OTHER FACILITIES IN TAUCAL

TAUCAL also has the ability to correct for distortions in the spatial domain which are caused by the detector used in the observations. This is accomplished by an analysis of the spacing of reseau marks on a test exposure. Once the distortions have been mapped out a two dimensional axis array is generated for each of the spacial axes for use in subsequent processing of the data. This is important for detectors which are geometrically distorted, such as the IPCS, since this distortion will result in poor determination of the phase map.

A facility for creating an instrumental data base also exists in TAUCAL. This data base is used for information on both the observations and the reduction results. It provides a useful central repository for data which the user may need and a means for programmes within the package to communicate with each other. It can be stylised to include information on an arbitrary number of instrumental setups.

Finally, TAUCAL includes a program which does simple gaussian and cauchy function fits to phase calibrated spectral line data cubes. It can be run either interactively or in an automated way. An editing facility allows the user to view and edit the fits. Also included are facilities to create fit parameter or velocity maps, to display two dimensional images and to do rough statistics of resulting fit parameters.

TAUCAL is written to run within the FIGARO environment and is available upon request from the authors. The algorithms used in the package are described in more detail by Lewis & Unger (1991).

REFERENCES

Lewis, J. R., & Unger, S. W. 1991, RGO publication

STATISTICAL ANALYSIS TOOLS FOR THE STUDY OF TURBULENCE IN HII REGIONS

E. BOILY
Département de Physique, Université Laval and Observatoire du mont Mégantic, Québec, CANADA, G1K 7P4

ABSTRACT We have developed two new IRAF packages; a *non-scanning* Fabry–Perot interferogram reduction package and a radial velocity analysis package. The latter is the analysis tool for the velocity fields produced by the Fabry–Perot package.

Since a stationary Fabry–Perot device was used, the spatial sampling on the HII region is non uniform and incomplete. The velocity package is specially designed to deal with this fact. It permits to graphically visualise different local statistical characteristics (sigma, mean, minimum, maximum, median) of the velocity field, to isolate or remove systematic velocity trends and to compute different statistical functions like the autocorrelation or the structure functions in 1D or 2D for the turbulence study.

We also present some results from the analysis.

INTRODUCTION

The presence of turbulence in different components of the interstellar medium is now well accepted. It's existence in molecular clouds has been clearly demonstrated (Kleiner & Dickman 1987). However, if there are strong evidences for the presence of turbulence in HII regions (Castañeda 1988; Roy & Joncas 1985; O'Dell 1986; Joncas & Roy 1986), there is no true confirmation. To eliminate this uncertainty, new observations were obtained to improve the database and an ameliorated software package was developed for data reduction and analysis of this database.

OBSERVATIONS

We observed the Galactic HII regions S158 (NGC 7538) in the Hα and [OIII]λ5007 lines, and M17 and Orion in [SII]$\lambda\lambda$6717, 6731. A servo-stabilized piezoelectrically step-scanned Fabry–Perot interferometer attached to the Université Laval $f/8 \rightarrow f/3.5$ focal reducer was used. This instrument was mounted on the 1.6 meter telescope of the Observatoire du mont Mégantic. The apparatus is described in Joncas & Roy (1984). The interferograms were recorded on a 320×512 RCA CCD camera directly attached to the focal reducer. We isolated the Hα and [OIII] lines with interference filters having a bandwidth of

10 Å. A 50 Å filter was used to observe the [SII] doublet. This way, both lines were measured at the same time, thus improving our sampling of the nebulae and our efficiency at the telescope.

Each stepped interferogram is composed of a calibration frame, at least 3 object frames and a post-object calibration frame. This way, the mechanical stability of the Fabry–Perot is verified as well as the quality of the data. The 3 object frames represent a minimum number of frames in order to reach an acceptable signal to noise ratio and to efficiently remove the cosmic rays.

DATA REDUCTION AND ANALYSIS

We developed two IRAF packages for the data reduction and analysis of our observations. The CCD frames are previously treated using the IRAF package CCDRED.

The Fabry–Perot Package

The radial velocity map of the object is obtained with the Fabry–Perot package from the comparison of the object and calibration interferograms. The data reduction is divided in two parts. The pre and post CCD calibration frames must be reduced with the Fabry–Perot package before beginning the reduction of the object interferograms. On each calibration, circles were fitted on the rings with a standard least-square method in order to get the center and the radius of each ring. For the object frames, an adjustable-width radial vector, starting at the derived center, follows the intensity variations along it's path. The whole frames are scanned by varying the vector's position angle. A peak detection algorithm is used to determine the local ring position along each vector. Each position on the object ring is compared to the appropriate calibration circle in order to compute the velocities.

The package provides an interactive graphical user interface for setting the multiple reduction parameters and for filtering out the noise. The package also corrects for the wavelength difference between the observed and the calibration lines and for the geometrical distortions caused by the optics of the focal reducer. Most of the distortions have a radial dependence and thus have a direct influence on the determined velocities.

The intermediate files are all STSDAS tables and the final output is a text file containing the coordinates and the velocity.

The Velocity Package

The velocity package has been specially designed to work with the velocity fields produced by the Fabry–Perot package. The goal of the package is to compute the statistical functions like the autocorrelation, the correlation and the second order structure function in 1D and 2D. We do not use Fourier transforms since our velocity fields are not uniformly sampled. The method we use (the direct method) is very CPU demanding but the accessibility to supercomputers keeps the calculation time within reasonnable limits. In order to compute the statistical functions, the velocity field is filtered out to remove the systematic velocity components by fitting a parametric surface. The package also provides a number of tools to graphically visualize and manipulate the data such as: production of a theoretical velocity field, display of local histograms, construction of velocity maps, computation of the statistical moments of the field, extraction of the systematic velocity component(s) (large-scale motions), display of the velocity cube in different velocity interval slices, and so on.

SCIENTIFIC RESULTS

We computed so far the autocorrelation and the second order structure functions for the velocity fields of the Galactic HII regions S158, S142 (Joncas & Roy 1984; Roy & Joncas 1985) and M17 (Joncas & Roy 1986) on which we measured respectively 5000, 22 000 and 11 000 velocity points representing 12.5 millions, 240 millions and 60 millions lag samples. The functions were computed on a Convex 220 supercomputer and a SPARC1+ workstation. *The three regions* have a correlation length of about 0.1*pc*. This is a clear indication that turbulent motions are present. Kleiner & Dickman (1987) observed Heile's cloud 2 in the Taurus molecular complex and they also found a correlation length of 0.1*pc*. If this is not fortuitous, there may be a physical process that limits the size of the biggest turbulent cells in the interstellar medium.

The computed structure functions give an indication that the Kolmogorov energy cascade model must be adapted for a compressible and supersonic medium. We found a Kolmogorov exponent which is closer to 1/3 — 1/4, the prediction of Henriksen (1984, 1988) for star cloud turbulence.

ACKNOWLEDGMENTS

I am grateful to Dr. Gilles Joncas for designing the Fabry–Perot procedure, paying my residence fees at the conference and revising this document... after all, he is my thesis advisor! I also wish to thank the Smithsonian Institute for paying my travel expenses and registration fees, Dr. Daniel Durand for writing most of the SPP code of the Fabry–Perot package and MM. Richard Martineau and Guy Plante for their help in preparing the poster.

REFERENCES

Castañeda, H. O. 1988, ApJS, 67, 93

Henriksen, R. N. 1984, ApJ, 287, 200

Henriksen, R. N. 1988, ApJ, 331, 359

Joncas, G., & Roy, J. R. 1984, PASP, 96, 263

Joncas, G., & Roy, J. R. 1986, ApJ, 307, 649

Kleiner, S. C., & Dickman, R. L. 1987, ApJ, 312, 837

O'Dell, C. R. 1986, ApJ, 304, 767

Roy, J. R., & Joncas, G. 1985, ApJ, 288, 142

SNOOPY: A DATA REDUCTION AND ANALYSIS SYSTEM FOR A MID-INFRARED ECHELLE SPECTROMETER

JEFFREY M. ACHTERMANN
Astronomy Department, University of Texas at Austin, Austin, TX 78712

INTRODUCTION

The Snoopy software system was written to reduce and analyze data taken with the University of Texas IR-Echelle, named Irshell (Lacy 1989). Irshell is a mid-infrared (5–25 μm), high resolution cryogenic echelle spectrometer that uses a 10×64 Si:As detector array to measure ten 64-point spectra. Snoopy's two major strengths are an intelligent parser that treats arrays as ordinary variables and a display program that allows multiple windows and data sets.

The Snoopy software system consists of two programs: Snoopy, responsible for manipulating the data; and Woodstock, the display program. Both programs are written in C and were designed to run on Sun Microsystem computers. There are two versions of Woodstock, one for Sun Microsystem's SunView interface and the other for the OpenLook interface of the X Window System. Snoopy and Woodstock communicate through TCP client–server connections.

SNOOPY'S CAPABILITIES

Command Line Parser

Snoopy is written around an expression parser that understands numbers, mathematical operators, variables, and functions. All operations involve either setting a variable equal to an expression or calling a function. The following interaction demonstrates how to subtract the continuum from a line plus continuum source as a way to show how the expression parser works.

(1) First, raw data is loaded, corrected for spikes, flat fielded, calibrated, and then stored in an array, which we'll name 'data' (but could be any name desired). 'Data' will be automatically created and dimensioned. This routine also implements the infrared observing technique of subtracting the off beam from the on beam. Assume 'data' is a 10×64 array that contains ten 64–point spectra.

```
:->data = raw f c rawdatafile
```

The f parameter turns on flat fielding and the c parameter will cause the data to be flux calibrated.

(2) The array 'data' is plotted for a quick peek. The 'a' specifies a plot window. Twenty-six (a-z) plot windows are available. The type of plot is selected by menus in the plot window.

```
:->plot data a
```

(3) Assume that pixels 54-64 of the 64-point spectra contain only continuum. An average value for the continuum in each of the ten spectra is calculated by binning these 11 data points and dividing by 11. 'Continuum' will be a 10×1 array. The bin function will add together 1 pixel in the x dimension and 11 pixels in the y dimension.

```
:->continuum = (bin data[,54:64] 1 11) / 11
```

(4) A simple line flux can be computed by binning the pixels containing the line and subtracting the continuum. Assume the line lies between pixels 20 and 35. The resulting array, 'lineflux', will have dimensions of 10×1 – ten flux measurements, one for each of the ten spectra.

```
:->lineflux = (bin data[,20:35] 1 16) - 16 * continuum
```

For another example 'data' will be corrected for airmass and divided by the normalized spectra of a star to get rid of atmospheric lines.

```
:->result = (data^1.2)/(norm(star))
```

The ^ raises the array to a power. The function 'norm' returns a normalized array.

Reduction Capabilities

Built into Snoopy are the reduction procedures required by infrared observing and several specialized for Irshell:

- Chopping the secondary : On source - Off source
- Nodding the telescope : On source - Off source
- Chopping the secondary while nodding the telescope
- Telescope scan - Rapidly scan the telescope and use off source observations on either end of the scan to correct for sky fluctuations.
- Telescope chop and scan - Chop the secondary while scanning the telescope.
- Grating scan - Piece together spectra that partially overlap in wavelength.

Several procedures are carried out when raw data is converted to a usable form. Spikes are detected by comparing neighboring pixels and corrected by interpolation. A noise array is calculated from the standard deviation of pixel variations with time. The data is flat fielded and flux calibrated.

Snoopy contains non-linear fitting routines to automatically calculate line shapes and fluxes. For more complicated numerical fitting Snoopy exchanges data with the excellent least squares fitting program GaussFit (Jefferys 1990).

To create publication quality graphs Snoopy produces input files for the Lick Mongo plotting program (developed at Lick Observatory). Possible output includes multiple spectra and contour plots. Snoopy will automatically scale the axes for cm^{-1}, μm, and RA (in hours).

Most complicated functional enhancements to Snoopy are performed by additions to the source code. Snoopy does, however, have a simple macro language which allows loops and functions.

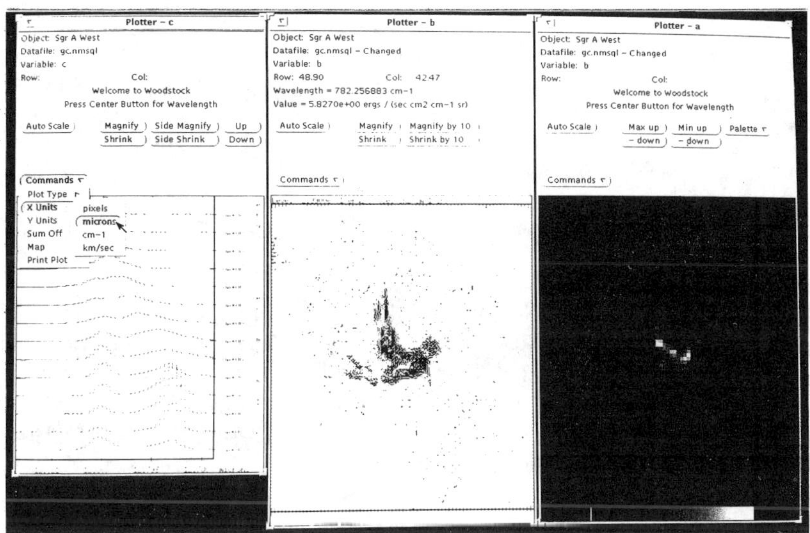

FIGURE I Multiple Instances of the Woodstock Plotting Program.

WOODSTOCK'S CAPABILITIES

One of the main advantages of the Woodstock plotting program is that multiple instances of the program may appear on screen simultaneously (Figure I). Each instance of the plotting program may display a different data set and use a different method of displaying the data. Through its menus and buttons Woodstock allows the control of many functions:

- Plot type: Multiple Spectra, Contour Map, Color Image, Movie (Scan through a 3–dimensional array).
- Autoscaling.
- Rescaling to a desired level.
- Send the plot to a printer.
- Click mouse button to determine cursor location, flux, and wavelength.
- Toggle on/off whether the screen pixels will be square.
- Select axes labels (cm^{-1}, μm, km/s, pixels, and flux units).
- Parameters for contour plotting, such as whether contours below or equal to zero will be plotted.
- Select a color palette.
- Allow fiddling of the color palette.

REFERENCES

Lacy, J. H., Achtermann, J. M., Bruce, D. E., Lester, F. F., Arens, J. F., Peck, M. C., & Gaalema, S. D. 1989, PASP, 101, 1166

Jefferys, W., McArthur B., & McCartney, J. 1990, BAAS, 22, 927

APPLICATIONS OF SYNTHETIC PHOTOMETRY

D. BAZELL
Space Telescope Science Institute, 3700 San Martin Dr., Baltimore, MD 21218

ABSTRACT The *Synphot* package developed at STScI allows HST users to model observations and calculate expected count rates using a database of HST optical component efficiencies and an internal spectrum/passband modeling language. The structure of the database is discussed and it is shown how other telescopes can make use of the same software by making changes to the database.

INTRODUCTION

Synthetic photometry currently is being used at STScI to provide a uniform method of calibrating photometric and spectrophotometric observations obtained by HST. The basic ideas behind synthetic photometry are discussed briefly below. We explain the concept of the passband generator that is used to assemble the appropriate sensitivity curve for any HST observation. We then describe the implementation of the passband generator as part of the *Synphot* package within IRAF. Synphot can be used as a driver for synthetic photometry calculations for any set of passbands in the database. Thus, if passband have been measured for any telescope, Synphot can be used to calculate expected count rates and flux for any planned observation.

PASSBAND GENERATION

The Synphot package was designed after the XCAL routines developed by Keith Horne at STScI. XCAL runs only under VMS and it was desired to have a portable version, hence Synphot was developed. One of the fundamental pieces of the Synphot package is the *passband generator.* The passband generator takes as input a keyword string that represents a possible optical path through the telescope. An example would be a string like foc,f/96,f195w,prism1. This is interpreted to mean: use the f/96 relay of the Faint Object Camera, the f195w filter and the far UV objective prism. Other components such as the throughput of the OTA and detector sensitivity are included by default. Each keyword in the string represents an optical component that has an associated efficiency as a function of wavelength, which is stored in the form of an STSDAS Table. Each instrument has an associated *Directed Graph* in the form of a graph table, which specifies the allowed combinations of optical components for that instrument. For each component in the graph table there is an associated keyword, an innode and an outnode. A path through the graph table is defined by a list of component keywords with the outnode of one component equal to the innode of another component with increasing node number.

The passband generator traces a path through the directed graph by going from innode to outnode. If an invalid observation mode string was specified, the software traces the path as far as possible and then prints a list of the possible keywords for the next node. If the mode string is valid then the passband generator multiples together the efficiencies of each component in the graph path and stores it in an array that is passed back to the calling subroutine.

Using this technique one can determine the efficiency of any valid HST observing mode. At STScI we also have made available numerous non-HST passbands (e.g. Johnson UBVRI) to allow for easy comparison with ground based observations. By creating a database and directed graph representing any set of optical components for any telescope, the same software can be used to dynamically generate passbands from observation mode keyword strings for that particular telescope.

CAPABILITIES OF THE SYNPHOT PACKAGE

In addition to dynamically generating passbands, the Synphot package within STSDAS has a number of other useful capabilities. We list some of these here without going into great detail.

The *CALCPHOT* task allows calculation of expected count rates and flux within a given passband for a particular object. The user chooses the output units and specifies the observation mode and input spectrum. The spectrum can be read from a file or generated using the Synphot mini-language.

The task *PLBAND* allows plotting of passbands, specified as either an observation mode string or using the Synphot mini-language. The *PLSPEC* task will plot passbands and spectra along with spectrophotometric and photometric observations. It will also calculate synthetic photometry based upon the specified observation mode.

Several other tasks exist for the comparison of data including *PLRATIO* which produces ratio plots of photometry and spectrophotometry, and *PLTRANS* which creates photometric transformation plots.

Two tasks are also included for fitting models to observations. These are *FITSPEC* and *FITBAND* which, as their names imply, fit models to spectra and to passbands. These have been used for modeling deviations of the expected throughput of filters from actual observations.

DISCUSSION

The techniques of synthetic photometry allow a user to model expected count rates and flux for specific observations. The Synphot package relies upon the passband generator to produce the desired passband from a database of efficiencies of individual optical components which includes filter, gratings, mirrors windows and detectors. For HST it is necessary to dynamically generate passbands because there are more than 1000 optical components producing over 10^9 possible observing mode combinations. The method of representing passbands using directed graphs is very general and applicable to other telescopes. Further information about the Synphot package can be obtained from the author. Consult the references for more detailed information on synthetic photometry.

REFERENCES

Horne, K. 1988, in New Directions in Astronomy, A. G. Davis Phillip, D. S. Hayes, & S. J. Adelman, eds., L. Davis Press, 145

Koornneef, J., Bohlin, R., Buser, R., Horne, K., & Turnshek, D. 1986, in Highlights in Astronomy, J.-P. Swings, ed. (Reidel: Dordrecht), 7, 833

THE ARCHIVAL REPROCESSING OF IUE DATA. I. IMAGE REGISTRATION AND PHOTOMETRIC CORRECTION

M.D. DE LA PEÑA†, R.A. SHAW‡, H.A. BUSHOUSE†, AND J.S. NICHOLS-BOHLIN†
†Computer Sciences Corporation, IUE Observatory, 10000-A Aerospace Road, Lanham, MD 20706
‡Space Telescope Science Institute, 3700 San Martin Drive, Baltimore, MD 21218

INTRODUCTION

A new image processing system has been developed by the NASA IUE Project in anticipation of the final archival reprocessing of the IUE dataset. We present the techniques employed by the new system (NEWSIPS) which produce a more accurate photometric correction of IUE data. The improved photometric accuracy is achieved through the combination of two critical components. The Intensity Transfer Functions (ITFs) (*i.e.*, graded ultraviolet floodlamp exposures used to linearize and normalize the camera response) are constructed in a manner which eliminates the necessity to resample, and consequently degrade, the data and preserves the inherent characteristics of the detector. In addition, a more precise image registration technique is utilized to align accurately the raw science data with the ITFs. Since the variability of the geometrical distortions, as introduced by the SEC Vidicon cameras, guarantees that the raw science images are never perfectly aligned with the ITFs, an accurate registration is crucial to avoid introducing systematic noise into the photometrically corrected data which will ultimately manifest itself as a reduced signal-to-noise ratio (SNR) in the extracted spectral data.

CONSTRUCTION OF THE ITFS

The modern epoch ITF series for each camera was obtained in a closely monitored and stable satellite configuration such that any parameters known to affect the image quality were carefully controlled. These improved exposure procedures ensured that the constituent ITF images at each exposure level share the same image geometry, allowing us to construct the ITFs in their own geometric space. The advantage of this construction method is the elimination of the necessity to resample, and therefore smooth, the data before co-adding the constituent images to produce each ITF mean level. Each level is then created using an enhanced averaging procedure which excludes transient artifacts and measures the noise characteristics independently for each pixel.

IMAGE REGISTRATION

A set of reseaux on the camera faceplates are intended to be used to rectify, and therefore align, the science and ITF images. However, they are difficult to detect at low exposure levels and many cannot be used if they fall on the exposed portion of a spectral order. Because of these severe limitations on the utility of the reseaux in an explicit registration method, predicted reseau positions are used in the current IUESIPS image reduction software for alignment. Unfortunately, these mean positions are imprecisely known, and the application of a mis-registered ITF (by more than about a few tenths of a pixel) manifests itself as a source of systematic noise in the photometrically corrected image, and ultimately in a degraded SNR in the extracted spectrum. Therefore, the primary goal of this registration method is to define a means to determine fiducial positions to sub-pixel accuracy such that the proper image alignment can be achieved.

Our new approach uses the "fixed pattern", the inherent pixel-to-pixel sensitivity variations found in all raw IUE images, as the registration fiducial, and a cross correlation algorithm as the registration mechanism. Such an approach has several advantages, perhaps the major one being that these fiducials can be located to a greater precision and on a finer spatial grid than the instrument fiducial marks for each raw science image. Moreover, the positions of the fixed pattern fiducials can be chosen to address better those regions where the optical aberrations (hence, the geometrical distortions) are the greatest. In contrast to the reseaux, the fixed pattern can be detected at even the lowest exposure levels, which eliminates the need for predicted information.

The linear cross correlation procedure incorporates several statistical verifications to secure the most accurate mapping between the raw images and the ITF. As it is the fixed pattern in the background portion of the images which is used for pattern matching, the presence of the exposed spectrum is successfully excluded from the correlation templates through an iterative exclusion algorithm where DN values which are larger than the median by a chosen factor, κ, times the square root of the median, σ, are ignored. Therefore, the spectral information does not inhibit the use of the fixed pattern in the surrounding background regions.

PHOTOMETRIC CORRECTION

The new photometric correction procedure accesses the *explicitly* determined displacements between the raw science image and the ITF to align the two images, in contrast to using predicted information. Because of the three critical components: the more accurately determined image alignment, the non-degraded ITF, and the ITF being resampled only once during its application to the raw data using a piece-wise, bi-cubic spline interpolation scheme, the photometrically corrected images produced by the new image processing system have a significantly improved SNR over the current data.

The improvement in the photometrically corrected data is evident in Figure I which is a plot of cross-cuts taken through both the photometrically corrected image (PI) produced by NEWSIPS and the current IUESIPS system.

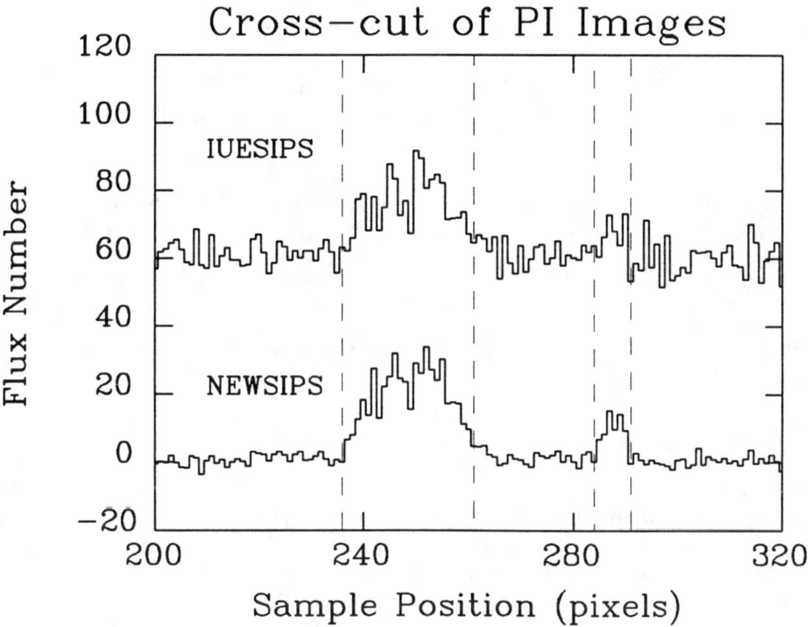

FIGURE I Spatial cut at Line=300 of a NEWSIPS PI image and the corresponding IUESIPS PI Image. The IUESIPS data has been displaced vertically for clarity. The dashed lines denote the approximate boundaries of the large and small apertures where there is detectable flux in this spectrum.

Both large and small aperture spectral information is present in these images. The improvement in the SNR in the background data is distinctive. More importantly, the small aperture data can now be clearly distinguished from the noise. The amount of improvement observed in the photometrically corrected data is ultimately realized as an increased SNR in the extracted spectrum. Proper photometric correction removes the large non-random component of noise, the fixed pattern, from the data and should permit a continuing increase in the SNR when co-adding spectra, limited only by the SNR in the ITF itself.

REFERENCES

De La Peña, M. D., Shaw, R. A., Linde P., & Dravins, D. 1990, Proc. Int. Symp. 'Evolution in Astrophysics', ESA SP-310, 617

Nichols-Bohlin, J., De La Peña, M. D., & Shaw, R. A. 1990, Proc. Int. Symp. 'Evolution in Astrophysics', ESA SP-310, 623

Shaw, R. A. 1990, Proc. Int. Symp. 'Evolution in Astrophysics', ESA SP-310, 621

THE ARCHIVAL REPROCESSING OF IUE DATA. II. GEOMETRIC CORRECTIONS AND SPECTRAL EXTRACTION

H. BUSHOUSE, M. DE LA PEÑA, AND J. NICHOLS-BOHLIN
Computer Sciences Corporation, IUE Observatory, 10000-A Aerospace Road, Lanham, MD 20706

R. SHAW
Space Telescope Science Institute, Baltimore, MD 21218

INTRODUCTION

Raw IUE images suffer from spatial distortions introduced by the SEC Vidicon detectors and the dispersion direction lies at an angle of about 45 degrees relative to the image axes. Furthermore, the dispersion is not linear within the spectral orders. The combination of these effects can make the task of spectral extraction and subsequent analysis very difficult. Therefore our goal is to create a geometrically corrected IUE image in which the echelle spectral orders are all parallel to one image axis and the dispersion is linear within each order, thus providing an image format that is best suited to scientific analysis. In this paper we describe the techniques that will be used to accomplish the geometric corrections, as well as the subsequent spectral extraction process and the use of "flag" images which contain information on problem conditions associated with each pixel of the science images.

GEOMETRIC CORRECTIONS

Our approach to geometrically correcting an IUE image is to construct a vector field that maps each pixel from its instrumental raw space to a geometrically rectified space. These vectors include the displacements between the science image and the Intensity Transfer Function (ITF), the displacement of the reseau pattern (camera fiducials) in the science image from its original grid due to the distortions introduced by the Vidicon cameras, the rotation of the spectral format to lie along pixel rows, the displaying of the orders for high-dispersion mode, and the small change of scale needed to linearize the dispersion. Low-dispersion observations of extended sources will receive an additional correction to account for the fact that the major axis of the oval-shaped spectrograph aperture is not exactly perpendicular to the dispersion direction, thus resulting in a slight artificial broadening (in wavelength space) of spatially extended features.

The dispersion linearization correction is characterized using a third-order Chebyshev polynomial. The use of a polynomial with orthogonal terms is necessitated by the fact that the dispersion relation is intimately tied to the geometric rectification. The form of the linearization correction is determined empirically by first geometrically correcting a set of wavelength calibration images using no linearization correction and then fitting a third-order Chebyshev polynomial to the nonlinear wavelength residuals. This function is then retro-fitted into the geometric correction process by adding the wavelength linearization vector to the other remapping vectors. This then forces the geometrically corrected images onto a linear wavelength scale.

The various displacement vectors necessary to accomplish all of these corrections are computed independently and then coadded to produce one set of vectors which are used to remap pixels from their coordinates in raw image space into the geometrically corrected space. In this way the image data are resampled only once throughout the entire processing sequence. The resampling of the image to its corrected space is accomplished using a modified Shepard technique. This technique conserves flux well and also minimizes smoothing of the data.

ERROR FLAG IMAGES

In addition to the spectral images, the IUE Final Archive will also maintain images containing error flags to complement the data images at each stage of the processing. These are 16-bit integer images with a zero value indicating no known problems and bit-encoded negative values indicating up to a maximum of fifteen unique problem conditions associated with a given pixel in the corresponding science image. Problem conditions that are accounted for by the processing system include such things as telemetry dropouts, permanent camera blemishes, reseau mark contamination, saturated pixels, bright spots, cosmic ray hits, and microphonics.

The flag image is first constructed in the raw space of the science image. It is then resampled to the space of the low- and high-dispersion geometrically corrected images using the same vector displacements calculated for the science data. However, instead of actually resampling the flag values (which would of course destroy their bit-encoded nature) they are reassigned to pixels in the geometrically corrected flag image based on the amount of flux from the original raw science image pixel that is being resampled into neighboring pixels in the resampled image.

SPECTRAL EXTRACTION

One-dimensional spectra are extracted from the two-dimensional geometrically corrected images using a weighted slit extraction technique as described by Kinney, Bohlin & Neill (1991). The advantages of this method as compared to a simple boxcar extraction are: (1) The extraction improves the signal-to-noise ratio of the spectrum while conserving total flux; (2) Most cosmic ray hits are automatically removed; (3) The output includes an error estimate for each extracted point.

The weighted-slit extraction empirically determines the cross-dispersion spectral profile shape by fitting high-order polynomials as a function of wavelength to the normalized spectral data. The routine then uses the profile information along with an empirically determined model of the detector noise to calculate a weighted sum of the cross-dispersion flux and an error estimate at

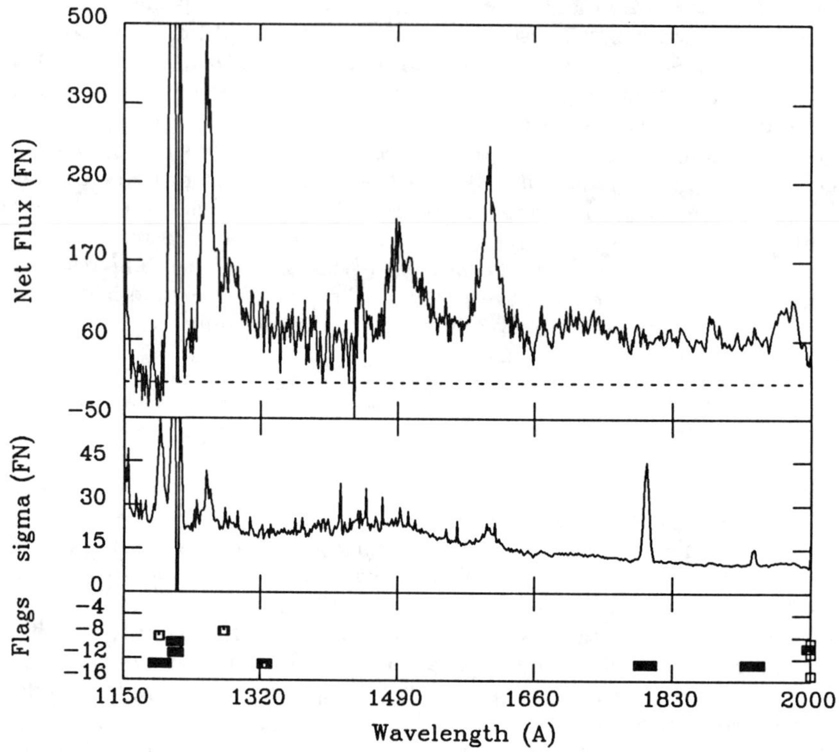

FIGURE I Extracted spectral data for an observation of 3C120, including the net flux, the sigma spectrum, and flag information. Some flag values of interest here are bright spots (-7), saturated pixels (-11), and reseau contamination (-13). Notice the increase in the estimated sigmas in the flagged regions.

each wavelength interval along the spectrum. Through the use of the profile, noise model, and flag image information the routine is able to identify bad data points (such as cosmic ray hits or reseaux) which are excluded from the calculation of the integrated flux at a given wavelength. The final extracted data file contains several one-dimensional records including the fitted background, the net flux spectrum, the absolutely calibrated net spectrum, the estimated error (or "sigma") spectrum, and a flag spectrum which contains information on problematic pixels lying along the spectrum as determined from the two-dimensional flag image (see Figure I).

REFERENCES

Kinney, A., Bohlin, R., & Neill, D. 1991, PASP, 103, 694

ANALYSIS OF *COBE FIRAS* SKY SPECTRA VIA *IDL*

R.E. EPLEE, JR. AND R.B. ISAACMAN
General Sciences Corporation, Code 685.3, NASA/Goddard Space Flight Center, Greenbelt, MD 20771

S.M. READ, K.A. JENSEN, AND L.P. ROSEN
Hughes STX Corporation, Code 685.9, NASA/Goddard Space Flight Center, Greenbelt, MD 20771

E.S. CHENG AND R.A. SHAFER
Code 685, NASA/Goddard Space Flight Center, Greenbelt, MD 20771

INTRODUCTION

The Far Infrared Absolute Spectrophotometer (*FIRAS*) on board the Cosmic Background Explorer (*COBE*) satellite collected over 2 million spectra of the sky during its 10-month orbital lifetime, resulting in the measurement of the spectrum of the Cosmic Microwave Background Radiation (CMBR) to an unprecedented accuracy. The spectra are binned into 6144 pixels on the sky using a quadrilateralized spherical cube format; the sky is divided into six faces, each composed of 32×32 2.6° pixels. Homogeneous spectra in each pixel are averaged and then calibrated. To achieve the ultimate sensitivity of the instrument, the calibrated spectra must be analyzed collectively over the entire sky. *IDL* is the software system used in this analysis. We have tailored the *IDL* environment for *FIRAS* use by developing or importing the *IDL* routines discussed below. There is an ongoing effort to develop tools common to the three *COBE* instruments (the Far Infrared Absolute Spectrophotometer, the Diffuse Infrared Background Experiment, and the Differential Microwave Radiometers).

INCORPORATION OF EXISTING *IDL* ROUTINES

Any astronomical data analysis system must perform the basic tasks of data input/output, data manipulation, and image processing. We have incorporated a number of existing *IDL* routines which perform these functions into the *COBE IDL* environment. Currently, we have imported various astronomical analysis tools, FITS format file input/output routines, and astronomical database tools from the Ultraviolet Imaging Telescope (on *ASTRO*) and Goddard High Resolution Spectrograph (on *HST*) Projects. These routines are available through the *IDL* Astronomy User's Library.

INTERFACES TO THE *COBE IDL* ENVIRONMENT

We have developed several Fortran subroutines, callable from *IDL*, that serve as interfaces to the *IDL* environment. These Fortran subroutines are dynamically linked to the *COBE IDL* environment by the *IDL* `Linkimage` procedure.

The *COBE* Project software is written in Fortran and runs under VAX/ VMS; its database system is built around VAX Datatrieve and binary data files whose formats are defined in the VAX Common Data Dictionary. Interfaces between the *COBE IDL* environment and the *COBE* database system are provided by two Fortran subroutines. One program provides *IDL* users with access to particular data fields in the *COBE* time-ordered data archives, while the other provides similar access to the *COBE* skymap-ordered data archives.

The *FIRAS* data processing software uses as its plotting utility the *PLT (PGPLOT)* software package from the X-Ray Analysis and Data Utilization (*XANADU*) system. *PLT* has a built-in least-squares curve fitting capability that includes both predefined and user-supplied models. In order to provide a common plot format for both the *FIRAS* processing software and the *COBE IDL* environment, and in order to provide *IDL* users with access to the *PLT* curve fitting routines, a Fortran subroutine serves as an interface between the *COBE IDL* environment and the *PLT* software package.

FIRAS-SPECIFIC *IDL* ROUTINES

Within the *COBE IDL* environment the *FIRAS* sky spectra are stored in the spherical cube format, each face of which constitutes a three dimensional dataset whose dimensions are 32×32×the number of frequencies. It is difficult to visualize what is happening to the CMBR over the entire sky from such a dataset. Accordingly, we have developed several *FIRAS*-specific *IDL* routines to manipulate the spectral data in this format, fitting models to the spectra on a pixel-by-pixel basis. Fitting routines solve for the following:

1) the temperature and optical depth of galactic dust for spectra spanning a frequency range of 20-100 wavenumbers (a wavelength range of 0.1-0.5 mm);

2) the temperature of the CMBR and the optical depth of galactic dust, using the dust temperature derived in 1), for spectra spanning a frequency range of 1-20 wavenumbers (a wavelength range of 0.5-10 mm); and

3) the dipole anisotropy of the CMBR temperature.

These fitting routines produce skymaps of spectrum (or temperature) residuals for each pixel. The residual skymaps are analyzed using other *IDL* routines to:

1) evaluate the quality of the instrument calibration and to identify systematic errors in the calibration;

2) model distortions of the CMBR spectrum from that of a blackbody;

3) construct atomic and molecular line emission maps of the galaxy; and

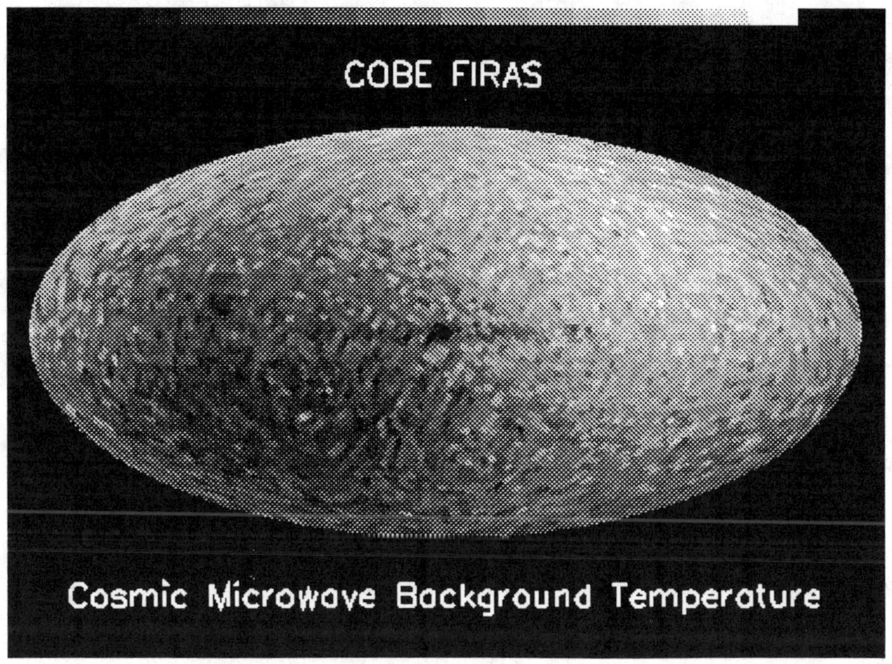

FIGURE I Hammer-Aitoff projection in galactic coordinates.

4) model spatial anisotropies of the CMBR.

The model parameters, line emissions, and average residuals produced by these routines can be displayed as two-dimensional skymaps. We have developed a number of *IDL* routines to display data that are stored in the spherical cube format. Sky data can be displayed in one of three projections (unfolded spherical cube (see White, 1991, these Proceedings), Hammer-Aitoff, or global sinusoidal) and in one of three coordinate systems (ecliptic, galactic, or equatorial). A map of the CMBR temperature is shown in Figure I.

COBE GUEST OBSERVER INTERFACE

The *COBE* Guest Observer Interface is currently under development and will be available to the astronomical community when the initial *COBE* datasets are delivered to the National Space Science Data Center in mid 1993. Among other capabilities, the Guest Observer Interface will allow users to manipulate and display *FIRAS* data using *IDL* in the *COBE IDL* environment.

ACKNOWLEDGMENTS

This research is supported by the Astrophysics Division of NASA's Office of Space Science and Applications.

SECTION 11

DATA ACQUISITION, PIPELINE PROCESSING AND QUALITY ASSURANCE

DATA REDUCTION FACILITIES FOR THE WILLIAM HERSCHEL TELESCOPE

S. W. UNGER
Royal Greenwich Observatory, Madingley Rd, Cambridge CB3 OEZ, UK

ABSTRACT A wide range of instruments are either available now or are planned for the William Herschel Telescope; CCD imaging, slit spectroscopy, echelle spectroscopy, Fabry-Perot spectroscopy, multi-object spectroscopy using aperture masks and fibres, spectropolarimetry. The provision of adequate data reduction facilities is essential (but often forgotten) if these instruments are to realise their potential.

In order to provide adequate facilities for on-line data reduction, RGO plan to interface a Sparcstation 2 to the existing VME based data acquisition system. The data analysis software which might be used on this workstation includes a commercial image analysis package (SunVision), as well as IRAF and FIGARO.

OVERVIEW OF THE WHT CONTROL SYSTEM

The instruments already commissioned on the William Herschel Telescope (WHT) include the intermediate–dispersion ISIS double spectrograph, the low–dispersion Faint Object Spectrograph, the Utrecht Echelle Spectrograph, the TAURUS Fabry–Perot interferometer, as well as various CCD imaging options. Multi–object spectroscopy using aperture masks (LDSS) will be commissioned early in 1992, whilst a fibre positioner (AUTOFIB) and dedicated fibre spectrograph (WYFFOS) is under construction. In order for these instruments to be used effectively, it is essential to provided adequate data reduction facilities, and in this paper I describe RGO's current plans in this area.

In order to set the question of on–line data reduction for the WHT into context, I first give an outline description of the existing control and data–acquisition system. A block diagram of the WHT computer and microprocessor system is shown in Figure I. This system differs substantially from the earlier systems used on the Isaac Newton and Jacobus Kapteyn Telescopes, in that all the instrument and detector controllers are microprocessor based. The telescope itself is controlled by a dedicated microVax II computer. We therefore have a highly distributed control system, in which the detailed control of individual mechanisms is delegated to the local controllers.

Overall supervisory control of the system is provided by a VAX 3600 computer running VAX ADAM. This is responsible for the coordinated operation of telescope, A&G unit, instrument and detector, the aim being to maximise observing efficiency. Communication between the VAX 3600 system computer and

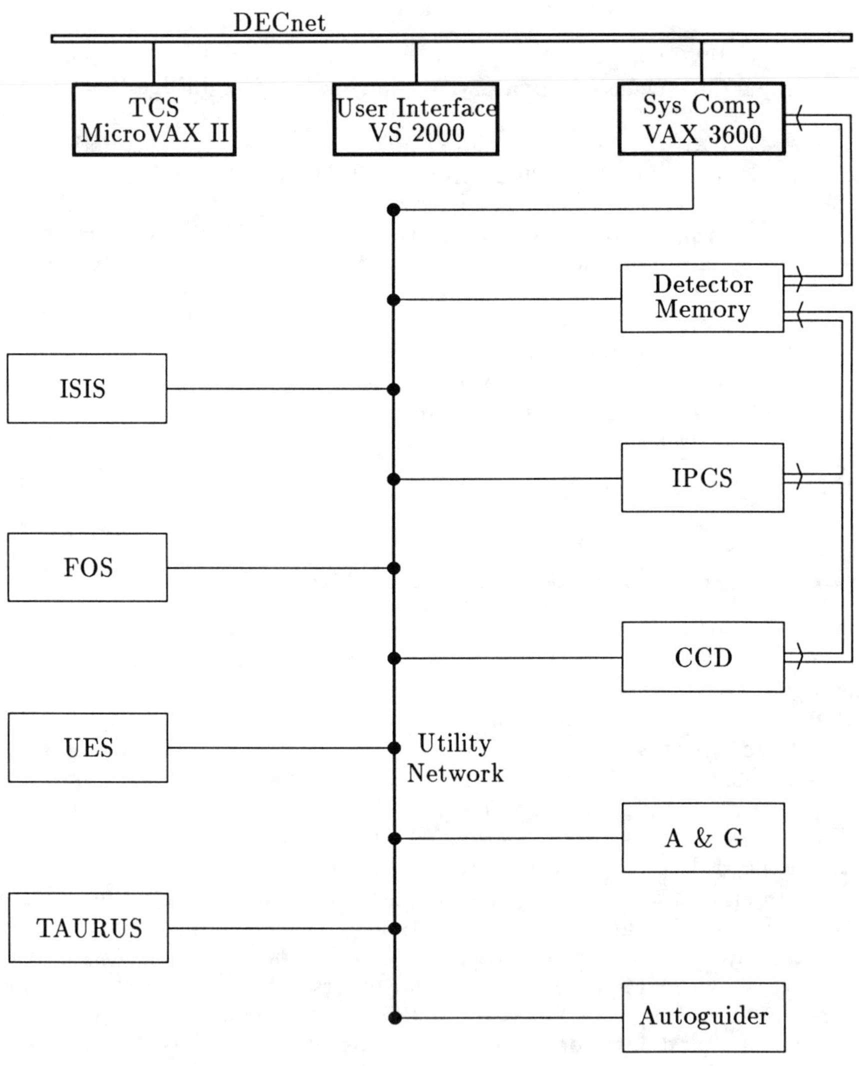

FIGURE I **WHT System Configuration**

the microprocessor-based local controllers is provided across an Ethernet based Utility Network, whilst communication with the Telescope Control System is via DECNet. The user interface is provided by a VAXstation 2000, which provides a means of entering commands to the system, and provides feedback to the user via a multi–page mimic display.

This paper is primarily concerned with that part of the system responsible for data–acquisition, and this appears at the top right–hand corner of Figure I. Data from CCD detectors and from the IPCS photon counting detector is collected by a Detector Memory System (DMS). This is a VME system based around a 68020 microprocessor, with 32 Mbyte of memory. The data acquired into the DMS memory can be displayed on a local colour image display, and some basic image processing applications (e.g. image statistics) are available for rapid data assessment. The data is then transferred over a high speed parallel link to the VAX 3600, where is is merged with header information into an HDS format file, and archived. It is then also possible to transfer the data over DECnet from the 3600 to one of several data–reduction VAX workstations (not shown in Figure I) for further analysis.

THE DATA REDUCTION PROBLEM

Over the last few years two major developments have greatly increased the load on the WHT data reduction facilities.

The first important development, noted by a number of speakers at this meeting, is the increasing availability of large CCD detectors. In Table I I summarise the formats of the CCD devices expected to be used on the ISIS and UES spectrographs, amd for prime focus imaging on the WHT. As a comparison I also list the device used for the last few years on the Intermediate Dispersion Spectrograph (IDS) on the Isaac Newton Telescope. I have calculated expected framesizes assuming the data is stored as R*4, and have estimated typical data–rates assuming an average of 50 exposures per night. It is clear that data–rates have increased by at least an order of magnitude, and are now close to 1 Gbyte/night.

TABLE I Typical data–rates

Instrument	CCD device	Format	Framesize (Mbyte)	Data–rate (Mbyte/night)
IDS	GEC P8603	385×578	.89	45
ISIS	2 × EEV 88300	2 × 1280×1152	11.8	590
UES	EEV 88500	2186×1152	10.1	504
Prime	Tektronix	2048×2048	16.8	840

The second major development is that not only is the volume of data increasing, but so is the data 'density'. A principal goal of most recent work on new spectrographs is to make full use of area detectors, by packing as much information as possible into a given area of silicon. Multiobject spectrographs such as LDSS and WYFFOS make it possible to observe up to 150 objects simultaneously, whilst cross-dispersed echelle spectrographs such as the UES provide a large wavelength coverage for individual objects by making it possible to observe a large number of interference orders simultaneously. Future observers will want to extract not one spectrum from each data frame, but a hundred !

The existing system for on-line data assessment and reduction on the WHT will not be able to handle these developments. The Detector Memory System itself is fast, but provides limited functionality. The VAXes have poor performance, and the process of getting data from the detector onto the VAXes is too slow for them to be useful for on–line data assessment.

THE PROPOSED SOLUTION

Our aim must clearly be to provide rapid access to data that has just been acquired, from a fast Unix workstation, running a standard data reduction package such as IRAF or FIGARO.

The hardware solution we have adopted is to purchase a SPARCstation 2, partially for hardware compatibility with the rest of the astronomical community, but also because it turns out to be possible to provide a very direct interface between the SPARCstation and our existing VME–based data–acquisition system. An SBus to VME adapter (Performance Technologies model PT–SBS915) provides transparent access to data held in the VME memory from the SPARCstation. It will therefore be possible to use software running on the SPARCstation to start analysing data as soon as its been acquired.

The software running on the SPARCstation will include a data acquisition monitor. This will provide a window in which the current status of each detector is displayed. Some basic image statistics and/or image display might also be provided automatically, as well as a means of exporting data into a standard reduction package. We have found the software provided by SUN for building graphical user interfaces extremely useful in designing the data acquisition monitor. It is clear that the use of this type of commercially available software can significantly reduce the costs of developing a new system.

Probably the most contentious issue for the new system is the question of what data reduction packages to support. Over the last few years the RGO has adopted the FIGARO package as the basis for all reduction software produced for La Palma, as have ROE for the telescopes on Hawaii and AAO for the AAT. A variety of applications have been produced targeted at particular WHT instruments, including software produced at RGO for the batch processing and optimal extraction of CCD spectroscopy, software produced at UCL for the analysis of echelle spectroscopy, software produced at Durham University for the analysis of multi-object spectroscopy, and software produced at RGO for the analysis of Fabry–Perot spectroscopy.

However, whilst the FIGARO package is extremely popular with users, the version currently used by UK astronomers has not yet been ported to Unix. Although a Unix version of FIGARO has been produced at Caltech, the relationship between the two versions of FIGARO is not clear. In any case, the level of support available internationally for the FIGARO system is rather low compared to that available for systems such as IRAF and MIDAS. IRAF is already being increasingly used in the UK as a result of the Space Telescope project, and it seems likely that the Gemini project (if funded) will over the next few years further increase interchange between the UK and US of data reduction software. Our current plans are therefore to make IRAF the principal (but not the only) package for on–line data reduction at the WHT.

One other option we have been exploring is the use of commercial image processing packages, and we currently plan to make the SUNVision package available to WHT observers. Commercial packages such as SUNVision and AVS show great potential, providing both general–purpose image–processing applications, and tools for building astronomy–specific applications. It is possible that by building on these commercial systems we will in the future be able to provide much more powerful data reduction systems than has been possible in the past.

THE LOOK AND FEEL OF THE CFHT PEGASUS SOFTWARE SYSTEM

J. KERR, B. GRUNDSETH, S. SMITH, AND J. WRIGHT
Canada-France-Hawaii Telescope Corp., P.O. Box 1597, Kamuela HI 96743

ABSTRACT Taking into account the diverse community of users as found at most observatories, a user interface needs to be quickly and easily mastered by the occasional user, and yet be able to provide the power and customization capability that the engineering staff demands. At CFHT, we believe that the PEGASUS system has met the challenge of providing such an interface.

INTRODUCTION

The PEGASUS software system is a graphical user interface (GUI) for acquisition systems developed to run on UNIX machines supporting the X window system and OSF / Motif. It is both a GUI style guide, providing a uniform "look and feel" for all the CFHT acquisition systems, and a complete programming environment to aid software developers in providing a GUI without having detailed knowledge of window programming.

Visiting astronomers can intuitively navigate through the PEGASUS system because all the tools they need are visually made available to them. Pointing with a mouse and selecting an option is much less demanding, especially at 14,000 feet, than trying to accomplish the same task with a command driven interface.

However, members of the CFHT staff sometimes need to bypass the graphical user interface software and manipulate the hardware more directly. By using the easily configured PEGASUS software, they are able to design their own customized instrumentation software and interact with it via windows, command-line processing, or UNIX shell programming.

THE GRAPHICAL USER INTERFACE

The user normally sees the session as depicted in Figure I. That is, across the top of the workstation screen is a menubar with pushbuttons on it, a feedback window at the bottom of the screen, and a large workspace in the middle of the screen in which user interface (UI) forms appear.

By conforming to a consistent style throughout all of the acquisition systems, we build software with which the users can quickly understand how to interact. In turn, this frees them to concentrate on the observing tasks at hand.

The user quickly learns that clicking on an object with the left mouse button

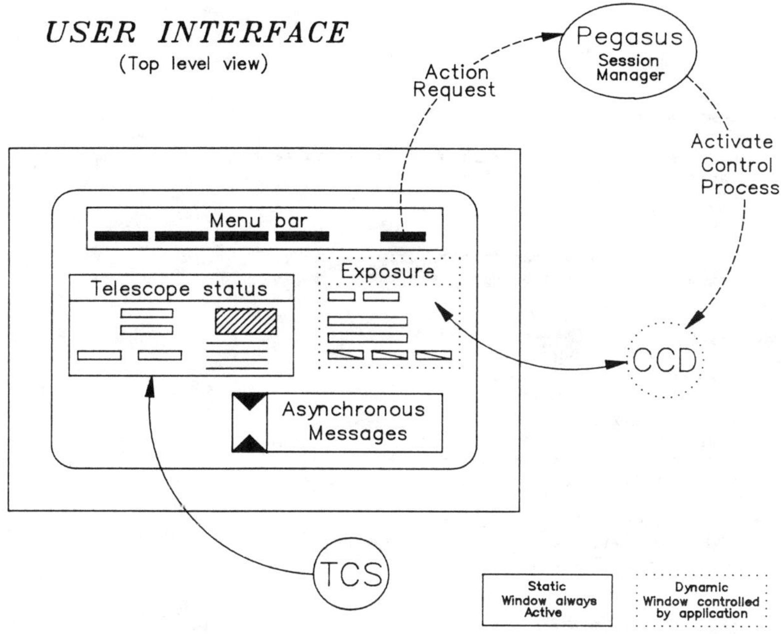

FIGURE I Session Overview

will result in some action being taken. In the case of the menubar, he knows that clicking on a button will cause a form to appear in the workspace. When the form is presented, the user enters information, closes the form, and then a process will be executed to perform some task. The task will do its job and terminate. The feedback window will keep the user informed of any issues which need attention and provide a running log of events as they unfold.

All of the forms which appear have the same set of action buttons across the bottom:

Accept – does the requested action and unmaps the form,

Save Values – saves the form parameters and unmaps the form,

Defaults – puts preset defaults into the form,

Cancel – forgets everything entered and unmaps the form,

Help – invokes the help system.

The consistency afforded by the PEGASUS environment also makes the job of the support staff much easier because there is less need to cross-train them on the various acquisition packages.

```
\#
\# PEGASUS Session Manager Configuration Section
\#

session -> focamsession

ccdsession = ccd raster focus fraster chip iqe modes image graph
    routefits suggs news aloha
    .
    .
    .
focamsession = ccd raster focus caf fraster chip focam iqe
      modes image graph routefits suggs news aloha

alohalabel = aloha
alohainvoke = aloha
alohahelp = Log out of this observing session.
ccdlabel = Expose
ccdinvoke = ccd -X
ccdhelp = Take a CCD exposure.
ccdsetlabel = Coude Setup
ccdsetinvoke = xform -p coudeseth.par -f -invoke
ccdsethelp = Setup of Coude gratings and Hartmann masks
iraflabel = IRAF
irafinvoke = xterm -t -n IRAF -geometry 80x24+1+160
            %800x600-50-125 -e runiraf
irafhelp = Start up an IRAF session.
```

FIGURE II Menubar Section of a Configuration File

SESSION CONFIGURATION

It is quite easy for the engineering staff to configure a session such as the one depicted in Figure I. Residing in each user's home directory is a localized, session-specific configuration file. It contains information about which buttons appear on the session menubar, what to do when a button is pressed, and what help text should be displayed. Figure II depicts an excerpt from a configuration file showing the ease with which the system may be defined.

In the excerpt of the configuration file shown in Figure II, the session line is automatically setup by the installation scripts. The PEGASUS Session Manager (PSM) will follow the session-pointer to the current observing session – in this case "focamsession".

Looking further down the file, you will see focamsession defined on the left side of an "=" sign. The string on the right side will be further dereferenced. PSM will parse the string, take a token, ccd, for instance, and append the strings "label", "invoke", and "help". For the case of ccd, we derive "ccdlabel", "ccdinvoke", and "ccdhelp".

Note that the program invocation lines may be anything which you can type at a shell prompt. The most common programs seen here are xform and the controller programs such as ccd. In the configuration file are also found instructions for exposure sequencing, which detector is being used, and other session-specific details. Without this file, nothing happens except an error message from the system.

It should be quite clear that configuring this system is easy and requires no window programming knowledge. Even changing the text of the system to another language, French, for instance, is as easy as editing the configuration file.

WINDOW DEFINITION

The average user interacts with the data acquisition system via windows; each window being a graphic representation of choices for the user. A variety of shapes and options, referred to as widgets, are available to X programmers. The PEGASUS system is built around a selection of Motif widgets and CFHT-developed "meta-widgets". By limiting ourselves to, and standardizing upon, these widgets we are able to provide a software development environment that separates the developers from the complexities of window programming and provides a consistent "look and feel" for every piece of software we develop.

A variety of PEGASUS widgets are available to the designer when creating a form window. Currently, the following "meta-widgets" may be used:

- title bar,
- radio button and radio button multiple,
- push button and push button invoke,
- check box,
- static text and static raster,
- text edit.

A configuration file (called a par file) must be created for each form. The par file contains the grammar explaining the position of each widget within the window as well as the labels and variables necessary for the proper functioning of the window. The file may also contain the programs invoked as a result of the choices selected. The UI routines in the support library parse through each par file and decipher its contents. Then an application, either a controller task or the program xform, paints the results in a window.

THE CFHT ACQUISITION METHODOLOGY

At CFHT, we use the PEGASUS system to do the user interface work and control the data acquisition processes. Within our software architecture, we have two types of acquisition programs – controllers and handlers.

Controllers do the more intensive UI work and provide sequencing of handlers, while the handlers are designed to isolate hardware knowledge from higher level software. All of these programs are command line invokable, making them perfect for shell programming and xform interfacing.

Figure III depicts an overview of the CFHT acquisition system. The data flow depicts a typical CCD exposure sequence. When the user presses the "Expose" button on the menu bar, the PSM retrieves the invoke line from the session

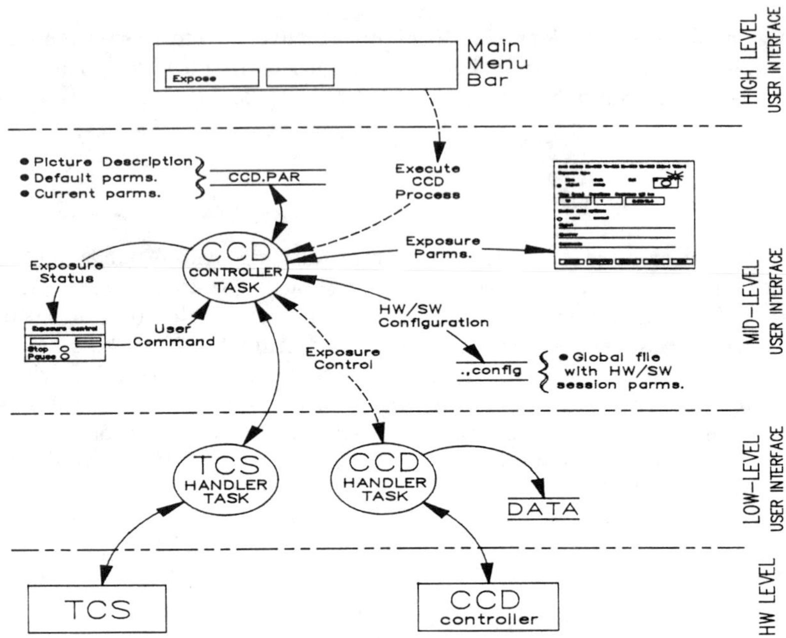

FIGURE III Dataflow Diagram for a CCD Exposure

configuration file in the current directory. It will then execute the process "ccd" which will in turn popup a dialog form. When the user is through modifying the pertinent parameters, he presses the "Accept" button on the form and the exposure sequence starts.

The CCD process may invoke several pre-exposure instrument-specific handlers, (just which ones to execute are defined in the session configuration file), popup an exposure control form, and execute the ccd handler process to acquire the data. When the exposure is completed, the CCD process will execute post-exposure handlers and then the process terminates.

CONCLUSIONS

The PEGASUS software system has allowed us to rapidly develop, and easily maintain approximately 15 different data acquisition packages during the first 18 months of its existence. All of the sessions have a common "look and feel" and all of the sessions have the same acquisition methodology. Visitors, as well as staff members, have found the system to be quite user-friendly and easy to learn. The software is now being used at several other institutions for data acquisition and prototyping graphical user interfaces and is available via ftp. We believe that is how success is defined.

AUTOMATIC ON-LINE DATA ANALYSIS FOR AN INFRARED LONG SLIT ARRAY SPECTROMETER

J. M. STEWART, S. M. BEARD, C. M. MOUNTAIN, D. A. PICKUP
Royal Observatory, Edinburgh, EH9 3HJ, U.K.

A. BRIDGER
Joint Astronomy Centre, Hilo, HI 96720

INTRODUCTION

CGS4 is a state-of-the-art long slit infrared spectrometer, developed at the Royal Observatory Edinburgh and installed on the United Kingdom Infrared Telescope on Mauna Kea, Hawaii in February 1991. It operates in the 1-5 μm region of the spectrum (Mountain et al. 1990), using a 58x62 InSb array. Integration of the various control, data acquisition and data reduction subsystems was facilitated through the use of the ADAM software environment (Kelly & Chipperfield 1991; Lawden 1989).

On-line automatic data analysis was made an integral part of the project from an early stage for two important reasons. First of all it is necessary to allow proper data assessment by the observer so that telescope time is used most effectively. Secondly in order to maximize the scientific productivity of a common user instrument, it is important that the observer leaves the telescope with fully reduced data.

DATA ACQUISITION SYSTEM

The data acquisition system moves the telescope, sets the instrument configuration, acquires the raw data, writes them to disk and informs the data reduction system that data are available for analysis. It is implemented as a set of ADAM tasks running on VAX/VMS computers with front-end array control and data reduction performed by a MicroPower Pascal PDP system. (Future systems will use transputer front-ends and UNIX workstations also running ADAM.)

The data acquisition system may be driven observation by observation by the user, but is normally driven from a sequential queue, the EXEC system, which allows an entire observing programme to be defined in advance. While it is being executed it may be interrupted ('broken') by the observer between its component operations. The observer may then choose to resume the queue at its next scheduled operation, to restart the queue from its beginning or to edit the upcoming operations in the queue using a text editor. This allows the observer to interact with the predefined observing programme in response, for example, to changing weather conditions.

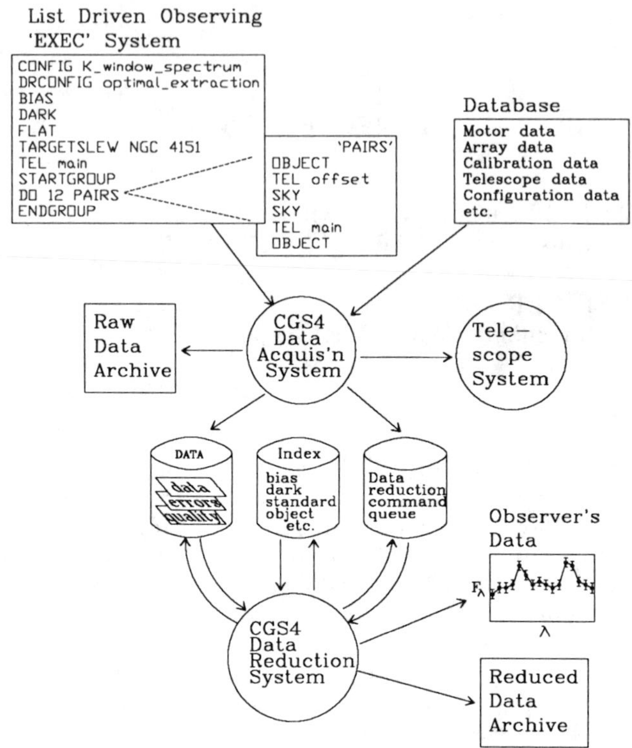

FIGURE I The CGS4 data acquisition and data reduction systems, connected solely by means of data files and the data reduction command queue, both on disk.

DATA REDUCTION SYSTEM

The CGS4 data reduction system automatically reduces and displays observations as they are made. Like the data acquisition system it consists of ADAM tasks running on VAX/VMS computers. The data acquisition system communicates with the data reduction system via a data reduction queue (Figure I), which contains a list of observations to be reduced in the order in which they were observed. This queue, which is distinct from the data acquisition queue driven from EXEC files, can also be used to communicate other instructions, such as a request to change to a new data reduction configuration. Communication between the two systems is handled entirely through disk files, which means that neither system requires the other to be present.

The disk files, in a format compatible with the FIGARO reduction system (Shortridge 1991), contain not only a data array, but also arrays indicating the quality (e.g. bad pixels) and statistical error of the data at each point. Data can also be exported using FITS.

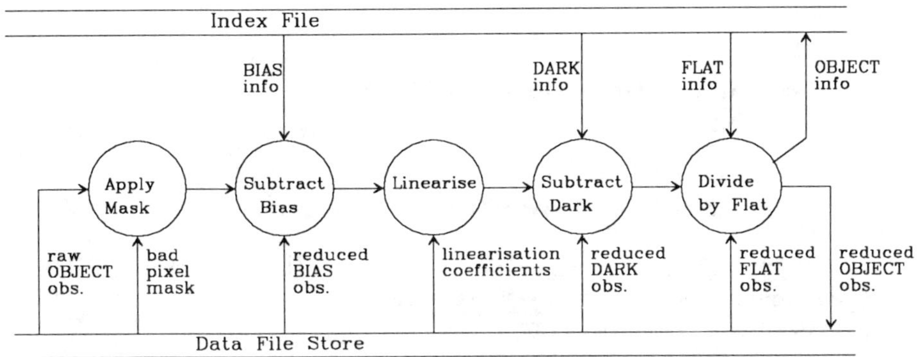

FIGURE II The flow of data through the CGS4 data reduction system for a simple OBJECT observation.

The stages in the reduction of a simple OBJECT observation are shown in Figure II. A BIAS frame is subtracted, the data are linearized, a DARK frame is subtracted, and the image is then flat-fielded by dividing it by a FLAT frame. Error and quality information is propagated at each stage.

A key feature which enables all these reduction steps to be carried out automatically is a database of previously-reduced observations, known as the 'index file'. Each time an observation, such as a DARK, is required the index file is searched for a reduced observation with suitable characteristics. Wavelength calibration data and standards used for the removal of atmospheric features are also accessed using the index file.

CONCLUSIONS

The scientific productivity of the recently commissioned CGS4 spectrometer has been greatly facilitated by making automatic on-line data reduction an integral part of the system design. The use of the ADAM software environment meant that the integration of the data acquisition system with the data reduction and telescope systems could be achieved relatively easily by a project team working on the other side of the world from the telescope.

REFERENCES

Kelly, B. D., & Chipperfield, A. J. 1991, ADAM - Guide to Writing Instrumentation Tasks, Starlink User Note 134

Lawden, M. D. 1989, ADAM - The Starlink Software Environment, Starlink Guide 4

Mountain, C. M., Robertson, D. J., Lee, T. J., & Wade, R. 1990, Instruments in Astronomy, SPIE, 1235, 25

Shortridge, K. 1991, Figaro User Guide, Anglo-Australian Observatory

IRAF DATA ACQUISITION SOFTWARE

SKIP SCHALLER
Steward Observatory, University of Arizona, Tucson, Arizona

ABSTRACT Data acquisition software written in IRAF's SPP/VOS environment is now in use at several telescopes. The advantages and success of this approach are discussed.

INTRODUCTION

This paper discusses an IRAF SPP task originally described two years ago (Schaller 1990). At that time it was running only in the laboratory. It is now used to control CCD cameras at the Multiple Mirror Telescope and at the Steward 90 inch telescope on Kitt Peak, where it has been in use for more than a year, and more recently at six NOAO telescopes on Kitt Peak.

DESIGN ADVANTAGES

The observing task is written as an IRAF SPP task, thereby immediately inheriting the rich facilities of the IRAF environment. This allows the user interface to be flexible enough to cover many different observing modes. The user is able to use both menu and command modes of operation for reviewing and altering observing parameters, and is able to specify parameters for which he should be prompted during each observation. IRAF provides this parameter manipulation capability, via hidden parameters, parameter sets, the *eparam* screen-oriented parameter editor, and the *cl* menu mode. The user is able to create custom sequences of data acquisition commands, intersperse them with data reduction commands, and run these sequences interactively or non-interactively, using *cl* scripts and background tasks. The data are recorded with all the header information needed to invoke the IRAF data reductions immediately. IRAF is used for data plotting and display, and for making hardcopies of the plots and displays. It is also used for archival recording of the data. The observing task can write any image format or data type supported by IRAF. One of the unique features of this software is the ability to deal with a variety of data types to be stored on disk or acquired from the detector, such as unsigned short, long, real, or complex. The IRAF environment also provides the tools to support networking and portability. The same source code will run on any machine that IRAF supports. This close integration of the data acquisition into the IRAF data reduction environment is one of the major advantages to this approach.

The software is written in a device independent manner. It uses a generalized interface to access the functions of each of the detector, instrument, and telescope devices. Thus, the software is able to interface easily not only to our existing CCD controllers, but also to new controllers which might be acquired in the future. The user interface always appears the same. The design uses a client-server architecture. The user interface is the client which makes requests to each of the three servers containing the actual interface to the real detector, instrument, and telescope hardware. The client and server modules can be combined into a single process or distributed among several processes. These processes can run together on one machine or be distributed among interconnected machines, possibly using dissimilar architectures or operating systems, thus allowing one or more of the above modules to run in an environment with more real-time capabilities than the others. This configuration lends itself well to remote observing.

RESULTS

This software was installed on two Steward telescopes in 1990 under the package name *ccdacq*. In 1991, NOAO adopted this software for their use at six telescopes on Kitt Peak. They made minor site-specific modifications, interfaced new server modules for their particular instruments and telescopes, and renamed it ICE. Together, a very large number of observers have used this software. This includes both resident staff and visitors, and those who were experienced with IRAF and those who were not. The software has been well received by these observers. They have found it to be simple and very easy to use.

Several other astronomical institutions have requested copies of this software. They are now in the process of adapting it to their hardware and, in a few cases, have already done so.

FUTURE

In collaboration with Doug Tody of NOAO, new device interfaces are being developed. This work will make even easier the job of interfacing new CCD controllers and other devices. It will also support the creation of more specialized client tasks, such as online photometry, drift scanning observations, laboratory calibrations, and others.

REFERENCES

Schaller, S., 1990, in "CCDs in Astronomy", ed. G. H. Jacoby, A. S. P. Conference Series No. 8 (San Francisco: A.S.P.), p. 206

CCDPHOT: AN IRAF BASED REAL TIME CCD PHOTOMETER

DOUG TODY AND LINDSEY E. DAVIS
National Optical Astronomy Observatories, Tucson, AZ 85726

INTRODUCTION

CCDPHOT is an IRAF program designed to do multi-filter photometry of single objects in real time. CCDPHOT performs all the functions of a photoelectric photometer, using a CCD in place of the photomultiplier, and software apertures in place of the aperture wheel.

The CCDPHOT project grew out of a request by the Kitt Peak scientific staff for a program which would allow CCDs to be used for single object photometry, providing a modern alternative to the aging KPNO photoelectric photometry systems. The goals of the project were to: 1) improve the efficiency of short-exposure multi-filter CCD observations of single objects, e.g. standard stars, 2) make better use of available bright time on small telescopes, and 3) explore the use of CCDs for some photometry programs previously done with a photoelectric photometer.

In practice the CCDPHOT project has developed into something which is not only a response to the request by the KPNO staff, but an important R&D project for IRAF as well, spurring progress in a number of areas. CCDPHOT is a vehicle for exploring new types of user interfaces, new program structures, and experimenting with the integration of telescope, detector, and instrument control with data reduction and quick look, in a single high level, device and system independent program. The techniques and approaches being developed for CCDPHOT will benefit IRAF programs currently under development as well as future IRAF programs, including general IRAF programs having nothing to do with data acquisition.

SCIENTIFIC REQUIREMENTS

The CCDPHOT program was designed to meet a diverse set of scientific and technical requirements. Chief among these was that CCDPHOT should function like a single-channel photoelectric photometer, producing raw photometry records containing instrumental counts and magnitudes as its primary output. However, unlike a classical photoelectric photometer, the observing and reduction algorithms should take advantage of the 2D information provided by the CCD, e.g. the focusing algorithm should be based on analysis of the 2D stellar profile, and the photometry algorithms should use digital centering techniques to automatically center the aperture on the object (and optionally autoguide the telescope). Image analysis should be used to to measure the object and sky counts simultaneously, to reject bad pixels from the sky region, and to detect bad pixels in the object region.

Effective data acquisition requires that the observer have constant feedback on the state of the instrument and telescope systems, and be able to continually monitor the quality of the data in order to guide the observations. CCDPHOT therefore had to be interactive, maintaining a live connection to the user, the telescope, and the CCD at all times. The observer had to be able to command, and receive back status from, the detector and telescope subsystems via facilities integrated into the CCDPHOT observing environment.

Most observers do not have time to learn a complex system, hence CCDPHOT had to be simple to setup and use. A typical CCDPHOT run, including instrument setup and calibration, should require knowledge of perhaps a couple dozen commands, of which only a few are used for routine observing. Nonetheless, to be flexible, CCDPHOT needed to be user programmable, i.e. the user had to be able to create new commands by combining existing commands. Although the program framework may be device independent, it should be possible to send device or system dependent commands to the detector controller or telescope system, in order to take full advantage of the capabilities of the local system to prepare special observing programs.

To provide an acceptable replacement for the old KPNO photometry programs, CCDPHOT had to be able to acquire, calibrate, reduce and record the photometry records in real time. A set of quick-look display and reduction commands were required to enable the user to evaluate the data in real time. The reduction algorithms should be able to detect, and reject and/or flag bad data automatically wherever possible. Although the image data would be reduced to raw photometry and quick look colors and magnitudes in real time, it was also necessary to be able to log all the raw image data in the event that it should later be necessary to verify the photometry.

Good program structure required that the program should call a well defined set of telescope and detector system interface routines. Knowledge of the telescope and CCD hardware should be encapsulated in device capability files or driver routines, which could be prepared for a given telescope system without need to change any of the CCDPHOT code (or the code for any similar client application). The bulk of the code should be system and device independent and usable with any telescope system or detector, once the low level system specific interfaces were in place.

BASIC DESIGN

CCDPHOT is a single program which runs under the IRAF command language. The user interface is a minilanguage, coupled with a forms-based parameter editor for editing parameter sets. An alias facility permits the user to create new commands by combining existing ones. Both IRAF commands and host level commands are available via an escape mechanism, along with detector or telescope system commands, local system permitting. In normal usage CCDPHOT is an interactive program, but it can run in batch mode, reading commands from a file or a user-supplied string.

CCDPHOT automatically saves its full run time context, including the values of all parameters and the contents of all image registers, in a binary file on disk, and restores it from one observing session to the next. It is possible to exit the program at any time and restart it later without affecting the program state.

Input to CCDPHOT consists of: 1) program and standard star lists supplied by the user, 2) a local site instrument database describing the characteristics of the local telescopes and detectors, 3) a local site photometry database describing the local filter systems and 4) default parameter sets which the user must edit at program startup and occasionally throughout the night.

The primary output from CCDPHOT is a binary file containing three types of records: 1) an observing record which describes the instrument setup, which is written each time the setup changes, 2) an object record, which is written each time the object being observed changes, and 3) the raw photometry records, which are written for each integration. Optional output includes 1) the raw and/or calibrated image data, 2) a log of the observing session, and 3) a summary of the quick-look results.

INTERNAL STRUCTURE

Each built-in CCDPHOT command is implemented using an action procedure of the form "procedure (ptr, argptrs, nargs)", where procedure is the name of the action procedure or subroutine, ptr is the pointer to the CCDPHOT data structure, argptrs is an array of pointers to the argument strings, and nargs is the number of arguments. CCDPHOT is implemented as a set of independent subsystems. These are 1) the general framework, e.g. show <parameter>, 2) command definition, e.g. alias cmd <value>, 3) star list control, e.g. cache <starlist>, 4) telescope control, e.g. preset, 5) imaging, e.g. mkzero, mkflat, radprof, 6) observing sequences, e.g. itimes <t1 t2 ...>, obs, 7) photometry, e.g. phot, and 8) quick-look reductions, e.g. reduce. During initialization the subsystems register their action procedures and parameter sets with the user interface as a set of "call-back" procedures, which are invoked by the user interface if a command is executed or a parameter is modified. The call-back mechanism isolates the subsystem code from the user interface, and should be well suited to the future addition of an optional window and panel user interface.

CCDPHOT communicates with the external telescope and detector control systems through two interfaces: 1) TCP, the telescope control system interface, and 2) DCP, the detector control system interface. CCDPHOT acts as a client, issuing commands to the telescope or detector server program. TCP is a general telescope control interface with facilities for setting the telescope position, offsetting the telescope, setting or querying the filter position, querying the telescope status, and so on. DCP is a general detector control interface with facilities for taking exposures using the full frame or any subregion, and returning data frames to the client. It is assumed that the telescope and detector servers are intelligent programs or subsystems in their own right, and that these will be unique to each telescope system or observatory. The DCP and TCP interfaces merely define a standard interface to these external control systems.

CCDPHOT's internal data structures include: 1) current values of all the parameters, 2) several image registers which hold the raw and calibrated data frames and the zero and flat field calibration frames, 3) 2 accumulation buffers which hold the current and total counts for sky and object in each filter and aperture, and 4) the default observing program.

CURRENT STATUS AND FUTURE PLANS

CCDPHOT currently supports single aperture photometry programs only, although multiple software apertures within a single region of the CCD may be defined. Both annular sky regions surrounding the object of interest, and offset sky regions are supported. Photometry is done by fractional pixel integration. Future algorithm improvements include support for optimal weighting (Stover & Allen 1987) and, depending on scientific interest and the availability of future controllers supporting the capability, support for multiple-region readout.

Successful observing runs with the CCDPHOT prototype code were made in June and November 1991, although substantial work still remains to be done on the telescope control and imaging sub-systems. Future software plans include modifications to the IRAF system to better support minilanguage programs like CCDPHOT, and the implementation of a window and panel user interface on top of the current CCDPHOT program. This will retain the current minilanguage structure, implementing the window user interface as a higher level layer.

Interested readers can find more detailed information about CCDPHOT in Tody, Davis & Valdes (1990).

ACKNOWLEDGMENTS

We are grateful to T. Kinman for his willingness to test the software and the telescope allocation committee for generous allotments of engineering time.

REFERENCES

Stover, R. J., & Allen, S. J. 1987, PASP, 99, 877

Tody, D. C., Davis, L. E., & Valdes, F. 1990, IRAF CCD Photometer Design, Draft

THE OSU PHASE II INSTRUMENT CONTROL SYSTEM

ARNE A. HENDEN AND RYLAND J. TRUAX
Astronomy Department, The Ohio State University, Columbus, OH 43210

ABSTRACT This paper presents the design for the Phase II Instrument Control System of The Ohio State University. Key points addressed are instrument transportability and a transparent remote observing interface.

INTRODUCTION

The Ohio State University (OSU) has developed an instrument control system that can operate standalone, provide debugging and setup capability, interface cleanly with a workstation computer for data archival and reduction, and provide remote observing capabilities.

The instrument control system described in Henden (1990) has been extended by adding a Data Reduction Computer (DRC, presently a Sun Sparcstation) with a Ethernet link to the AT workstation (WC) and forms the basis of our current Phase I system. It currently supports two direct imaging CCD cameras, a CCD long-slit spectrograph and an imaging Fabry-Perot spectrometer. An IR imaging system is under development and will see first light within the next few months. The system has been in operation for two years, long enough to work out most of the bugs and to gain adequate feedback from users about the features they like and dislike.

We plan to upgrade this system in gradual, phased implementations to meet changing needs. For example, hardware advances have provided new directions. Also, increases in instrumentation complexity have demanded a corresponding increase in on-line data reduction capability to provide sufficient real-time data to make informed scientific judgements.

REMOTE OBSERVING CONSIDERATIONS

A primary requirement of the OSU system is the ability to support remote observing. About 75 percent of our observing with the Perkins 1.8m telescope in Flagstaff is through a remote interface from Columbus. A night assistant on the telescope opens up at the beginning of the night, controls the movement of the telescope from object to object, and shuts down when observing is finished. The Columbus observer controls the instrument, sees the acquisition and data images, and has the same functionality as if he/she were present in the observatory warm room with the night assistant.

We have found that the night assistant must be included in the control loop to ensure instrument safety and to expedite data acquisition. We accomplish this by using a "Y" interface, where both a local and a remote keyboard can enter commands and the results are displayed simultaneously to both observers.

A further consideration is image transmission to the remote observer. Transmitting a full 512x512x8 image over our current 14Kbps phone line takes 4 minutes, which is unacceptably slow for real time observing. When ISDN is available the data rate will increase to 56Kbps, but even this speed is insufficient for unprocessed image transmission. Therefore, data compression and massaging will continue to be required to transmit the image quickly over any long-distance carrier.

UPGRADE PATHS TO PHASE II

Based on our positive experience with the Phase I system we are planning an upgrade to a Phase II system which will meet the increased needs of the observer.

One possible upgrade is to control the instrument completely from the DRC. This approach would eliminate both the WC and the Instrument Computer (IC). However, Unix is not a real-time operating system and this solution would require at least a full-image buffer between the detector and the DRC.

A less radical upgrade is to embed the instrument control software into IRAF on the DRC. This approach would eliminate the WC and have the Detector Control System (DCS) talk directly to either the Instrument Control System (ICS) or the DCS subsystems. Schaller (1990) proposed and implemented such a system for both Steward and KPNO instruments.

This approach has several advantages. The data is written in IRAF format immediately, and all IRAF data types are supported. Scripting is built into IRAF and this provides an easy interface for performing data reduction after a frame is acquired. All of the interface tools are available such as: plotting, fetch/store, command-line interpreter, and editing. This means that moving the instrument control system to a new computer just requires rewriting the low-level drivers that interface to the instrument.

However, there are disadvantages to this solution. The difficulty of interfacing to IRAF requires an IRAF guru to be available each time an instrument is added or a new release of IRAF is installed. IRAF has a poor user interface for real-time control since it does not provide updated status displays, windows, pull-down menus, etc. IRAF lacks remote-observing support. This solution requires that all users must know and use IRAF, even though they may come from some other data analysis background (such as VISTA) or intend to write their own reduction package for an instrument not available at NOAO. The system would not be highly portable; i.e., if you want to use your instrument at some other observatory you must bring your own complete solution rather than use their workstations.

We are encouraged by the Schaller approach and intend to adopt a similar system in the future. However, we think that our needs are different enough to suggest the alternative solution described below and shown in Figure I.

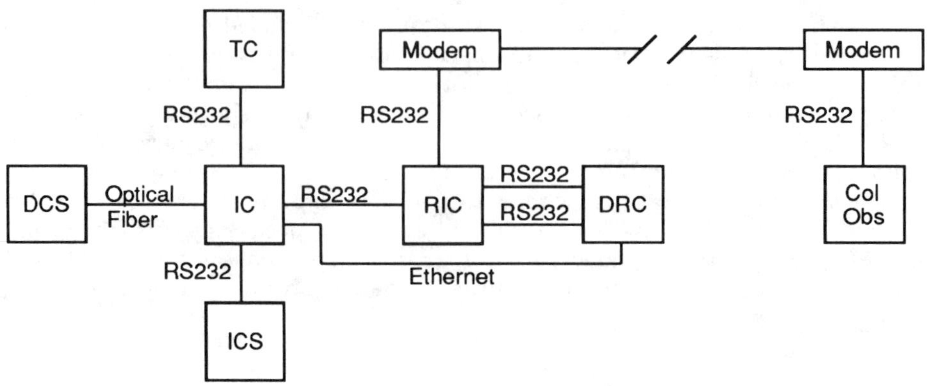

FIGURE I Proposed Phase II Layout.

PROPOSED LAYOUT

In our Phase II system, the ICS and DCS remain as in Phase I. The IC is provided with high-speed data transfer capability to the DRC. In this configuration adding an external WORM to the IC for archival storage allows the IC to operate as a standalone unit. Raw data is displayed on the IC image display; status is shown in a VT-100 window on the DRC via the RS232 link, and X-Y cuts through the image are created using Tektronix 4010 emulation on the DRC.

The AT-class WC becomes the Remote Interface Computer (RIC). The RIC provides the serial connection between the IC and the DRC, as well as the 'Y' connection data path between the remote observer and the IC. The RIC also provides a separate serial link from the remote observer to the data reduction program on the DRC (remote login), and performs the packetized image transfer to the remote observer. The RIC also acts as a backup computer system for the IC.

Files are transferred to the DRC in FITS format. For IRAF, a small task can be written in the DRC that takes each file as it is transferred and converts it into an IRAF format if desired. Or, it is feasible for the IC to write IRAF file formats directly. Files can be written asynchronously via Ethernet at the same rate as an incoming image, or a dedicated high-speed IC-DRC link can be implemented. For example, we are using a 48Mbps fiberoptic link between the IC and the DCS; a similar system could be custom-built for the IC-DRC or a FDDI-based Ethernet could be implemented with commercial products.

Note that in this configuration, one could remove the RIC completely with no software modifications if remote observing was not required. The Instrument Computer remains a standalone system that can be set up in the lab with ease. While the IC has more functionality in the sense that it must display the raw collected data and be able to archive that data in emergencies, it does not have to handle remote observing. The instrument can be moved to any observatory that has a workstation with Ethernet and a graphics terminal interface. No additional software need be developed for the workstation.

The remote observing interface is a major part of our Phase II design. By making it separate from either the DRC or the IC we achieve maximum flexibility and readily allow new types of implementation to be studied. For example, one could examine Xwindows for the local and remote observing interfaces. Also, inexpensive AT-based image compression hardware implementing the JPEG, MPEG, or wavelet algorithms can be evaluated for image transmission.

CONCLUSIONS

We have proposed a system which meets our instrument control needs by first designing an instrument that can be used in a standalone mode, then interfacing that instrument with standard ASCII and Ethernet protocols to a Unix workstation. Such a solution requires that the user learn the command language for controlling the instrument and for data reduction on the workstation, but provides great flexibility for instrument development and portability.

REFERENCES

Henden, A. A. 1990, in CCDs in Astronomy, ed. G. H. Jacoby (ASP Conference Series Vol. 8), 201

Schaller, S. 1990, in CCDs in Astronomy, ed. G. H. Jacoby (ASP Conference Series Vol. 8), 206

SOFTWARE FOR HYBRID TV SYSTEM FOR ASTRONOMICAL INVESTIGATIONS ON SMALL TELESCOPES.

LEONID Y. KOKOURINE
Institute of Astronomy, USSR Acad. Sci., 48 Pyatnitskaya st., 109017, Moscow, USSR

PETER A. GANGNUS
Institute of Astronomy, Moscow State Univ., 1 Universitetsky prosp. Moscow, USSR

INTRODUCTION

This CCD-based system provides high-speed real-time data reduction for solving non-extreme astronomical tasks which do not require extremely high accuracy and sensitivity. The system performs simultaneously real-time photometry and astrometry and can also be used as an applied guiding system for another instrument. We use the either PDP-11/23 or IBM PC/386 as data acquisition computer, and the special software performs all the real-time calculations.

THE HARDWARE

The double-frame surface-illuminated TV-mode CCD we use, has dimensions of 360×576 pixels. One of two CCD's frames (360×288 pixels) is insensitive and is used as a frame-storage buffer. In operation an image is formed on the exposed half of the CCD. At the end of desired exposure the CCD controller quickly clocks the accumulated charge to the frame buffer for reading it out while making next exposure, so there is almost no loss of the observing time because of the readout. The video signal in analogue form is transmitted to the 8-bit digitizer and then to the video memory, which is large enough to store 4 CCD frames. The video memory controller provides visualization of the video memory contents on the TV-monitor, synchronizes the computer access requests. The interface built in the computer, provides fast data link to video memory. The CCD exposure time range is 0.04...10 sec. The Direct Quantum Efficiency of the detector is about factor of 0.015.

THE SOFTWARE

Actually, the software is represented by a main program module with major functions such as hardware control, real-time measurements algorithms built-in, and a couple of external transient algorithms, which could be called from the main module. This technique simplifies the design and modifications of data processing algorithms for the system.

The main module includes following built-in functions:

a) Hardware and CCD control;

b) Real-time photometry and coordinates measurements of objects including "object guiding" facility;

c) Direct disk recording of images (or pieces of images) with four-frame buffering to reduce or exclude observation data loss.

The transient procedures are:

a) The flat field determination procedure for image correction;

b) The image tube field distortions measurement algorithm;

c) Algorithm for coordinates measurements of trailing stars; (useful for artificial satellites astrometrical observations only).

SOME ALGORITHMS IN DETAIL

The Flat Field produced by one of the external procedures is in fact a coefficients' matrix - three coefficients for each pixel: two of them represent the linear function of the dark signal in dependence on the integration time. Thus the dark signal also includes the bias (zero point) for each pixel. The third coefficient is for the sensitivity correction.

Field Distortions Determination external procedure. To calculate the distortions, we expose the artificial light sources matrix (usually 4×4) with accurately pre-measured relative coordinates onto the image tube cathode, then we determine each light source's coordinates within the CCD frame with the astrometrical algorithm (described below) and reduce the resulting coordinate differences by two-dimensional polynomial approximation. Thus, the two-dimensional matrix of the polynomial coefficients we obtain is the measure of field distortions. This procedure is periodically being run during observations to re-calculate distortions matrix.

Real-Time Measurements resident algorithm. Here we accept the following limitation: typically just few objects are to be investigated, so that most of the frame area doesn't contain significant information. So that real-time algorithms process only small apertures of the image, which contain objects. Typically stars have full width at half intensity of about 3...6 pixels or less, and we use square or rectangular apertures 20...40 pixels on a side to insure that essentially all the light is included. One important facility of this program is the ability to process moving objects, adjusting the apertures' coordinates for each subsequent frame. This "object guiding" could be useful for accurate telescope positioning according to the data issued by this procedure when observing stars as well as artificial satellites. The steps of processing in this mode are as follows:

- Sky background and dark signal correction within apertures.
- Resizing of the apertures so that they closely follow objects' shapes.
- Determination of coordinates and brightnesses of objects.
- Coordinates correction with the distortions correction matrix.
- Adjustment of each aperture coordinates so that the target appears at the center of aperture in the next frame.

The brightness is calculated simply by summing reduced data, the actual object's magnitude is determined via comparison with the artificial source of light, pre-calibrated on some standard star. Coordinate are calculated by determining the object's "mass center". Although this method provides reliable results, it is good enough to provide the accuracy of 0.1...0.2 of a pixel. The photometrical accuracy is better than 0.04. The described procedure takes no more than 0.5 sec per aperture and is repeated cyclically for each subsequent CCD frame. If the exposure time is no more than total data reduction time for all apertures, there would be no observation data loss.

The Trailing Star Track Centering algorithm. It is designed specifically for astrometrical observations of artificial satellites. It determines the coordinates of trailing star track center to find out the absolute satellite's sky coordinates. The algorithm assumes that the brightness has variations along the star track, so we should make some special calculation, but not simply find the "mass center" of the target. First of all, algorithm filters the CCD frame, then it scans frame for objects. If it finds some, it surrounds the object with an aperture that closely follows its shape. Then the object's axis along the star track is estimated, cross sections are made, and the "mass center" for each is found. Once this is done, the function for brightness distribution along the track is constructed. This function's derivation is examined for extremes, and the bending point of the brightness function is calculated as the "mass center" of the derivation function in the vicinity of each extreme. The star track center is the geometric center between two bending points.

MAKING AN OBSERVATION

CCD-based systems bring some additional complexity to an instrument setup for observation, so the computer software must help the astronomer manage them. The software system we describe has a flexible user interface with help system and macros sequencer included, which is very useful for setting up for observations and hardware control. For example to make real-time photometry setup, user should run the appropriate macro, which includes the following steps:

- Initialization of the hardware.
- Loading pre-estimated flat field file (or recalculating it)
- Standardization of the artificial source of light (LED) on some known star.
- Setting up the desired sizes and initial positions of apertures within the frame and defining their individual functionality (static positioned or guiding).
- Result file and result form definition.
- Starting observation.

The observation process is illustrated by Figure I. Similar macros are written for other observation modes, specifically real-time astrometry, field photometry, etc. Moreover, it is possible to switch between different modes with saving/restoring current state.

FUTURE DEVELOPMENT

We plan to improve the performance of the system by adding the array processor to the system when it becomes available, so that we can process the entire CCD frame in real time. Also we are going to add the FITS files processing facility to the system. The corresponding software is being already designed. These mod-

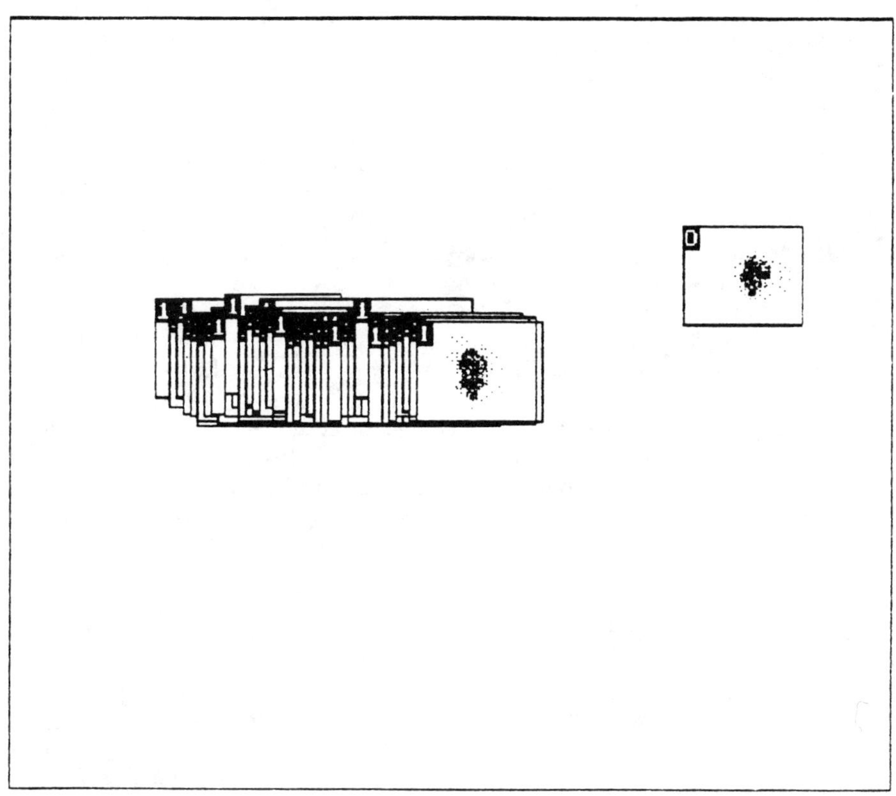

FIGURE I The observation process using “Guiding Facility”.

ifications give us the possibility to perform real-time field photometry and some additional functionality. Furthermore, we plan to develop a new fast scanning mode CCD camera with specifical CCD controller and fast 14-bit digitizer.

OVERVIEW OF THE DATA ANALYSIS SYSTEM FOR THE EXTREME ULTRAVIOLET EXPLORER PROJECT

P.W. VEDDER, C.A. DOBSON, H.L. MARSHALL, W. MARCHANT, V. SABA, A. HOPKINS, E. OLSON, J. FLANAGAN, I. HAWKINS, F. GIROUARD, F. MACHI, AND J. LEWIS
Center for EUV Astrophysics, Space Sciences Laboratory, University of California, Berkeley

ABSTRACT We describe the data analysis system (called the End-to-End System, or EES) for handling data from the Extreme Ultraviolet Explorer, which is scheduled for launch in March, 1992. The EES is nearing completion and was developed in stages in parallel with the development of the science payload. It is designed to contain nearly all components of operations, data processing and data analysis, from data simulation to the production of source catalogs and all-sky maps suitable for publication.

The Extreme Ultraviolet Explorer (EUVE) satellite will perform a six month all-sky survey in the extreme ultraviolet (EUV, 70 – 700 Å) in four bandpasses, along with a simultaneous deep survey of the ecliptic in two bandpasses. Following the all-sky survey, EUVE will conduct EUV spectroscopic observations of the brightest sources as part of a Guest Observer program. All of the hardware and software for the EUVE mission was developed at the University of California, Berkeley (UCB).

Data from the EUVE all-sky survey will be processed from raw telemetry into skymaps and catalogs at UCB. At the same time, the science operations of EUVE, which includes the health and safety monitoring of the science payload, instrument commanding, mission science planning and telemetry archiving, will be conducted from Berkeley. Data processing for the pointed spectroscopy observations is handled using a separate IRAF-based system.

The EUVE End-to-End System (EES) was developed concurrently with the flight hardware and incorporates simulation, operations and analysis functions. This parallel software development proved to be extremely successful. It allowed the software to incrementally evolve during the hardware development and enabled software tools developed for instrument testing to be easily incorporated.

Major releases of the EES have occurred approximately yearly, with software modules evolving from functional to preliminary to final versions. The EES makes extensive use of parameter files which contain the important parameters controlling the data processing. These parameters can be controlled separately from the software to incorporate calibration and testing results. The EES also utilizes several internal data formats, and is currently in Release 7.3, which will be the flight version of the software.

The End-to-End System contains four major modules (see Figure I): En-

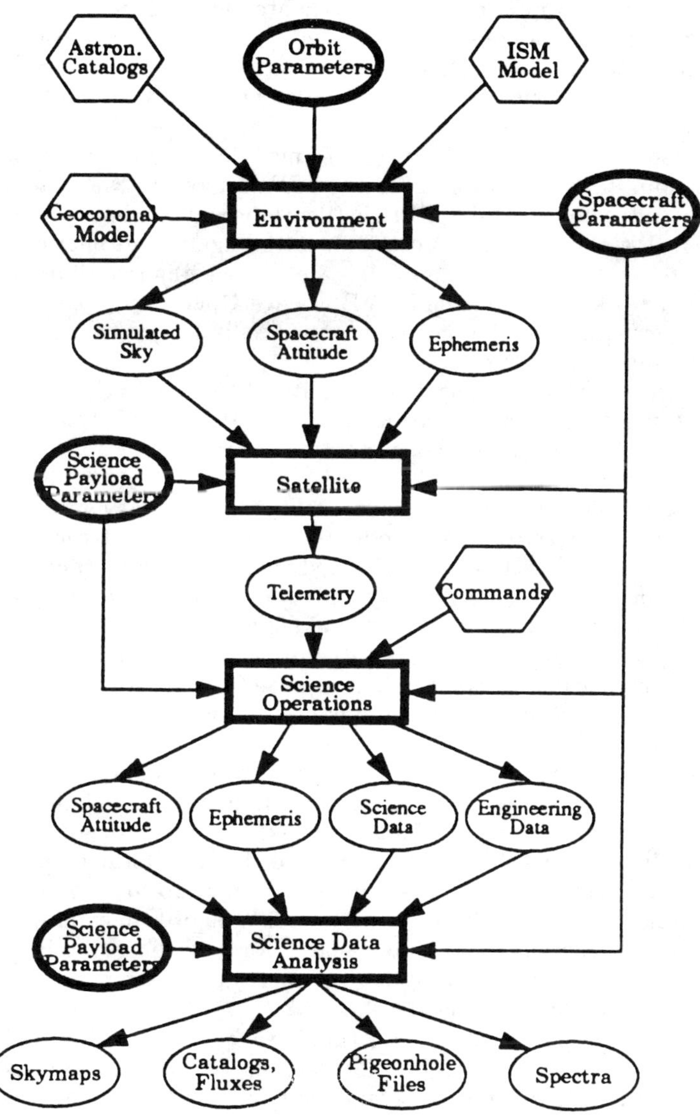

FIGURE I Overview of the End-to-End System

vironment, Satellite, Science Operations and Science Data Analysis. The two simulation modules (Environment and Satellite) use models of likely sources, models of the interstellar medium, attitude and orbit information to produce simulated satellite telemetry. Photon data from sources are ray-traced through extensive models of the instruments, which has proved invaluable for evaluating and interpreting calibration data results.

The Science Operations software performs telemetry reception and archiving, health and safety monitoring of the EUVE payload via automatic limit checking, science planning using SPIKE (developed at the Space Telescope Science Institute), and enables instrument commanding and automatic verification. Instrument telemetry monitors and limits are displayed in real time using X windows on color workstations in the EUVE Science Operations Center (ESOC) at UCB. Engineering limit violations are automatically logged to a file and flagged with yellow and red markers.

The science data processing functionality of the EES is split between two independent networks: an Operations network and an Analysis network. A data archive system straddles the two networks, which are not interconnected for security reasons.

The EES Pipeline runs autonomously on the Operations network processing each packet of data received from Goddard. Automatic limit checking and updating of the history database are performed along with remapping the detected photons back onto the sky. Photon sky addresses, based upon the quadrilateralized sphere used by COBE, are computed along with source crossing files for use in computing the exposure. Time-tagged photon data are saved in "pigeonhole" files, where each pigeonhole is a small pre-defined region around the position of a likely source. Intermediate data products are passed to the analysis network via the data archive.

Continued data analysis on the Analysis network is divided into two pieces: one which is run daily and one which is run on demand. The daily processing accumulates data to produces strip maps and exposure files for source detections. The intermediate data files are them discarded. The on-demand processing searches for new sources in the strip maps and plots exposure corrected skymaps for the different bandpasses. The source detection algorithm employs a model of the point spread function to optimally search for new sources. Pigeonhole files are created automatically for newly found sources and passed to the operations network for use by the EES Pipeline.

The pigeonhole files used in the EES allow for a relatively small subset of the data ($\sim$10%) to be kept on-line for detailed analysis by the science staff. These files contain data for small regions on the sky containing suspected or actual sources, and eliminate the need to search through the entire time-ordered data set to pull out data for a given spatial region.

ACKNOWLEDGMENTS

This work was supported by NASA contracts NAS 5-29298 and NAS 5-30180.

EUVE REAL-TIME TELEMETRY PROCESSING AND DATA ANALYSIS SOFTWARE

J. FLANAGAN
University of California Center for EUV Astrophysics, Berkeley, CA 94720

INTRODUCTION

The End-to-End System (EES) data analysis pipeline is in three parts: The ESOC Pipeline, the SDAF Daily Pipeline, and the SDAF Demand Pipeline. Each section of the EES Pipeline is a Bourne shell script comprising several modules, each one of which can be activated separately.

The ESOC Pipeline runs on the EUVE Science Operations Center (ESOC) computer network at U.C. Berkeley, where telemetry packets are received from the Packet Processor (PACOR) at Goddard Space Flight Center. Each PACOR message is archived and then processed with the ESOC Pipeline, which extracts engineering information, produces engineering stripcharts, updates the engineering history database, and extracts and maps photons back to their source positions on the sky. The photons are stored in two formats: a compact, 5-byte/photon representation for binning into sky maps, and a more detailed, 22-byte/photon representation which retains timing and detector position information. The 22-byte photons are fed to the *do_pigeonholing* software module, which saves the photons associated with locations on the sky which have been deemed to be interesting.

Once a day, the merged products from each ESOC Pipeline run are transferred to the Science Data Analysis Facility (SDAF) network at U.C. Berkeley. Here, the SDAF Daily Pipeline is run to bin the photons are binned into sky maps, and to accumulate exposure maps. Once a week, the SDAF Demand Pipeline is run to detect point sources in the accumulated sky maps, and to calculate fluxes from the "pigeonhole" catalogue sources. New detections are added to the pigeonhole catalogue for future processing.

SPEED AND LOAD DISTRIBUTION

The principle speed constraints are on the ESOC Pipeline. Currently, it runs between 2 and 3 times faster than real time (*i.e.*, 1 hour of telemetry is processed in 20-30 minutes). Since startup overhead is a significant part of the ESOC Pipeline's run time, and several programs need to read in a large number of engineering parameters, the engineering parameters which describe the locations of engineering monitors in the telemetry are loaded into a shared memory segment for instant access.

The SDAF Daily Pipeline runs somewhere between 3 and 6 times faster than real time, and the SDAF Demand Pipeline runs in a few hours.

ESOC pipelines can be run in parallel, the products being merged before

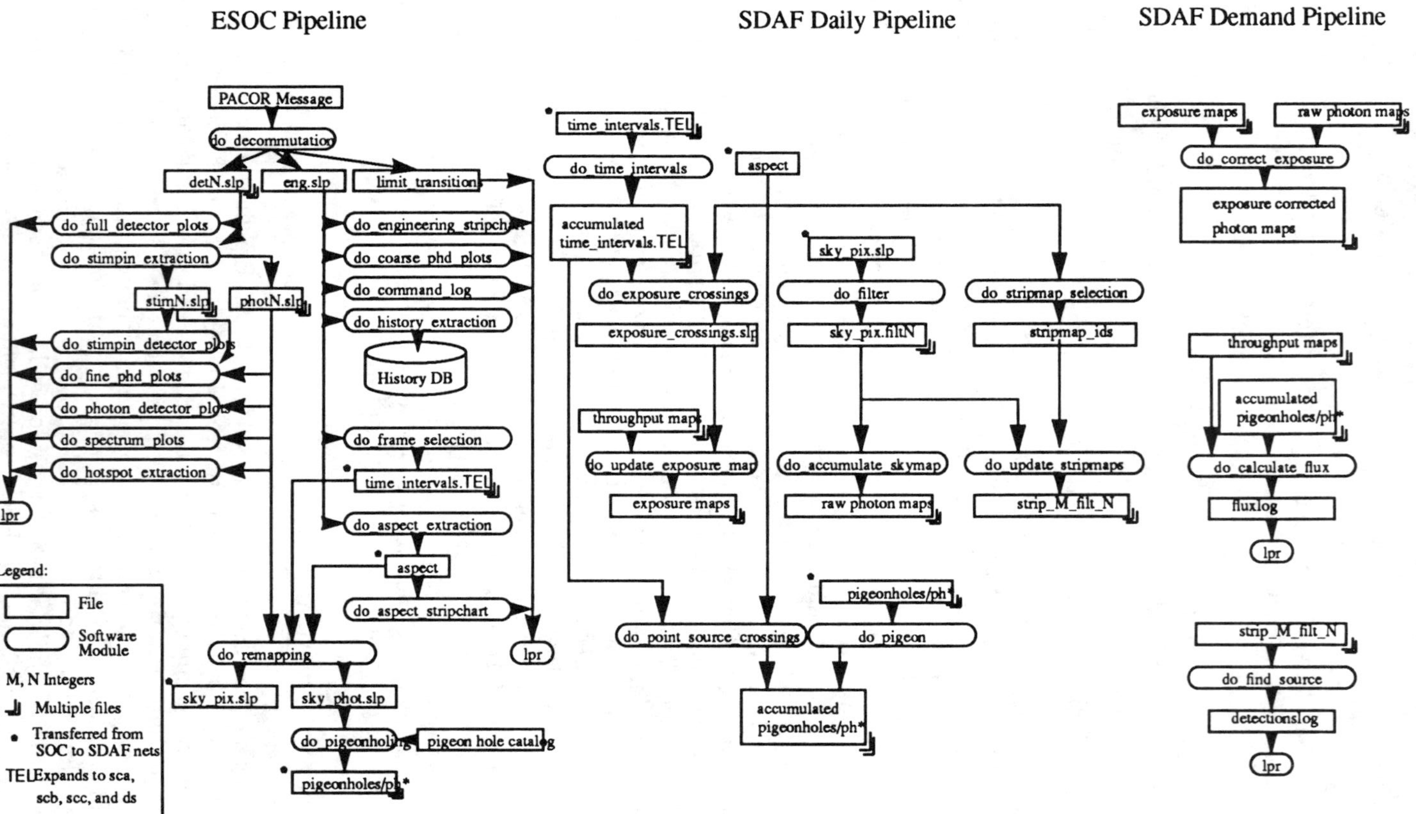

ESOC Pipeline
SDAF Daily Pipeline
SDAF Demand Pipeline
PACOR Message
do_decommutation
detN.slp
eng.slp
limit_transition
do_full_detector_plots
do_stimpin_extraction
stimN.slp
photN.slp
do_stimpin_detector_plots
do_fine_phd_plots
do_photon_detector_plots
do_spectrum_plots
do_hotspot_extraction
lpr
do_engineering_stripchart
do_coarse_phd_plots
do_command_log
do_history_extraction
History DB
do_frame_selection
time_intervals.TEL
do_aspect_extraction
aspect
do_aspect_stripchart
lpr
do_remapping
sky_pix.slp
sky_phot.slp
do_pigeonholing
pigeon hole catalog
pigeonholes/ph*
Legend:
File
Software Module
M, N Integers
Multiple files
Transferred from SOC to SDAF nets
TEL Expands to sca, scb, scc, and ds
time_intervals.TEL
do_time_intervals
aspect
accumulated time_intervals.TEL
do_exposure_crossings
exposure_crossings.slp
throughput maps
do_update_exposure_maps
exposure maps
sky_pix.slp
do_filter
sky_pix.fltN
do_accumulate_skymap
raw photon maps
do_stripmap_selection
stripmap_ids
do_update_stripmaps
strip_M_flt_N
pigeonholes/ph*
do_point_source_crossings
do_pigeon
accumulated pigeonholes/ph*
exposure maps
raw photon maps
do_correct_exposure
exposure corrected photon maps
throughput maps
accumulated pigeonholes/ph*
do_calculate_flux
fluxlog
lpr
strip_M_flt_N
do_find_source
detectionslog
lpr

transfer to the SDAF net. If added functionality should lower the processing speed of the ESOC Pipeline to below an acceptable level, time can be gained by splitting up incoming PACOR messages and processing the pieces simultaneously on different machines, or in parallel on one multi-CPU machine.

SOFTWARE MANAGEMENT

Software version control is handled via the Source Code Control System (SCCS). Each source directory in the EES file tree contains files used to build an application, stored in an SCCS database in that directory. To release a new version of the EES, a new directory hierarchy is created which mirrors the development tree, with links from the development SCCS directories into the released source directories. Version lists are used to extract the correct revisions of the source files in the released hierarchy from the original SCCS databases.

The use of parallel directory hierarchies also permits the support of different machine architectures. For example, a Sun 3 hierarchy is built using the sources from the Sun 4 development hierarchy.

PARAMETER HANDLING

Each program in the EES reads its runtime parameters from parameter files which are accessed through the *ui* library. The *ui* parameter file for a program is located by searching the directories listed in the PFPATH environment variable, which is set by the EES setup script. Parameters may be redirected to other parameter names and files, which must be located somewhere in the parameter file path. Parameters which characterize the instrument (dimensions, calibration data) are kept under separate revision control from the EES, as the Science Baseline Document (SBD). The SBD parameters are accessed by EES programs via parameter file redirection.

DATA FORMATS

Data files are stored in a machine-independent binary format. A library of I/O routines, *binio*, handles translation of file record formats to and from machine-dependent in-memory data structures.

The records in binary data files are organized into Self-Labelled Packets (SLP), each one of which has a corresponding in-memory structure definition. Each SLP record comprises an identifying header and either a fixed-length body, or multiple bodies of the same type. An SLP file can contain any combination of SLP record types. The *binio* library provides routines for identifying and manipulating the different SLP record types in a file.

As photon records are processed through the EES Pipeline, informational SLP records are passed into the data stream by each program to provide a processing history. An EES program passes all records from input to output, except for those which are of special interest to the program. For example, the *remap* program passes through all records except for detector photon records, which it replaces in the output stream with remapped sky photon records.

THE ALEXIS DATA PROCESSING PACKAGE: AN IDL-BASED SYSTEM

J. J. BLOCH, B. W. SMITH, AND B. C. EDWARDS
Space Astronomy and Astrophysics Group,
Los Alamos National Laboratory,
Los Alamos, NM, 87545

INTRODUCTION

The ALEXIS experiment (Array of Low Energy X-ray Imaging Sensors; Priedhorsky et al. 1990) is a mini-satellite containing six wide angle EUV/ultrasoft x-ray telescopes. Its purpose is to map out the sky in three narrow (5%) bandpasses around 66, 71, and 93 eV. The 66 and 71 eV bandpasses are centered on intense Fe emission lines which are characteristic of million-degree plasmas such as the one thought to produce the soft x-ray background. The 93 eV bandpass is not near any strong emission lines and is more sensitive to continuum sources. The mission will be launched on the Pegasus Air-Launched Vehicle in the first quarter of 1992 into a 400-nautical-mile, high-inclination orbit and will be controlled entirely from a small ground station located at Los Alamos. The project is a collaborative effort between Los Alamos National Laboratory, Sandia National Laboratory, and the University of California-Berkeley Space Sciences Laboratory.

The six telescopes are arranged in three pairs. As the satellite spins twice a minute they scan the entire anti-solar hemisphere. Each f/1 telescope consists of a spherical, multilayer-coated mirror with a curved, microchannel plate detector located at the prime focus. The multilayer coatings determine the bandpasses of the telescopes. The field of view of each telescope is 30 degrees with a spatial resolution of 0.5 degree, limited by spherical aberration.

The data processing requirements for ALEXIS are large. Each event in one of the six telescopes is telemetered to the ground with its time of arrival and position on the detector. This information must be folded with the aspect solution for the satellite to reconstruct the direction on the sky from which the photon came. Because of the way the six telescopes scan the sky, the effective exposure calculation is also very compute-intensive. ALEXIS may generate up to 100 megabytes of raw data per day, which are converted into a gigabyte per day of processed data. The entire analysis system is built on a set of SPARCstation platforms.

SOFTWARE OVERVIEW

While the processing job for ALEXIS is sizable, the programming staff is small. To maximize programming efficiency, and to make the best use of tools available in the public domain, we chose Research Systems Incorporated's IDL package as our software development platform. IDL was used from the start of instrument development through flight. We use IDL as a top-level executive for the processing tasks (replacing Unix shell scripts), as a device-independent graphics engine, as a database manager, and as a final data manipulator. IDL routines spawn special-purpose C programs to perform detailed telemetry deconvolution and other specialized functions.

ALEXIS Data Streams and the End-to-End Philosophy

Early in the ALEXIS project, a uniform standard for all binary data files and streams was adopted. This standard, a self-describing record format called General Data Format or GDF, was applied to ground as well as flight software. The data streams are made up of variable-length records, each consisting of a header section comprised of three 16-bit integer words followed by a data section consisting of a variable number of bytes. The three words in the header contain a record ID, a data type, and the number of bytes in the data section, respectively. The record ID indicates the source of data. As an example it might identify with which of the six ALEXIS telescopes a record of photon data originated. The data type indicates the structure or type of data found in the data section of the record. Examples of this include: an array of integers or real numbers, or a heterogeneous data structure. The last of the three integers indicates the actual length of the data section of the record, in bytes. This record format allows heterogeneous record streams to be read by software that may not know about all of the records in the stream and can therefore pick out only those records of interest, skipping over the rest.

This uniform data standard allowed us to adopt an end-to-end system test and development philosophy. Simulation software for the ALEXIS instrument produced GDF data streams just as the flight system does. When we tested individual ALEXIS telescopes in the laboratory, the ground test equipment (GSE) also generated GDF data streams that looked as they would in flight. In this way, software developed to analyze experiment simulations could later be used for instrument testing, and then for flight operations, with little or no modifications. This aided our software development efforts a great deal. We did not have to re-write existing software to match new data formats that could have arisen at each phase of experiment planning and integration.

Use of IDL

The ALEXIS experiment will generate up to 100 Megabytes of data per day that must be automatically processed and reduced. At the program's start, we began writing software for the project from the bottom up, writing, in the C language, Unix "filter" programs first that read gdf streams from standard input and write new ones to standard output. We debated for quite some time as to how to tie all the programs together into an automated system for processing, archiving, and plotting the flight data. Our initial choice was to write Unix shell scripts, and use a plotting package such as MONGO for graphics. After a demonstration of the IDL data processing/graphics package, we obtained a license and began experimenting with it. Considering IDL's operating system access and process-spawning capabilities, we came to the conclusion that IDL could exceed shell scripts in versatility and provide a device-independent graphics capability as well. IDL would also provide a high-level, array-oriented, data manipulation functionality. Another advantage of IDL was an existing library of astronomy routines that was available from the UIT project at Goddard Spaceflight Center, an effort funded by the NASA Astrophysics Data Program.

Currently, our software design has control, or "Glue," routines in IDL at the top level. These, in turn, call specialized IDL functions which perform specific data processing tasks. These functions in turn spawn Unix pipelines of C-language data filters to do the majority of the telemetry data manipulation. Temporary files produced by these Unix filters, or attached to the Unix pipe output of the spawned processes, can be accessed easily within IDL to obtain transformed data for final results. In the next section we will discuss how the Glue routines interact with an IDL database package for production analysis control.

Databases

All the ALEXIS database work is built on the UIT IDL astronomy library database routines. We realized that if we used a database to control the flow of production processing we could gain a great deal of efficiency in managing the data. All of the Glue routines interact with IDL databases that keep track of what has been done to each file as well as record specific characteristics of each archived telemetry file. For off-line data processing, the IDL database inquiry routines then allow easy recall of the names of specific files that match any user specified characteristics.

An example is the IDL procedure to control routine full processing of flight data. The procedure receives as input a parameter file setting a number of flags, the path names of the raw telemetry file and the aspect file, and the locations where output files may be placed. The database is queried for information on which, if any, of the functions have already been performed on the designated telemetry file. Sky coverage in the file is entered into the database if it has not been entered previously. Point sources that are covered in the file are extracted from the database, and sub-procedures are started to extract count data from sky locations near these sources. Appropriate sub-procedures are started to tag all counts with their origin in sky coordinates and to add them to cumulative sky maps. At the end of processing, an entry is made in the database registering the tasks completed.

ACKNOWLEDGMENTS

This work was supported by the Department of Energy.

REFERENCES

Priedhorsky, W. C., Bloch, J. J., Cordova, F., Smith, B. W., Ulibarri, M., Chavez, J., Evans, E., Siegmund, O. H. W., Marshall, H., Vallerga, J., & Vedder, P. 1990, in Extreme Ultraviolet Astronomy, ed. R. F. Malina, & S. Bowyer, (New York: Pergammon Press)

THE SCIENTIFIC DATA PROCESSING SYSTEM AT THE OPERATIONAL CONTROL CENTER OF THE SAX SATELLITE

F.D'ALESSANDRO, A.COLETTA, L.NATALUCCI
Telespazio SPA, Via Tiburtina 965, Rome (Italy)

M.B.NEGRI
Agenzia Spaziale Italiana, Viale Regina Margherita 202, Rome (Italy)

OUTLINE OF THE MISSION

SAX is an astronomical satellite designed to perform detailed spectral and timing studies of celestial X-ray sources in the wide energy range $0.1 \div 200$ *keV* (Butler & Scarsi 1991). It is a major program of the Italian Space Agency (ASI), involving a collaboration with the Netherlands Agency for Space Programs (NIVR). The *SAX* payload (see Figure I) consists of six separate instruments operating simultaneously which can be grouped into:

- 4 *Narrow Field Instruments (NFI)*, comprising two imaging concentrator systems (*LEC/S, MEC/S*) and two collimated detectors (*HP-GSPC, PDS*).
- 2 *Wide Field Cameras (WFC)*, each consisting of a coded mask-detector system.

INFORMATION TO BE PROCESSED

At the *SAX Operational Control Center (OCC)* (Manca et al. 1990; Manca & D'Alessandro 1991) in addition to the usual activities aimed to spacecraft and payload control, a *Quick Look (QL)* analysis of the scientific telemetry (TLM) information will be performed in almost real time. A total of 500 *Mbit* TLM information is received at the *OCC* every 96 min, coming from the equatorial Ground Station located at Malindi (Kenya). The packets containing scientific data and housekeeping (HK) information, along with the relevant attitude and orbital parameters, are stored in a dedicated archive, named *RRD (Reformatted Raw Data)*, organized as a relational database. The on-line *RRD* contains the data of the last week observations, while older data is available on optical disks (D'Alessandro et al., this volume; D'Alessandro et al. 1990). For each instrument, the contents of the data field in the in the scientific packets TLM depends on the particular mode chosen. Operative modes can be classified into two categories: a) Direct modes, in which relevant data sets for each valid event are transmitted, b) Indirect modes, in which on-board accumulation is performed prior to TLM channeling.

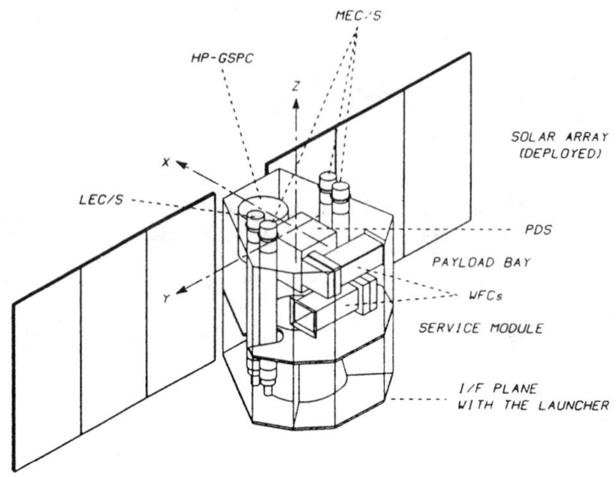

FIGURE I The satellite configuration

H/W RESOURCES

At the *OCC* in a LAN connects a VAX-Cluster including 3 VAX 6510 with 5 DecStations 5000 and a Terminal Server to 3 DEC X25 Routers. A VAX 6510 in the cluster is devoted to a series of activities related to the scientific data telemetry, such as the Scientific *Quick Look.* In addition 2 DECStations 5000-200 with their disks interface the Duty Scientists in charge of the *QL* permitting to increase the system potentialities. A 6.55 *GBytes* Optical Disk device allows data storing for final archiving. Standard output devices as printers and graphic terminals complete the configuration.

QUICK-LOOK OPERATIONAL PROPOSAL

A schematic view of the overall *Scientific Quick Look* is proposed in Figure II. The activities to be performed at the *OCC* concern processing of the payload HK data (both real-time and delayed) and of the scientific information previously filed into the *RRD* database. In the Short Term Scientific Data Processing (ST-SDP) performed at the *OCC*, data is typically processed on a orbit-by-orbit basis in order to retain fast feedback on satellite (within 2-3 orbits). Within this environment the user can merge and process any of the on-line *RRD* data and perform trend analysis on a relatively short time scale (max. 1 week). The scientific data files accumulated are stored on a dedicated magnetic disk within a database environment to be retrieved for further quick look processing. The above Quick Look system is made up of two main components:

• a Brief Quick Look, mainly consisting of a set of standard procedures running with a low level of interactivity. The purpose of this type of quick look is to ensure a fast, reliable overview of the payload performances.

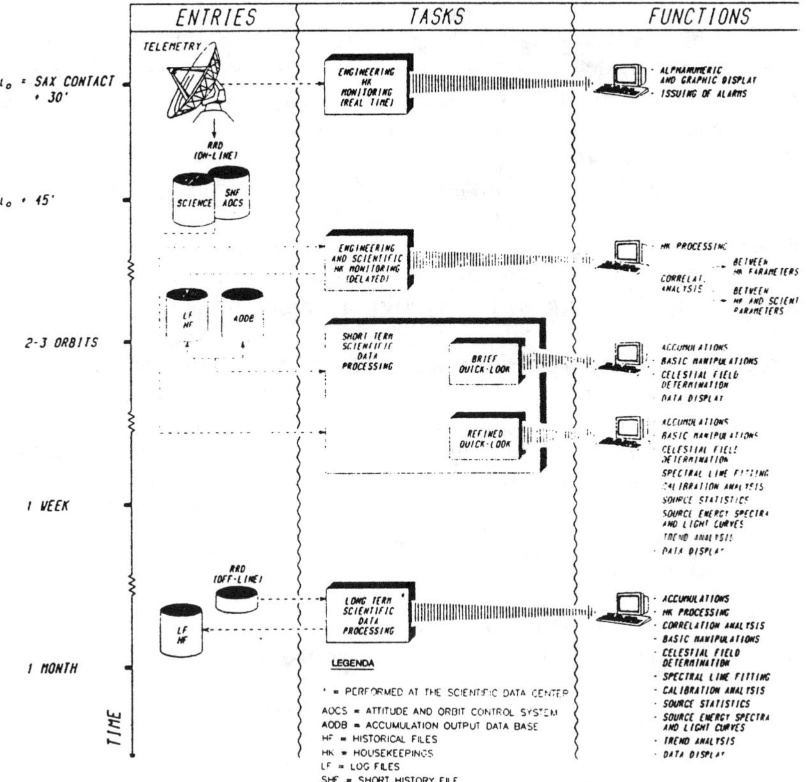

FIGURE II The Quick Look processing system

• a Refined Quick Look, allowing more specific and deeper analysis, characterized by a high level of interactivity. The HK data processing and the ST-SDP constitute a unique environment with common tools for user interface purposes. Within this environment, the user can create, modify and run its own commands and procedures in a number of different modalities. Once the data has been transferred to the *Scientific Data Center (SDC)* a Long Term Scientific Data Processing can be performed having as input the off-line archive available on optical disks. This type of quick look includes all the tasks accomplished by the ST-SDP, plus more sophisticated processing with the possibility of trend analysis on a longer time scale (up to 1 month).

USER INTERFACE

The Duty Scientist will perform a Quick Look session by means of a window-based, mouse-driven Man Machine Interface (MMI) system dedicated to scientific applications. We have developed an example MMI based on the DECwindows environment that provides a suitable approach to our operational requirements. The user will access the above specified system from the general *OCC*

MMI. An Application Window will control the choice of different activities (e.g. Brief and Refined Quick Look, editing of Quick Look procedures, Observation Scheduling) along with some control/command functions. As regards the ST-SDP we consider two different operative modalities:
(*a*) a fully interactive approach,
(*b*) an approach based on command procedures characterized by a low interactivity level.
In the former case by "clicking" with the mouse on the proper menu entry of the Application Window the user will select the wanted instrument in the related pull down menu. The available tasks are grouped in functional classes. Each task foreseen by the *QL* is controlled by a complex input form containing different areas to be filled (Classes Area, Functions Area, Parameters Area). The user is also provided with facilities allowing to generate a "command string" containing information on the current input form, that is, containing all the selections for the related function and parameter values in an abbreviate and readable form. The user has the possibility to edit such command strings in order to rapidly recall and modify them.

A different way of operating is the use of *Quick Look* Procedures (case b). These procedures are available for editing and executing under a dedicated environment accessible from the Application Window. The user will have the possibility to build his own procedure, consisting of a set of commands which can be inserted, modified or deleted in the course of a session.

EXPECTED RESULTS

The *Quick Look* Scientific Data Processing at the *SAX OCC* has been designed taking into account the basic requirements for fast data processing and immediate interpretation of the results. The interpretation of the results will significantly impact on the execution of the short term operation plan, including the scheduling of targets of opportunity detected by the *SAX* instruments. The flexibility and friendliness of the system is guaranteed by a high-level user interface, based on the well-established windowing approach.

REFERENCES

Butler, R. C., & Scarsi, L. 1991, Adv. Space Res., 11, N. 8, 265

D'Alessandro, F., Coletta, A., Negri, B., & Trifoglio, M. 1990, in Proceedings of ESA Symp. on Ground data systems for spacecraft control, ESA SP-308

D'Alessandro, F., Lupi, M. C., & Negri, B. 1992, this volume

Manca, M., Appolloni, A., Negri, B., & Trifoglio, M. 1990, in Proceedings of ESA Symp. on Ground data systems for spacecraft control, ESA SP-308

Manca, M., & D'Alessandro, F. 1991, Journal of the British Interplanetary Society, 44, 431

DESIGNING AND PROTOTYPING THE RAW DATA ARCHIVE FOR THE SAX MISSION

F.D'ALESSANDRO, M.C.LUPI
Telespazio, Via Tiburtina 965, Rome (Italy)

M.B.NEGRI
Agenzia Spaziale Italiana, Viale R.Margherita 202, Rome (Italy)

INTRODUCTION

SAX is an astronomical satellite designed to investigate celestial x-ray sources both inside and outside the galaxy. It is a major program of the Italian Space Agency (ASI), involving a collaboration with the Netherlands Agency for Space Programs (NIVR). The satellite will be launched at the end of 1993 into a 600 *km* circular orbit at 3 degrees inclination, with an operative life of at least 2 years. The orbit period will be of about 96 minutes.
The description of the *SAX* Ground Support System and the *Operational Control Centre (OCC)* is given by Manca et al. (1990). The volume of the telemetry data, transmitted during each contact period (about 9 minutes), is up to 550*Mbit*.
One of the three CPU at the *OCC* will perform: scientific quick look in almost real time; production of the observations plan; management of the scientific raw data archive, including the possibility of producing optical disks and final tapes.

MISSION REQUIREMENTS AND CONSTRAINTS

Five types of instruments, eight separated units, make up the *SAX* payload.

The *MEC* and the *LEC*,(respectively the Medium and the Low Energy Concentrators) the *HP-GSPC* (High Pressure Gas Scintillation Proportional Counter) and the *PDS* (Phoswich Detector System) are the so-called *Narrow Field Instruments (NFI)* having an average Field Of View (FOV) of 30 *arcmin*. They always will point to a target at the same time, covering the wide energy range $0.1 \div 200$ *keV*. The wide field instruments will be arranged orthogonally to the *NFI* pointing direction and consist of two *Wide Field Cameras (WFCs)*. Their FOVs are 20 degrees square looking in opposite directions in an energy range of $2 \div 30$ *keV*. (Butler & Scarsi 1991)

During a given *Pointing*, it is possible to identify several *Observations* characterized by an *Experiment Configuration* (coming from the On Board Data Handling processor), an attitude and an instruments Operative Mode. The total data volume will rise to 550 *Mbit* considering the data collected and transmitted during the contact period; the scientific data will be the 90% of such amount and the 10% will be *Housekeepings (HK)*.
At the *OCC* the main requirement on data archiving come from the need of

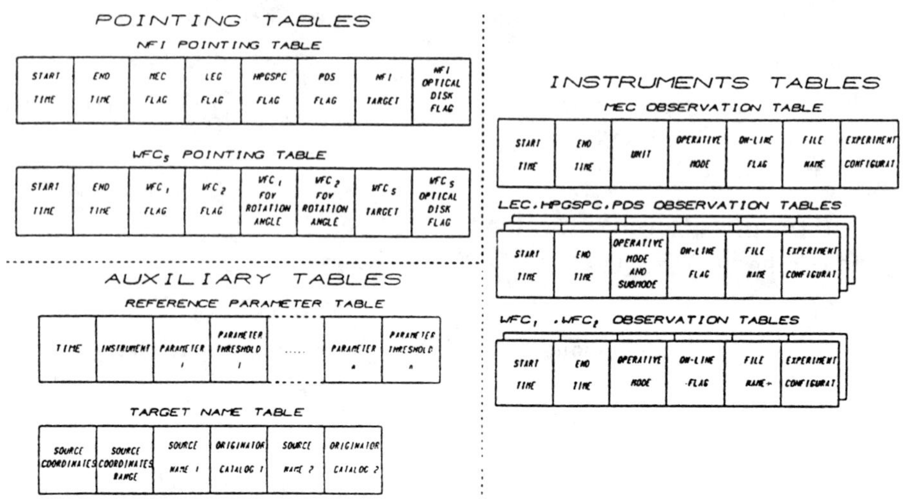

FIGURE I Database description

performing a scientific *Quick Look (QL)* by 2 orbits, as to allow an almost real time control on experiments, so the archiving process will not exceed 30 minutes. Moreover all data has to remain on line for a week and copied into Optical Disks (OD) after 48 hours from its acquisition. All data pertaining to a given *Pointing* will be contained on a single OD, so *NFI* and *WFCs* will have separated ODs. Data belonging to an *Observation* and coming from each instrument will be stored in separated data sets.

THE REFORMATTED RAW DATA ARCHIVE DESIGN

Typical functions to be performed in QL session are the production and integration of spatial images, energy spectra and time profiles for each instrument (D'Alessandro et al., this volume). Such a complexity has been faced by organizing the Reformatted Raw Data (*RRD*)in three parts:

- an on-line Data Base (DB) managed by a commercial relational Data Base Management System, in which are stored all information for retrieving raw data,
- an on-line File System in which raw data are stored by Observation,
- an off-line set of ODs containing the definitive RRD to be sent to the *SDC*.

An application software will control and coordinate these parts.
The non-normalized DB, outlined in Figure I, consists of three groups of tables. The first group concerns activities of *NFI* and *WFCs* during their *Pointing* period, with an occurrence for each pointed target. Start and End Time are the primary key. The Flags indicate if a given instrument is ON/OFF during a *Pointing*. The FOV rotation angle for the *WFCs* is useful for searching a target in wide square FOVs. All the previous information comes from On Board Data Handling or Attitude and Orbital Control System data evaluation. The OD Id. identifies the OD in which are copied data of a given *Pointing* (if already done) and it is generated by the OD production process. The second group of tables

separates information about the 6 instruments. There is an occurrence for every *Observation*. Start and End Time are the primary key. The Operative Mode and Submodes indicate the data type and their format. *Experiment Configuration* is a code extracted from the payload HK analysis. Another group consists of 2 relations, named Auxiliary Tables, containing collateral information not extracted from telemetry but separately entered. One of them is the Target Name Table, that is a restricted catalogue of the targets to be observed with their names commonly used in astronomy. The other one is a table containing scientific parameter (e.g. energy thresholds, peak positions) liable to vary during the mission.
The on-line File System is composed by the Scientific Raw Data Files, each one referring to a single *Observation* of an instrument and it is characterized by one data format, depending on the Operative Mode and Submode. Also Historical Files generated by the CPU A are part of the *RRD* structure. They contain Verified Commands, Orbital and Attitude Data, all the HKs.
The off-line part of the *RRD* archive is made up of ODs. Due to the unrelated pointing of *NFI* and *WFCs*, two sets of ODs are present.

The average volume of raw data is 62 $MBytes$ and the volume of the on line raw data can reach 6 $GBytes$ in a week. Each file can vary from $4 \div 300$ $Mbytes$ of data. On the other hand the DB volume can vary in a week from a minimum of 6 $KBytes$ to a maximum of 192 $KBytes$.

DESCRIPTION OF THE PROTOTYPE

The prototype of the *RRD* archive tests its feasibility mainly from the point of view of the temporal performances both for DB load and data retrieval. These critical processes are studied by four different data simulations, both typical cases and extreme ones. The DB (except for the Auxiliary Tables) is built following the above described structure using the VAX-VMS Rdb. The application programme is written using SQL FORTRAN embedded.
In the prototype the whole telemetry packets are simulated but only the needed fields are considered: the primary and secondary headers including all information allowing data archiving. Once the different types of data packets are generated, they are divided by Application Identifier (AP.ID.) and filed, as in the reality. The AP.ID. is a code included in the primary header of each packet as foreseen by the ESA telemetry standard; it contains information about the data type, the instrument and its operative mode.
In order to distinguish the different Observations, the kernel of the application programme scans the data packet secondary header, extracting Start and End Time, and the packet format of each *Observation*. So this kernel can perform the DB updating and the raw data filing. The Figure II shows the characteristics of the four simulated cases and the related results for the DB loading. The tests were performed on a VAX 6510 completely dedicated to the present activity. All cases satisfy the temporal requirements, being the elapsed time much lower than 30 minutes. The second case represents a contingency situation, but also in this condition there isn't any problem in the DB loading performance. In order to test the archive, different querying programmes are implemented. The queries can allow to verify,in real time, the number of *Observations* included in

	CASE 1	CASE 2	CASE 3	CASE 4
POINTING PERIOD	5760 s	10000 s	5760 s	5760 s
NUMBER OF OBSERVATIONS	2 OBS. => 3000 s 2760 s	4 OBS. => 3000 s 3000 s 3000 s 1000 s	30 OBS. x 192 s	20 OBS.: 19 OBS. x 300 s 1 OBS. x 60 s
INSTRUMENT DATA RATE (kb/s)	MEC => 44 LEC => 15 PDS => 7 WFC1 => 22	MEC => 30 LEC => 20 HPGSPC => 10 PDS => 10 WFC1 => 30	LEC => 88	MEC => 14.5 LEC => 14.5 HPGSPC => 14.5 PDS => 14.5 WFC1 => 14.5 WFC2 => 14.5
ELAPSED TIME (h:m:s.)	00:10:08.10	00:20:20.73	00:10:12.54	00:09:32.16
CPU TIME (h:m:s.)	0:06:20.12	0:12:08.22	0:06:35.89	0:03:12.88

FIGURE II Data simulation and DB loading performances

a *Pointing period*, and if data concerning an *Observation* are on magnetic disk or on OD; to know the lasting or the main features (e.g the active instruments, the Operative Mode) of an *Observation*; to search a data packet corresponding to a fixed time or all data packets with a defined delay from the beginning or the end of an *Observation*. Moreover some additional functionalities are under study to get better performances: to increase computer memory and/or to write in a parallel way on DB using six different processes, respectively related to the six instruments.

REFERENCES

Butler, R. C., & Scarsi, L. 1991, Adv. Space Res., 11, N. 8, 265

D'Alessandro, F., Coletta, A., Natalucci, L., & Negri, B. 1992, this volume

D'Alessandro, F., Coletta, A., Negri, B., & Trifoglio, M. 1990, in Proceedings of ESA Symp. on Ground data systems for spacecraft control, ESA SP-308

Manca, M., Appolloni, A., Negri, B., & Trifoglio, M. 1990, in Proceedings of ESA Symp. on Ground data systems for spacecraft control, ESA SP-308

Manca, M., & D'Alessandro, F. 1991, Journal of the British Interplanetary Society, 44, 431

CONVERSION OF THE EINSTEIN OBSERVATORY LEVEL 1 PROCESSING SYSTEMS

J. S. ORSZAK, D. W. KIM, N. KRONENFELD, K. R. MANNING, J. MCSWEENEY, F. A. PRIMINI, S. ROMAINE, J. SCHWARZ
Smithsonian Astrophysical Observatory, Cambridge, Massachusetts 02138

ABSTRACT The Einstein Observatory Level 1 Processing Systems for the Imaging Proportional Counter (IPC Rev. 1B) and High Resolution Imager (HRI Rev. 1) are being converted from Fortran V and Fortran 77 programs running on an antiquated Data General M600 minicomputer to Fortran 77 programs running on SAO's network of Sun workstations. When completed, the processing systems will be archived at SAO along with the raw data, and as such, will represent one of the few examples of a concurrent archive of functioning Level 1 software and raw data available to the general astronomical community. We will discuss the conversion effort with particular emphasis on the methods used to convert large amounts of source code, and the approach taken in testing the validity of the converted code. Difficulties encountered in the conversion effort and their resolution will also be addressed. We will discuss the relevance of this project to the general problem of porting large software projects to multiple environments and suggest guidelines for the development of such systems which will minimize the difficulties inherent in such ports.

INTRODUCTION

The Einstein Observatory Level 1 Processing Systems are used to reduce the raw telemetry data and to perform some basic scientific analysis of x-ray events of the IPC and HRI. The Processing Systems apply the aspect, energy and position corrections to the x-ray events. The data are screened according to predetermined engineering and orbital criteria, an x-ray image generated, and basic scientific analysis performed on that image.

We are converting the Level 1 Processing Systems to operate on a network of Sun workstations, thus allowing us to retire the M600 computer. This conversion will maintain the ability to process the raw data into the indefinite future.

METHODS USED IN CODE CONVERSION

We have developed a set of tools which facilitate the conversion of large amounts of source code. These include a set of four UNIX scripts and a nawk program that accomplish the first level of FORTRAN conversion. These perform the common conversion tasks, such as maintaining data type and changing library call names.

These scripts complete about 90% of the code changes from M600 FORTRAN 77 and 80% of the code changes from FORTRAN V to Sun FORTRAN 77.

In addition, we have written several programs that check the results of converted programs. Programs were written to port original, binary M600 data, view this data in ASCII format, and take the difference between two data files. These programs use templates of the data structure to accomplish the binary translation between M600 format, Sun (IEEE) format, and ASCII. Also, programs were written to test specific segments of the conversion.

These scripts and programs, and the large amount of code converted, lead to a standard method of conversion. The first stage is to run the code conversion scripts on the original M600 FORTRAN source code. After this each program is debugged by hand, comparing the results on the Sun system with those produced on a test sequence from the M600. Finally, the program is integrated into the complete Processing System and verification of the system results occurs.

Table I outlines the conversion that is completed. The code conversion took $>$ 2 person years of programming effort.

TABLE I The converted Level 1 Processing Systems.

System	Number of Programs	Subprogram Modules	Approximate Lines of Code
Input, Pre-aspect, Aspect	12	161	15,300
IPC	44	596	43,400
HRI[a]	15	128	11,500
HEAO Library		57	3,100
Control Macros[b]	2	15	2,000

[a]There are two HRI programs which remain to be converted.

[b]The Level 1 System Processing Macros have been rewritten as KornShell scripts.

DIFFICULTIES IN THE CODE CONVERSION

During the effort to convert the Level 1 Processing systems we found many recurring difficulties and common problems.

The most troublesome difficulties are due to the conversion from a 16 bit integer computer to a 32 bit integer computer. This is especially difficult within arrays. It was also difficult because one of the major libraries had been previously rewritten to take advantage of 32 bit integers, thus requiring that we change type definition on calls. Much of this is handled by the automatic conversion routines.

Bugs in the original M600 code, that only appeared due to the careful inspection the code is now undergoing, were difficult to differentiate from conversion issues.

A large amount of effort addressed the need to replace machine specific functionality. This was achieved by rewriting all Assembly language code in C, and AOS (M600 Operating System) Macros in the KornShell. M600 databases were integrated into the existing INGRES system on the Suns. Contour plotting functionality was written utilizing the IRAF/PROS system and strip plot functionality is being written in new programs that utilize the SM plotting package.

The inherent differences in floating point arithmetic on different platforms causes difficulties which can become amplified as the data moves further through processing. Packing of bits and bytes, and word alignments in common blocks, that are machine dependent also caused problems.

A serious difficulty is the lack of accurate documentation to resolve all issues.

TESTING THE VALIDITY OF THE CONVERSION EFFORT

Throughout the conversion of the Level 1 Processing Systems we maintained a set of standards for preserving the validity of the original processing results.

The validity of our system was tested by reprocessing approximately 70 randomly selected sequences through the complete IPC Processing System and comparing the results between the M600 and the Sun system. This program will also be carried out with the HRI system. UNIX scripts, nawk programs and IRAF/PROS macros were developed to simplify the data comparison during the test phase of these 70 IPC sequences.

The testing and operations procedure took approximately 2 person years to date. This was time in addition to that spent on programming.

SOME LESSONS TO BE APPLIED TO FUTURE SYSTEMS

There are some valuable lessons from this conversion effort that, if applied from the beginning, can ease the porting of other systems.

- Using software standards (e.g., ANSI) can reduce many of the problems. When this is not possible, coding standards will ease the process of making the conversion automatic.
- Avoid using machine-dependent, proprietary, or external packages, that might be difficult to port.
- When using machine-dependent, proprietary, or external packages keep them as isolated as possible to ensure that replacing them is straightforward.
- Document all details of the system completely. Documentation should especially address those parts of the system that are most unusual.
- Avoid packing of bits and bytes, which is machine specific.
- Maintain libraries as a part of the system and not as an external resource.

It is important that the issue of converting Processing Systems is addressed early due to the rapid advances in computing and the extended lengths of missions and data analysis.

SOLAR DISK IMAGE PROCESSING FOR SPATIAL AND TEMPORAL FREQUENCIES FOR THE GONG PROJECT

W. E. WILLIAMS
National Optical Astronomical Observatory/National Solar Observatory

ABSTRACT The performance of the prototype data processing system developed to reduce the GONG solar disk observations is evaluated. Current test results indicate that data processing can keep pace with data acquisition.

INTRODUCTION

The GONG (Global Oscillation Network Group) project was devised to study the internal structure of the sun by long-term, continuous observations of solar acoustic oscillation modes. The solar acoustic oscillation modes are studied by spatial and temporal analysis of radial velocity maps of the sun. Basic helioseismology is described by Leibacher et al. (1985). Duvall & Harvey (1986), among others, discuss analysis of helioseismology data.

The performance of the prototype data reduction package developed for the GONG (Global Oscillation Network Group) solar disk observations is examined. The data reduction is of interest because a large volume of data must be processed in a timely fashion in order to keep pace with data acquisition using both I/O and cpu intensive tasks.

GONG DATA OVERVIEW

In order to make continuous observations of the sun, data will be collected from six observing sites in a worldwide network. This data will be acquired continuously for three or more years totalling to a terabyte or more of output data. A site survey indicates that two to four sites will be on line simultaneously, with a network duty cycle of >90%.

The network will generate images totalling 750 to 1500 Mbytes per day, to be processed into time series totalling 360 Mbytes per day. The data processing must keep pace with the data acquisition. As a consequence, the data must be processed two or three times faster than it is acquired to allow for errors, equipment down time, processing of anomalous but salvagable data and post processing procedures. Figure I shows the daily data rates for the GONG network and the data products.

Currently GRASP, an IRAF package, is used to analyze prototype instrument data. The GRASP package was developed by the GONG Data Reduction and Analysis Team in cooperation with the solar community. The GRASP pack-

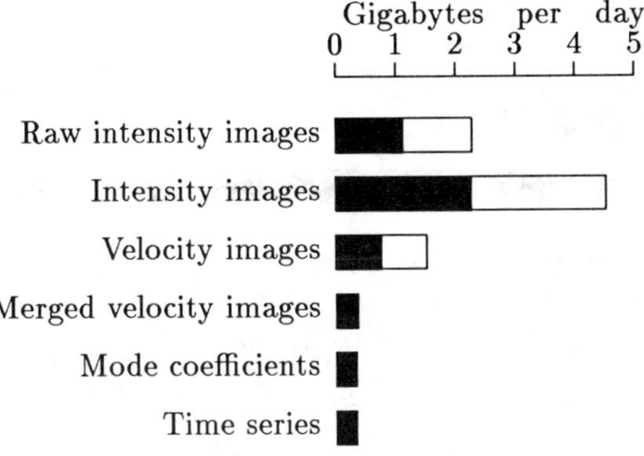

FIGURE I Distribution of data in the GONG data reduction stream. Filled bars indicate the data rate for two simultaneous sites. Empty bars indicate the data rate for four simultaneous sites.

age includes the operations described in Table I. Parts of the package, such as the spherical harmonic transform are cpu intensive, with cpu utilization at >80% of elapsed (wall clock) time. The initial data corrections and final time transpose are I/O intensive, with cpu utilization at <20% of elapsed time. The GRASP package also includes analysis of the time series and the temporal frequencies for each spatial mode. The final analysis of the time series and generation of the temporal mode frequencies is not discussed here, as it is still under development (Anderson et al. 1990).

TABLE I Operations during Data Reduction

Operation	% cpu	Algorithm description
dark, clamp correction	27	addition, multiplication
image restoration	92	deconvolution using 2-D fft and inverse
velocity calibration	81	vector subtraction, scaling
temporal filter, remap	65	running mean removal, coordinate remap
multisite merge	63	weighted sum of images
spatial transformation	86	Legendre and Fourier transforms
time transpose	20	extensive data reordering

The first three steps–the *upstream data reductions*–are dependent on the instrument, and have evolved with instrument development. These steps are the least optimized. The scattering and seeing correction are the newest functions in the data reduction and are in the earliest stages of development.

The temporal filtering through time transpose steps–the *downstream data reductions*–are the most fully optimized and developed parts of the data reduction. These parts are in the SUNTRANS subpackage of GRASP, described in detail by Williams (1991). The only part of the downstream data reduction still in active development is the multisite merging, which is envisioned as a weighted average of simultaneous images. Development work is concentrated on

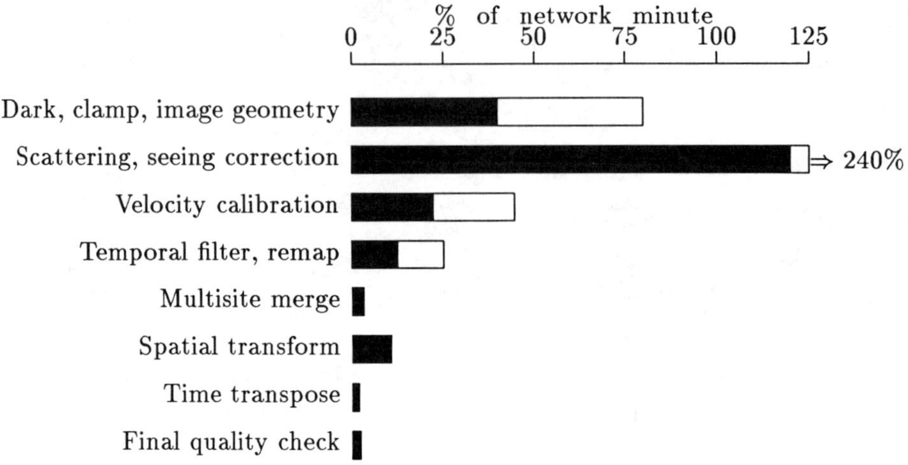

FIGURE II Processing versus acquisition time for the GONG prototype data reduction package. Filled bars indicate processing/network time ratio for two simultaneous sites. Empty bars indicate processing/network time ratio for four simultaneous sites. Tests were run on a Sun Sparc2 workstation.

the computation of the weights.

Test runs with 8-12 hours of data have been performed on currently available work stations. Figure II shows the elapsed times for the segments of the data reduction in Table I. The times are for a Sun Sparc2, the fastest workstation currently owned by the GONG data reduction team. The times in Figure II indicate that keeping pace with the data acquisition is achievable using desk-top workstations for the SUNTRANS part of the data reduction. Further reductions in processing time for the dark and clamp corrections are anticipated shortly, as the instrument is nearing completion and the codes can now be optimized.

The remaining serious timing problem is the scattering and seeing correction. Currently, this is envisioned as a deconvolution by two dimensional Fourier transform of the intensity images and the scattering function, a multiplication in Fourier space and an inverse transform of the product. Alternative approaches are being actively investigated.

ALGORITHM AND HARDWARE STRATEGIES

It is enlightening to look at execution times for the well-developed downstream part of the processing. Figure III shows the evolution of the time consumption by the downstream data reduction codes on a DECstation 3100. Although functionality has increased, the amount of cpu time has remained essentially constant. The elapsed time, however, has dropped dramatically over three years of code development.

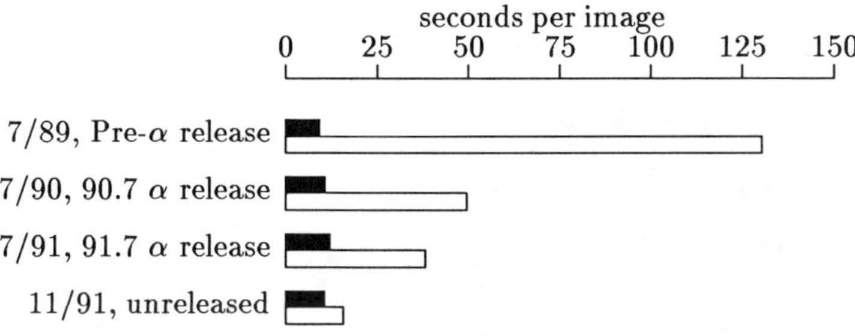

FIGURE III Time consumption as a function of algorithm development. Filled bars indicate cpu seconds. Empty bars indicate elapsed time. Tests were run on a DECstation 3100, a 1989 RISC machine.

Achievement of the eight-fold drop in elapsed time is attributed to a number of algorithmic improvements. The largest single improvement, a factor of 66, resulted from the strategy of performing the time transpose from large files of mode coefficients to large time series files by use of one set of interim smaller files. Further improvements in performance resulted from simultaneous reorientation and heliographic remapping of the images. A mask was used to remove spikes in the temporally filtered images, which result from interpolating in data too near the limb. Piecing together time series files of about 700 time slots into longer time series files was found to be more efficient than adding data for smaller amounts of time into long time series.

In order to keep pace with the data acquisition, the workstations used must meet arithmetic requirements of 7-50 Mflops and I/O requirements of some ten Gigabytes per day. Of the two requirements, the I/O requirement is the most restrictive. Bottlenecks are most apparent for the removable media I/O during the reductions. A substantial part of the I/O requirement is invested in the use of data product tapes in subsequent reduction steps to establish the integrity of the removable media library.

Figure IV shows the evolution of execution time for the downstream data reduction as a function of hardware improvements. The codes used are those in the GRASP-SUNTRANS 91.7 software release. Hardware improvements over the past six years have allowed a factor of eight drop in elapsed time for the same SUNTRANS code release. A further factor of two drop in cpu time has been noted in tests of a pre-release desktop workstation temporarily loaned to NOAO for testing.

It is possible to keep pace with the network's data acquisition using multiple desk-top workstations for the upstream and downstream data reduction steps. Asynchronous processing on multiple workstations allows bottlenecks to be addressed dynamically. Such processing, with most workstations being able to do any task in the data reduction, also minimizes the impact of hardware failure. Heterogeneous UNIX workstations and software allow the maximal use of upgrades during the project to take advantage of system innovations.

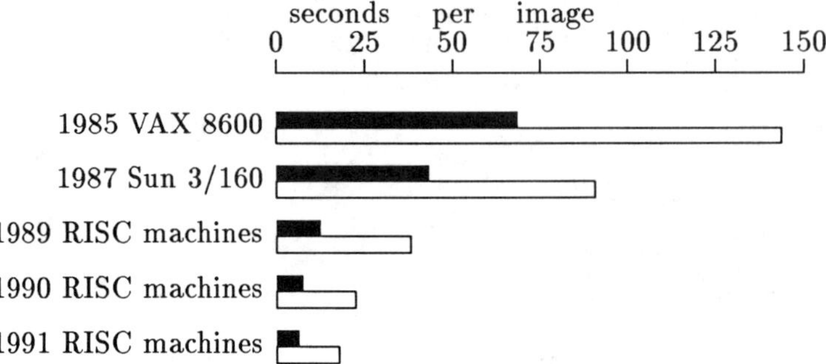

FIGURE IV Time consumption as a function of hardware development. Filled bars indicate cpu seconds. Empty bars indicate elapsed time. Tests were run using the GRASP-SUNTRANS 91.7 software release.

REMAINING PROBLEMS

A number of problems remain. The most acute algorithmic need is a faster scheme for image restoration. The most pressing hardware needs are associated with faster (and still reasonably priced) I/O capabilities. Automation of initial image quality assessment and resolution of details in the merging algorithm are also undergoing further study.

The drop in elapsed time for data reduction due to both algorithmic and hardware improvements has been most encouraging, and bodes well for achieving the stated goal of processing the GONG project data in cadence with acquisition.

ACKNOWLEDGMENTS

I wish to thank J. Pintar, F. Hill, D. Armet and R. Toussaint of NOAO/NSO for supporting information.

REFERENCES

Anderson, E. R., Duvall, T. L., Jr., & Jefferies, S. M. 1990, ApJ, 364, 699

Duvall, T. L., & Harvey, J. W. 1986, in Seismology of the Sun and the Distant Stars, D. Gough, Dordrecht: Reidel, 1986, 105

Leibacher, J. W., Noyes, R. W., Toomre, J., & Ulrich, R. K. 1985, Sci. Amer., 253, no. 3, 48

Williams, W. E. 1991, "SUNTRANS: Spherical Harmonic Transpose and Time Transpose Package" in GONG Alpha GRASP Version 91.7, J. Pintar, NOAO/NSO GONG project document

AN ALGORITHM FOR EXTRA-SOLAR SYSTEM PLANET DETECTION

JOHN R. ENGEL
Western Research Co., 2127 E. Speedway, Suite 209, Tucson, AZ 85719

ROY A. TUCKER
Tucker and Associates, 5500 W. Nebraska, Tucson, AZ 85746

ERIC R. CRAINE
Western Research Co., 2127 E. Speedway, Suite 209, Tucson, AZ 85719

BACKGROUND

We propose to conduct a statistically significant extra-solar planet search program using a photometric technique to detect dips in stellar light curves caused by planetary transits. Although use of this technique is not a new idea, the requisite observational intensity of a program of long-term, continuous photometric monitoring of a large sample ($\sim 10^3$) of stars has in the past precluded this approach. Technology is now available which makes observationally intensive photometric monitoring of a large number of stars possible. Indeed, Crawford (1989) describes the structure of the Global Network of Automated Telescopes (GNAT) which we will use as the basis for our observing system.

To increase the chances of detecting transits we will create a program star list from the catalog of cool dwarf stars which we are currently developing (Craine et al. 1991). These stars potentially have favorable orbital geometries and photometric properties which would make planetary transit detection feasible with modest-sized telescopes.

PROJECT REQUIREMENTS

To provide continuous coverage of stars, the global network will consist of three or four sites separated by 120 or 90 degrees of longitude, respectively. Each site will be in communication with a base site which will have a computer system to coordinate the observing program, edit the remote site observing schedules, communicate schedules and data via modem with remote sites, archive data, and perform additional data processing.

Each observing site will house 5 to 6 automated, one-meter class telescope systems, comparable to the existing Mt. Hopkins remote observing facility. Each automated telescope will have a microcomputer to conduct its observing program and perform photometric data reduction.

Photometric data will be obtained with a CCD imaging detector. Single field differential photometry will be conducted to obtain stellar light curve data. This technique will provide better than 0.01 mag accuracy necessary to detect the expected 0.03 mag drops in brightness due to a planetary transit.

We estimate that less than 5 minutes will be required to observe each star. The stars will be observed once every 6 hours in order to detect minimum duration transits based on our planetary system models. Assuming 10 hours of observing time each night, about 120 stars can be monitored by one system per year. Six systems could monitor about 720 stars per year, or 3600 stars over the expected 5 year lifetime of the project.

OBSERVATION PROCEDURE

Each site will conduct a routine observation procedure, largely without manual intervention. Site personnel will be responsible for powering the system on and off, opening and closing the dome, visually inspecting the hardware and handling emergency shut-downs. Automated procedures include:

- Initialize the system: boot the system control software, run hardware diagnostics, set system clock, load observing schedule
- Perform calibrations: set focus, locate and observe standard star fields, check positioning accuracy, obtain bias, dark and flat field images
- Perform observations and reduce data (see Figure I)
- Shut down the system: backup data, contact base system, send data

DATA REDUCTION AND ANALYSIS

Image data reduction will be performed by the remote systems in real-time, meaning that each star field image will be processed before acquisition of the next image has been completed. Consequently, the system can respond quickly to possible transit events and will not be burdened with an enormous image archive. The computer will process a star field image while the telescope is moving to the next field and while the CCD is integrating on that field. Major data reduction steps include:

- Flat-field the digitized star field image using the standard technique of division by sky flat frames following bias and dark frame subtraction.
- Determine star locations using Gaussian fit to stellar profiles. A priori knowledge of the star fields may be used to optimize star finding.
- Perform aperture photometry by calculating the integrated flux inside a circular aperture after sky subtraction.
- Verify that correct star field has been acquired with a pattern matching algorithm which uses known locations of brighter stars in the field.
- Analyze the program star light curve by comparing the magnitude difference of each program/comparison star pair with its historical value.

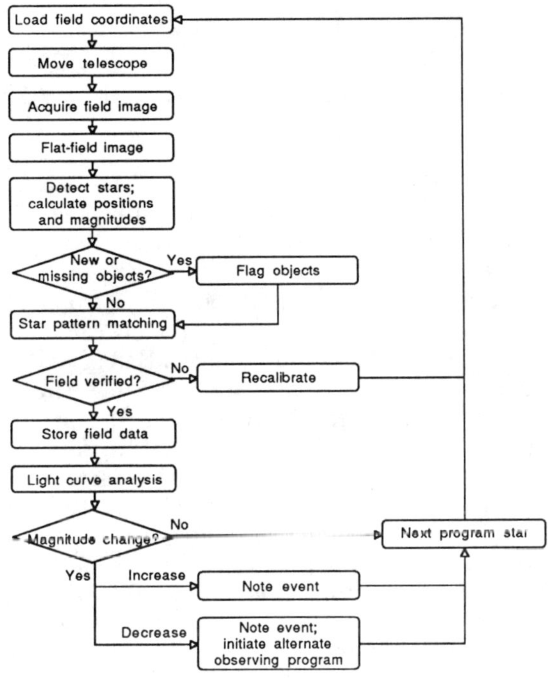

FIGURE I Observation and Data Reduction Flowchart

SUMMARY

A search for extra-solar planetary systems using a global network of modestly-priced, automated telescopes is a viable and scientifically interesting undertaking. It presents no great technological hurdles, all hardware for this project currently exists and the total project cost is very low. The program star selection process is already well underway. There is a high probability of obtaining statistically significant (even if null) results and we expect that much useful collateral data will be acquired. We also believe there are good prospects for serendipitous findings.

REFERENCES

Craine, E. R., Tucker, R. A., & Engel, J. R. 1991, in Cool Stars, Stellar Systems, and the Sun, Volume 7 (in publication)

Crawford, D. L., Genet, R. M., & Hayes, D. S. 1989, in Automatic Small Telescopes, ed. D. S. Hayes & R. M. Genet, Fairborn Press, Mesa, Arizona p.115

THE *COBE* FIRAS PROJECT PIPELINE SOFTWARE SYSTEM

S.M. READ[1], S. ALEXANDER[1], D.B. BOULER[1], N.G. GONZALES[1], K.A. JENSEN[1], L.P. ROSEN[1], F.G. SHUMAN[1], H.Y. WANG[1], R.E. EPLEE[2], R.B. ISAACMAN[2], A.R. TRENHOLME [2]

ABSTRACT The Far Infrared Absolute Spectrophotometer (FIRAS) on *COBE* was designed to measure extremely small deviations of the cosmic microwave background radiation (CMBR) from a blackbody spectrum. The data, taken by a rapid scan polarizing Michelson interferometer, cover a frequency range of 1 to 100 cm^{-1}. Data are taken by four detectors, two at 1 to 20 cm^{-1} and two at 20 to 100 cm^{-1}, at two stroke lengths and two scan speeds of the mirror transport mechanism. The two low frequency detectors are used for CMBR and cosmological studies. The two high frequency detectors are used for study of the galactic plane. On orbit calibration is performed with a movable, high precision, external blackbody calibrator. The FIRAS Project Pipeline (Figure I) is a complex software system which takes the time ordered, raw interferograms (IFGs) and produces skymap data sets of calibrated spectra. The pipeline software is designed to handle unique problems such as the asynchronous relation of science and engineering data. The software is designed so that intermediate products can be created for analysis and validation purposes.

PRELIMINARY PROCESSING AND QUALITY CHECKING

The Preprocessor is the first program in the pipeline processing of FIRAS data. Its primary function is to compute the midpoint of the collection time of the IFGs. Because science data from the detector are transmitted from the spacecraft asynchronously with engineering and instrument state data, the latter must be correlated with the science data. An IFG is valid only if status data remain constant during the entire IFG collection. The program reports telemetry gaps, finds the engineering and instrument state data associated in time with the science data, and extracts the stroke length, scan speed, and status data. Using the extracted data, the Preprocessor computes the midpoint of collect time and flags anomalous results.

The Data Qualify Program sets quality flags which are used to make the primary decision on whether the raw IFG record will be passed on to the remain-

[1]Hughes STX Corp. 4400 Forbes Blvd., Lanham, MD 20706

[2]General Sciences Corp. Code 685.3, NASA/GSFC Greenbelt, MD 20771

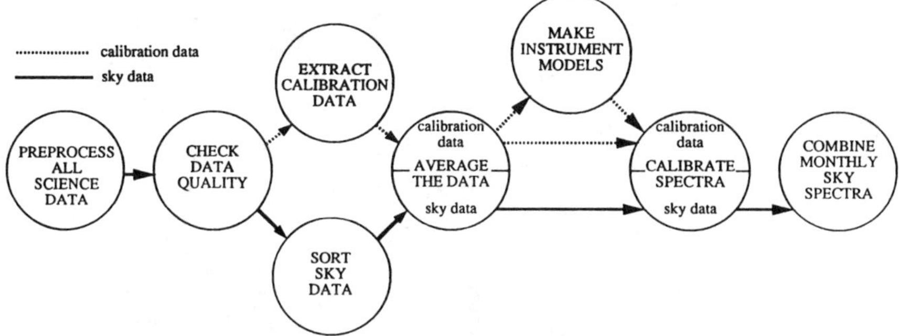

FIGURE I The *COBE* FIRAS Project Pipeline Software System

der of the pipeline for further processing. This program is designed to handle unique problems such as the asynchronous relation of each detector's data to the engineering data and the correlation of the FIRAS line-of-sight with the observation. Using the midpoint of collect time, derived attitude parameters are computed from the line-of-sight. The science records are aligned between two bracketing housekeeping records for interpolation followed by conversion to engineering units. Sets of the four detector IFGs are associated with the engineering data and engineering statistics are computed by orbital period. Finally, the program performs limit checking on engineering and attitude parameters and sets quality flags accordingly.

AVERAGING OF THE DATA FOR CALIBRATION AND SKY BRANCHES

After the data quality checking, the FIRAS Pipeline splits into two processing streams; one stream is for calibration and the other for sky data. Both branches of data feed back into the program which does the averaging.

During calibration, commands were sent to set and control several components of the instrument at temperatures over a range of 2° to 25° K. The Extract Calibration Program identifies time ranges during which the temperatures remain within a specified tolerance for time-ordered groups of calibration IFGs. In contrast, the sky data must be rearranged into groups of IFGs from the same location in the sky so that the data can be averaged. Sky IFGs have been binned into 6144 pixels in the sky using a quadrilateralized spherical cube format. Data are sorted by pixel, scan mode and internal calibrator temperature. Other optional selections are available based on galactic latitude, position within the *COBE* orbit, and whether associated engineering data meet specified criteria.

Ensembles of IFGs are averaged to improve signal-to-noise. The Average Data Program performs this function separately for both sky and calibration data. Ensembles of IFGs are checked for consistent shapes and instrument states. Following this, a pattern-recognition technique is used to remove cosmic ray hits on the detectors. All IFGs in the ensemble which pass the consistency checks are averaged and a quartic baseline is fitted and subtracted. This is the first skymap data set (Figure II). Weights and covariance matrices are calculated for statistical purposes. Several intermediate products are created at this stage

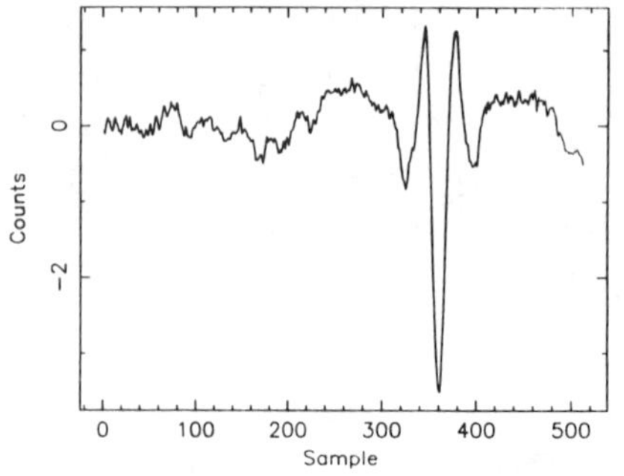

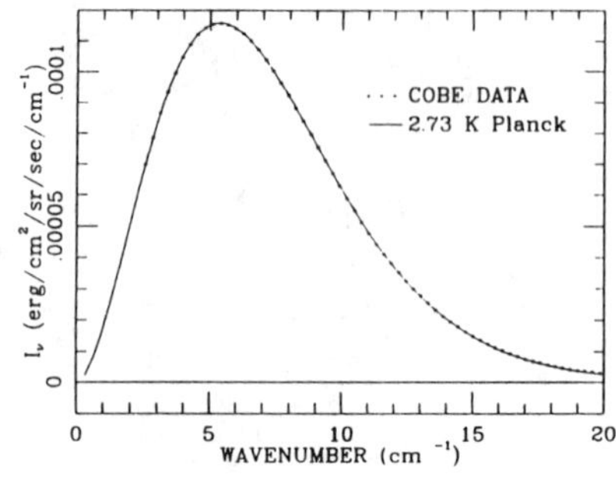

FIGURE II Averaged Sky Interferogram and Calibrated Sky Spectrum

of the pipeline. Templates made from midaveraging the IFGs can be used to study contributions of galactic gas and dust. Analysis of residual IFGs can demonstrate that cosmic ray effects are successfully removed.

CALIBRATION

The FIRAS calibration model is composed of a detector model used to convert the spectra from volts to ergs/sec and an optical model defined by a set of complex emissivities that characterize the thermal emission of various components of the instrument. The Make Instrument Models Program solves iteratively for the calibration model using a polytopic search algorithm. Input to the program consists of the averaged calibration data and laboratory measured values for the detector parameters. Detector parameters and emissivities are varied to minimize the least square differences between calibrated spectra and Planck curves for the instrument component temperature. For each of the detectors and instrument scan modes, a corresponding model solution is produced.

The Calibrate Spectra Program uses the calibration models to produce monthly calibrated spectra from either the sky-averaged or calibration-averaged IFGs. Rotation about the peak position, followed by apodization and a Fourier transform, creates the spectrum, which is then normalized by the electronics transfer function. The detector responsivity and time constant for the spectrum are computed from the bolometer parameters and the detector temperature and voltage. Finally, the program applies the calibration model to the spectra. This is the second skymap data set (Figure II).

FIRAS observes each part of the sky twice a year. In order to have a final product of a single calibrated spectrum for each scan mode and pixel in the sky, the Combine Monthly Sky Spectra Program uses a statistical weighting scheme to combine the monthly spectra.

ACKNOWLEDGMENTS

This research is supported by the Astrophysics Division of the Office of Space Science and Applications at NASA Headquarters.

THE DIRBE PROJECT PIPELINE SOFTWARE SYSTEM

J.A.J. SKARD[1], H.T. FREUDENREICH[1], R.J. HOLLENHORST[2], T. KELSALL[2], A.R. PANITZ[1], K.P. TEWARI[1], AND G.N. TOLLER[3]
[1]Hughes STX Corporation, Lanham, Md. 20706
[2]NASA/Goddard Space Flight Center, Greenbelt, Md. 20771
[3]General Sciences Corporation, Laurel, Md. 20707

INTRODUCTION

The Diffuse Infrared Background Experiment (DIRBE) is a ten-band infrared photometer on the Cosmic Background Explorer (COBE) spacecraft. The instrument is designed to perform all-sky, absolute, photometric measurements in the wavelength range from 1 to 300 micrometers (Hauser et al. 1990).

Data from the DIRBE instrument are stripped from the telemetry stream and become inputs to the 'Project Pipeline' processing. The primary tasks for the DIRBE Project Pipeline are to provide insight into the performance of the DIRBE instrument and the nature and quality of the data, to remove all 'instrument signatures' prior to the execution of major scientific investigations, and to produce the DIRBE 'Project Data Sets' for guest investigator use and delivery to the National Space Science Data Center (NSSDC).

In the following we describe the concept, the implementation, and the operation of the DIRBE 'Project Pipeline', along with a discussion of the lessons learned in the process of constructing and utilizing this software.

THE 'PIPELINE' CONCEPT

For removal of instrument signatures the DIRBE data must go through an extensive process of filtering, linearization, and calibration. Characteristic of the DIRBE Pipeline concept is that the process is broken up into a sequence of 'software facilities', i. e. independent S/W modules, each with a separate, well-defined function, through which the data proceed in an essentially linear fashion. The interfaces between the facilities are defined through the record structures of the intermediate data sets.

The pipeline model lends itself to extensive Quality Assurance during the processing via analysis of the pipeline byproducts, such as printed report files, as well as plots and tables generated on the basis of the intermediate data sets.

Run modes and parameters determining the operation of each facility are set in configured ASCII-formatted 'script' files and binary reference files, which are read by the facility at run-time. This system greatly facilitates the creation of non-standard 'off-line' runs, utilizing the standard, configured S/W in combination with personalized scripts for analysis purposes.

DEVELOPMENT OF THE PIPELINE FACILITIES

The pipeline is designed for operation in a VAX/VMS environment (Cheng 1992). Drivers for the facilities are all written in VAX DCL, while the facilities themselves are coded in FORTRAN in order to ease the S/W maintenance and validation activities. We have adhered to strict coding standards, and emphasis has been placed on consistency between the different DIRBE facilities. Wherever possible we have made use of the extensive 'Core' S/W libraries. The most significant component of the Core S/W is 'COBETRIEVE', a large subroutine package for data set archiving and retrieval.

The development of the DIRBE pipeline S/W is made up of of several distinct phases: Overall design of the pipeline and its internal and external interfaces, detailed requirements specifications for each facility, structured analysis resulting in detailed pseudocode (PDL), coding, testing, and implementation, and, finally, enhancements resulting from experience with real data. The entire DIRBE Pipeline S/W system will upon completion amount to approximately 140000 lines of code, in the following categories: indexing, quality checking and data merging, engineering and noise trends, internal calibration and calibration trends, linearization, celestial calibration, production of sky maps, and production of calibrated time-ordered data.

OPERATIONAL CONSIDERATIONS

The DIRBE data processing is performed by independent operations staff. This fact has led to several significant design features in the DIRBE S/W: The pipeline was given a menu-driven interface, and the standard pipeline processing was broken up into 3 sequential 'menu-selections', each of which resulting in the submission of a 'root' batch job. This 'root' batch job contains all the sequential pipeline activities in that processing segment, and submits as separate batch jobs the various processing chores which can be run in parallel, such as plot production. Minimal operator interaction is necessary for each step, as each menu selection requires only the specification of a time range.

In order to prevent crashes due to anomalies in the data, all S/W facilities were coded to take a 'sensible' course of action when anomalies occur (often by simply skipping the bad data), while signaling the condition in a clear and unambiguous way. Subsequent off-line analysis can then be performed in order to determine whether or not the action taken by the S/W was appropriate.

TRACKING AND QUALITY ASSURANCE

The processing is performed according to a detailed processing script. The Data Base Administration (DBA) staff perform frequent archive backups and consistency checks. The ancestry of all files in the archive is automatically maintained in an Archive History File.

Processing of each data week results in a 'weekly folder', containing diagnostic reports from each facility, as well as plots, tables, and grey scale and contour plots of the sky (Freedman et al. 1992). These folders are inspected by developers and validators. Powerful graphics and analysis tools, such as IDL

(Interactive Data Language) and IOPS (IGSE Offline Processing System) are available for investigation of anomalies. Analysis of the sky map products (Toller et al. 1992) also gives valuable insight into the nature of the data.

LESSONS LEARNED

A number of lessons have been learned during the course of the S/W development for the DIRBE pipeline. PDL proved to be an extremely efficient tool for defining procedures at a high level as well as at more detailed levels. 'Stubs' in the S/W in preparation for handling unexpected, but possible, difficulties in the data turned out to be well worth the effort. A SIMPLE user interface and ROBUST S/W has been critical in order to allow processing to be conducted by non-developers. The design, whereby most run parameters are set by ASCII-formatted 'scripts', has allowed the same, configured S/W to be used in a flexible analysis environment and in the strictly controlled production environment.

The large data expansion factor for DIRBE (=15x) should have been taken into account from the beginning, starting with the initial system design.

This type of project leads to an extremely high S/W maintenance load as experience is gained with the S/W and the data. This should be considered when drawing up a resource plan. Furthermore, good communication between developers and end users is invaluable during the design and development phase, as it is impossible to have foreseen and defined everything in advance.

Continuity in staffing is important in such a long-term project. Developers should become familiar with the instrument and the science goals in order to contribute to the coherency of the product. Good documentation is essential for quickly bringing new personnel 'up to speed'. Sufficient resources should be allocated for documentation, starting early in the project.

ACKNOWLEDGMENTS

The contributions to the DIRBE Pipeline S/W by A. Ahearn, W. Daffer, I. Freedman, A. Ramsey, T. Roegner, and V. Snowel are gratefully acknowledged.

This research is supported by the Astrophysics Division of the Office of Space Science and Applications at NASA Headquarters.

REFERENCES

Cheng, E. 1992, this volume

Freedman, I. F., Skard, J. A. J., & Kelsall, T. 1992, this volume

Hauser, M. G., et al. 1990, AIP Conf. Proceedings, 222, 161

Toller, G. N., et al. 1992, this volume

DAILY QUALITY ASSURANCE SOFTWARE FOR A SATELLITE RADIOMETER SYSTEM

P.B. KEEGSTRA[1], G.F. SMOOT[2], C.L. BENNETT[3], J. AYMON[2], C. BACKUS[1], G. DE AMICI[2], G. HINSHAW[4], P.D. JACKSON[1], A. KOGUT[4], C. LINEWEAVER[2], L.A. ROKKE[1], J. SANTANA[1], AND L. TENORIO[1]

ABSTRACT Six Differential Microwave Radiometers (DMR) on the *COBE* (Cosmic Background Explorer) measure the large-angular-scale isotropy of the cosmic microwave background (CMB) at 31.5, 53, and 90 GHz. In addition to basic plotting and limit-checking, we implement a collection of more specialized programs to perform data quality assurance and long-term trending. Quality assurance of DMR data poses challenging requirements. The data are differential, so a single bad point can affect a large region of the sky, yet lengthy integration times ($>$ 1 year) are required to limit potential CMB anisotropies.

PRE-ATTITUDE SOFTWARE

Stability Analysis

Tests are run on the daily DMR data to assure that the instrument is functioning properly and that the data are scientifically usable. These tests do not require spacecraft attitude solutions so they can be run immediately. One such test investigates whether the noise in the data is consistent with a Gaussian distribution (see Figure I). This is done by averaging groups of 2^N data points and then calculating the scatter in the groups as a function of N. If the scatter were entirely due to Gaussian noise, the resulting log-log plot would be linear with a slope of $-1/2$. One expects variation from this due to real signals seen at the spin and orbit period, but unexpected variations would be cause for concern.

Fourier Analysis

An additional two tests on the daily DMR data confirm the presence of the signal from the cosmological dipole and check for any spurious signals of comparable amplitude (see Figure II). These tests exploit the fact that the interaction of the

[1]Hughes STX Corp. 4400 Forbes Blvd., Lanham, MD 20706

[2]Lawrence Berkeley Lab. and Space Sciences Lab. MS 50-232, UC Berkeley, CA 94720

[3]Lab. for Astronomy and Solar Physics, NASA/GSFC, Code 685, Greenbelt, MD 20771

[4]Universities Space Research Assoc. Code 610.3, NASA/GSFC, Greenbelt, MD 20771

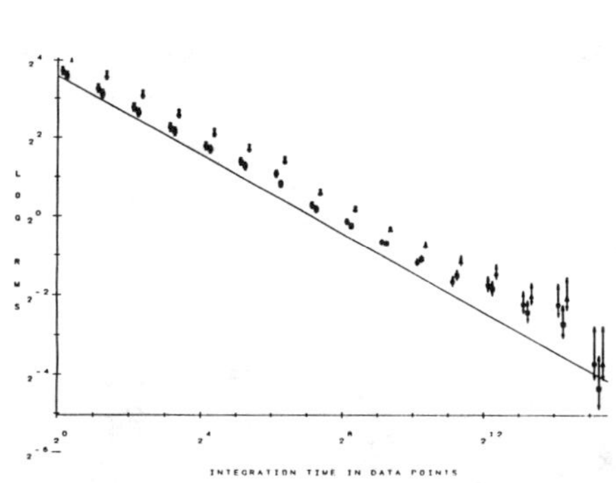

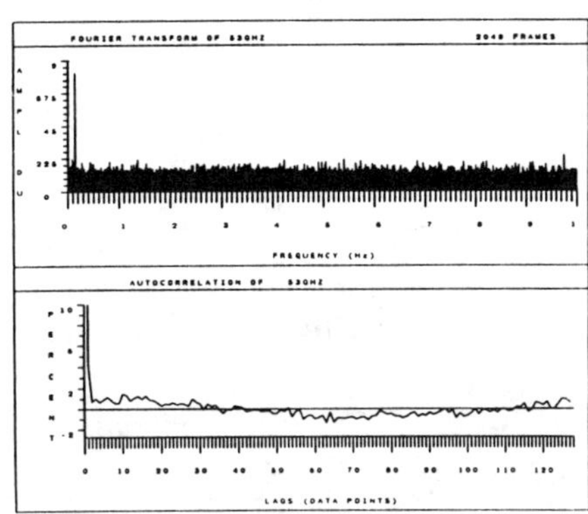

FIGURE I This shows the scatter in the daily data as a function of the number of points averaged together. The solid line shows the expected scaling, offset for visibility. Squares represent the standard deviation, triangles represent the RMS step size between adjacent blocks, and circles represent the mean absolute value of the step size. The flattening at $N = 2^{14}$ corresponds to celestial signals (dipole + Galaxy) sampled at the spacecraft orbital period.

FIGURE II The upper plot is a Fourier transform of the daily data. The cosmological dipole causes a modulation of the DMR signal which is visible as a peak in the FFT at the spin frequency. The lower plot shows the correlation between the original data and the data shifted by the indicated number of half-second datapoints. The peak at a shift of one is an instrumental memory effect. The gentle sinusoid is a correlation induced by the dipole; the minimum at a shift of 75 corresponds to half the spin frequency.

spacecraft spin with the cosmological dipole gives rise to a modulation in the time domain. These tests also reveal the presence of the signal from the Moon during the 2 weeks per month that it is visible in the data.

POST-ATTITUDE SOFTWARE

Data Binning

Additional tests are run after the DMR data has been calibrated and merged with spacecraft attitude to provide a more detailed analysis. The first of these sorts the data into bins according to basic spacecraft quantities such as spin angle and orbit angle. The purpose of data binning is to detect systematic effects correlated with the spacecraft orbit or spin. A sample is shown in Figure III.

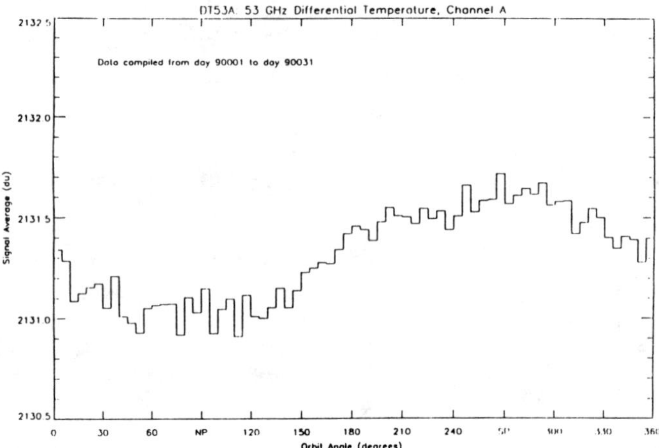

FIGURE III The variation of the time-ordered differential temperature signal from the 53 GHz radiometer over an orbit is shown. This variation is due to an instrument susceptibility that causes the signal to be modulated by the Earth's magnetic field. We model this effect and apply a correction to the data.

Long-term Trending

The purpose of long-term trending is to track instrumental behavior on time scales ranging from a few orbits to the entire mission. Trends in the data are investigated by averaging the quantities of interest over one spacecraft orbit ($\sim$ 102 minutes). The motivation for this is to reduce the noise level in the data, and to average over any systematic orbit modulation in the signal.

ACKNOWLEDGMENTS

COBE is managed by the Goddard Space Flight Center and sponsored by NASA's Astrophysics Division, Office of Space Science and Applications, Washington, DC.

REFERENCES

Bennett, C., et al. 1992, ApJ, in press

Smoot, G., et al. 1990, ApJ, 360, 685

Smoot, G., et al. 1991, ApJ, 371, L1

SOFTWARE CONFIGURATION MANAGEMENT AND QUALITY ASSURANCE IN THE COSMIC BACKGROUND EXPLORER (COBE) PROJECT

A. C. RAUGH, K. L. BABST, C. A. BURKE, L. A. HERREID, S. R. KALTENBAUGH, P. W. MCCASLIN, N. L. WOODSON AND S. H. YOM
Hughes STX, Cosmology Data Analysis Center, 7601 Ora Glen Drive, Greenbelt, MD 20771

ABSTRACT The COBE project software, consisting of over 700,000 lines of executable code running on a VAX/VMS cluster comprised of 8000-series VAXes and VAX workstations, will ultimately produce the project datasets. It is the task of the configuration management subsystem to ensure that each new release of software contains only code which has been formally tested and approved. This paper describes the configuration management and quality control procedures being applied to assure the integrity of the software prior to production of the project datasets.

HISTORICAL PERSPECTIVE

The COBE software has been released to the Project in progressive stages, identified as "builds". Build 0, in October 1986, established the system structure used for subsequent deliveries. Build 1.0, January 1987, introduced the constraints on software dependencies described below. Build 2.0, September 1987, saw the introduction of the subsystem hierarchy, while the dataset archive structure was formalized with Build 3.2, May 1988. Build 4.2, March 1989, contained all launch-critical capabilities.

Following launch of the spacecraft in November 1989, the build schedule has been accelerated so that the production software is now updated on a weekly basis with the latest deliveries and required fixes. This shortening of the build cycle has been accomplished without substantial change to the existing configuration management/quality assurance structures.

Throughout this time, there has been a steady growth in the amount of code under configuration management (CM).

SYSTEM HIERARCHY

In order to limit both the implicit (common areas, e.g.) and explicit (included files, e.g.) links within the software, specific rules were laid out in the earliest stages of development defining the ways in which various levels of code may depend on each other.

The software divisions and allowed dependencies, from lowest level to highest, are as follows:

facility A single directory containing a set of related modules (a module is any source file, such as a FORTRAN file, a text file included in other files during compilation, a command file, etc.). Facilities generally produce one or more executable images or reference libraries.

No two facilities may depend on each other, except as noted below.

subsystem A set of facilities relevant to a specific aspect of the COBE project. For example, there is an Attitude Subsystem to handle attitude determination and processing; a Mission Planning Subsystem, which handles command generation; a Data Management Subsystem, to deal with archiving software; and so on. The build procedure is partitioned along subsystem boundaries.

Each subsystem may have one facility consisting of general utilities which may be referred to by other facilities of that subsystem.

system-level partition These partitions (analogous to subsystems) build facilities which may be used by any and all other subsystems. These include things like system-wide libraries and shareable images. Because these facilities may be referred to by other subsystems, they must be built first.

Any subsystem facility may depend on any module in a system-level partition.

PARALLEL USER ENVIRONMENTS

One of the main objectives for configuration management was to provide a system in which development, problem resolution and dataset production could continue simultaneously without risk of compromising the integrity of the final data product. This has been achieved by taking advantage of the VMS logical name and search list capabilities. Logical names allow the user to point to identically structured but physically separate environments by simply redefining one or a few key logical names. Search lists are used for layering environments on top of each other during the build process, which creates the executable images and libraries comprising each environment. New versions of modules located at top levels of the search list supersede older versions lying farther down in the list

There are three concurrent environments running at any given time on the cluster. Level-1, or **L1**, is the production environment. This code is the most rigorously controlled and is the software used to generate and validate the project datasets. **L2** is a transitory test environment, created by layering new, approved software deliveries on top of the existing L1 code via search lists. Regression testing is then performed here and, when the results have been reviewed and approved, a change of logical pointers turns this environment into the new **L1.** **L3** is the integration testing environment, where new and modified software is layered on top of the **L2** and **L1** code and tested for the first time within the full COBE environment.

IMPLEMENTING SOFTWARE CHANGES AND ENHANCEMENTS

Enhancements, modifications and bug fixes all follow basically the same procedure for configuration, although for purposes of budgeting and scheduling enhancements require additional approval by the Project prior to the start of development.

1. The new software is first coded and tested in the user's local area.

2. It is then copied into the L3 area, the facility and subsystem are rebuilt using the new code, and the software is tested in the new environment.

3. The changes made and results of the integration test are then reviewed by a software engineering review board, which must approve the code for inclusion in the next build of L2.

4. Following the L2 build, regression testing ensures that the code not only delivers the desired results, but does it without affecting any existing functionality.

5. Once the regression test results are approved and the software accepted, a simple change of pointers turns the old L2 into the new L1.

Required approvals are always collected prior to modifying L1, but in emergency cases approval may be obtained by telephone or proxy.

The problem reports, review and sign-off forms, and test reports are all maintained and tracked in on-line databases which may be read by the entire user community. These systems have been set up using a combination of in-house code, DCL command files, and vendor-supplied utilities.

IN SUMMARY

The configuration management/quality assurance system has achieved its major objectives of: 1) allowing software building, verification testing, development and production to continue simultaneously without significant interruption; 2) continually maintaining a verified software environment for production; 3) providing flexibility to handle critical changes with a short turn-around time without compromising software integrity; and 4) tracking on-line all modifications and enhancements made to the system.

ACKNOWLEDGMENTS

This work is supported by the Astrophysics Division of the Office of Space Science and Applications at NASA Headquarters, under contracts NAS5-30726 and NAS5-30440.

AUTHOR INDEX

SUBJECT INDEX

COLOPHON

These Proceedings were prepared with LaTeX using a style file customized for the PASP series. Authors submitted manuscripts electronically, most using the electronic mail facilities of the Internet. We also accepted papers on floppy disk, and a very small number of articles had to be entered by hand. Of the 123 papers in this volume, 119 were submitted with usable LaTeX markup by the authors themselves. We received 107 of the submissions by the deadline date, most flooding in via email during the final two days.

Final proof pages for the entire book (from page v on) were produced at NOAO on an Apple LaserWriter IIntx. Tomas Rokicki's *dvips* program was used to translate the device-independent output from LaTeX. This combination of PostScript printer and driver program enabled us to take advantage of the PostScript language to merge figures into the pages as they were printed.

Several authors (13 papers) submitted graphical material as Encapsulated PostScript (EPS) files, and others provided PostScript that we were able to encapsulate. Most of the EPS files were incorporated into the pages by the *dvips* program, saving us the tedium of scaling originals and pasting up figures. The EPS standard is immature at this date, however, and considerable technical effort was required in place of the photocopying and gluing time; some of the files could not be incorporated by the software and the figures had to be inserted by hand. The automatic inclusion of EPS graphics is advantageous and produces a superior product, and it is our hope that graphics-producing software will evolve toward a common implementation of the EPS standard.

Most of the front and back matter was generated mechanically from the material submitted by the authors. A database of pertinent information about each paper was maintained, and the table of contents, author index, etc. were derived from it. The overall pagination of the volume was determined by software after the papers were edited and ordered, and the running heads and folio numbers were applied by LaTeX using standard markup commands. The software that was developed for these chores is generalized so that it can be used for other projects, and it is available to other editors through the PASP Conference Series Office.